Structural Design of Steelwork
to
EN 1993 and EN 1994

Third edition

Structural Design of Steelwork to EN 1993 and EN 1994

Third edition

L.H. Martin Bsc, PhD, CEng FICE
J.A. Purkiss Bsc (Eng), PhD

AMSTERDAM • BOSTON • HEIDELBERG • LONDON • NEW YORK • OXFORD
PARIS • SAN DIEGO • SAN FRANCISCO • SINGAPORE • SYDNEY • TOKYO
Butterworth-Heinemann is an imprint of Elsevier

Butterworth-Heinemann is an imprint of Elsevier
Linacre House, Jordan Hill, Oxford OX2 8DP, UK
30 Corporate Drive, Suite 400, Burlington, MA 01803, USA

First edition published by Edward Arnold 1984
Second edition 1992
Third edition 2008

British Library Cataloguing in Publication Data
A catalogue record for this book is available from the British Library

Library of Congress Cataloging-in-Publication Data
A catalog record for this book is available from the Library of Congress

ISBN13: 978-0-7506-5060-1

For information on all Butterworth-Heinemann publications
visit our web site at books.elsevier.com

Typeset by Charon Tec Ltd (A Macmillan Company), Chennai, India
www.charontec.com

Printed and bound in Great Britan

07 08 09 10   10 9 8 7 6 5 4 3 2 1

Contents

CHAPTER 8

Frames and Framing 282

CHAPTER 9

Trusses 341

CHAPTER 10

Composite Construction 358

CHAPTER 11

Cold-formed Steel Sections 413

Preface

This book conforms to the latest recommendations for the design of steel and composite steel–concrete structures as described in Eurocode 3: Design of steel structures and Eurocode 4: Design of composite steel–concrete structures. References to relevant clauses of the Codes are given where appropriate. Note that for normal steelwork design, including joints, three sections of EN 1993 are required:

- Part 1–1 General rules and rules for buildings
- Part 1–5 Plated structural elements
- Part 1–8 Design of joints

Additionally if design for cold formed sections is carried out from first principles then Part 1–3 Cold formed thin gauge members and sheeting is also required.

Whilst it has not been assumed that the reader has a knowledge of structural design, a knowledge of structural mechanics and stress analysis is a prerequisite. However, as noted below certain specialist areas of analysis have been covered in detail since the Codes do not provide the requisite information. Thus the book contains detailed explanations of the principles underlying steelwork design and provides appropriate references and suggestions for further reading.

The text should prove useful to students reading for engineering degrees at University, especially for design projects. It will also aid designers who require an introduction to the new Eurocodes.

For those familiar with current practice, the major changes are:

(1) There is need to refer to more than one part of the various codes with calculations generally becoming more extensive and complex.
(2) Steelwork design stresses are increased as the gamma values on steel are taken as 1,0, and the strength of high yield reinforcement is 500 MPa albeit with a gamma factor of 1,15.
(3) A deeper understanding of buckling phenomena is required as the Codes do not supply the relevant formulae.
(4) Flexure and axial force interaction equations are more complex, thus increasing the calculations for column design.
(5) The checking of webs for in-plane forces is more complex.
(6) Although tension field theory (or its equivalent) may be used for plate girders, the calculations are simplified compared to earlier versions of the Code.
(7) Joints are required to be designed for both strength and stiffness.
(8) More comprehensive information is given on thin-walled sections.

Acknowledgements

Lawrence Martin and John Purkiss would like to thank Long-yuan Li and Xiao-ting Chu for writing Chapter 11 on Cold-formed steel sections.

The second author would like to thank Andrew Orton (Corus) for help with problems over limiting critical lengths for lateral–torsional buckling of rolled sections.

The authors further wish to thank the following for permission to reproduce material:

- Albion Sections Ltd for Fig. 11.1
- www.access-steel.com for Figs 11.24 and 11.25
- Karoly Zalka and the Institution of Civil Engineers for Fig. 8.13
- BSI

	BSI Ref.	Book Ref.
BS 5950: Part 1: 1990	Tables 15–17	Annex A7
BS 5950: Part 1: 2000	Table 14	Table 5.3
EN 1993-1-3	Fig. 5.6	Fig. 11.8
	Fig. 5.7a	Fig. 11.9
	Fig. 10.2	Fig. 11.20
EN 1994-1-1	Fig. 9.8	Fig. 10.3

British Standards may be obtained from BSI Customer Services, 389 Chiswick High Road, London W4 4AL. Tel: +44 (0)20 8996 9001. e-mail: cservices@bsi-global.com

Principal Symbols

Listed below are the symbols and suffixes common to European Codes

LATIN UPPER AND LOWER CASE

A	accidental action; area
a	distance; throat thickness of a weld
B	bolt force
b	breadth
c	outstand
d	depth of web; diameter
E	modulus of elasticity
e	edge distance; end distance
F	action; force
f	strength of a material
G	permanent action; shear modulus of steel
H	total horizontal load or reaction; warping constant of section
h	height
i	radius of gyration
I	second moment of area
k	stiffness
L	length; span; buckling length
l	effective buckling length; torsion constant; warping constant
M	bending moment
N	axial force
n	number
p	pitch; spacing
Q	variable action; prying force
q	uniformly distributed action
R	resistance; reaction
r	radius; root radius; number of redundancies
S	stiffness
s	staggered pitch; distance; bearing length
T	torsional moment
t	thickness
uu	principal major axis
vv	principal minor axis

V shear force; total vertical load or reaction

v shear stress

W section modulus

w deflection

GREEK LOWER CASE

α coefficient of linear thermal expansion; angle; ratio; factor

β angle; ratio; factor

γ partial safety factor

δ deflection; deformation

ε strain; coefficient $(235/f_y)^{1/2}$ where f_y is in MPa

η distribution factor; shear area factor; critical buckling mode; buckling imperfection coefficient

θ angle; slope

λ slenderness ratio; ratio

μ slip factor

ν Poisson's ratio

ρ unit mass; factor

σ normal stress; standard deviation

τ shear stress

ϕ rotation; slope; ratio

χ reduction factor for buckling

ψ stress ratio

SUFFIXES

Ed   design strength

el   elastic

f   flange

j   joint

o   initial; hole

p   plate

pl   plastic

Rd   resistance strength

t   torsion

u   ultimate strength

v   shear

w   web; warping

x x-x axis

y y-y axis; yield strength

z z-z axis

Chapter 1 / General

1.1 DESCRIPTION OF STEEL STRUCTURES

1.1.1 Shapes of Steel Structures

The introduction of structural steel, circa 1856, provided an additional building material to stone, brick, timber, wrought iron and cast iron. The advantages of steel are high strength, high stiffness and good ductility combined with relative ease of fabrication and competitive cost. Steel is most often used for structures where loads and spans are large and therefore is not often used for domestic architecture.

Steel structures include low-rise and high-rise buildings, bridges, towers, pylons, floors, oil rigs, etc. and are essentially composed of frames which support the self-weight, dead loads and external imposed loads (wind, snow, traffic, etc.). For convenience load bearing frames may be classified as:

(a) Miscellaneous isolated simple structural elements (e.g. beams and columns) or simple groups of elements (e.g. floors).
(b) Bridgeworks.
(c) Single storey factory units (e.g. portal frames).
(d) Multi-storey units (e.g. tower blocks).
(e) Oil rigs.

A real structure consists of a load bearing frame, cladding and services as shown in Fig. 1.1(a). A load bearing frame is an assemblage of members (structural elements) arranged in a regular geometrical pattern in such a way that they interact through structural connections to support loads and maintain them in equilibrium without excessive deformation. Large deflections and distortions in structures are controlled by the use of bracing which stiffens the structure and can be in the form of diagonal structural elements, masonry walls, reinforced concrete lift shafts, etc. A load bearing steel frame is idealized, for the purposes of structural design, as center lines representing structural elements which intersect at joints, as shown in Fig. 1.1(b). Other shapes of load bearing frames are shown in Figs 1.1 (c) to (e).

Structural elements are required to resist forces and displacements in a variety of ways, and may act in tension, compression, flexure, shear, torsion or in any combination of

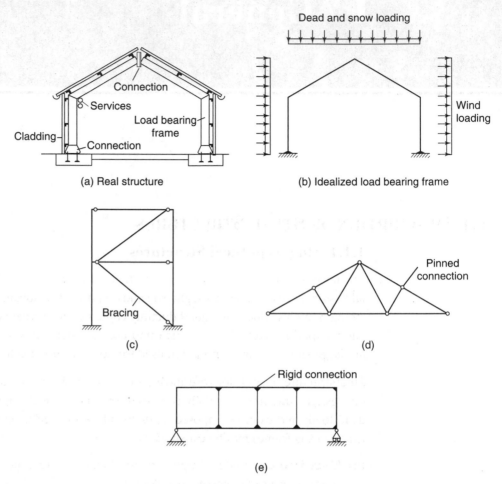

FIGURE 1.1 Typical load bearing frames

these forces. The structural behaviour of a steel element depends on the nature of the forces, the length and shape of the cross section of the member, the elastic properties and the magnitude of the yield stress. For example a tie behaves in a linear elastic manner until yield is reached. A slender strut behaves in a non-linear elastic manner until first yield is attained, provided that local buckling does not occur first. A laterally supported beam behaves elastically until a plastic hinge forms, while an unbraced beam fails by elastic torsional buckling. These modes of behaviour are considered in detail in the following chapters.

The structural elements are made to act as a frame by connections. These are composed of plates, welds and bolts which are arranged to resist the forces involved. The connections are described for structural design purposes as pinned, semi-rigid and rigid, depending on the amount of rotation, and are described, analysed and designed in detail in Chapter 7.

1.1.2 Standard Steel Sections

The optimization of costs in steel construction favours the use of structural steel elements with standard cross-sections and common bar lengths of 12 or 15 m. The billets of steel are hot rolled to form bars, flats, plates, angles, tees, channels, I sections and hollow sections as shown in Fig. 1.2. The detailed dimensions of these sections are given in BS 4, Pt 1 (2005), BSEN 10056-1 (1990), and BSEN 10210-2 (1997).

Where thickness varies, for example, Universal beams, columns and channels, sections are identified by the nominal size, that is, 'depth × breadth × mass per unit length × shape'. Where thickness is constant, for example, tees and angle sections, the identification is 'breadth × depth × thickness × shape'. In addition a section is identified by the grade of steel.

To optimize on costs steel plates should be selected from available stock sizes. Thicknesses are in the range of 6, 8, 10, 12,5, 15 mm and then in 5 mm increments. Thicknesses of less than 6 mm are available but because of lower strength and poorer corrosion resistance their use is limited to cold formed sections. Stock plate widths are in the range 1, 1,25, 1,5, 2, 2,5 and 3 m, but narrow plate widths are also available. Stock plate lengths are in the range 2, 2,5, 3, 4, 5, 6, 10 and 12 m. The adoption of stock widths and lengths avoids work in cutting to size and also reduces waste.

The application of some types of section is obvious, for example, when a member is in tension a round or flat bar is the obvious choice. However, a member in tension may

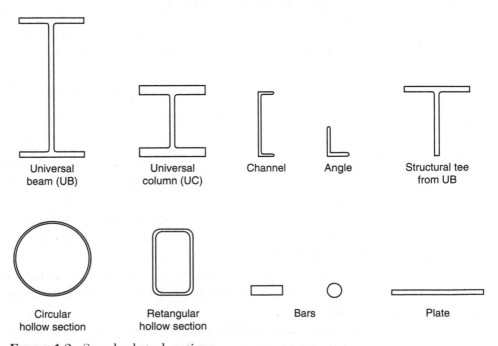

FIGURE 1.2 Standard steel sections

be in compression under alternative loading and an angle, tee, or tube is often more appropriate. The connection at the end of a bar or tube, however, is more difficult to make.

If a structural element is in bending about one axis then the 'I' section is the most efficient because a large proportion of the material is in the flanges, that is, at the extreme fibres. Alternatively, if a member is in bending about two axes at right angles and also supports an axial load then a tube, or rectangular hollow section, is more appropriate.

Other steel sections available are cold formed from steel plate into a variety of cross sections for use as lightweight lattice beams, glazing bars, shelf racks, etc. Not all these sections are standardized because of the large variety of possible shapes and uses, however, there is a wide range of sections listed in BSEN 10162 (2003). Local buckling can be a problem and edges are stiffened using lips. Also when used as beams the relative thinness of the material may lead to web crushing, shear buckling and lateral torsional buckling. Although the thickness of the material (1–3 mm) is less than that of the standard sections the resistance to corrosion is good because of the surface finish obtained by pickling and oiling. After degreasing this surface can be protected by galvanizing, or painting, or plastic coating. The use in building of cold formed sections in light gauge plate, sheet and strip steel 6 mm thick and under is dealt with in BSEN 5950 (2001) and EN 1993-1-1 (2005).

1.1.3 Structural Classification of Steel Sections (cl 5.5. EN 1993-1-1 (2005))

A section, or element of a member, in compression due to an axial load may fail by local buckling. Local buckling can be avoided by limiting the width to thickness ratios (b/t_f or d/t_w) of each element of a cross-section. The use of the limiting values given in Table 5.2, EN 1993-1-1 (2005) avoids tedious and complicated calculations.

Depending on the b/t_f or d/t_w ratios standard or built-up sections are classified for structural purposes as:

- *Class 1*: Low values of b/t_f or d/t_w where a plastic hinge can be developed with sufficient rotation capacity to allow redistribution of moments within the structure.
- *Class 2*: Full plastic moment capacity can be developed but local buckling may prevent development of a plastic hinge with sufficient rotation capacity to permit plastic design.
- *Class 3*: High values of b/t_f and d/t_w, where stress at the extreme fibres can reach design strength but local buckling may prevent the development of the full plastic moment.
- *Class 4*: Local buckling may prevent the stress from reaching the design strength. Effective widths are used to allow for local buckling (cl 5.5.2(2), EN 1993-1-1 (2005)).

1.1.4 Structural Joints (EN 1993-1-8 (2005))

Structural elements are connected together at joints which are not necessarily at the ends of members. A structural connection is an assembly of components (plates, bolts, welds, etc.) arranged to transmit forces from one member to another. A connection may be subject to any combination of axial force, shear force and bending moment in relation to three perpendicular axes, but for simplicity, where appropriate, the situation is reduced to forces in one plane.

There are other types of joints in structures which are not structural connections. For example a movement joint is introduced into a structure to take up the free expansion and contraction that may occur on either side of the joint due to temperature, shrinkage, expansion, creep, settlement, etc. These joints may be detailed to be water-tight but do not generally transmit forces. Detailed recommendations are given by Alexander and Lawson (1981). Another example is a construction joint which is introduced because components are manufactured to a convenient size for transportation and need to be connected together on site. In some cases these joints transmit forces but in other situations may only need to be waterproof.

1.2 DEVELOPMENT, MANUFACTURE AND TYPES OF STEEL

1.2.1 Outline of Developments in Design Using Ferrous Metals

Prior to 1779, when the Iron Bridge at Coalbrookdale on the Severn was completed, the most important materials used for load bearing structures were masonry and timber. Ferrous materials were only used for fastenings, armaments and chains.

The earliest use of cast iron columns in factory buildings (circa 1780) enabled relatively large span floors to be constructed. Due to a large number of disastrous fires around 1795, timber beams were replaced by cast iron with the floors carried on brick jack arches between the beams. This mode of construction was pioneered by Strutt in an effort to attain a fire proof construction technique.

Cast iron, however, is weak in tension and necessitates a tension flange larger than the compression flange and consequently cast iron was used mainly for compression members. Large span cast iron beams were impractical, and on occasions disastrous as in the collapse of the Dee bridge designed by Robert Stephenson in 1874. The last probable use of cast iron in bridge works was in the piers of the Tay bridge in 1879 when the bridge collapsed in high winds due to poor design and unsatisfactory supervision during construction.

In an effort to overcome the tensile weakness of cast iron, wrought iron was introduced in 1784 by Henry Cort. Wrought iron enabled the Victorian engineers to produce the following classic structures. Robert Stephenson's Brittania Bridge was the first box girder bridge and represented the first major collaboration between engineer, fabricator (Fairburn) and scientist (Hodgkinson). I.K. Brunel's Royal Albert Bridge at Saltash combined an arch and suspension bridge. Telford's Menai suspension bridge used wrought iron chains which have sine been replaced by steel chains. Telford's Pont Cysyllte is a canal aqueduct near Llangollen. The first of the four structures was replaced after a fire in 1970. The introduction of wrought iron revolutionized ship building and enabled Brunel to produce the S.S. Great Britain.

Steel was first produced in 1740, but was not available in large quantities until Bessemer invented the converter in 1856. The first major structure to use the new steel exclusively was Fowler and Baker's railway bridge at the Firth of Forth. The first steel rail was rolled in 1857 and installed at Derby where it was still in use 10 years later. Cast iron rails in the same position lasted about 3 months. Steel rails were in regular production at Crewe under Ramsbottom from 1866.

By 1840 standard shapes in wrought iron, mainly rolled flats, tees and angles, were in regular production and were appearing in structures about 10 years later. Compound girders were fabricated by riveting together the standard sections. Wrought iron remained in use until around the end of the nineteenth century.

By 1880 the rolling of steel 'I' sections had become widespread under the influence of companies such as Dorman Long. Riveting continued in use as a fastening method until around 1950 when it was superseded by welding. Bessemer steel production in Britain ended in 1974 and the last open hearth furnace closed in 1980. Further information on the history of steel making can be found in Buchanan (1972), Cossons (1975), Derry and Williams (1960), Pannel (1964) and Rolt (1970).

1.2.2 Manufacture of Steel Sections

The manufacture of standard steel sections, although now a continuous process, can be conveniently divided into three stages:

(1) Iron production
(2) Steel production
(3) Rolling.

Iron production is a continuous process and consists of chemically reducing iron ore in a blast furnace using coke and crushed limestone. The resulting material, called cast iron, is high in carbon, sulphur and phosphorus.

Steel production is a batch process and consists in reducing the carbon, sulphur and phosphorus levels and adding, where necessary, manganese, chromium, nickel, vanadium, etc. This process is now carried out using a Basic Oxygen Converter, which consists of a vessel charged with molten cast iron, scrap steel and limestone through which oxygen is passed under pressure to reduce the carbon content by oxidation. This is a batch process which typically produces about 250–300 tons every 40 min. The alternative electric arc furnace is in limited use (approximately 5% of the UK steel production), and is generally used for special steels such as stainless steel.

From the converter the steel is 'teemed' into ingots which are then passed to the rolling mills for successive reduction in size until the finished standard section is produced. The greater the reduction in size the greater the work hardening, which produces varying properties in a section. The variation in cooling rates of different thicknesses introduces residual stresses which may be relieved by the subsequent straightening process. Steel plate is now produced using a continuous casting procedure which eliminates, ingot casting, mould stripping, heating in soaking pits and primary rolling. Continuous casting permits, tighter control, improved quality, reduced wastage and lower costs.

1.2.3 Types of Steel

The steel used in structural engineering is a compound of approximately 98% iron and small percentages of carbon, silicon, manganese, phosphorus, sulphur, niobium and vanadium as specified in BS 4360 (1990). Increasing the carbon content increases strength and hardness but reduces ductility and toughness. Carbon content therefore is restricted to between 0,25% and 0,2% to produce a steel that is weldable and not brittle. The niobium and vanadium are introduced to raise the yield strength of the steel; the manganese improves corrosion resistance; and the phosphorus and sulphur are impurities. BS 4360 (1990) also specifies tolerances, testing procedure and specific requirements for weldable structural steel.

Steels used in practice are identified by letters and number, for example, S235 is steel with a tensile yield strength of 235 MPa (Table 3.1, EN 1993-1-1 (2005)).

1.3 STRUCTURAL DESIGN

1.3.1 Initiation of a Design

The demand for a structure originates with the client. The client may be a private person, private or public firm, local or national government, or a nationalized industry.

In the first stage preliminary drawings and estimates of costs are produced, followed by consideration of which structural materials to use, that is, reinforced concrete, steel, timber, brickwork, etc. If the structure is a building, an architect only may be involved at this stage, but if the structure is a bridge or industrial building then a civil or structural engineer prepares the documents.

If the client is satisfied with the layout and estimated costs then detailed design calculations, drawings and costs are prepared and incorporated in a legal contract document. The design documents should be adequate to detail, fabricate and erect the structure.

The contract document is usually prepared by the consultant engineer and work is carried out by a contractor who is supervised by the consultant engineer. However, larger firms, local and national government, and nationalized industries, generally employ their own consultant engineer.

The work is generally carried out by a contractor, but alternatively direct labour may be used. A further alternative is for the contractor to produce a design and construct package, where the contractor is responsible for all parts and stages of the work.

1.3.2 The Object of Structural Design

The object of structural design is to produce a structure that will not become unserviceable or collapse in its lifetime, and which fulfils the requirements of the client and user at reasonable cost.

The requirements of the client and user may include any or all of the following:

(a) The structure should not collapse locally or overall.
(b) It should not be so flexible that deformations under load are unsightly or alarming, or cause damage to the internal partitions and fixtures; neither should any movement due to live loads, such as wind, cause discomfort or alarm to the occupants/users.
(c) It should not require excessive repair or maintenance due to accidental overload, or because of the action of weather.
(d) In the case of a building, the structure should be sufficiently fire resistant to, give the occupants time to escape, enable the fire brigade to fight the fire in safety and to restrict the spread of fire to adjacent structures.

The designer should be conscious of the costs involved which include:

(a) The initial cost which includes fees, site preparation, cost of materials and construction.
(b) Maintenace costs (e.g. decoration and structural repair).
(c) Insurance chiefly against fire damage.
(d) Eventual demolition.

It is the responsibility of the structural engineer to design a structure that is safe and which conforms to the requirements of the local bye-laws and building regulations. Information and methods of design are obtained from Standards and Codes of Practice and these are 'deemed to satisfy' the local bye-laws and building regulations. In exceptional circumstances, for example, the use of methods validated by research or testing, an alternative design may be accepted.

A structural engineer is expected to keep up to date with the latest research information. In the event of a collapse or malfunction where it can be shown that the engineer has failed to reasonably anticipate the cause or action leading to collapse, or has failed to apply properly the information at his disposal, that is, Codes of Practice, British Standards, Building Regulations, research or information supplied by the manufacturers, then he may be sued for professional negligence. Consultants and contractors carry liability insurance to mitigate the effects of such legal action.

1.3.3 Limit State Design (cl 2.2, EN 1993-1-1 (2005))

It is self-evident that a structure should be 'safe' during its lifetime, that is, free from the risk of collapse. There are, however, other risks associated with a structure and the term safe is now replaced by the term 'serviceable'. A structure should not during its lifetime become 'unserviceable', that is, it should be free from risk of collapse, rapid deterioration, fire, cracking, excessive deflection, etc.

Ideally it should be possible to calculate mathematically the risk involved in structural safety based on the variation in strengths of the material and variation in the loads. Reports, such as the CIRIA Report 63 (1977), have introduced the designer to elegant and powerful concept of 'structural reliability'. Methods have been devised whereby engineering judgement and experience can be combined with statistical analysis for the rational computation of partial safety factors in codes of practice. However, in the absence of complete understanding and data concerning aspects of structural behaviour, absolute values of reliability cannot be determined.

It is not practical, nor is it economically possible, to design a structure that will never fail. It is always possible that the structure will contain material that is less than the required strength or that it will be subject to loads greater than the design loads. If actions (forces) and resistance (strength of materials) are determined statistically then the relationship can be represented as shown in Fig. 1.3. The design value of resistance (R_d) must be greater than the design value of the actions (A_d).

It is therefore accepted that 5% of the material in a structure is below the design strength, and that 5% of the applied loads are greater than the design loads. This does not mean therefore that collapse is inevitable, because it is extremely unlikely that the weak material and overloading will combine simultaneously to produce collapse.

The philosophy and objectives must be translated into a tangible form using calculations. A structure should be designed to be safe under all conditions of its useful life

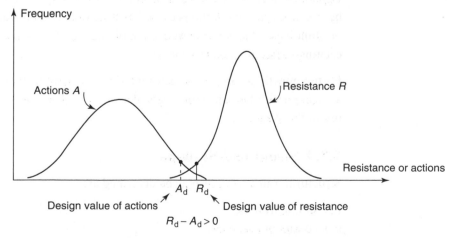

FIGURE 1.3 Statistical relationship between actions and resistance

and to ensure that this is accomplished certain distinct performance requirements, called 'limit states', have been identified. The method of limit state design recognizes the variability of loads, materials, construction methods and approximations in the theory and calculations.

Limit states may be at any stage of the life of a structure, or at any stage of loading and are important for the design of steelwork. To reduce the number of load cases to be considered only serviceability and ultimate limit states are specified. Each of these sections is subdivided although some may not be critical in every design. Calculations for limit states involve loads and load factors (Chapter 3), and material factors and strengths (Chapter 2).

Stability, an ultimate limit state, is the ability of a structure, or part of a structure, to resist overturning, overall failure and sway. Calculations should consider the worst realistic combination of loads at all stages of construction.

All structures, and parts of structures, should be capable of resisting sway forces, for example, by the use of bracing, 'rigid' joints, or shear walls. Sway forces arise from horizontal loads, for example, winds, and also from practical imperfections, for example, lack of verticality. The sway forces from practical imperfections are difficult to quantify and advice is given in cl 5.3.3, EN 1993-1-1 (2005).

Also involved in limit state design is the concept of structural integrity. Essentially this means that the structure should be tied together as a whole, but if damage occurs, it should be localized.

Deflection is a serviceability limit state. Deflections should not impair the efficiency of a structure, or its components, nor cause damage to the finishes. Generally the worst realistic combination of unfactored imposed loads is used to calculate elastic deflections. These values are compared with empirical values related to the length of a member or height.

Dynamic effects to be considered at the serviceability limit state are vibrations caused by machines, and oscillations caused by harmonic resonance, for example, wind gusts on buildings. The natural frequency of the building should be different from the exciting source to avoid resonance.

Fortunately there are few structural failures and when they do occur they are often associated with human error involved in design calculations, or construction, or in the use of the structure.

1.3.4 Structural Systems

Structural frame systems may be described as:

(a) simple frames,
(b) continuous frames,
(c) semi-continuous frames.

These titles refer to the types of joints and whether bracing is included.

Simple design assumes that 'pin joints' connect the members and joint rotations are prevented by bracing. Historically this method was popular because parts of the structure could be designed in isolation and calculations could be done by hand. With the advent of the computer calculations are less onerous but the method is still in use.

Continuous frames assume that the connections between members are rigid and therefore the angles between members can be maintained without the use of bracing. Calculations for the design of members and connections are more complicated and a computer is generally used. Global analysis of the frame is based on elastic, plastic, or elastic–plastic analysis assuming full continuity.

Semi-continuous frames acknowledges that in reality, end moments and rotations exist at the connections. Global analysis using the computer is based on the moment–rotation and force displacement characteristics of the connections. Bracing is often necessary for this type of frame to reduce sway.

1.3.5 Errors

The consequences of an error in structural design can lead to loss of life and damage to property, and it is necessary to appreciate where errors can occur. Small errors in design calculations can occur in the rounding off of figures but these generally do not lead to failures. The common sense advice is that the accuracy of the calculation should match the accuracy of the values given in the European Code.

Errors that occur in structural design calculations and which affect structural safety are:

(1) Ignorance of the physical behaviour of the structure under load and which consequently introduces errors in the basic assumptions used in the theoretical analysis.
(2) Errors in estimating the loads, especially the erection forces.
(3) Numerical errors in the calculations. These should be eliminated by checking, but when speed is paramount checks are often ignored.
(4) Ignorance of the significance of certain effects (e.g. residual stresses, fatigue, etc).
(5) Introduction of new materials, or methods, which have not been proved by tests.
(6) Insufficient allowance for tolerances or temperature strains.
(7) Insufficient information (e.g. in erection procedures).

Errors that can occur in workshops or on construction sites are:

(1) Using the wrong grade of steel, and when welding using the wrong type of electrode.
(2) Using the wrong weight of section. A number of sections are the same nominal size but differ in web or flange thickness.
(3) Errors in manufacture (e.g. holes in the wrong position).

Errors that occur in the life of a structure and also affect safety are:

(1) Overloading
(2) Removal of structural material (e.g. to insert service ducts)
(3) Poor maintenance.

1.4 FABRICATION OF STEELWORK

1.4.1 Drawings

Detailed design calculations are essential for any steel work design but the sizes of the members, dimensions and geometrical arrangement are usually presented as drawings. Initially the drawings are used by the fabricator and eventually by the contractor on site. General arrangement drawings are often drawn to scale of 1:100, while details are drawn to a scale of 1:20 or 1:10. Special details are drawn to larger scales where necessary.

Drawings should be easy to read and should not include superfluous detail. Some important notes are:

(a) Members and components should be identified by logically related mark numbers, for example, related to the grid system used in the drawings.
(b) The main members should be presented by a bold outline (0,4 mm wide) and dimension lines should be unobtrusive (0,1 mm wide).
(c) Dimensions should be related to centre lines, or from one end; strings of dimensions should be avoided. Dimensions should appear once only so that ambiguity cannot arise when revisions occur. Fabricators should not be put in the position of having to do arithmetic in order to obtain an essential dimension.
(d) Tolerances for erection purposes should be clearly shown.
(e) The grade of steel to be used should be clearly indicated.
(f) The size, weight and type of section to be used should be clearly stated.
(g) Detailing should take account of possible variations due to rolling margins and fabrication variations.
(h) Keep the design and construction as simple as possible. Where possible use simple connections, avoid stiffeners, use the minimum number of sections and avoid changes in section along the length of a member.
(i) Site access, transport and use of cranes should be considered.

1.4.2 Tolerances (cl 3.2.5, EN 1993-1-1 (2005))

Tolerances are limits places on unintentional inaccuracies that occur in dimensions which must be allowed for in design if structural elements and components are to fit together. In steelwork variations occur in the rolling process, marking out, cutting and drilling during fabrication, and in setting out during erection.

In the rolling process the allowable tolerances for length, width, thickness and flatness for plates are given in BS 4360 (1990). Length and width tolerances are positive while those for thickness and flatness are negative and positive. The dimensional and weight tolerances for sections are given in BS 4 Pt 1 (2005), or BS 10056-1 (1990), as appropriate.

During fabrication there is a tendency for members and components to increase rather than reduce, and the tolerance is therefore often specified as negative; it is often cheaper and simpler to insert packing rather than shorten a member, provided that the packing is not excessive. Where concrete work is associated with steelwork variations in dimensions are likely to be greater. When casting concrete, for example, errors in dimensions may arise from shrinkage or from warping of the shuttering, especially when it is re-used. Therefore, by virtue of the construction method, larger tolerances are specified for work involving concrete.

To facilitate erection all members and connections should be provided with the maximum tolerance that is acceptable from structural and architectural considerations. A typical example is a connection between a steel column and a reinforced concrete base. It would produce great difficulties if the base were set too high and a tolerance of approximately 50 mm is often included in the design, with provision for grouting under the base. Tolerances are also provided to allow lateral adjustment of the foundation bolts. Tolerances between concrete and steelwork are also important because two different contractors are involved.

1.4.3 Fabrication, Assembly and Erection of Steelwork

The drawings produced by the structural designer are used first by the steel fabricator and later by the contractor on site.

The steel fabricator obtains the steel either direct from the rolling mills or from the steel stockiest, and then cuts, drills and welds the steel components to form the structural elements as shown on the drawings. In general, for British practice, the welding is confined to the workshop and the connections on site are made using bolts. In American, however, site welding is common practice.

When marking out, the measurements of length for overall size, position of holes, etc. can be done by hand, but if there are several identical components then wooden or cardboard templates are made and repeated measurements avoided. Now automatic machines, controlled by a computer, or punched paper tape, are used to cut and drill standard sections. When completed, the steel work should be marked clearly and manufactured to the accepted tolerances.

When fabricated, parts of the structure are delivered to the site in the largest pieces that can be transported and erected. For example a lattice girder may be sent fully assembled to a site in this country, but sent in pieces to fit a standard transport container for erection abroad. All components should be assembled within the tolerances and cambers specified, and should not be bent or twisted or otherwise damaged.

On site the general contractor may be responsible for the assembly, erection, connections, alignment and leveling of the complete structure. Alternatively the erection work may be done by the steel fabricator, or sublet to a specialist steel erector. The objective of the erection process is to assemble the steelwork in the most cost-effective method whilst maintaining the stability of individual members, and/or part or complete structure. To do this it may be necessary to introduce cranes and temporary bracing which must also be designed to resist the loads involved.

During assembly on site it is inevitable that some components will not fit, despite the tolerances that have been allowed. A typical example is that the faying surfaces for a friction grip joint are not in contact when the bolts are stressed. Other examples are given by Mann and Morris (1981). The correction of some faults and the consequent litigation can be expensive.

1.4.4 Testing of Steelwork (cl 2.5, EN 1993-1-1 (2005))

Steel is routinely sampled and tested during production to maintain quality. However, occasionally new methods of construction are suggested and there may be some doubt as to the validity of the assumptions of behaviour of the structure. Alternatively if the structure collapses there may be some dispute as to the strength of a component, or member, of the structure. In such cases testing of components, or part of the structure, may be necessary. However it is generally expensive because of, the accuracy required, cost of material, cost of fabrication, necessity to repeat tests to allow for variations and to report accurately.

Tests may be classified as:

(a) acceptance tests – non-destructive for confirming structural performance,
(b) strength tests – used to confirm the calculated capacity of a component or structure,
(c) tests to failure – to determine the real mode of failure and the true capacity of a specimen,
(d) check tests – where the component assembly is designed on the basis of tests.

The size, shape, position of the gauges, and method of testing of small sample pieces of steel is given in BS 4360 (1990) and BSEN 10002-1 (2001). The tensile test is most frequently employed, and gives values of, Young's modulus, limit of proportionality, yield stress or proof stress, percentage elongation and ultimate stress. Methods of destructive testing fusion welded joints and weld metal in steel are given in numerous Standards.

The Charpy V-notch test for impact resistance is used to measure toughness, that is, the total energy, elastic and plastic, which can be absorbed by a specimen before fracture. The test specimen is a small beam of rectangular cross section with a 'V' notch at mid-length. The beam is fractured by a blow from a swinging pendulum, and the amount of energy absorbed is calculated from the loss of height of the pendulum swing after fracture. Details of the test specimen and procedure are given in BS 4360 (1990) and BSEN 10045-1 (1990). The Charpy V-notch test is often used to determine the temperature at which transition from brittle to ductile behaviour occurs.

Structures which are unconventional, and/or method of design which are unusual or not fully validated by research, should be subject to acceptance tests. Essentially these consist of loading the structure to ensure that it has adequate strength to support, for example, 1 (test dead load) + 1,15 (remainder of dead load) + 1,25 (imposed load).

Where welds are of vital importance, for example, in pressure vessels, they should be subject to non-destructive tests. The defects that can occur in welds are: slag inclusions, porosity, lack of penetration and sidewall fusion, liquation, solidification, hydrogen cracking, lamellar tearing and brittle fracture.

A surface crack in a weld may be detected visibly but alternatively a dye can be sprayed onto the joint which seeps into the cracks. After removing any surplus dye the weld is resprayed with a fine chalk suspension and the crack then shows as a coloured line on the white chalk background. A variant of this technique is to use fluorescent dye and a crack then shows as a bright green line in ultra violet light. A surface crack may also be detected if the weld joint area is magnetized and sprayed with iron powder. The powder congregates along a crack, which shows as a black line.

Other weld defects cannot be detected on the surface and alternative methods must be used. Radiographic methods use an X-ray, or gamma-ray, source on one side of the weld and a photographic film on the other. Rays are absorbed by the weld metal, but if there is a hole or crack there is less absorbtion which shows as a dark area on the film. Not all defects are detected by radiography since the method is sensitive to the orientation of the flaw, for example, cracks at right angles to the X-ray beam are not detected. Radiography also requires access to both sides of the joint. The method is therefore most suitable for in-line butt weld for plates.

An alternative method to detect hidden defects in welds uses ultrasonics. If a weld contains a flaw then high frequency vibrations are reflected. The presence of a flaw can therefore be indicated by monitoring the reduction of transmission of ultrasonic vibrations, or by monitoring the reflections. The reflection method is extremely useful for welds where access is only possible from one side. Further details can be obtained from Gourd (1980).

REFERENCES

Alexander S.J. and Lawson R.M. (1981). *Movement design in buildings.* Technical Note 107. Construction Industry Research and Information Association Publication.

BS 4 (2005). *Specification for hot rolled sections Pt1.* BSI.

BSEN 10002-1 (2001). *Methods for tensile testing of metals.* BSI.

BSEN 10045-1 (1990). *The Charpy V-notch impact test on metals.* BSI.

BSEN 10162 (2003). *Cold rolled sections.* BSI.

BS 4360 (1990) *Specification for weldable structural steels.* BSI.

BSEN 10056-1 (1990). *Hot rolled structural steel sections.* BSI.

BSEN 10210-2 (1997). *Hot rolled structural steel hollow sections.* BSI.

BSEN 5950 (2001 and 1998). *Structural use of steel work in buildings.* BSI.

EN 1993-1-1 (2005). *General rules and rules for buildings.* BSI.

EN 1993-1-8 (2005). *Design of joints.* BSI.

Buchanan R.A. (1972). *Industrial archeology in Britain.* Penguin.

CIRIA (1977). *Rationalisation of safety and serviceability factors in structural codes.* Report 63. Construction Industry Research and Information Association Publication.

Cossons N. (1975). *The BP book of industrial archeology.* David and Charles.

Derry T.K. and Williams T.I. (1960). *A short history of technology.* Oxford University Press.

Gourd L.M. (1980). *Principles of welding technology.* Edward Arnold.

Mann A.P. and Morris L.J. (1981). *Lack of fit in steel structures*, Technical Report 87. Construction Industry Research and Information Association Publication.

Pannel J.P.M. (1964). *An illustrated history of civil engineering.* Thames and Hudson.

Rolt, L.T.C. (1970). *Victorian engineering.* Penguin.

Chapter 2 / Mechanical Properties of Structural Steel

2.1 VARIATION OF MATERIAL PROPERTIES

All manufactured material properties vary because the molecular structure of the material is not uniform and because of inconsistencies in the manufacturing process. The variations that occur in the manufacturing process are dependent on the degree of control. Variations in material properties must be recognized and incorporated into the design process.

The material properties that are of most importance for structural design using steel are strength and Young's modulus. Other properties which are of lesser importance are hardness, impact resistance and melting point.

If a number of samples are tested for a particular property, for example, strength, and the number of specimens with the same strength (frequency) plotted against the strength, then the results approximately fit a normal distribution curve as shown in Fig. 2.1.

This curve can be expressed mathematically by the equation shown in Fig. 2.1 which can be used to define 'safe' values for design purposes.

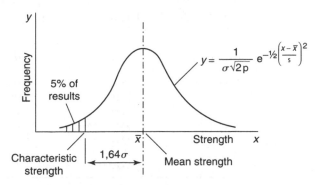

$$y = \frac{1}{\sigma\sqrt{2p}}\, e^{-\frac{1}{2}\left(\frac{x-\bar{x}}{s}\right)^2}$$

FIGURE 2.1 Variation in material properties

2.2 Characteristic Strength (cl 3.1, EN 1993-1-1 (2005))

A strength to be used as a basis for design must be selected from the variation in values shown in Fig. 2.1. This strength, when defined, is called the characteristic strength. If the characteristic strength is defined as the mean strength, then clearly from Fig. 2.1, 50% of the material is below this value. This is not acceptable. Ideally the characteristic strength should include 100% of the samples, but this is also impractical because it is a low value and results in heavy and costly structures. A risk is therefore accepted and it is therefore recognized that 5% of the samples fall below the characteristic strength.

The characteristic strength is calculated from the equation

$$f_y = f_{mean} - 1{,}64\sigma$$

where for n samples the standard deviation

$$\sigma = \left[\frac{\Sigma(f_{mean} - f)^2}{(n - 1)} \right]^{1/2}$$

Nominal values of the characteristic yield strengths and ultimate tensile strengths are given in Table 3.1, EN 1993-1-1 (2005) and some examples are given in Table 2.1.

TABLE 2.1 Some nominal values of yield strength for hot rolled steel (Table 3.1, EN 1993-1-1 (2005)).

Standard EN 10025-2: Grade	Nominal thickness of material			
	$t \leq 40\,mm$		$40 < t \leq 80\,mm$	
	$f_y\ (MPa)$	$f_u\ (MPa)$	$f_y\ (MPa)$	$f_u\ (MPa)$
S235	235	360	215	360
S275	275	430	255	410
S355	355	510	335	470

TABLE 2.2 Some partial safety factors (cl 6.1, EN 1993-1-1 (2005) and Table 2.1, EN 1993-1-8 (2005)).

Situation	Symbol	Value
Buildings	γ_{M0}	1,00
	γ_{M1}	1,00
	γ_{M2}	1,25
Joints	γ_{M3}	1,25
	γ_{M4}	1,00
	γ_{M5}	1,00
	γ_{M7}	1,10

2.3 DESIGN STRENGTH (cl 6.1, EN 1993-1-1 (2005))

The characteristic strength of steel is the value obtained from tests at the rolling mills, but by the time the steel becomes part of a finished structure this strength may be reduced (e.g. by corrosion or accidental damage). The strength to be used in design calculations is therefore the characteristic strength divided by a partial safety factor (γ_M) (Table 2.2).

2.4 OTHER DESIGN VALUES FOR STEEL (cl 3.2.6, EN 1993-1-1 (2005))

The elastic modulus for steel (E) is obtained from the relationship between stress and strain as shown in Fig. 2.2. This is a material property and therefore values from a set of samples vary. However, the variation for steel is very small and the European Code assumes $E = 210$ GPa.

The elastic shear modulus (G) is related to Young's modulus by the expression

$$E = 2G(1 + v)$$

where Poissons ratio $v = 0{,}3$ in the elastic range and is used in calculations involving plates.

The thermal coefficient of expansion for steel is given as $\alpha = 12\text{E-}6/K$ for $T \leq 100°\text{C}$ and is used in calculations involving temperature changes.

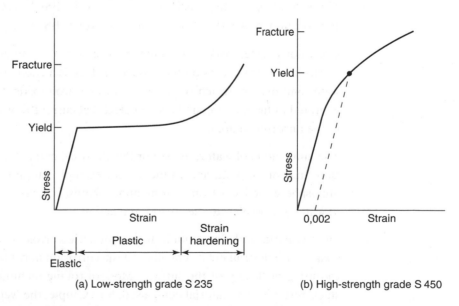

(a) Low-strength grade S 235   (b) High-strength grade S 450

FIGURE 2.2   Tensile stress–strain relationships for steel

Hardness is material property that is occasionally of importance in structural steel design. It is measured by the resistance the surface of the steel offers to, the indentation of a hardened steel ball (Brinell test), a square-based diamond pyramid (Vickers test) or a diamond cone (Rockwell test). Higher strength often correlates with greater hardness but this relationship is not infallible.

Ductility may be described as the ability of a material to change its shape without fracture. This is measured by the percentage elongation, that is, $100 \times$ (change in length)/(original length). Values of 20% can be obtained for mild steel but it is less for high-strength steel. A high value is advantageous because it allows the redistribution of stresses at ultimate load and the formation of plastic hinges.

2.5 CORROSION AND DURABILITY OF STEELWORK (cl 4, EN 1993-1-1 (2005))

Durability is a service limit state and the following factors should be considered at the design stage:

(a) Environment,
(b) Degree of exposure,
(c) Shape of the members and details,
(d) Quality of workmanship and control,
(e) Protective measures,
(f) Maintenance.

Methods of protecting steel work are given in BSEN ISO 12944 (1998) and the specification for weather resistant steel is given is BS 7668 (1994).

Corrosion of steel work reduces the cross-section of members and thus affects safety. Corrosion, which occurs on the surface of steel, is a chemical reaction between iron, water and oxygen, which produces a hydrated iron oxide called rust. Electrons are liberated in the reaction and a small electrical current flows from the corroded area to the uncorroded area.

The elimination of water, oxygen or the electrical current, reduces the rate of corrosion. In contrast pollutants in the air, for example, sulphur dioxides from industrial atmospheres and salt from marine atmospheres, increase the electrical conductivity of water and accelerate the corrosion reaction.

Steel is particularly susceptible to atmospheric corrosion which is often severe in coastal or industrial environments and the corrosion may reduce the section size due to pitting or flaking of the surface. Modern rolling techniques and higher-strength steels result in less material being used, for example, the web of an 'I' section may be only 6 mm thick. Generally in structural engineering 8 mm is the minimum thickness used for exposed steel, and 6 mm for unexposed steel. For sealed hollow sections these limits are reduced to 4 and 3 mm, respectively.

Corrosion of steel usually takes the form of rust which is a complex oxide of iron. The rust builds up a deposit on the surface and may eventually flake off. The coating of rust does not inhibit corrosion, except in special steels, and corrosion progresses beneath the rust forming conical pits and the thickness of the metal is reduced. The conical pits can act as stress raisers, that is, centres of high local stress, and in cases where there are cyclic reversals of stress, may become the initiating points of fatigue cracks or brittle fracture.

The corrosion resistance of unprotected steel is dependent on its chemical composition, the degree of pollution in the atmosphere, and the frequency of wetting and drying. Low-strength carbon steels are inexpensive but are particularly susceptible to atmospheric corrosion which is often greatest in industrial or coastal environments. High-strength low-alloy steels (Cr–Si–Cu–P) do not pit as severely as carbon steels and the rust that forms becomes a protective coating against further deterioration. These steels therefore have several times the corrosion resistance of carbon steels.

The longer steel remains wet the greater the corrosion and therefore the detailing of steelwork should include drainage holes, avoid pockets and allow the free flow of air for rapid drying.

The most common, and cheapest form of protection process is to clean the surface by sand or shot blasting, and then to paint with a lead primer, generally in the workshop prior to delivery on site. Joint contact surfaces need not be protected unless specified. On site the steel is erected and protection is completed with an undercoat and finishing coat, or coats, of paint.

In the case of surfaces to be welded steel should not be painted, nor metal coated, within a suitable distance of any edges to be welded, if the paint specified or the metal coating is likely to be harmful to welders or impair the quality of the welds. Welds and adjacent parent metal should not be painted prior to de-slagging, inspection and approval.

Encasing steel in concrete provides an alkaline environment and no corrosion will take place unless water diffuses through the concrete carrying with it SO_2 and CO_2 gases from the air in the form of weak acids. The resulting corrosion of the steel and the increase in pressure spalls the concrete. Parts to be encased in concrete should not be painted nor oiled, and where friction grip fasteners are used protective treatment should not be applied to the faying surfaces.

A more expensive protection is zinc, or aluminium spray coating which is sometimes specified in corrosive atmospheres. Further improvements are hot dip zinc galvanizing, or the use of stainless steels. These and other forms of protection are described in BSEN ISO 12944 (1998). Recently zinc coated highly stressed steel has been shown to be susceptible to hydrogen cracking.

2.6 BRITTLE FRACTURE (cl 3.2.3, EN 1993-1-1 (2005))

Brittle fracture is critical at the ultimate limit state. Evidence of brittle fracture is a small crack, which may or may not be visible, and which extends rapidly to produce a sudden failure with few signs of plastic deformation. This type of fracture is more likely to occur in welded structures (Stout *et al.*, 2000).

The essential conditions leading to brittle fracture are:

(a) There must be a tensile stress in the material but it need not be very high, and may be a residual stress from welding.
(b) There must be a notch, or defect, or hole in the material which produces a stress concentration.
(c) The temperature of the material must be below the transition temperature (generally below room temperature). At low temperatures crack initiation and propagation is more likely because of lower ductility.

The mechanism of failure is that the notch, defect or hole raises the local tensile stresses to values as high as three times the average tensile stress. The material which generally fails by a shearing mechanism now tends to fail by a brittle fracture cleavage mechanism which exhibits considerably less plastic deformation. A drop in the temperature encourages the cleavage failure. A ductile material which has an extensive plastic range is more likely to resist brittle fracture and a test used as a guide to resistance to brittle fracture is the Charpy V-notch impact test (BS 7668 (1994)).

The importance of brittle fracture was shown by the failure of the welded 'liberty' cargo ships mass produced by the USA during the Second World War. The ships broke apart in harbour and at sea during the cold weather.

Brittle fracture is considered only where tensile stresses exist. The mode of failure is mainly dependent on the following:

(a) Steel strength grade
(b) Thickness of material
(c) Loading speed
(d) Lowest service temperature
(e) Material toughness
(f) Type of structural element.

No further check for brittle fracture need to be made if the conditions given in EN 1993-1-10 (2005) are satisfied for lowest temperature. For further information see NDAC (1970).

2.7 RESIDUAL STRESSES

Residual stresses are present in steel due to uneven heating and cooling. The stresses are induced in steel during, rolling, welding which constrains the structure to a particular geometry, force fitting of individual components, lifting and transportation. These stresses may be relieved by subsequent reheating and slow cooling but the process is expensive. The presence of residual stresses adversely affects the buckling of columns, introduces premature yielding, fatigue resistance and brittle fracture.

Welding raises the local temperature of the steel which expands relative to the surrounding metal. When it cools it contracts inducing tensile stresses in the weld and the immediately adjacent metal. These tensile stresses are balanced by compressive stresses in the metal on either side.

During rolling the whole of the steel section is initially at a uniform temperature, but as the rolling progresses some parts of the cross-section become thinner than others and consequently cool more quickly. Thus, as in the welded joint, the parts which cool last have a residual tensile stress and the parts which cool first may be in compression. Since the cooling rate also affects the yield strength of the steel, the thinner sections tend to have a higher yield stress than the thicker sections. A tensile test piece cut from the thin web of a Universal Beam will probably have a higher yield stress than one cut from the thicker flange. The residual stress and yield stress in rolled sections are also affected by the cold straightening which is necessary for many rolled sections before leaving the mills.

Residual stresses are not considered directly in the European Code but are allowed for in material factors. For further information see Ogle (1982).

2.8 FATIGUE

This is an ultimate limit state. The term fatigue is generally associated with metals and is the reduction in strength that occurs due to progressive development of existing small pits, grooves or cracks when subject to fluctuating loads. The rate of development of these cracks depends on the size of the crack and on the magnitude of the stress variation in the material and also the metallurgical properties. The number of stress variations, or cycles of stress, that a material will sustain before failure is called fatigue life and there is a linear experimental relationship between the log of the stress range and the log of the number of cycles. Welds are susceptible to a reduction in strength due to fatigue because of the presence of small cracks, local stress concentrations and abrupt changes of geometry.

Research into the fatigue strength of welded structures is described by Munse (1984). Other references are BS 5400 (1980), Grundy (1985) and ECSS.

All structures are subject to varying loads but the variation may not be significant. Stress changes due to fluctuations in wind loading need not be considered, but wind-induced oscillations must not be ignored. The variation in stress depends on the ratio of dead load to imposed load, or whether the load is cyclic in nature, for example, where machinery is involved. For bridges and cranes fatigue effects are more likely to occur because of the cyclic nature of the loading which causes reversals of stress.

Generally calculations are only required for:

(a) Lifting appliances or rolling loads,
(b) Vibrating machinery,
(c) Wind-induced oscillations,
(d) Crowd-induced oscillations.

The design stress range spectrum must be determined, but simplified design calculations for loading may be based on equivalent fatigue loading if more accurate data is not available. The design strength of the steel is then related to the number and range of stress cycles. For further information see EN 1993-1-9 (2005).

2.9 STRESS CONCENTRATIONS

Structural elements and connections often have abrupt changes in geometry and also contain holes for bolts. These features produce stress concentrations, which are localized stresses greater than the average stress in the element, for example, tensile stresses adjacent to a hole are approximately three times the average tensile stress. If the average stress in a component is low then the stress concentration may be ignored, but if high then appropriate methods of structural analysis must be used to cater for this effect. The effect of stress concentrations has been shown to be critical in plate web girders in recent history. Stress concentrations are also associated with fatigue and can affect brittle fracture. Formulae for stress concentrations are given in Roark and Young (1975).

2.10 FAILURE CRITERIA FOR STEEL

The structural behaviour of a metal at or close to failure may be described as ductile or brittle. A typical brittle metal is cast iron which exhibits a linear load–displacement relationship until fracture occurs suddenly with little or no plastic deformation. In contrast mild steel is a ductile material which also exhibits a linear load–displacement relationship, but at yield large plastic deformations occur before fracture.

The nominal yield strength is a characteristic strength in the European Code and is therefore an important failure criterion for steel. The tensile yield condition can be related to various stress situations, for example, tension, compression, shear or various combinations of stresses.

There are four generally acceptable theoretical yield criteria as follows:

(1) The maximum stress theory, which states that yield occurs when the maximum principal stress reaches the uniaxial tensile stress.

(2) The maximum strain theory, which states that yield occurs when the maximum principal tensile strain reaches the uniaxial tensile strain at yield.

(3) The maximum shear stress theory, which states that yield occurs when the maximum shear stress reaches half of the yield stress in uniaxial tension.

(4) The distortion strain energy theory, or shear strain energy theory, which states that yielding occurs when the shear strain energy reaches the shear strain energy in simple tension. For a material subject to principal stresses σ_1, σ_2 and σ_3 it is shown (Timoshenko, 1946) that this occurs when

$$(\sigma_1 - \sigma_2)^2 + (\sigma_2 - \sigma_3)^2 + (\sigma_3 - \sigma_1)^2 = 2f_y^2 \tag{2.1}$$

This theory was originally developed by Huber, Von-Mises and Hencky.

Alternatively Eq. (2.1) can be expressed in terms of direct stresses σ_b, σ_{bc} and σ_{bt}, and shear stress τ on two mutually perpendicular planes. It can be shown from Mohr's circle of stress that the principal stresses

$$\sigma_1 = \frac{(\sigma_b + \sigma_{bc})}{2} - \left[\frac{(\sigma_b - \sigma_{bc})^2}{4} + \tau^2 \right]^{1/2} \tag{2.2}$$

and

$$\sigma_2 = \frac{(\sigma_b - \sigma_{bc})}{2} + \left[\frac{(\sigma_b - \sigma_{bc})^2}{4} + \tau^2 \right]^{1/2} \tag{2.3}$$

If Eqs (2.2) and (2.3) are inserted in Eq. (2.1) with $\sigma_3 = 0$ and f_y is equal to the design stress f_y/γ_M then

$$(f_y/\gamma_M)^2 = \sigma_{bc}^2 + \sigma_b^2 - \sigma_{bc}\sigma_b + 3\tau^2 \tag{2.4}$$

If σ_{bc} is replaced by σ_{bt} with a change in sign then

$$(f_y/\gamma_M)^2 = \sigma_{bc}^2 + 3\tau^2 \tag{2.5}$$

This equation is a yield criteria which is applicable in some design situations, for example the design of welds (EN 1993-1-8 (2005)).

REFERENCES

BS 7668 (1994), BSEN 10029 (1991), BSEN 10113-2 (1993), BSEN 10113-3 (1993) and BS 10210-1 (1994). *Specification for weldable structural steels*. BSI.

BS 5400 (1980). *Code of practice for fatigue, Pt 10*. BSI.

BSEN ISO 12944-1 to 8 (1998) and BSEN ISO 14713 (1999). *Code of practice for protection of iron and steel structures*. BSI.

ECCS (1981). *Recommendations for fatigue design of steel structures*. European Convention for Structural Steelwork. Construction Press.

EN 1993-1-1 (2005). *General rules and rules for buildings*. BSI.

EN 1993-1-8 (2005). *Design of joints*. BSI.

EN 1993-1-9 (2005). *Design of steel structures: Fatigue strength of steel structures*. BSI.

EN 1993-1-10 (2005). *Design of steel structures: Selection of steel for fracture toughness and through thickness properties*. BSI.

Grundy, P. (1985). *Fatigue limit design for steel structures*. Civil Engineering Transactions, Institution of Civil Engineers, Australia, CE27, No. 1.

Munse, W.H. (1984). *Fatigue of welded structures,* Welding Research Council.

NDAC (1970). Brittle fracture in steel structures, *Navy Department Advisory Committee on Structural Steels* (ed G.M. Boyd). Butterworth.

Ogle, M.H. (1982). *Residual stresses in a steel box-girder bridge*. Tech Note 110. Construction Industry Research and Information Association Publication.

Roark, J.R. and Young, W.C. (1975). *Formula for Stress and Strain* (5th edition). McGraw-Hill.

Stout, R.D., Tor, S.S. and Ruzek, J.M. (1951). The effect of fabrication procedures on steels used in pressure vessels, *Welding Journal* **30**.

Timoshenko, S. (1946). *Strength of Materials, Pt II*. D. Van Nostrand.

Chapter 3 / Actions

3.1 DESCRIPTION

Actions are a set of forces (loads) applied to a structure, or/and deformations produced by temperature, settlement or earthquakes (EN 1990 (2002)).

Values of actions are obtained by determining characteristic or representative values of loads or forces. Ideally, loads applied to a structure during its working life, should be analysed statistically and a characteristic load is determined. The characteristic load might then be defined as the load above which no more than 5% of the loads exceed, as shown in Fig. 3.1. However, data is not available and the characteristic value of an action is given as a mean value, an upper value or a nominal value.

3.2 CLASSIFICATION OF ACTIONS

Actions are classified as:

(1) Permanent
(2) Variable
(3) Accidental
(4) Seismic

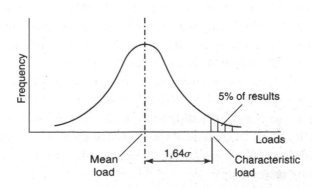

FIGURE 3.1 Variation in loads

In addition actions can be classified by:

(1) Variation in time:
    (a) Permanent actions (G), for example, self-weight and fixed equipment.
    (b) Variable actions (Q), for example, imposed loads, wind actions or snow loads.
(2) Spacial variation:
    (a) Fixed actions, for example, structures sensitive to self-weight.
    (b) Free actions which result in different arrangements of actions, for example, movable imposed loads, wind actions and snow loads.

3.3 ACTIONS VARYING IN TIME

3.3.1 Permanent Actions (G)

Permanent actions are due to the weight of the structure, that is, walls, permanent partitions, floors, roofs, finishes and services. The actual weights of the materials (G_k) should be used in the design calculations, but if these are unknown values of density in kN/m^3 may be obtained from EN 1991-1-1 (2002). Also included in this group are water and soil pressures, prestressing force and the indirect actions such as settlement of supports.

3.3.2 Variable Actions (Q)

(a) Imposed floor loads are variable actions and for various dwellings are given in EN 1991-1-1 (2002). These loads include a small allowance for impact and other dynamic effects that may occur in normal occupancy. They do not include forces resulting from the acceleration and braking of vehicles or movement of crowds. The loads are usually given in the form of a distributed load or an alternative concentrated load. The one that gives the most severe effect is used in design calculations.

When designing a floor it is not necessary to consider the concentrated load if the floor is capable of distributing the load and for the design of the supporting beams the distributed load is always used. When it is known that mechanical stacking of materials is intended, or other abnormal loads are to be applied to the floor, then actual values of the loads should be used, not those obtained from EN 1991-1-1 (2002). In multi-storey buildings the probability that all the floors will simultaneously be required to support the maximum loads is remote and reductions to column loads are therefore allowed.

(b) Snow roof loads are variable actions and are related to access for maintenance. They are specified in EN 1991-1-3 (2002) and, as with floor loads, they are expressed as a uniformly distributed load on plan, or as an alternative concentrated load. The magnitude of the loads decrease as the roof slope increases and in special situations, where roof shapes are likely to result in drifting snow, then loads are increased.

(c) Wind actions are variable but for convenience they are expressed as static pressures in EN 1991-1-4 (2002). The pressure at any point on a structure is related to the shape of the building, the basic wind speed, topography and ground roughness. The effects of vibration, such as resonance in tall buildings must be considered separately.

(d) Thermal effects need to be considered for chimneys, cooling towers, tanks, hot and cold storage, and services. They are classed as indirect variable actions. Elements of structures which are restrained or highly redundant introduce stresses which need to be determined (EN 1991-1-5 (2003)).

(e) EN 1991-1-2 (2002) covers the actions to be taken into account in the structural design of buildings which are required to give adequate performance in fire.

3.3.3 Accidental Actions (A)

(a) Accidental actions during execution include scaffolding, props and bracing (EN 1991-1-6 (2002)). These may involve consideration of construction loads, instability and collapse prior to the completion of the project.

Erection forces are of great importance in steelwork construction because pre-fabrication is normal practice. Compression members which will be restrained in a completed structure may buckle during erection when subject to relatively minor forces. Joints which are rigid when fully bolted may, during erection, act as a pin and induce collapse of the structure. Suspension points for members or parts of structures may have to be specified to avoid damage to components. It is extremely difficult to anticipate all possible erection forces and the contractor is responsible for erection which should be carried out with due care and attention. Nevertheless a designer should have knowledge of the most likely method of erection and design accordingly. If necessary temporary stiffening or supports should be specified, and/or instructions given.

(b) Accidental actions include impact and explosions which are covered in EN 1991-1-7 (2004). No structure can be expected to resist all actions but it must be designed so that it does not suffer extreme damage from probable actions, for example, vehicle collisions in a multi-storey car park. Local damage from accidental actions is acceptable.

(c) When designing for earthquakes the inertial forces must be calculated as described in EN 1998-8 (2004). This is not of major importance in the UK. Actions induced by cranes and machinery are dealt with in EN 1991-3 (2004).

3.4 DESIGN VALUES OF ACTIONS

Partial safety factors allow for the probability that there will be a variation in the effect of the action, for example, a variable action is more likely to vary than a permanent action. The values also allow for inaccurate modelling of the actions, uncertainties in the assessment of the effects of actions, and uncertainties in the assessment of the limit state considered.

The design value of an action is obtained by multiplying the characteristic value by a partial safety factor, for example, for a permanent action the design value $G_d = \gamma_G G_k$. For a variable action the design value $Q_d = \psi_0 Q_k$ or $\psi_1 Q_k$ or $\psi_2 Q_k$. These represent combination, frequency and quasi-permanent values. The combination value ($\psi_0 Q_k$) allows for the reduced probability that unfavourable independent actions occur simultaneously at the ultimate limit state. The frequency value ($\psi_1 Q_k$) involves accidental actions and reversible ultimate limit states. The quasi-permanent value ($\psi_2 Q_k$) also involves accidental actions and reversible serviceability limit states. Recommended values of ψ_0, ψ_1 and ψ_2 are given in EN 1990 (2002).

3.4.1 Combination of Design Actions

For the ultimate limit state three alternative combinations of actions, modified by appropriate partial safety factors (γ), must be investigated:

(a) *Fundamental*: a combination of all permanent actions including self-weight (G_k), the prestressing action (P), the dominant variable action (Q_k) and combination values of all other variable actions ($\psi_0 Q_k$).
(b) *Accidental*: a combination value of the dominant variable actions ($\psi_0 Q_k$). This combination assumes that accidents (explosions, fire or vehicular impact) of short duration have a low probability of occurrence.
(c) *Seismic*: reduces the permanent action partial safety factors (γ_G) with a reduction factor (ξ) between 0,85 and 1.

For the serviceability limit state three alternative types of combination of actions must be investigated:

(a) The characteristic rare combination occurring in cases when exceeding a limit state causes permanent local damage or deformation.
(b) The combination which produces large deformations or vibrations which are temporary.
(c) Quasi-permanent combinations used mainly when long-term effects are important.

A combination of actions can be symbolically represented for design purposes, for example, for one of three conditions at the ultimate limit state:

$$\sum \gamma_G G_k + \gamma_P P + \sum \gamma_Q \psi_0 Q_k$$

Similar equations can be formed for the other two conditions at ultimate limit state and for the three conditions at the serviceability limit state (EN 1990 (2002)).

3.5 ACTIONS WITH SPACIAL VARIATION

3.5.1 Pattern Loading

All possible actions relevant to a structure should be considered in design calculations. The actions should be considered separately and in realistic combinations to determine which is most critical for strength and stability of the structure.

For continuous structures, connected by rigid joints or continuous over the supports, vertical actions should be arranged in the most unfavourable but realistic pattern for each element. Permanent actions need not be varied when considering such pattern loading, but should be varied when considering stability against overturning. Where horizontal actions are being considered pattern loading of vertical actions need not be considered.

For the design of a simply supported beam it is obvious that the critical condition for strength is when the beam supports the maximum permanent action and maximum variable action at the ultimate limit state. The size of the beam is then determined from this condition and checked for deflection at the serviceability limit state.

A more complicated structure is a simply supported beam with a cantilever as shown in Fig. 3.2(a).

Assuming that the beam is of uniform section and that the permanent actions are uniformly applied over the full length of the beam, it is necessary to consider various combinations of the variable actions as shown in Figs. 3.2(b)–(d). Although partial

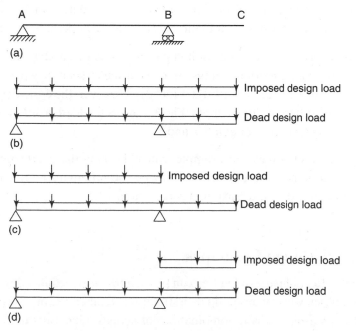

FIGURE 3.2 Pattern loading

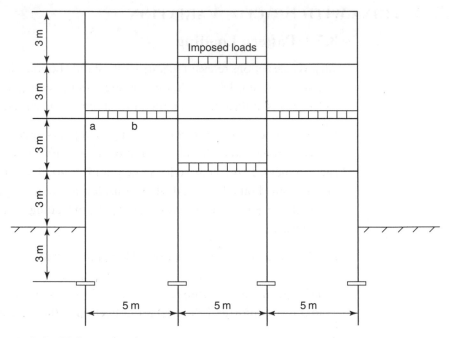

FIGURE 3.3   Multi-storey frame

loading of spans is possible this is not generally considered except in special cases of rolling actions (e.g. a train on a bridge span).

For a particular section it is not immediately apparent which combination of actions is most critical because it depends on the relative span dimensions and magnitude of the actions. Therefore calculations are necessary to determine the condition and section for maximum bending moment and shear force at the ultimate limit state.

Analysis of a multi-storey building is more complicated as shown by Holicky (1996). Where loads on other storeys affect a particular span they may be considered as uniformly distributed (EN 1991-1-1 (2002)). However, the critical positioning of the loads may be different, for example, Fig. 3.3 shows the load positions for verification of the bending resistance at points a and b.

In other situations, for example, when checking the overturning of a structure, the critical combination of actions may be the minimum permanent action, minimum imposed action and maximum wind action.

3.5.2 Design Envelopes

The effect of pattern loading can be seen by constructing a design envelope. This is a graph showing, at any point on a structural member, the most critical effect that results from various realistic combinations of actions. Generally the most useful envelopes

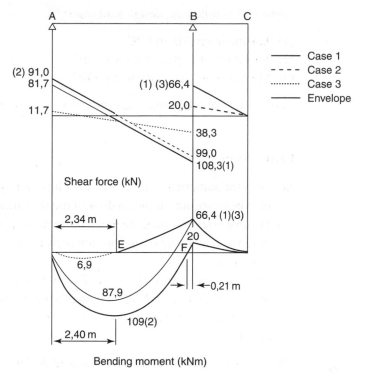

FIGURE 3.4 Example: Design envelope

are for shear force and bending moment at the ultimate limit state. The formation and use of a design envelope is demonstrated by the following example.

EXAMPLE 3.1 Example of a design envelope. The beam ABC in Fig. 3.4-carries the following characteristic loads:

Dead load $G_k = 10\,\text{kN/m}$ on both spans;

Imposed loads $Q_k = 15\,\text{kN/m}$ on span AB, $12\,\text{kN/m}$ on span BC.

Sketch the design envelope for the bending moment and shear force at the ultimate limit state. Indicate all the maximum values and positions of zero bending moment (points of contraflexure).

The maximum and minimum design loads on the spans are:

Maximum on AB $= \gamma_G G_k + \gamma_Q Q_k = 1{,}4 \times 10 + 1{,}6 \times 15 = 38\,\text{kN/m}$.

Maximum on BC $= \gamma_G G_k + \gamma_Q Q_k = 1{,}4 \times 10 + 1{,}6 \times 12 = 33{,}2\,\text{kN/m}$.

Minimum on AB or BC $= \gamma_G G_k = 1{,}0 \times 10 = 10\,\text{kN/m}$.

Consider the following design load cases:

(1) Maximum on AB and BC.
(2) Maximum on AB, minimum on BC.
(3) Minimum on AB, maximum on BC.

The bending moment and shear force diagrams are shown in Fig. 3.4.

Comments

(a) Only the numerical value of the shear force is required in design, the sign however, may be important in the analysis of the structure.
(b) Positive (sagging) bending moments indicate that the bottom of the beam will be in tension; negative (hogging) moments indicate that the top of the beam will be in tension.
(c) The envelope, shown as a heavy line, indicates the maximum values produced by any of the load cases. Note that on AB the envelope for shear force changes from case (2) to case (1) at the point where the numerical values of the shear force are equal.

This process is tedious and an experienced designer knows the critical action combinations and the positions of the critical values and avoids some of the work involved. Alternatively the diagrams can be generated from input data using computer graphics. In more complicated structures and loading situations, envelopes are useful in determining where a change of member size could occur and where splices could be inserted. In other situations wind action is a further alternative to combinations of permanent and variable actions.

REFERENCES

EN 1990 (2002). *Basis of structural design*. BSI.
EN 1991-1-1 (2002). *Densities, self weight and imposed loads*. BSI.
EN 1991-1-2 (2002). *Actions on structures exposed to fire*. BSI.
EN 1991-1-3 (2002). *Snow loads*. BSI.
EN 1991-1-4 (2002). *Wind loads*. BSI.
EN 1991-1-5 (2003). *Thermal actions*. BSI.
EN 1991-1-6 (2002). *Actions during execution*. BSI.
EN 1991-1-7 (2004). *Accidental actions due to impact and explosions*. BSI.
EN 1998-8 (2004). *Design of structures for earthquake resistance*. BSI.
EN 1991-3(2004). *Actions induced by cranes and machinery*. BSI.
Holicky, M. (1996). *Densities, self weight and imposed loads on buildings*. Czech Technical University, Prague.

Chapter 4 / Laterally Restrained Beams

4.1 STRUCTURAL CLASSIFICATION OF SECTIONS (CL 5.5, EN 1993-1-1 (2005))

Chapters 4 and 5 are concerned with the design of members which are predominantly in bending, that is, where axial loads, if any, are small and transverse shear forces are not excessive. Chapter 4 contains basic theoretical work on section properties and the design of laterally restrained beams using Class 1 standard sections and plastic methods of analysis.

Sections of steel beams in common use are shown in Fig. 4.1. The rolled sections shown at (a) are used most often and of these the 'I' section is used widely. Some sections are of uniform thickness while others are of different thickness for the web and flange. The rolled sections are generally in stock, are lowest in cost, require less design and connections are straightforward. Hollow sections are not as efficient in bending but corrosion resistance is better and aesthetically they may be more acceptable. Cold formed sections are thinner and are therefore more susceptible to corrosion unless protected, however they are very economical for use as puriins. Fabricated sections are used when a suitable rolled section is not available, but costs are higher and delivery times are longer. Castellated sections are used for large spans with relatively low loads and where transverse shear forces are not excessive. Tapered beams are efficient in resisting bending moments but must be checked for shear forces. Composite steel–concrete sections are used for floors.

The four classes of cross-section of steel 'I' beams are described in cl 5.5.2, EN 1993-1-1 (2005). To allow for flange buckling sections are reduced to effective sections.

All members subject to bending should be checked for the following at critical sections:

(a) A combination of bending and shear force
(b) Deflection
(c) Lateral restraint
(d) Local buckling
(e) Web bearing and buckling.

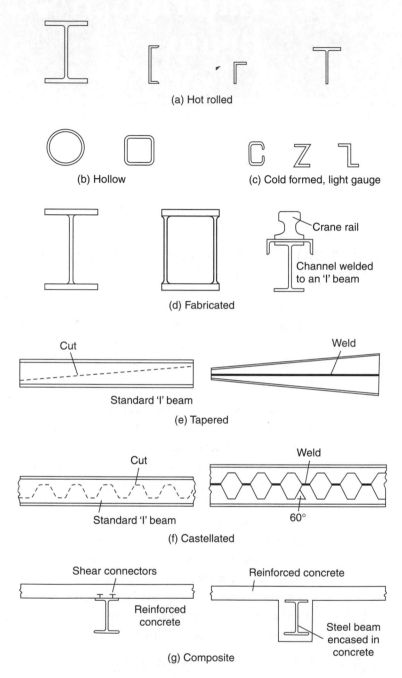

(a) Hot rolled

(b) Hollow

(c) Cold formed, light gauge

Crane rail

Channel welded
to an 'I' beam

(d) Fabricated

Cut

Weld

Standard 'I' beam

(e) Tapered

Cut

Weld

Standard 'I' beam

60°

(f) Castellated

Shear connectors

Reinforced concrete

Reinforced
concrete

Steel beam
encased in
concrete

(g) Composite

FIGURE 4.1 Types of steel beams

This chapter is concerned with members which are predominantly subject to bending and where lateral torsional buckling and local buckling of the compression flange are prevented. It is important to recognize the characteristics of these two forms of buckling shown in Fig. 4.2. Lateral torsional buckling exhibits vertical movement (bending about about the y–y axis), lateral displacement (bending about the z–z axis),

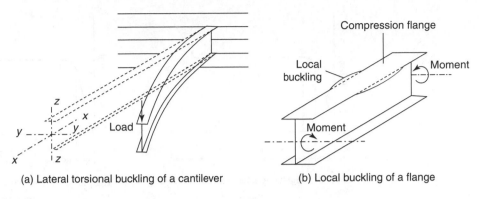

(a) Lateral torsional buckling of a cantilever   (b) Local buckling of a flange

FIGURE 4.2 Buckling of beams

and torsional rotation (rotation about the x–x axis). Local buckling exhibits local deformation of an outstand, for example, a flange of an 'I' beam.

Lateral torsional buckling occurs when the buckling resistance about the z–z axis and the torsional resistance about the x–x axis are low. The buckling resistance about the z–z axis can be improved by lateral restraints, for example, transverse members which prevent lateral movement of the compression flange. Local buckling occurs when the flange outstand to thickness ratio (b/t_f) is high and is avoided by choosing Class 1 sections (cl 5.5.2, EN 1993-1-1 (2005)).

Where both types of buckling are prevented, as for Class 1 and Class 2 sections, then the section can be stressed to the maximum design stresses in bending, i.e. plastic methods of analysis and design can be used. If Class 3 or Class 4 sections are used the plastic moment capacity is reduced to an elastic moment capacity.

If a steel 'I' section is used as a simply supported beam and loaded with a uniformly distributed load then the bending moment distribution varies parabolically. If the section is bent about a major axis then the stress distribution at centre span at various stages of loading is shown in Fig. 4.3(c). In the early stages of loading the stress distribution is elastic, then elastic–plastic and finally fully plastic. The corresponding moment curvature relationship is shown in Fig. 4.3(b).

The fully plastic stage corresponds to the condition for the tensile stress–strain relationship for the steel shown in Fig. 4.3(a). Theoretically the load cannot be increased beyond this plastic condition but strain hardening occurs and this increases the resistance. Note that for full plasticity large strains occur, of the order of 20%, which makes mild steel ideal, while other steels with less plastic strain behave in a more brittle fashion.

Although bending is the predominant design criteria checks must be made for the magnitude of the shear stresses. Shear stresses are introduced from vertical shear forces, or torsion moments (cls 6.2.6 and 6.2.7, EN 1993-1-1 (2005)).

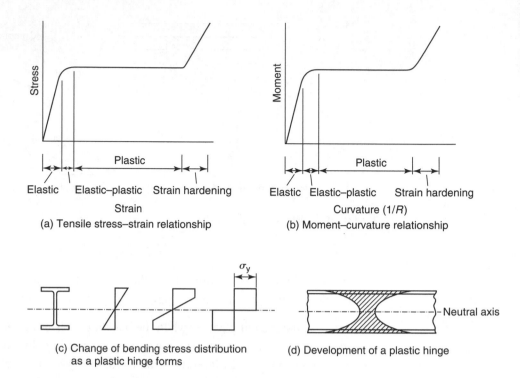

FIGURE 4.3 Development of a plastic hinge

For the design of beams calculations are required in the elastic stage of behaviour, for example, stresses and deflections, and also at the fully plastic stage (e.g. collapse load). The calculations involve certain basic section properties.

4.2 ELASTIC SECTION PROPERTIES AND ANALYSIS IN BENDING

4.2.1 Sectional Axes and Sign Conventions

For all standard sections rectangular centroidal axes y–y and z–z are defined parallel to the main faces of the section, as shown in Fig. 4.4. The position of these axes is given in Section Tables. For angles, and other sections where the rectangular and principal axes do not coincide, the principal axes are denoted by u–u and v–v. The major axis u–u is conventionally inclined to the y–y axis by an angle α, as shown in Figs 4.4(e) and (f). For equal angles, $\alpha = 45°$.

For problems involving simple uniaxial or biaxial bending of symmetrical sections a strict sign convention is not necessary, but for the solution of complex problems it is desirable. In this chapter the positive conventions of sagging curvature and downward deflections are adopted; and the direction of the angle α is anti-clockwise, consistent with the Section Tables. Fig. 4.5(a) shows the coordinates of a point P in the positive

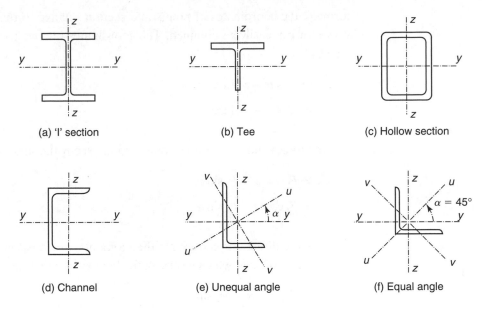

FIGURE 4.4 Sectional axes

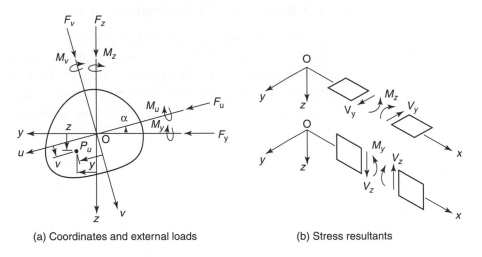

(a) Coordinates and external loads (b) Stress resultants

FIGURE 4.5 Sign conventions

quadrant of a section and the positive directions for the externally applied forces and couples. The positive directions of the corresponding stress resultants (shear forces and bending moments) are shown for the horizontal and vertical planes in Fig. 4.5(b). Positive directions relative to the u–u and v–v axes can be inferred. The convention for the moments has been chosen so that positive moments give tensile stresses in the positive quadrant of the section.

Normally the coordinates of points in a section relative to the rectangular axes are known, or can easily be obtained. The coordinates relative to the principal axes are given by

$$u = y \cos \alpha + z \sin \alpha$$
$$v = z \cos \alpha - y \sin \alpha \tag{4.1}$$

External forces and shear forces transform in exactly the same way, thus

$$F_u = F_y \cos \alpha + F_z \sin \alpha$$
$$F_v = F_z \cos \alpha - F_y \sin \alpha \tag{4.2}$$

However the directions chosen for the moments are consistent with the rules for a right hand set of axes, which gives rise to changes in sign, thus

$$M_u = M_y \cos \alpha - M_z \sin \alpha$$
$$M_v = M_z \cos \alpha + M_y \sin \alpha \tag{4.3}$$

4.2.2 Elastic Second Moments of Area

This property is derived from the simple theory of elastic bending (Croxton and Martin Vol 1 (1987 and 1989)). In design it is used to calculate stresses and deflections in the elastic stage of behaviour, that is, at service loads. Second moments of area for all standard sections are given in Section Tables but for fabricated sections they must be calculated. The procedure involves application of the theorems of parallel axes which, for the single element of area A in Fig. 4.6, can be stated as follows:

$$I_y = I_a + Az^2$$
$$I_z = I_b + Ay^2 \tag{4.4}$$
$$I_{yz} = I_{ab} + Ayz$$

where

I_y = elemental second moment of area about y–y

I_z = elemental second moment of area about z–z

I_{yz} = elemental product moment of area about y–y and z–z

a–a and b–b are centroidal axes through the element, parallel to y–y and z–z, respectively.

For the determination of I_{yz}, which can be either positive or negative, the correct signs must be allocated to the coordinates y and z. The positive directions are indicated by arrows in Fig. 4.6.

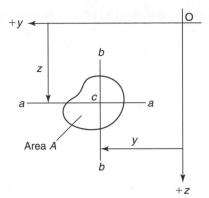

FIGURE 4.6 Parallel axes for an element

When second moments of area about the rectangular axes have been computed, the direction of the principal axes can be obtained from

$$\tan 2\alpha = 2I_{yz}/(I_z - I_y)$$
(4.5)

The principal second moments of area are then give by

$$I_u = I_y \cos^2 \alpha + I_z \sin^2 \alpha - I_{yz} \sin 2\alpha$$
$$I_v = I_y \sin^2 \alpha + I_z \cos^2 \alpha + I_{yz} \sin 2\alpha$$
(4.6)

If I_y is arranged to be greater than I_z, then α will be less than $45°$ and I_u will be the major principal second moment of area. A negative result for Eq. (4.5) indicates that α is to be measured clockwise from y–y.

4.2.3 Elastic Section Moduli

These values are derived from the second moments of area by dividing by the distance to the extreme fibres (i.e. $W_{el} = I/z$). Values of section moduli are given in Section Tables. For structural tees two values of W_{el} are given, referring to the extreme fibres in the table and the stalk.

4.2.4 Elastic Bending of Symmetrical Sections

When either of the rectangular axes is an axis of symmetry the normal bending stress at any point in the section is given by

$$\sigma = \frac{M_x z}{I_y} + \frac{M_z y}{I_z}$$
(4.7)

If the directions of the bending moments and the coordinates are in accordance with the sign convention of Fig. 4.5, a positive result indicates that the stress is tensile.

For simple bending about the y–y axis, which is the most common event, the stress in the extreme fibres

$$\sigma_{max} = \frac{M_y z_{max}}{I_y} = \frac{M_y}{W_{el_y}} \qquad (4.8)$$

Similarly for bending about the z–z axis

$$\sigma_{max} = \frac{M_y z_{max}}{I_y} = \frac{M_y}{W_{el_y}} \qquad (4.9)$$

4.2.5 Elastic Bending of Unsymmetrical Sections

When a section is subject to bending about an axis which is not a principal axis the effect is the same as if the section were subject to the components of the bending moment acting about the principal axis. In other words the bending is biaxial.

For standard rolled angles the principal second moments of area and the directions of the principal axes are given in Section Tables. Transforming bending moments and coordinates to the principal axes by means of Eqs (4.3) and (4.1). Bending stress

$$\sigma = \frac{M_u v}{I_u} + \frac{M_v u}{I_v} \qquad (4.10)$$

This is the same as Eq. (4.7) but with all the terms treated to the principal axes. If the sign convention of Fig. 4.5 is observed, a positive result indicates tension.

In other cases the additional calculations required for the solution of problems by principal axes can be avoided by the use of 'effective bending moments'. These are modified bending moments which can be considered to act about the rectangular axes of the section. The bending stress is then given by an expression having exactly the same from as Eq. (4.7)

$$\sigma = \frac{M_{ey} z}{I_y} + \frac{M_{ez} y}{I_z} \qquad (4.11)$$

where M_{ey} and M_{ez} are effective bending moments about y–y and z–z axes, respectively and are given by

$$M_{ey} = \frac{M_y - M_z I_{yz}/I_z}{1 - I_{yz^2}/(I_y I_z)}$$

$$M_{ez} = \frac{M_z - M_y I_{yz}/I_y}{1 - I_{yz^2}/(I_y I_z)} \qquad (4.12)$$

These expressions are derived from the application of conventional elastic bending theory to curvature in both the yx and zx planes. One such derivation is given by Megson (1980).

By successive differentiation with respect to x, the longitudinal dimension, similar expressions for the effective shear force and effective load intensity can be obtained, thus

$$V_{ey} = \frac{V_y - V_z I_{yz}/I_y}{1 - I_{yz^2}/(I_y I_z)}$$

$$V_{ez} = \frac{V_z - V_y I_{yz}/I_z}{1 - I_{yz^2}/(I_y I_z)} \qquad (4.13)$$

and

$$f_{ey} = \frac{(f_y - f_z I_{yz}/I_y)}{1 - I_{yz^2}/(I_y I_z)}$$

$$f_{ez} = \frac{(f_z - f_y I_{yz}/I_z)}{1 - I_{yz^2}/(I_y I_z)} \qquad (4.14)$$

It should be noted that the quantities I_y and I_z are interchanged in Eqs (4.13) and (4.14). This is because the expressions for the shear force and the load intensity in the y direction are obtained by successive differentiation of bending moments along the z–z axis and vice versa.

All bending moment problems with unsymmetrical sections can be solved simply by replacing ordinary loads, shears, and bending moments by their effective counterparts. Note however that these effective counterparts have values related to both the y–y and z–z axes, even if the section is only loaded in the direction of one of the rectangular axes.

EXAMPLE **4.1**   Principal axes for an unequal angle section. Find the directions of the principal axes and the values of the principal second moments of area for the angle section in Fig. 4.7(a).

For the calculation of section properties the work is simplified considerably, with insignificant loss of accuracy, by using the dimensions of the section profile, that is, the shape formed by the centre line of the elements, as shown in Fig. 4.7(b).

The position of the centroid O is found by taking moments of the area about the centre lines of each leg in turn

Areas	mm^2
A'B'	$140 \times 20 = 2800$
A'C'	$290 \times 20 = 5800$
	Total $= 8600$

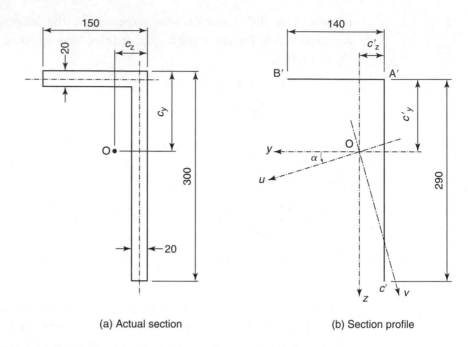

(a) Actual section   (b) Section profile

FIGURE 4.7 Example: principal axes for an unequal angle

Taking moments about A′B′

$$8600c_y' = \frac{5800 \times 290}{2} \quad \text{hence } c_y' = 97{,}8\,\text{mm}$$

Taking moments about A′C′

$$8600c_z' = \frac{2800 \times 140}{2} \quad \text{hence } c_z' = 22{,}8\,\text{mm}$$

Hence for the full section (Fig. 4.7(a))

$$c_y = 107{,}8\,\text{mm} \quad \text{and} \quad c_z = 32{,}8\,\text{mm}$$

Coordinates of the centroids of the legs AB and AC are therefore given by

$$\text{Leg A′B′} \quad y = \frac{140}{2} - c_z' = 47{,}2\,\text{mm}$$

$$z = -c_y' = -97{,}8\,\text{mm}$$

$$\text{Leg A′C′} \quad y = -c_z' = -22{,}8\,\text{mm}$$

$$z = \frac{290}{2} - c_y' = 47{,}2\,\text{mm}$$

The second moments of area about the rectangular axes are obtained in the usual way be applying the parallel axes formula to each leg.

$$I_y(\text{leg}) = \frac{bh^3}{12} + A(\text{leg})z^2$$

where b and h are dimensions of the leg parallel to the y–y and z–z axes, respectively.

$$\text{Leg A}'\text{B}' \quad 140 \times \frac{20^3}{12} + 2800 \times (-97{,}8)^2 = 26{,}87\text{E6 mm}^4$$

$$\text{Leg A}'\text{C}' \quad 20 \times \frac{290^3}{12} + 5800 \times 47{,}2^2 = 53{,}57\text{E6 mm}^4$$

$$I_y = 80{,}44\text{E6 mm}^4$$

$$I_z(\text{leg}) = \frac{hb^3}{12} + A(\text{leg})y^2$$

$$\text{Leg A}'\text{B}' \quad 20 \times \frac{140^3}{12} + 2800 \times 47{,}2^2 = 10{,}81\text{E6 mm}^4$$

$$\text{Leg A}'\text{C}' \quad 290 \times \frac{20^3}{12} + 5800 \times (-22{,}8)^2 = 3{,}21\text{E6 mm}^4$$

$$I_z = 14{,}02\text{E6 mm}^4$$

The product moment of area I_{yz} is obtained by applying the parallel axis formula to each leg

$$I_{yz}(\text{leg}) = I_{ab} + A(\text{leg})yz$$

For each leg the term I_{ab} is equal to zero, because the parallel axes through the centroid of the leg are principal axes.

$$\text{Leg A}'\text{B}' \quad 2800 \times 42{,}7 \times (-97{,}8) = -12{,}93\text{E6 mm}^4$$

$$\text{Leg A}'\text{C}' \quad 5800 \times (-22{,}8) \times 47{,}2 = -6{,}24\text{E6 mm}^4$$

$$I_{yz} = -19{,}17\text{E6 mm}^4$$

Direction of the principal axes from Eq. (4.5)

$$\tan 2\alpha = \frac{2I_{yz}}{I_z - I_y}$$

$$2\alpha = \arctan\left[\frac{2 \times (-19,17)}{14{,}02 - 80{,}44}\right] \quad \text{hence } \alpha = 15°$$

Principal second moments of area from Eq. (4.6)

$$I_u = I_y \cos^2 \alpha + I_z \sin^2 \alpha - I_{yz} \sin 2\alpha$$
$$I_v = I_y \sin^2 \alpha + I_z \cos^2 \alpha + I_{yz} \sin 2\alpha$$

Substituting values

$$I_u = 85{,}58\text{E6 mm}^4 \quad \text{and} \quad I_v = 8{,}88\text{E6 mm}^4.$$

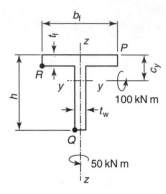

FIGURE 4.8 Example: structural tee in biaxial bending

As a check on the transformation

$$I_y + I_z = I_u + I_v \text{ which is correct.}$$

EXAMPLE 4.2 Structural tee in biaxial bending. Calculate the maximum extreme fibre stresses in a standard $292 \times 419 \times 113$ kg structural tee cut from a Universal Beam. The tee is loaded by two moments as shown in Fig. 4.8.

From Section Tables

$b_f = 293,8$ mm, $h = 425,5$ mm, $t_w = 16,1$ mm, $t_f = 26,8$ mm, $C_y = 108$ mm, $I_y = 246,6E6$ mm^4, $I_z = 56,76E6$ mm^4, W_{ely}(flange) $= 2,277E6$ mm^3, W_{ely}(toe) $= 0,7776E6$ mm^3, $W_{elz} = 0,3865E6$ mm^3.

By inspection the maximum compressive stress occurs at a point P because the stresses from both moments are compressive.

$$\sigma = \frac{M_y}{W_{ely}(\text{flange})} + \frac{M_z}{W_{elz}}$$

$$= \frac{100E6}{2,277E6} + \frac{50E6}{0,3865E6} = 173 \text{ MPa}$$

The maximum tensile stress can occur at point Q or point R, depending on the relative magnitude of the bending moments. It is necessary to check both points.

Using the sign convention of Fig. 4.5 both bending moments are positive and the coordinates of the points are given by

Point Q, $y = \dfrac{t_w}{2} = \dfrac{16,1}{2} = 8,05$ mm, $z = h - c_y = 425,5 - 108 = 317,5$ mm.

Point R, $y = \dfrac{b_f}{2} = \dfrac{293,8}{2} = 146,9$ mm, $z = t_f - c_y = 26,8 - 108 = -81,2$ mm.

From Eq. (4.7) the stresses

$$\sigma = \frac{M_y z}{I_y} + \frac{M_z y}{I_z}$$

$$\sigma_Q = \frac{100 \times 317,5}{246,6} + \frac{50 \times 8,05}{56,76} = 135,8 \text{ MPa}$$

$$\sigma_R = \frac{100 \times (-81,2)}{246,6} + \frac{50 \times 146,9}{56,76} = 96,5 \text{ MPa}$$

Summarizing, the maximum stresses are:

At Q, $\sigma_Q = +135,8$ MPa (tension)

At P, $\sigma_P = -173$ MPa (compression)

EXAMPLE 4.3 Bending stresses in an unequal angle section. Calculate the bending stresses in the angle section shown in Fig. 4.9 where

$$I_y = 80,44\text{E}6 \quad I_z = 14,02\text{E}6 \quad I_{yz} = -19,17\text{E}6 \text{ mm}^4$$

Effective moments from Eq. (4.12)

$$M_{ey} = \frac{(M_y - M_z I_{yz}/I_z)}{(1 - I_{yz^2}/I_y I_z)} = 74,92 \text{ kNm}$$

$$M_{ez} = \frac{(M_z - M_y I_{yz}/I_y)}{(1 - I_{yz^2}/I_y I_z)} = 32,86 \text{ kNm}$$

Maximum compressive stress at A

$$\sigma_A = \frac{M_{ey} z_A}{I_y} + \frac{M_{ez} y_A}{I_z}$$

$$= \frac{74,92\text{E}6 \times (-107,8)}{80,44\text{E}6} + \frac{32,86\text{E}6 \times (-32,8)}{14,02\text{E}6}$$

$$= -177,3 \text{ MPa (compression)}$$

Check for maximum tensile stress at B, and at C

$$\sigma_B = \frac{M_{ey} z_B}{I_y} + \frac{M_{ez} y_B}{I_z}$$

$$= \frac{74,92\text{E}6 \times (-87,8)}{80,44\text{E}6} + \frac{32,86\text{E}6 \times 117,2}{14,02\text{E}6}$$

$$= +192,9 \text{ MPa (tension)}$$

$$\sigma_C = \frac{M_{ey} z_C}{I_y} + \frac{M_{ez} y_C}{I_z}$$

$$= \frac{74,92\text{E}6 \times 192,2}{80,44\text{E}6} + \frac{32,86\text{E}6 \times (-12,8)}{14,02\text{E}6}$$

$$= +149,0 \text{ MPa (tension)}$$

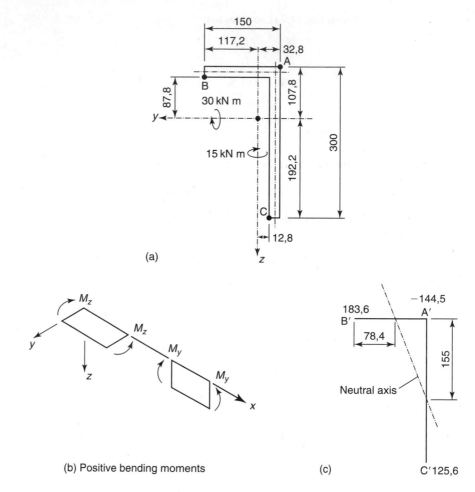

(a)

(b) Positive bending moments

(c)

FIGURE 4.9 Example: bending stresses in an unequal angle

Summarizing, the maximum stresses are:

At B, $\sigma_B = +192{,}9\,\text{MPa}$ (tension)

At A, $\sigma_A = -177{,}3\,\text{MPa}$ (compression)

Note that although the position of the centroid and the values of the second moment of area can be calculated without significant error from the profile dimensions of the section, the same is not true of the stresses.

EXAMPLE 4.4 Bending about principal axes of an angle section. Recalculate the stresses at points A, B and C in the previous example considering bending about the principal axes.

The coordinates of the points are transformed in accordance with Eq. (4.1),

$$u = y \cos \alpha + z \sin \alpha \quad \text{and} \quad v = z \cos \alpha - y \sin \alpha.$$

$$\sin \alpha = 0,2588 \quad \text{and} \quad \cos \alpha = 0,9659,$$

Point	y/z axes		u/v axes	
	y	z	u	v
A	−32,8	−107,8	−59,6	−95,6
B	117,2	−87,8	90,5	−115,1
C	−12,8	192,2	37,5	189,0

Bending moments transform in accordance with Eq. (4.3)

$$M_u = M_y \cos \alpha - M_z \sin \alpha = 25,10 \,\text{kNm}$$

$$M_v = M_z \cos \alpha + M_y \sin \alpha = 22,25 \,\text{kNm}$$

Bending stresses from Eq. (4.10)

$$\sigma = \frac{M_u v}{I_u} + \frac{M_v u}{I_v}, \quad \sigma_A = -177,3 \quad \sigma_B = 192,9 \quad \sigma_C = 149,0 \,\text{MPa}$$

which are the same as obtained previously.

4.2.6 Elastic Analysis of Beams

The elastic analysis of simply supported beams with examples of shear force and bending moment diagrams and deflection calculations are given in many introductory books on structural analysis (e.g. Croxton and Martin Vol. 1 (1987 and 1989)). The analysis of continuous beams is more complicated but there is a choice of methods such as area–moment, moment distribution, slope deflection, and matrix methods. These methods are also covered by Croxton and Martin Vol. 2 (1987 and 1989). with a computer program for analysis using matrix methods.

4.2.7 Elastic Deflections of Beams (cl 7.2.1, EN 1993-1-1 (2005))

The deflections under serviceability loads of a building or part should not impair the strength or efficiency of the structure or its components or cause damage to the finishes. When checking for deflections the most adverse realistic combination and arrangement of serviceability loads should be assumed, and the structure may assumed to be elastic.

The theory and the methods of calculating deflections for static and hyperstatic structures are given in Croxton and Martin Vols. 1 and 2 (1987 and 1989). For simple beams standard cases can be superimposed and some useful cases are shown in Figs 4.10 and 4.11.

Maximum deflection (at free end)

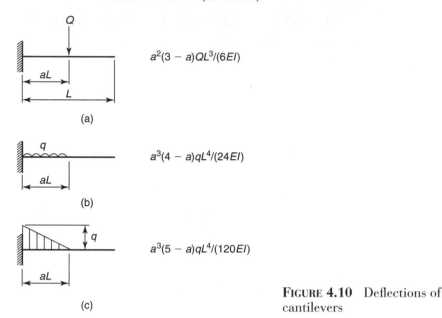

(a) $a^2(3 - a)QL^3/(6EI)$

(b) $a^3(4 - a)qL^4/(24EI)$

(c) $a^3(5 - a)qL^4/(120EI)$

FIGURE 4.10 Deflections of cantilevers

For simply supported beams the central deflection rather than the maximum is given, so that deflections from individual cases can be added. For most loading cases the central deflection only differs by a small percentage from the maximum. In case (a) of Fig. 4.11, for example, the difference is always within 2,5%. A notable exception is the case of equal end moments acting in the same direction, when the central deflection is zero. However in such a case the deflection at other points along the beam are likely to be small. A more accurate analysis can be formed if it is suspected that the deflection is likely to exceed the limit. Recommendations for limiting values for deflections are given in cl 7.2.1, EN 1993-1-1 (2005).

EXAMPLE 4.5 Deflections for a hyperstatic structure. The size of the members for the symmetrical structure shown in Fig. 4.12 have been determined and the structure requires to be checked for deflections in the elastic range of behaviour. The imposed variable characteristic loads are shown in Fig. 4.12 and the second moment of area is $I_y = 127{,}56\mathrm{E}6\ \mathrm{mm}^4$. The bending moments (positive clockwise) at the joints are given in the following table.

Joint	Span	Moment (kNm)
B	AB	+60
B	BC	−60
C	BC	−6
C	CG	+72
C	CD	−66
G	CG	+36

Central deflection, and
rotation at supports

(a)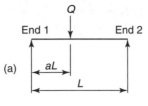

For $a \leq 0.5$
$$\delta = (3a - 4a^3)QL^3/(48EI)$$
$$\theta_1 = (2a - 3a^2 + a^3)QL^2/(6EI)$$
$$\theta_2 = (a^3 - a)QL^2/(6EI)$$

(b)

$$\delta = 5qL^4/(384EI)$$
$$\theta_1 = qL^3/(24EI)$$
$$\theta_2 = -\theta_1$$

(c)

For $a \leq 0.5$
$$\delta = (3a^2 - 2a^4)qL^4/(96EI)$$
$$\theta_1 = (a^4 - 4a^3 + 4a^2)qL^3/(24EI)$$
$$\theta_2 = (a^4 - 2a^2)qL^3/(24EI)$$

(d)

$$\delta = qL^4/(120EI)$$
$$\theta_1 = 5qL^3/(192EI)$$
$$\theta_2 = -\theta_1$$

(e)

$$\delta = (M_1 - M_2)L^2/(16EI)$$
$$\theta_1 = (2M_1 - M_2)L/(6EI)$$
$$\theta_2 = (2M_2 - M_1)L/(6EI)$$

FIGURE 4.11
Displacements of simply
supported beams

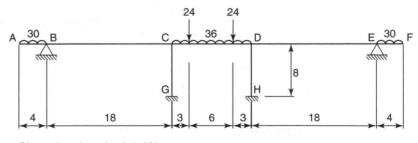

Dimensions in m; loads in kN.

FIGURE 4.12   Example: deflections of a symmetrical continuous structure

For all beams

$$EI_y = 210E3 \times 127{,}56E6 = 26{,}79E12 \, \text{N mm}^2$$

Span CD

1. Uniform load (Fig. 4.11(b))

$$\partial_{ul} = \frac{5qL^4}{384EI_y} = \frac{5QL^3}{384EI_y}$$

$$= \frac{5 \times 36E3 \times 12E3^3}{384 \times 26{,}79E12} = 30{,}2 \, \text{mm}$$

2. Concentrated loads (Fig. 4.11(a))

$$\partial_{cl} = \frac{2[3a - 4a^3]QL^3}{48EI_y}$$

$$= \frac{2[3 \times (3/12) - 4 \times (3/12)^3] \times 24E3 \times 12E3^3}{48 \times 26{,}79E12} = 44{,}3 \, \text{mm}.$$

3. End moments (Fig. 4.11(e))

$$\partial_M = L^2 \frac{(M_1 - M_2)}{16EI_y}$$

$$= 12E3^2 \times \frac{(-66 - 66)E6}{16 \times 26{,}79E12} = -44{,}3 \, \text{mm (upwards)}$$

Total deflection $= \partial_{ul} + \partial_{cl} + \partial_M = 30{,}2 + 44{,}3 + (-44{,}3) = 30{,}2 \, \text{mm}$

A limit for beams with plaster finish $= L/350 = 12E3/350 = 34{,}3 \, \text{mm}$

Span BC

End moments (Fig. 4.11(e))

$$\partial_3 = L^2 \frac{(M_1 - M_2)}{16EI_y}$$

$$= 18E3^2 \times \frac{(-60 + 6)E6}{16 \times 26{,}79E12} = -40{,}8 \, \text{mm (upwards)}$$

An accurate analysis gives $-42{,}4$ mm at 7,35 m from B,

A limit for beams with plaster finish $= L/350 = 18E3/350 = 51{,}43 \, \text{mm}$

Cantilever span AB

For this span the deflection is due to the flexure of the cantilever, assuming the beam is horizontal at B, plus the effect of the anti-clockwise rotation of the beam at B, that is, $-\theta_1$ for span BC.

End moments for span BC (Fig. 4.11(e))

$$\theta_1 = L \times \frac{(2M_1 - M_2)}{6EI_y}$$

$$= 18\text{E}3 \times \frac{(-2 \times 60 + 6)\text{E}6}{6 \times 26{,}79\text{E}12} = -0{,}01277 \text{ rad}$$

1. Deflection at A due to rotation

$$\partial_R = -L\theta_1 = -4\text{E}3 \times (-0{,}01277) = 51{,}07 \text{ mm}$$

2. Deflection due to load (Fig. 4.10(b))

$$\partial_{ul} = \frac{a^3(4-a)qL^4}{24EI} = \frac{QL^3}{8EI_y}$$

$$= \frac{30\text{E}3 \times 4\text{E}3^3}{8 \times 26{,}79\text{E}12} = 8{,}96 \text{ mm}$$

Total deflection $= \partial_R + \partial_{ul} = 51{,}1 + 9{,}0 = 60{,}1$ mm

A limit for a cantilever beam with plaster finish

$$\frac{2L}{350} = \frac{2 \times 4\text{E}3}{350} = 22{,}9 \text{ mm}$$

If the deflection of the cantilever exceeds the limit and stiffening is required then increase the size of the section, or add flange plates. It is also necessary to stiffen spans AB and BC because the deflection is dependent on both.

4.2.8 Span/Depth Ratios for Simply Supported Beams

An initial estimate for the depth of a simply supported 'I' beam carrying a uniformly distributed load can be obtained by using the deflection limit. If σ_{max} is the maximum elastic bending stress at service load then from elastic bending theory

$$\sigma_{max} = \frac{M_z}{I_y} = \frac{QL}{8} \times \frac{h/2}{I_y}$$

rearranging

$$Q = \frac{16\sigma_{max}I_y}{Lh} \tag{i}$$

Assuming a deflection limit for beams with plaster finish of

$$\frac{L}{350} = \frac{5QL^3}{384EI_y} \tag{ii}$$

Eliminating Q by combining Eqs (i) and (ii) and putting $E = 210$ GPa, the span depth ratio

$$\frac{L}{h} = \frac{2880}{\sigma_{max}}$$

If a beam is laterally restrained so that lateral torsional buckling does not occur, then if the maximum bending stress is approximately $f_y/1,5$, the span/depth ratios for different grades of steel and different limits are:

TABLE 4.1 Span/depth ratios for beams.

Grade	Characteristic stress (f_y)	Span/depth ratio	
		L/350	L/250
S235	235	18,4	25,7
S275	275	15,7	22,0
S355	355	12,2	17,0

Note that since Youngs modulus (E) is a constant for all grades of steel, the stiffness of a beam does not increase with a higher grade of steel. If a design is governed by deflection then there is no advantage in using a higher grade of steel.

4.3 ELASTIC SHEAR STRESSES

4.3.1 Elastic Shear Stress Distribution for a Symmetrical Section

When a beam is bent elastically by a system of transverse loads, plane sections no longer remain plane after bending, but are warped by shear strains. In most cases the effect is small and the errors introduced in the use of conventional bending theory are negligible. Important exceptions are discussed briefly in Section 4.3.2. Formulae for the calculation of shear stresses in an elastic beam are derived by considering the variation in bending stresses along a short length of beam.

Consider the very short length of beam in Fig. 4.13(a). At a point S in the web the shear stresses on the vertical and the horizontal section are complementary and are given by the established formula (Eq. (6.20), EN 1993-1-1 (2005))

$$v_s = \frac{VAz_c}{It} \tag{4.15}$$

where

$V =$ the vertical shear force on the section

$A =$ the hatched area, that is, the part of the section between point S and the extreme fibres

$z_c =$ the distance from the centroid of the area A to the neutral axis

$I =$ the second moment of area of the whole section about the neutral axis

$t =$ the thickness of the section at the point S.

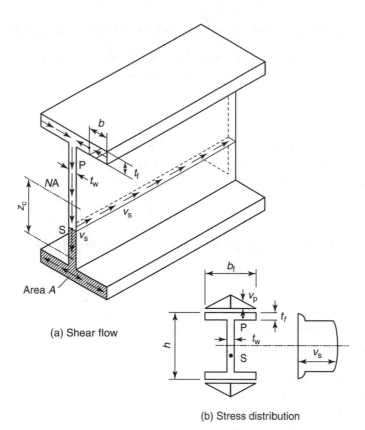

(a) Shear flow

(b) Stress distribution

FIGURE 4.13 Shear stresses in an 'I' beam

The formula cannot be used to obtain the vertical shear stress in the outstanding parts of the flange. However as this must be equal to zero at the top and bottom faces, it must be very small. In fact the resistance of the section to vertical shear is provided almost entirely by the web.

The resultant of the longitudinal shear stress in the web is in equilibrium with the change in the normal tensile force on the area A due to the variation in bending moment along the beam. Similar longitudinal stresses exist in the flanges and give rise to horizontal complementary stresses in the directions shown. For example, at point P in the top flange.

$$A = b_f t_f, \quad t_f = t, \quad z_c = \frac{(h - t_f)}{2}$$

and Eq. (4.15) becomes

$$v_p = \frac{Vb(h - t_f)}{2I} \tag{4.16}$$

This expression is linear with respect to the variable b, and v_p has a maximum value at the centre of the flange where $b = b_f/2$, that is

$$v_{p(max)} = \frac{Vb_f(h - t_f)}{4I} \tag{4.17}$$

The complete distribution of shear stress on the cross-section is shown in Fig. 4.13(b).

Equation (4.15) can be expressed in terms of the shear flow, which is the product of the shear stress and the thickness of the section, thus

$$v = t v_s = \frac{V A_{zc}}{I} \tag{4.18}$$

In the longitudinal sense the shear flow is equal to the shear force per unit length of beam, and is a convenient quantity for the calculation of the shear force to be resisted by bolts or welds in a fabricated section.

4.3.2 Elastic Shear Stresses in Thin Walled Open Sections

The shear stress distribution for a rectangular cross-section subject to a transverse shear force is shown in Fig. 4.14(a) and can be calculated using Eq. (4.15). The shear stress distribution for the open cross-section is different as shown in Fig. 4.14(b). the distribution for an 'I' section is shown in Fig. 4.14(c). Steel sections are usually composed of relatively thin elements, for which the analysis can be simplified by:

(a) referring all dimensions to the profile of the section;
(b) assuming that the shear stress does not vary across the thickness;
(c) ignoring any shear stresses acting at right angles to the section profile. As these are equal to zero at each outside surface they must always be very small in a thin walled section.

If it is further assumed that the load is applied in such a way that no twisting of the beam occurs, the shear flow at a point S on the profile of the section is given by

$$v_s = v_o - \left(\frac{V_{ez}}{I_y}\right) \int_o^s tz \, ds - \left(\frac{V_{ey}}{I_z}\right) \int_o^s ty \, ds \tag{4.19}$$

here V_{ey} and V_{ez} are the effective shear forces obtained from Eq. (4.13) or by applying the effective loads obtained from Eq. (4.14). The variable s is the distance around the profile to the point of interest, starting from any point at which the shear flow

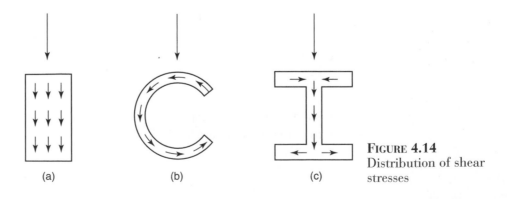

(a)   (b)   (c)

FIGURE 4.14
Distribution of shear stresses

v_0 is known. At any open end, such as the end of a flange, the value of v_0 is zero. The direction of s can be chosen arbitrarily and, provided that the sign convention of Fig. 4.5 is adopted, a positive sign for v_s indicates that the shear flow is in the direction chosen for s.

4.3.3 Elastic Shear Stresses in Thin Walled Closed Sections

The shear stress and shear flow in a symmetrical closed section can be obtained directly from Eqs (4.15) and (4.18), respectively. For an unsymmetrical section Eq. (4.19) can be used, but analysis is complicated by the fact that that v_0 is not known at any point. The problem can be solved by first cutting the section at some point and finding the position of the shear centre in the resulting open section. The shear flow in the closed cross-section then results from the combined action of the applied shear loads transferred to the shear centre of the cut (open) section, and the torque on the closed section due to the transference of loads. Examples of shear stress distribution are given in Fig. 4.14. A similar approach is used to find the position of the shear centre of the closed section.

4.3.4 Elastic Shear Lag (cl 6.2.2.3, EN 1993-1-1 (2005))

The simple theory of bending is based on the assumption that plane sections remain plane after bending. In reality shear strains cause the section to warp. The effect in the flanges is to modify the bending stresses obtained by the simple theory, producing higher stresses near the junction of a web and lower stresses at points remote from it as shown in Fig. 4.15. This effect is described as 'shear lag'. The discrepancies produced by shear lag are minimal in rolled sections, which have relatively narrow and thick flanges. However in plate girders, or box sections, having wide thin flanges the effects can be significant when they are subjected to high shear forces, especially in the vicinity of concentrated loads where the sudden change in shear force produces highly incompatible warping distortions. Shear lag effects can be allowed for by using effective widths (cl 3, EN 1993-1-5 (2003)).

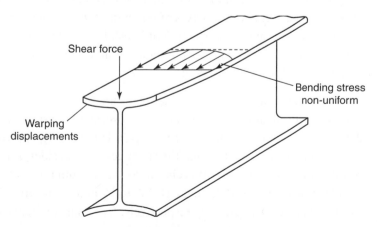

FIGURE **4.15** Shear lag effects for an 'I' section

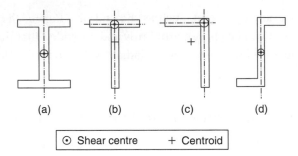

FIGURE **4.16**   Position of shear centre

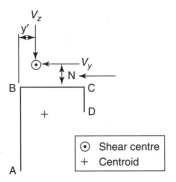

FIGURE **4.17**   Shear centre–unsymmetrical section

4.3.5 Elastic Shear Centre

Equation (4.19) is only valid if no twisting of the beam occurs at the section considered. Torsion in a section can be generated by a transverse load if the resultants of the shear stresses in the elements of the section produce a torque. To counteract this the line of action of the applied load must pass through the shear centre. In a symmetrical section the shear centre lies on an axis of symmetry, and loads applied along such an axis do not cause twisting. In some sections the position of the shear centre can be inferred from the direction of the shear flow (Fig. 4.16). In (a) the shear center lies at the intersection of the two axes of symmetry and is coincident with the centroid; in (b) and (c) it lies at the intersection of lines of shear flow; in (d), if the flanges are of the same size, the shear stresses in them set up opposing torques about the centroid which is therefore the shear centre.

For the general case of an unsymmetrical thin walled open section subject to biaxial bending, with shear forces V_y and V_z, the position of the shear centre can be found by determination of the shear flow from Eqs (4.18) or (4.19), applying V_y and V_z in turn, assuming that they pass through the shear centre. Consider, for example, the section profile in Fig. 4.17. If point B is chosen as the fulcrum it is only necessary to find the resultant shear forces in the leg CD due to V_y and V_z in turn. These forces produce torques equal to $V_y z'$ and $V_z y'$, respectively. By taking moments about B the values of

z' and y' can be obtained. There is no need to calculate the shear stresses in AB or BC because their lines of action pass through the point B, and generate no moment. The resultant shear forces in CD are obtained by integrating the shear stresses obtained by Eq. (4.19) along the leg.

The above process is tedious since, for each value of V_y and V_z, the corresponding effective shear forces V_{ey} and V_{ez} must be calculated and applied. If there is an axis of symmetry Eq. (4.18) can be used and the analysis is simplified.

EXAMPLE 4.6   Distribution of shear stresses for an angle section. Calculate the shear stresses in the simply supported angle section shown in Fig. 4.18(a). $I_y = 80{,}44E6$, $I_z = 14{,}02E6$, $I_{yz} = -19{,}17E6\,\text{mm}^4$.

To the left of mid-span the shear forces are

$$V_y = 5 \quad \text{and} \quad V_z = 10\,\text{kN}$$

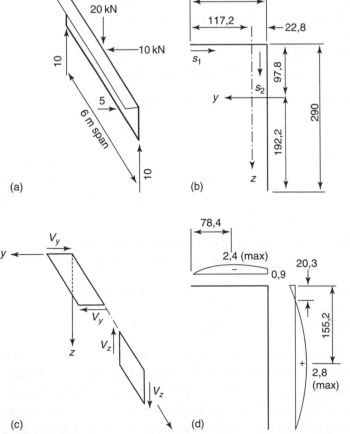

FIGURE 4.18
Example: distribution of shear stresses for an unsymmetrical section

Effective shear forces from Eq. (4.13)

$$V_{ey} = \frac{(V_y - V_z I_{yz}/I_y)}{1 - I_{yz^2}/I_y I_z} = 10{,}95 \text{ kN}$$

$$V_{ez} = \frac{(V_z - V_y I_{yz}/I_z)}{1 - I_{yz^2}/I_y I_z} = 24{,}97 \text{ kN}$$

Shear flow from Eq. (4.19)

$$v_s = v_0 - \left(\frac{V_{ez}}{I_y}\right) \int_o^s tz \, ds - \left(\frac{V_{ey}}{I_z}\right) \int_o^{s} ty \, ds \qquad \text{(i)}$$

For the horizontal leg starting from the left hand end

$$v_0 = 0, \quad s = s_1, \quad y = (117{,}2 - s_1) \quad \text{and} \quad z = -97{,}8 \text{ mm}$$

Substituting these values in (i) and integrating

$$v_{s_1} = 0{,}00781 s_1^2 - 1{,}224 s_1 \qquad \text{(ii)}$$

This equation shows that $v_{s_1} = 0$ only when $s_1 = 0$. Differentiating with respect to s_1 and equating to zero gives a turning point at $s_1 = 78{,}4$ mm. Hence from (ii) $v_{s_1(\max)} = -47{,}96$ N/mm, and $s_1 = 140$ mm, $v_{s_1} = -18{,}28$ N/mm. The negative signs indicate that the shear flow is in the opposite direction to s_1.

For the vertical leg

$$s = s_2, \quad y = -22{,}8 \text{ mm}, \quad z = (s_2 - 97{,}8) \text{ mm}, \quad v_0 = v_A = -18{,}28 \text{ N/mm}.$$

Substitution of these values into (i) gives

$$v_{s_2} = -18{,}28 + 0{,}9633 s_2 - 0{,}003104 s_2^2 \qquad \text{(iii)}$$

Solving (iii) for s_2 shows that when $v_{s_2} = 0$, $s_2 = 20{,}3$ mm. There is also a turning point at $s_2 = 155{,}2$ mm. Hence from (iii) $v_{s_2(\max)} = 56{,}46$ N/mm. The positive sign indicates that the shear flow is in the direction of s_2. As a check, putting $s_2 = 290$ mm gives $v_{s_2} = 0$, which is correct.

The shear stresses in MPa are obtained by dividing the shear flows by the thickness, that is, 20 mm, and are plotted for the whole section in Fig. 4.18(d).

This example demonstrates the method of analysis but the shear stresses are very low and do not justify such a detailed treatment. As a rough guide as to whether detailed analysis is required the shear forces are divided by the area of the appropriate leg.

If the shear stress is only required at particular points in the section Eq. (4.15) can be used with the effective shear forces, taking each axis in turn and superimposing the results. Integration is avoided but the directions of the stresses have to be found by inspection. It can be seen from the position of the neutral axis in Example 4.3 that the maximum shear stresses occur where the neutral axis intersects the profile of the

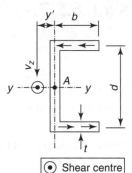

FIGURE 4.19 Example: shear centre of a channel

section, as in a symmetrical section. If these points have previously been found, then the maximum shear stress can be calculated directly as above, using Eq. (4.15).

EXAMPLE 4.7 Shear centre for a channel section. Find the position for the shear centre for the channel shown in Fig. 4.19 which has a uniform thickness.

As the shear centre lies on the axis of symmetry y–y, there is no need to consider V_y. If point A at the intersection of axis y–y with the centre line of the web is taken as the fulcrum, then only the shear force in the flanges need to be considered since the resultant shear force in the web produces no moment about A. Equation (4.18) is

$$v_s = \frac{VAz_c}{I}$$

The distribution of shear flow in the flanges in linear with zero at the ends and maximum at the web centre line. Hence, for maximum shear flow

$$A = bt, \quad V = V_z, \quad z_c = \frac{d}{2} \quad \text{and} \quad I = I_y, \text{ which gives}$$

$$v_{s(\text{max})} = \frac{V_z btd}{2I_y}$$

The resultant shear force is equal to half the maximum shear flow multiplied by the flange width, that is

$$\text{Force} = \frac{V_z b^2 td}{4I_y}$$

The torque about point A from both flanges is equivalent to the torque produced by the applied shear force when it passes through the shear centre, thus

$$V_z y' = \frac{V_z b^2 td^2}{4I_y}$$

from which

$$y' = \frac{b^2 d^2 t}{4I_y} \tag{4.20}$$

4.4 ELASTIC TORSIONAL SHEAR STRESSES (CL 6.2.7, EN 1993-1-1 (2005))

4.4.1 Elastic Torsion

Generally torsion is not a major problem in the design of beams except for special cases. Torsional shear stresses arise from a variety of causes including, beams cranked or curved in plan, and loads whose line of action does not pass through the shear centre of the section. It is important to realize that in torsional problems the lever arm for the torque is measured from the shear centre, not the centroid. In particular, the distributed loads on unsymmetrical sections, such as angles or channels, usually act through the centroid of the section, as in the case of self-weight, although generally self-weight does not produce large torsional stresses. However, if other loads are offset from the shear centre the effect can be considerable, for example, for a 381×102 channel section the lever arm for loads acting through the centroid is approximately 50 mm.

The total torsional moment at any section is the sum of the elastic torsion (St. Venant) plus the internal warping torsion. The warping torsion consists of a bi-moment plus warping.

4.4.2 Uniform and Non-uniform Elastic Torsion

In general the cross-sections of members subject to torsion do not remain plane, but tend to warp. Warping is the change in geometric shape of the member so that cross-sections do not remain plane as shown in Fig. 4.20. The degree of warping that takes place depends on the shape of the section, and is most pronounced in thin walled channels. In some sections, such as angles and tees, solid and hollow circular sections and square box sections of uniform thickness, warping is virtually non existent, while in others, such as closed box sections of general shape, its effect is small. In 'I' sections most of the warping takes place in the flanges and its effect on the web is very small and can be ignored. If torque is applied only at the ends of the member and warping is not restrained, the flanges remain virtually straight and maintain their original shape as shown in Fig. 4.20(a). The result is that the sectional planes of the flanges rotate in opposite directions, producing warping displacements which are constant along the whole length of the member. Under these circumstances the member is said to be in the state of uniform, or St. Venant, torsion.

If warping is prevented, for example, by rigid supports at the ends of the member, the torsional stiffness is increased and longitudinal tensile and compressive stresses are induced. In practice warping is not so obvious but may arise from the action of structural connections, or from the incompatibility of warping displacements that occur when the torque is not uniform along the length of the member. Warping restraint increases the torsional stiffness of a member, and at any point along its length the applied torque is resisted by the two components, one due to the St. Venant torsion,

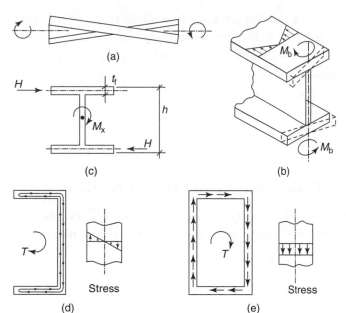

FIGURE 4.20 Torsion of thin walled sections

and the other to warping torsion from the effects of the restraints. The proportions of the two components depend on the type of loading and the distance from a restraint.

Both components of torsion produce shear stresses parallel to the walls of the section, and their combined effects can locally be greater than the effect of St. Venant alone. However in members other than channels with very thin walls the increase in shear stresses can usually be ignored in design. In beams the maximum shear stress can be obtained approximately by combining the effects of transverse shears, using Eq. (4.15), with the shear stresses from torsion, assuming that the whole of the applied torque results in St. Venant torsion.

A more significant effect of warping restraint in the design of beams is the introduction of longitudinal stresses. The effect is illustrated for an 'I' beam in Fig. 4.20(b). In this case the warping displacements are confirmed to the flanges, whose positions, if warping were allowed to occur freely, are shown by dotted outlines. Bi-moments M_b are induced in the planes of the flanges when warping is restrained, and these give rise to tensile and compressive stresses as shown.

A full treatment of the analysis of members subject to warping torsion is beyond the scope of this book and the reader is referred to Zbirohowski-Koscia (1967). The analysis is tedious, but for 'I' beams a conservative estimate of the longitudinal stresses due to warping torsion can be obtained by assuming that each flange acts independently and is bent in its own plane by an analogous system of lateral loads which replace the applied torques, as in Fig. 4.20(c). The value of the equal and opposite lateral loads H, analogous to the applied torque M_x is

$$H = \frac{M_x}{h - t_f} \tag{4.21}$$

The ends of the flanges can be assumed to be either fixed or simply supported, depending on whether or not warping is restrained by the structural connections. The results obtained by this method are conservative because in reality the warping stresses are produced only by the warping component of the applied torque, not the whole torque as assumed. However, Eq. (4.21) can be useful in preliminary designs where it is necessary to assess whether the effects of torsion are likely to be significant.

4.4.3 Elastic Torsion of Circular Sections

The elastic (St. Venant) theory of torsion of prismatic members with solid or hollow sections can be expressed by the well established formula

$$\frac{T}{I_t} = \frac{v}{r} = \frac{G\theta}{L} \tag{4.22}$$

where

T is the torsional moment

I_t is the torsion constant which, for a circular section, is equal to the polar second moment of area

v is the shear stress at radius r

G is the shear modulus

θ is the angle of twist

L is the length of the member

The polar second moment of area for a solid section of radius R is

$$I_t = \frac{\pi R^4}{2} \tag{4.23}$$

For a thin walled tube of mean radius R_m and wall thickness t, an approximate formula is

$$I_t = 2\pi R_m^3 t \tag{4.24}$$

The error is below 3% for t/R_m ratios of 1/3 or less, and is on the safe side. The polar second moment of area (I_t) is twice the second moment of area (I) about a diameter, values for which are given in Section Tables.

The elastic distribution of shear stress along the radius of a solid circular section is linear, with zero at the centre and a maximum at the outside surface. For a thin walled tube, the stress varies linearly across the wall thickness and in the range of standard structural tubes unsafe errors in the shear stress of up to about 18% are introduced by the use of R_m in Eq. (4.22) instead of the outside radius.

4.4.4 Elastic Torsion of Thin Walled Open Sections

The torsional constant I_t for a thin rectangle of width b and thickness t is

$$I_t = \frac{bt^3}{3} \tag{4.25}$$

This formula is accurate when b/t is infinite and gives unsafe errors of 6% when $b/t = 10$, and 10% when $b/t = 6$. The b/t ratios are typical of the flanges of Universal Column and Beam sections.

Most sections in steelwork design are composed of thin rectangles, and for a complete section the torsional constant can be obtained by summing the torsional constants for each rectangular element, that is

$$I_t = \frac{\sum bt^3}{3} \tag{4.26}$$

Values of I_t are given in Section Tables. In standard rolled sections the root fillets at the junctions of the web and the flanges give additional torsional stiffness.

The shear stresses in an open section under St. Venant torsion vary from zero on the centre line of the wall to a maximum on the outside surface (Fig. 4.20(d)), and their direction is reversed on each side of the centre line. The shear flow constitutes a closed loop. The maximum stress in any element of thickness t is

$$v = \frac{Tt}{I_t} \tag{4.27}$$

The maximum shear stress in a section therefore occurs in the thickest element. At re-entrant corners the flow lines are crowded together, giving rise to very high stress concentrations. The effect is reduced by fillet radii. The shear stress is zero at the outside corners.

The angle of twist

$$\theta = \frac{TL}{GI_t} \tag{4.28}$$

where I_t is calculated from Eq. (4.26).

4.4.5 Elastic Torsion of Thin Walled Closed Sections

The shear stress distribution for closed sections is shown in Fig. 4.20(e). The flow is unidirectional with respect to the profile, contrasting with open sections. Variations in stress across the thickness of the section are ignored. The shear flow is constant at all points on the profile and is given by

$$v = \frac{T}{2A} \tag{4.29}$$

where A is the cross-sectional area of the profile.

The shear stress is a maximum in the thinnest part of the section and is obtained by dividing the shear flow by the thickness. This is in direct contrast to the open section where maximum shear stress occurs in the thickest part.

Angle of twist

$$\theta = \frac{TL}{4A^2G} \int \frac{ds}{t} \tag{4.30}$$

As in open sections additional stresses are introduced when warping is restrained. Equations (4.29) and (4.30) are derived from the Bredt–Batho hypothesis in which it is assumed that the shape of the section remains unchanged. To ensure that this assumption remains valid it may be necessary to stiffen the section with internal diaphragms at intervals along its length, and at points where concentrated loads are applied.

EXAMPLE 4.7   Torsion and transverse shear in a box section. Find the maximum shear stress in the box section shown in Fig. 4.21 which is subject to a torque $T = 200\,$kNm and a shear force $V = 500\,$kN. Assume that the section is adequately stiffened to prevent distortion of the profile and ignore the effects of warping restraints. Calculate the angle of twist per metre length.

The torsion and shear may be considered separately and the resulting shear flows are shown in Figs 4.21(a) and (b). For both calculations the profile dimensions shown in Fig. 4.21(c) can be used. The shear modulus is 80 GPa.

First consider torsion. The area enclosed by the profile

$$A = 790 \times 380 = 0{,}3002\text{E6}\,\text{mm}^2.$$

From Eq. (4.29) the shear flow

$$v = \frac{T}{2A} = \frac{200\text{E6}}{2 \times 0{,}3002\text{E6}} = 333\,\text{N/mm}$$

The shear stress is maximum in the web where the section is thinnest

$$v_t = \frac{v}{t} = \frac{333}{10} = 33{,}3\,\text{MPa}$$

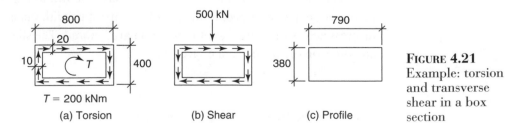

FIGURE 4.21
Example: torsion and transverse shear in a box section

From Eq. (4.30) the angle of twist

$$\theta = \frac{TL}{4A^2G} \int \frac{ds}{t}$$

$$= \frac{200E6 \times 1E3}{(4 \times 0,3002^2 \times 1E12 \times 80 \times 1E3)} \times 2 \left(\frac{790}{20} + \frac{380}{10} \right)$$

$$= 0,00107 \, \text{rad/m}$$

This is small which shows that the box is very stiff torsionally.

Now consider the direct shear force. The maximum shear stress is in the webs at the neutral axis, and the first moment of areas

$$Az_{c(\text{flange})} = 20 \times 790 \times \frac{380}{2} = 3,002E6 \, \text{mm}^3$$

$$Az_{c(\text{webs})} = 2 \times 10 \times \frac{380}{2} \times \frac{380}{4} = 0,361E6 \, \text{mm}^3$$

$$\text{Total} = 3,363E6 \, \text{mm}^3$$

Second moment of area

$$I = 2 \left(10 \times \frac{380^3}{12} + 20 \times 790 \times 190^2 \right) = 1232E6 \, \text{mm}^4$$

Direct shear stress from Eq. (4.15)

$$v_s = \frac{VAz_c}{It} = \frac{500E3 \times 3,363E6}{1232E6 \times 2 \times 10} = 68,2 \, \text{MPa}$$

Combing torsional and direct shear stresses

$$v_{(\text{combined})} = v_t + v_s = 33,3 + 68,2 = 101,5 \, \text{MPa}$$

Information on combining shear and torsion in design calculations is given in cl 6.2.7, EN 1993-1-1 (2005).

4.5 PLASTIC SECTION PROPERTIES AND ANALYSIS

4.5.1 Plastic Section Modulus

Plastic global analysis may be used in the design of steel structures for class 1 cross-sections (cl 5.6, EN 1993-1-1 (2005)). For the general case of a steel section symmetrical about the plane of bending, the stress distributions in the elastic and fully plastic state are shown in Fig. 4.3(c). For equilibrium of normal forces, the tensile and compressive forces must be equal. In the elastic state, when the bending stress varies from zero at the neutral axis to a maximum at the extreme fibres, this condition is achieved when the neutral axis passes through the centroid of the section. In the fully plastic state,

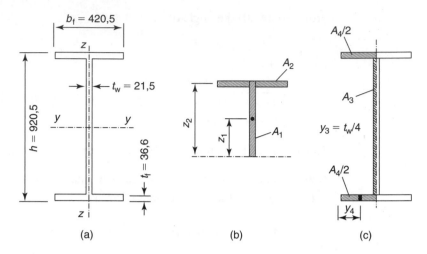

FIGURE 4.22 Example: plastic section modulus for an 'I' section

because the stress is uniformly equal to the yield stress, equilibrium is obtained when the neutral axis divides the section into two equal areas.

$$M_{pl} = \text{(first moment of area about the plastic NA)} f_y \qquad (4.31)$$

EXAMPLE 4.8 Plastic section moduli for an 'I' section. Determine the plastic section moduli about the y–y and z–z axes for the 'I' section shown in Fig. 4.22(a). The section is for a $914 \times 419 \times 388$ kg Universal Beam with the root radius omitted.

To determine the plastic section modulus about the y–y axis divide the section into A_1 and A_2 as shown in Fig. 4.22(b) where

$$A_1 = \left(\frac{h}{2}\right) t_w = \left(\frac{920,5}{2}\right) 21,5 = 9895,375 \text{ mm}^2$$

$$A_2 = (b_f - t_w) t_f = (420,5 - 21,5) 36,6 = 14603,5 \text{ mm}^2$$

and

$$z_1 = \frac{h}{4} = \frac{920,5}{4} = 230,125 \text{ mm}$$

$$z_2 = \frac{h}{2} - \frac{t_f}{2} = \frac{920,5 - 36,6}{2} = 441,95 \text{ mm}$$

Plastic section modulus

$$W_{ply} = 2(A_1 z_1 + A_2 z_2) = 2(9895,375 \times 230,125 + 14603,4 \times 441,95)$$
$$= 17,462 \text{E6 mm}^3$$

The value obtained from Section Tables is $17,657 \text{E6 mm}^3$ which is slightly greater because of the additional material at the root radius.

Similarly for the plastic section modulus about the z–z axis divide the section into areas A_3 and A_4 as shown is Fig. 4.22(c) where

$$A_3 = \frac{(h - 2t_f)t_w}{2} = \frac{(920,5 - 2 \times 36,6)21,5}{2} = 9108,475 \text{ mm}^2$$

$$A_4 = 2\left(\frac{b_f}{2}\right)t_f = 2\left(\frac{420,5}{2}\right)36,6 = 15390,3 \text{ mm}^2$$

and

$$y_3 = \frac{t_w}{4} = \frac{21,5}{4} = 5,375 \text{ mm}$$

$$y_4 = \frac{b_f}{4} = \frac{420,5}{4} = 105,125 \text{ mm}$$

Plastic section modulus

$$W_{plz} = 2(A_3y_3 + A_4y_4) = 2(9108,475 \times 5,375 + 15390,3 \times 105,125)$$

$$= 3,334\text{E}6 \text{ mm}^3.$$

The value obtained from Section Tables is $3,339\text{E}6 \text{ mm}^3$ which is slightly greater because of the additional material at the root radius.

From Section Tables the ratio of plastic/elastic section modulus (shape factor) for this section about the x–x axis is

$$W_{ply}/W_{ely} = 17,657\text{E}6/15,616\text{E}6 = 1,1307$$

The value of the shape factor for bending about the y–y axis generally quoted for 'I' sections in current use in design is 1,15.

The corresponding value of the shape factor for bending about the z–z axis is $W_{plz}/W_{elz} = 3,339\text{E}6/2,160\text{E}6 = 1,55$ which is typical for an 'I' section.

EXAMPLE 4.9 Plastic section moduli for a channel section. Determine the plastic section moduli about the y–y and z–z axes for the channel section shown in Fig. 4.23(a). The section is for a $432 \times 102 \times 65,54$ kg channel with the root radius omitted.

To determine the plastic section modulus about the y–y axis divide the section into A_1 and A_2 and shown in Fig. 4.23(b) where

$$A_1 = \left(\frac{h}{2}\right)t_w = \left(\frac{431,8}{2}\right)12,2 = 2633,98 \text{ mm}^2$$

$$A_2 = (b_f - t_w)t_f = (101,6 - 12,2)16,8 = 1501,92 \text{ mm}^2$$

and

$$z_1 = \frac{h}{4} = \frac{431,8}{4} = 107,95 \text{ mm}$$

$$z_2 = \frac{h}{2} - \frac{t_f}{2} = \frac{431,8 - 16,8}{2} = 207,5 \text{ mm}$$

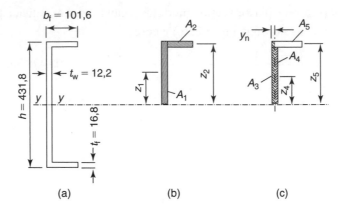

FIGURE 4.23 Example: plastic section modulus for a channel section

Plastic section modulus

$$W_{ply} = 2(A_1z_1 + A_2z_2) = 2(2633,98 \times 107,95 + 1501,92 \times 207,5)$$
$$= 1,192\text{E}6\,\text{mm}^3.$$

The value obtained from Section Tables is $1,207\text{E}6\,\text{mm}^3$ which is slightly different because of the additional material at the root radius and the fact that the flanges taper.

Similarly for the plastic section modulus about the z–z axis divide the section into areas A_3, A_4 and A_5 and shown in Fig. 4.23(c). It is first necessary to determine the position of the neutral axis z–z.

Since the axis z–z divides the total area into two equal parts

$$y_n = \frac{A_1 + A_2}{h} = \frac{2633,98 + 1501,92}{431,8} = 9,578\,\text{mm}$$

Areas

$$A_3 = \left(\frac{h}{2}\right)y_n = \left(\frac{431,8}{2}\right)9,578 = 2067,89\,\text{mm}^2$$

$$A_4 = \left(\frac{h}{2} - t_f\right)(t_w - y_n) = \left(\frac{431,8}{2} - 16,8\right)(12,2 - 9,578) = 522,04\,\text{mm}^2$$

$$A_5 = (b_f - y_n)t_f = (101,6 - 9,578)16,8 = 1545,97\,\text{mm}^2$$

and

$$y_3 = \frac{y_n}{2} = \frac{9,578}{2} = 4,789\,\text{mm}$$

$$y_4 = \frac{t_w - y_n}{2} = \frac{12,2 - 9,578}{2} = 1,311\,\text{mm}$$

$$y_5 = \frac{b_f - y_n}{2} = \frac{101,6 - 9,578}{2} = 46,011\,\text{mm}$$

Plastic section modulus

$$W_{pz} = 2(A_3 y_3 + A_4 y_4 + A_5 y_5)$$
$$= 2(2067,89 \times 4,789 + 522,04 \times 1,311 + 1545,97 \times 46,011)$$
$$= 0,1634E6 \, mm^3$$

The value obtained from Section Tables is $0,1531E6 \, mm^3$ which is slightly different because of the additional material at the root radius and the taper on the flanges.

4.5.2 Plastic Methods of Analysis

A plastic collapse mechanism depends on the formation of a plastic hinge(s) and this will now be considered in detail. The tensile stress–strain curve for mild steel is shown in Fig. 4.3(a). The curve is idealized into three stages namely elastic, plastic, and strain hardening stages of deformation. The moment–curvature for a beam made of the same material is shown in Fig. 4.3(b) with the corresponding distribution of stress at various loading stages shown in Fig. 4.3(c). The spread of the plastic hinge along the length of the beam is shown in Fig. 4.3(d).

The amount of rotation that can take place at a plastic hinge is determined by the length of the yield plateau shown in Fig. 4.3(b). For mild steel the length is considerable and the work hardening stage is ignored. For higher grade steels work hardening occurs immediately after yielding and there is no plateau. A plastic hinge does form but with an increasing moment of resistance. In design this increase in resistance due to work hardening is ignored which errs on the side of safety.

Plastic methods of analysis and design consider a structure at collapse when sufficient plastic hinges have formed to produce a mechanism. Examples of collapse mechanisms are shown in Fig. 4.24.

In simple situations, are shown in Fig. 4.24(a), the position of the plastic hinge is obvious and it is simple to calculate the collapse load using the method of virtual work.

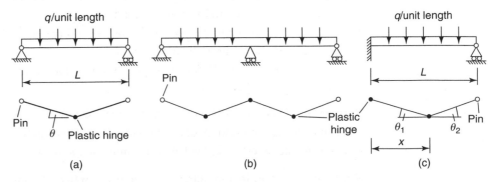

FIGURE 4.24 Examples of plastic collapse mechanisms

The method must only be applied to structures where the material becomes plastic at the yield stress and is capable of accommodating large plastic deformations. It must therefore not be applied to brittle materials such as cast iron. However it can be applied to reinforced concrete because the steel reinforcement behaves plastically at collapse but care must taken to check the rotation at the hinge for heavily reinforced sections.

The plastic method can be seen as a more rational method for design because all parts of the structure can be given the same safety factor against collapse. In contrast for elastic methods the safety factor varies. Intrinsically the plastic method of analysis is simpler than the elastic method because there is no need to satisfy elastic strain compatibilty conditions. However calculations for instability and elastic deflections require careful consideration when using the plastic method, but nevertheless it is very popular for the design of some structures (e.g. beams and portal frames).

The method of analysis demonstrated in this chapter is based on the principle of virtual work. This states that if a structure, which is equilibrium, is given a set of small displacements then the work done by the external loads on the external displacements is equal to the work done by the internal forces on the internal displacements. More concisely, external work equals internal work. The displacements need not be real, they can be arbitrary, which explains the use of the word 'virtual'. However the external and internal geometry must be compatible.

It is tacitly assumed that collapse is due to the formation of plastic hinges at certain locations and that other possible causes of failure, for example, local or general instability, axial or shear forces, are prevented from occurring. It is also important to understand that at collapse:

(a) the structure is in equilibrium, that is, the forces and moments, externally and internally, balance,
(b) no bending moment exceeds the plastic moment of resistance of a member,
(c) there are sufficient hinges to form a collapse mechanism.

These three conditions lead to three theorems for plastic analysis.

(1) *Lower bound theorem*: if only conditions (a) and (b) are satisfied then the solution is less than or equal to the collapse load.
(2) *Upper bound theorem*: if only conditions (a) and (c) are satisfied then solution is greater than or equal to the collapse load.
(3) *Uniqueness theorem*: if conditions (a), (b) and (c) are satisfied then the solution is equal to the collapse load.

Settlement of the supports has no effect on the solution at collapse because the only effect is to change the amount of rotation required. This is in contrast to elastic methods of analysis where settlement calculations must be included.

Plastic hinges form in a member at the maximum bending moment. However at the intersection of two members, where the bending moment is the same, the hinge forms in the weaker member. Generally the locations of hinges are at restrained ends, intersection of members and at point loads. The hinges may not form simultaneously as loading increases but this is not important for calculating the final collapse load. Generally the number of plastic hinges

$$n = r + 1$$

where

r is the number of redundancies.

However there are exceptions, for example, partial collapse of a beam in a structure.

Equating external to internal work for the member shown in Fig. 4.24(a)

$$wL\left(\frac{L}{4}\right)\theta = M_{\text{pl}}(2\theta) \quad \text{hence } M_{\text{pl}} = wL^2/8$$

In other conditions, as shown in Fig. 4.24(c), the position of the plastic hinge for minimum collapse load is not obvious and the calculations are more complicated. Equating external to internal work for the member shown in Fig. 4.24(c).

$$qx\left(\frac{x\theta_1}{2}\right) + q(L-x)^2\frac{\theta_2}{2} = M_{\text{pl}}(2\theta_1 + \theta_2) \tag{i}$$

From geometry

$$x\theta_1 = (L-x)\theta_2 \tag{ii}$$

Combining (i) and (ii)

$$qL = 2M_{\text{pl}}\left(\frac{2}{x} - \frac{1}{L-x}\right) \tag{iii}$$

Differentiating (iii) to determine the value of x for which q is a minimum

$$x^2 - 4Lx + 2L^2 = 0 \quad \text{hence } x = L(2 - \sqrt{2}) \tag{iv}$$

Combining (iii) and (iv)

$$M_{\text{pl}} = (1,5 - \sqrt{2})qL^2 \tag{v}$$

The method can be applied to a variety of structures. Further explanation and examples are given in Moy (1981), and Croxton and Martin Vol. 2 (1987 and 1989).

4.6 EFFECT OF A SHEAR FORCE ON THE PLASTIC MOMENT OF RESISTANCE (CLS 5.6 AND 6.2.6, EN 1993-1-1 (2005))

In general the effect of a transverse shear force is to reduce the plastic moment of resistance but the reduction for an 'I' section is small and may be ignored (cl 6.2.8, EN 1993-1-1 (2005)) if

$$\frac{V_{\text{Ed}}}{V_{\text{pl,Rd}}} \le 0,5 \tag{4.32}$$

where V_{Ed} is the design shear force, and the plastic shear resistance and (Eq. (6.18), EN 1993-1-1 (2005))

$$V_{pl,Rd} = \frac{A_v \left(f_y / 3^{1/2} \right)}{\gamma_{M0}} \tag{4.33}$$

The areas resisting shear (A_v) for various sections are given in cl 6.2.6(3), EN 1993-1-1 (2005) and in Section Tables.

The area resisting shear for an 'I' section is

$$A_v = A - 2bt_f + (t_w + 2r)t_f \tag{4.34}$$

This is a slight increase on the web area (ht_w) which has been used in the past.

The maximum shear stress of $f_y / 3^{1/2}$ is based on the failure criterion expressed in Eq. (2.5).

There is no reduction in the plastic moment of resistance for plastic or compact sections provided that the design shear force does not exceed 50% of the plastic shear resistance. This recommendation is related to the work of Morris and Randall (1979) who stated that shear can be ignored unless the average shear stress in the web exceeds $f_y/3$, or $f_y/4$ when the ratio of the overall depth to flange width (h/b_f ratio) exceeds 2.5.

Where the design shear force exceeds 50% of the plastic shear resistance the European Code recommends that the yield strength is reduced to $(1 - \rho)f_y$ where for shear (Eq. (6.29), EN 1993-1-1 (2005))

$$\rho = \left(2 \frac{V_{Ed}}{V_{pl,Rd}} - 1 \right)^2 \tag{4.35}$$

and for torsion

$$\rho = \left(2 \frac{V_{Ed}}{V_{pl,T,Rd}} - 1 \right)^2 \tag{4.35a}$$

These recommendations may be compared with theory by Horne (1971). If a shear force is applied to an 'I' section most of the shear force is resisted by the web. If it is assumed that all of the shear force is resisted by the web the effect on the plastic moment of resistance is shown in Fig. 4.25(a). The outer fibres in bending are at yield while the inner fibres are linear elastic. It is not possible to resist shear forces on the outer fibres and thus the inner fibres resist all of the shear force.

Plastic moment of resistance of the web

$$M_{plw} = \left[t_w \frac{d^2}{4} - t_w \frac{d_p^2}{4} + t_w \frac{d_p^2}{6} \right] f_y = \left[t_w \frac{d^2}{4} - t_w \frac{d_p^2}{12} \right] f_y \tag{i}$$

From the parabolic distribution of shear stress

$$V = \frac{2}{3} t_w d_p \tau y \tag{ii}$$

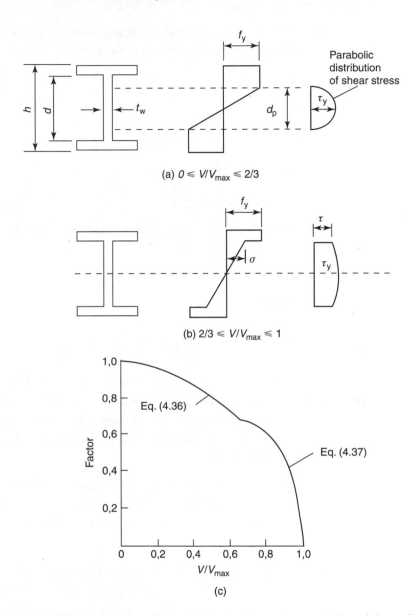

(a) $0 \leqslant V/V_{max} \leqslant 2/3$

(b) $2/3 \leqslant V/V_{max} \leqslant 1$

(c)

FIGURE 4.25 Effect of a shear force on the plastic section modulus of the web of an 'I' section

Defining

$$V_{max} = t_w h \tau_y \qquad \text{(iii)}$$

Combining (i) to (iii)

$$M_{plw} = \left(\frac{t_w d^2 f_y}{4}\right)\left[1 - \left(\frac{3}{4}\right)\left(\frac{V}{V_{max}}\right)^2\right] \qquad (4.36)$$

This expression is valid for $0 \leq V/V_{max} \leq 2/3$.

When $2/3 \leq V/V_{max} \leq 1$ then assuming the stress distributions shown in Fig. 4.25(b) the plastic moment of resistance of the web

$$M_{plw} = \frac{t_w d^2 \sigma}{6} \tag{i}$$

Applied shear force

$$V = dt_w \tau + \frac{2}{3} dt_w (\tau_y - \tau) \tag{ii}$$

If $V_{max} = dt_w \tau_y$, ratio

$$V/V_{max} = \frac{1}{3}\left(2 + \frac{\tau}{\tau_y}\right) \tag{iii}$$

Adopting a failure criterion of the form shown Eq. (2.5)

$$\left(\frac{\sigma}{f_y}\right)^2 + \left(\frac{\tau}{\tau_y}\right)^2 = 1 \tag{iv}$$

Combining (i) to (iv)

$$M_{plw} = \frac{t_w d^2 f_y}{4} \times \left[\frac{2}{3}\sqrt{1 - \left(\frac{3V}{V_{max}} - 2\right)^2}\right] \tag{4.37}$$

The expressions in the square brackets in Eqs (4.36) and (4.37) are plotted in Fig. 4.25(c) expressed as $M_{plw} = factor(t_w d^2 f_y/4)$. This theory is approximate and conservative but it does give a general appreciation of the effect of a shear force on the plastic moment of resistance. A fuller description and less conservative theories are given in Horne (1958).

In addition when the ratio $h_w/t_w > 72\varepsilon/\eta$ (cl 6.2.6, EN 1993-1-1 (2005)) the web should be checked for shear buckling. The limiting value of h_w/t_w is related to experimental work by Longbottom and Heyman (1956), and later work by Horne (1958).

EXAMPLE 4.10 Plastic section modulus reduced by shear. Determine the plastic modulus for a $762 \times 267 \times 197$ kg UB grade S275, using (a) Horne's method, (b) European Code method, when subject to a design value of shear force $V_{Ed} = 1145$ kN. Ignore the material factor for this example.

(a) Horne's method:

$$\frac{V_{Ed}}{V_{max}} = \frac{V_{Ed}}{ht_w f_y/3^{1/2}} = \frac{1145E3}{769,6 \times 15,6 \times 275/3^{1/2}} = 0,6$$

According to Horne, dividing Eq. (4.36) by f_y

$$W_{pl(web)} = \frac{td^2}{4}\left[1 - \frac{3}{4}\left(\frac{V_{Ed}}{V_{max}}\right)^2\right]$$

$$= 0,73(td^2/4), \quad \text{that is 27\% loss in the web.}$$

Reduced plastic section modulus of the complete section

$$W_{\text{pl(reduced)}} = W_{\text{pl(whole)}} - W_{\text{pl(web)}}(1 - \text{factor})$$

$$= 7{,}164\text{E}6 - \left(\frac{15{,}6 \times 685{,}8^2}{4}\right)(1 - 0{,}73) = 6{,}672\text{E}6 \,\text{mm}^3$$

Percentage reduction in plastic modulus for the whole section

$$= \frac{100(7{,}167\text{E}6 - 6{,}672\text{E}6)}{7{,}167\text{E}6} = 6{,}91\%$$

(b) EN 1993-1-1 (2005) method:

$$\frac{V_{\text{Ed}}}{V_{\text{pl,Rd}}} = \frac{V_{\text{Ed}}}{(A_v f_y/3^{1/2})}$$

$$= \frac{1145\text{E}3}{(12{,}7\text{E}3 \times 275/3^{1/2})} = 0{,}568$$

where $A_v = 12{,}7\text{E}3$ is obtained from Section Tables.

According to cl 6.29, EN 1993-1-1 (2005) there is a reduction in the plastic modulus if

$$\frac{V_{\text{Ed}}}{V_{\text{pl,Rd}}} > 0{,}5$$

From Eq. (4.35a)

$$\rho = \left(2\frac{V_{\text{Ed}}}{V_{\text{pl,Rd}}} - 1\right)^2 = (2 \times 0{,}568 - 1)^2 = 0{,}0185$$

and

$$W_{\text{pl(reduced)}} = \frac{M_{v,\text{Rd}}}{f_y} = W_{\text{pl}} - \frac{\rho A_v^2}{4t_w}$$

$$= 7{,}164\text{E}6 - 0{,}0185 \times \frac{12{,}7\text{E}3^2}{4 \times 15{,}6}$$

$$= 7{,}116\text{E}6 \,\text{mm}^3$$

$$\text{Percentage reduction} = \frac{(7{,}164\text{E}6 - 7{,}116\text{E}6) \times 100}{7{,}164\text{E}6} = 0{,}67\%.$$

4.7 Lateral Restraint (cl 6.3.5, EN 1993-1-1 (2005))

The full rotation required at a plastic hinge in a beam may not be realized unless lateral support is provided at the hinge position. It may also be necessary to provide lateral support at other points along the span to ensure that lateral torsional buckling does not occur. Lateral torsional buckling is considered to be prevented if the compression flange is prevented from moving laterally, either by an intersecting member, or by frictional restraint from intersecting floor units.

4.8 RESISTANCE OF BEAMS TO TRANSVERSE FORCES

The first check for transverse forces is the shear stress in the web at the neutral axis in the elastic stage of behaviour (cl 6.2.6(4), EN 1993-1-1 (2005)). In the plastic stage of behaviour the strength of the web is determined using values given in cl 6.2.6(2), EN 1993-1-1 (2005). Also it is necessary to consider web plate shear buckling at the ultimate limit state by checking if $72\varepsilon/\eta > h_w/t_w$ (cl 5.1(2), EN 1993-1-5(2003)).

Design calculations are also required for concentrated transverse forces applied to girders from supports, cross beams, columns, etc. (Fig. 4.26). The concentrated loads are dispersed through plates, angles and flanges to the web of the supporting girder. The deformations that occur to the supporting beam are shown in Fig. 4.27 and include yielding of the flange and local buckling of the web as shown in experiments (Astill *et al.* (1980)).

The design resistance is expressed simply as (cl 6.2, Eq. (6.1), EN 1993-1-5(2003))

$$F_{Rd} = \frac{f_{yw}L_{eff}t_w}{\gamma_{M1}}$$

The effective bearing length (L_{eff}) is an extension of the stiff bearing length (s_s) which assumes a 45° dispersion through plates, flanges and angles as shown in Fig. 4.26. The root radius of a section increases the length of the stiff bearing by $(2 - 2^{1/2})r$.

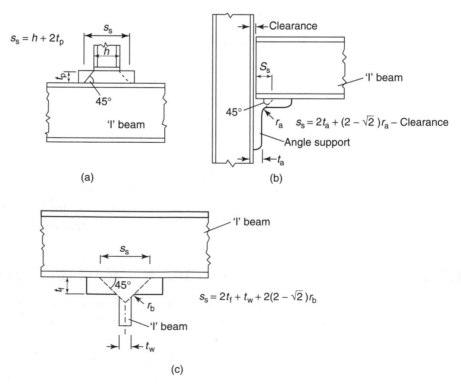

FIGURE 4.26 Stiff bearing lengths

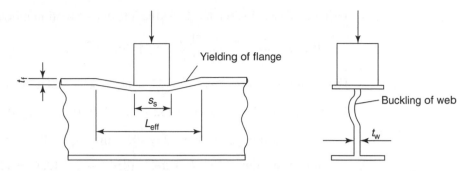

FIGURE 4.27 Transverse concentrated load

The extension of the stiff bearing length is based on theory of flange yielding and buckling of the web as shown in Examples 4.11 and 4.12.

If the transverse resistance of an unstiffened web is insufficient stiffeners are required (cl 9.4(1), EN 1993-1-5 (2003)) designed according to cl 6.3.3 or 6.3.4, EN 1993-1-1 (2005) with a buckling length of not less than $0,75h_w$ and using buckling curve c (Fig. 6.4, EN 1993-1-1 (2005)).

EXAMPLE 4.11   Simply supported beam carrying a uniformly distributed load and laterally restrained. The floor of an office building consists of 125 mm precast concrete units, with a mass of 205 kg/m^2, topped with a 40 mm concrete screed and 20 mm wood blocks. Lightweight partitions supported by the floor are equivalent to a superficial load of 1,0 kN/m^2 and the suspended ceiling has a mass of 40 kg/m^2. The floor rests on the top flanges of simply supported steel beams spanning 8 m and at a pitch of 3,75 m.

Characteristic loads	kg/m^2	kN/m^2
Dead load		
Self-weight of steel beam (assumed)	20	
125 mm precast units	205	
40 mm concrete screed $2400 \times 0,04$	96	
20 mm wood blocks $900 \times 0,02$	18	
Suspended ceiling	40	
	$379 \times 9,81/1E3 =$	3,72
Lightweight partitions		1,00
Total dead load		4,72
Imposed load for an office (EN 1991-1-1 (2002))		2,50
Maximum design bending moment at mid-span		kNm
Permanent load BM $= \gamma_G QL/8 = 1,35 \times 4,72 \times 8^2 \times 3,75/8 =$		191,2
Variable load BM $= \gamma_Q QL/8 = 1,5 \times 2,5 \times 8^2 \times 3,75/8 =$		112,5
Total BM		303,7

Using grade S275 steel, $f_y = 275$ MPa, the plastic section modulus required

$$W_{ply} = \frac{M}{f_y/\gamma_M} = \frac{303,7E6}{275/1,1} = 1,215E6 \text{ mm}^3.$$

From Table 4.1 for $L/250$, $L/h = 22$, hence $h = 8E3/22 = 364$ mm.

From Section Tables, try $457 \times 152 \times 60$ kg UB, $W_{ply} = 1,284E6$ mm^3,

$t_f = 13,3$ mm, $t_w = 8$ mm, $d = 407,7$ mm, $A_v = 3890$ mm^2, $b_f = 152,9$ mm,

$h = 454,7$ mm, $h_w = (h - 2t_f) = (454,7 - 2 \times 13,3) = 428,1$ mm,

$r = 10,2$ mm.

Check for buckling of web (Table 5.2 (sheet 1), EN 1993-1-1 (2005))

$$\frac{d}{t_w} = \frac{407,8}{8} = 50,96$$

$$72\varepsilon = 72 \left(\frac{235}{f_y}\right)^{1/2} = 72 \left(\frac{235}{275}\right)^{1/2} = 66,6 > 50,96 \text{ satisfactory.}$$

Check flange buckling (Table 5.2 (sheet 2), EN 1993-1-1 (2005))

$$\frac{c}{t_f} = \frac{[(B - t_w)/2 - r]}{t_f} = \frac{[(152,9 - 8)/2 - 10,2]}{13,3} = 4,67$$

$$9\varepsilon = 9 \times \left(\frac{235}{f_y}\right)^{1/2} = 9 \left(\frac{235}{275}\right)^{1/2} = 8,32 > 4,67 \text{ satisfactory.}$$

This is a Class 1 section and calculations can be reduced by using Section Tables.

Check deflection at the service limit state

Permanent and variable load on span $= 2,5 \times 8 \times 3,75 = 75$ kN

$$\text{Maximum deflection} = \delta_1 + \delta_2 = \frac{5QL^3}{384EI}$$

$$= \frac{5 \times 75 \times (8E3)^3}{384 \times 210 \times 254,64E6} = 9,35 \text{ mm}$$

Deflection limit from Table 4.1

$$\delta_{max} = \frac{L}{250} = \frac{8E3}{250} = 32 > 9,35 \text{ mm satisfactory.}$$

Check shear resistance of web at the ultimate limit state.

Design shear force at the support

$$V_{Ed} = \frac{QL}{2} = \frac{(1,35 \times 4,72 + 1,5 \times 2,5)8 \times 3,75}{2} = 151,8 \text{ kN.}$$

Area of web (cl 6.2.6(3), EN 1993-1-1 (2005))

$$A_v = A - 2bt_f + (t_w + 2r)t_f = 7580 - 2 \times 152,9 \times 13,3 + (8 + 2 \times 10,2)13,3$$

$$= 3890 \text{ mm}^2, \text{ or obtain value from Section Tables.}$$

Design plastic shear resistance (cl 6.2.6(1), EN 1993-1-1 (2005))

$$V_{\text{pl,Rd}} = \frac{A_v \tau_y}{\gamma_{M0}} = \frac{A_v(f_y/3^{1/2})}{\gamma_{M0}}$$

$$= 3890 \times \frac{(275/3^{1/2})}{1,1 \times 1E3} = 561,5\,\text{kN}$$

$$\frac{V_{\text{Ed}}}{V_{\text{pl,Rd}}} = \frac{151,8}{561,5} = 0,27 < 1,0 \text{ satisfactory.}$$

Check web plate buckling from shear at the ultimate limit state (cl 5.1 (2), EN 1993-1-5 (2003))

$$\frac{72\varepsilon}{\eta} = \frac{72(235/275)^{1/2}}{1,2} = 55,5$$

$$\frac{h_w}{t_w} = \frac{428,1}{8} = 53,5 < 55,5 \text{ satisfactory.}$$

Assuming $150 \times 75 \times 10$ mm angle supports at the ends of the beam (Fig. 4.26) the transverse shear buckling load (cl 6.2, Eq. (6.1), EN 1993-1-5 (2003))

$$F_{\text{Rd}} = \frac{f_{yw}L_{\text{eff}}t_w}{\gamma_{M1}} = \frac{275 \times 77,53 \times 8}{1,0 \times 1E3}$$

$$= 170,6 > V_{\text{Ed}} = 151,8\,\text{kN satisfactory no stiffener required.}$$

The effective length (L_{eff}) in the previous equation is obtained as follows:

The stiff bearing length (cl 6.3, Fig. 6.2, EN 1993-1-5 (2003)), for a $150 \times 75 \times 10$ mm angle support (Fig. 4.26)

$$s_s = 2t_a + (2 - 2^{1/2})r_a - \text{clearance}$$

$$= 2 \times 10 + (2 - 2^{1/2}) \times 11 - 3 = 23,44 < h_w = 428,1\,\text{mm}$$

Buckling co-efficient for the load application (cl 6.1(4), Type(c), EN 1993-1-5 (2003))

$$k_F = 2 + \frac{6(s_s + c)}{h_w} = 2 + \frac{6(23,44 + 0)}{428,1} = 2,33 < 6 \text{ satisfactory.}$$

Effective load length (cl 6.5, Eq. (6.13), EN 1993-1-5 (2003))

$$l_e = \frac{k_F E t_w^2}{2f_{yw}h_w}$$

$$= \frac{2,33 \times 210E3 \times 8^2}{2 \times 275 \times 428,1}$$

$$= 133,0 > (s_s + c = 23,44 + 0) \text{ use } 23,44\,\text{mm.}$$

Force (cl 6.4 Eq. (6.5) EN 1993-1-5 (2003))

$$F_{\text{cr}} = \frac{0,9\,k_F E t_w^3}{h_w} = \frac{0,9 \times 2,33 \times 210E3 \times 8^3}{(428,1 \times 1E3)} = 526,7\,\text{kN.}$$

The dimensionless parameters (cl 6.5, Eq. (6.8), EN 1993-1-5 (2003))

$$m_1 = \frac{f_{yf}b_f}{f_{yw}t_w} = \frac{275 \times 152,9}{275 \times 8} = 19,11$$

and (cl 6.5, Eq. (6.9), EN 1993-1-5 (2003))

$$m_2 = 0,02 \left(\frac{h_w}{t_f}\right)^2 = 0,02 \left(\frac{428,1}{13,3}\right)^2 = 20,72$$

Yield length (cl 6.5, Eq. (6.12), EN 1993-1-5 (2003))

$$l_y = l_e + t_f(m_1 + m_2)^{1/2} = 23,44 + 13,3(19,11 + 20,72)^{1/2} = 107,4 \, \text{mm}$$

An alternative yield length (cl 6.5, Eq. (6.11), EN 1993-1-5 (2003))

$$l_y = l_e + t_f \left[\frac{m_1}{2} + \left(\frac{l_e}{t_f}\right)^2 + m_2\right]^{1/2}$$

$$= 23,44 + 13,3 \left[\frac{19,11}{2} + \left(\frac{23,44}{13,3}\right)^2 + 20,72\right]^{1/2}$$

$$= 100,3 \, \text{mm use minimum.}$$

An alternative yield length (cl 6.5, Eq. (6.10), EN 1993-1-5 (2003))

$$l_y = s_s + 2t_f[1 + (m_1 + m_2)^{1/2}]$$

$$= 23,44 + 2 \times 13,3[1 + (19,11 + 20,72)^{1/2}] = 217,9 \, \text{mm.}$$

Factor (cl 6.4, Eq. (6.4), EN 1993-1-5 (2003))

$$\overline{\lambda}_F = \left(\frac{l_y t_w f_{yw}}{F_{cr}}\right)^{1/2} = \left(\frac{100,3 \times 8 \times 275}{526,7E3}\right)^{1/2} = 0,647$$

Reduction factor (cl 6.4, Eq. (6.3), EN 1993-1-5 (2003))

$$\chi_F = \frac{0,5}{\overline{\lambda}_F} = \frac{0,5}{0,647} = 0,773$$

The effective length (cl 6.2, Eq. (6.2), EN 1993-1-5 (2003))

$$L_{eff} = \chi_F l_y = 0,773 \times 100,3 = 77,53 < h_w = 428,1 \, \text{mm.}$$

Check self-weight of steel beam $= 60/3,75 = 16 < 20 \, \text{kg/m}^2$ assumed in loading calculations, acceptable.

EXAMPLE 4.12 Support for a conveyor. Part of the support for a conveyor consists of a pair of identical beams as shown in Fig. 4.28. Each beam is connected to a stanchion at end A by a cleat and is supported on a cross beam at D by bolting through the connecting flanges. Lateral restraint is provided by transverse beams at A, B and E connected to rigid supports. The loads shown are at the ultimate limit state.

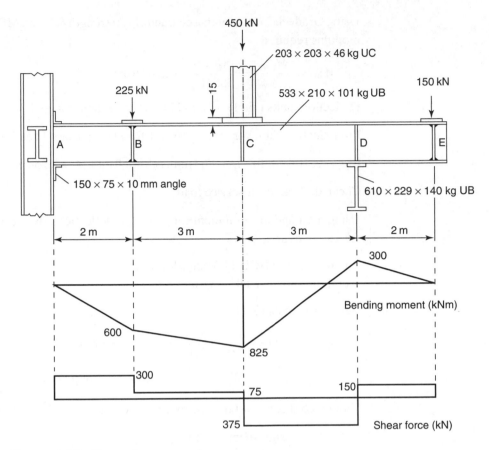

FIGURE 4.28 Example: support for a conveyor

Assuming pin joints at A and D.

Reactions are determined by taking moments about A, then about D.

$$R_D = \frac{2 \times 225 + 5 \times 450 + 10 \times 150}{8} = 525\,\text{kN}$$

$$R_A = \frac{6 \times 225 + 3 \times 450 - 2 \times 150}{8} = 300\,\text{kN}$$

Important bending moments are

$$M_B = R_A \times 2 = 300 \times 2 = 600\,\text{kNm}$$

$$M_C = R_A \times 5 - 225 \times 3 = 300 \times 5 - 225 \times 3 = 825\,\text{kNm}$$

$$M_D = -150 \times 2 = -300\,\text{kNm}$$

The shear force and bending moment diagrams at the ultimate limit state are shown in Fig. 4.28.

Using Grade S355 steel with a characteristic strength $f_y = 355$ MPa. The plastic section modulus required

$$W_{ply} = \frac{M_{max}}{f_y/\gamma_{M0}} = \frac{825E6}{355/1,1} = 2,56E6 \text{ mm}^3.$$

Deflection limits (Table 4.1), $L/250$, $L/h = 17$, hence $h = 8E3/17 = 470$ mm.

From Section Tables try $533 \times 210 \times 101$ kg UB, a Class 1 section,

$W_{ply} = 2,619E6$ mm^3, $t_f = 17,4$ mm, $t_w = 10,9$ mm, $A_v = 6250$ mm^2, $d = 476,5$ mm.

Check deflections at service load

Using area and area moment methods the deflections due to the imposed loads at service loads are:

mid-span in ABCD 11,3 mm (downwards)
end of cantilever DE −6,5 mm (upwards)

Deflection limits (Table 4.1)

$$\text{span ABCD} = \frac{L}{250} = \frac{8E3}{250} = 32 > 11,3 \text{ mm satisfactory.}$$

$$\text{span DE} = \frac{2L}{250} = \frac{2 \times 2E3}{250} = 16 > 6,5 \text{ mm satisfactory.}$$

Check web shear resistance between C and D at the ultimate limit state.

$$\text{Design shear force } V_{Ed} = 375 \text{ kN.}$$

$$\text{Plastic shear resistance } V_{pl,Rd} = \frac{A_v(f_y/3^{1/2})}{\gamma_M}$$

$$= \frac{6240 \times (355/3^{1/2})}{1,1 \times 1E3}$$

$$= 1163 > 375 \text{ kN satisfactory.}$$

Shear area (cl 6.2.6, EN 1993-1-1 (2005))

$$A_v = A - 2bt_f + (t_w + 2r)t_f$$

$$= 129,3 \times 100 - 2 \times 210,1 \times 17,4 + (10,9 + 2 \times 12,7) \times 17,4$$

$$= 6250 \text{ mm}^2 \text{ or obtain a value from Section Tables.}$$

$V_{Ed}/V_{pl,Rd} = 375/1163 = 0,322 < 0,5$ therefore no reduction in the plastic section modulus (cl 6.2.8, EN 1993-1-1 (2005)).

Check for buckling of web (Table 5.2 (sheet 1), EN 1993-1-1 (2005))

$$\frac{c}{t_w} = \frac{476,5}{10,9} = 43,7$$

$$72\varepsilon = 72\left(\frac{235}{f_y}\right)^{1/2} = 72\left(\frac{235}{355}\right)^{1/2} = 58,6 > 43,7 \text{ satisfactory.}$$

Check flange buckling (Table 5.2 (sheet 2), EN 1993-1-1 (2005))

$$\frac{c}{t_f} = \frac{[(B - t_w)/2 - r]}{t_f} = \frac{[(210,1 - 10,9)/2 - 12,7]}{17,4} = 4,99$$

$$9\varepsilon = 9 \times \left(\frac{235}{f_y}\right)^{1/2} = 9 \times \left(\frac{235}{355}\right)^{1/2} = 7,32 > 4,99 \text{ satisfactory.}$$

This is a Class 1 section and calculations can be reduced by using Section Tables.

Assuming a $150 \times 75 \times 10$ mm angle support at A at the end of the beam (Fig. 4.26) the transverse shear buckling strength (cl 6.2, Eq. (6.1), EN 1993-1-5 (2003)) is determined as shown previously in Example 4.11.

At D (Fig. 4.28) two UBs intersect which have the following dimensions.

From Section Tables the $533 \times 210 \times 101$ kg upper load carrying beam ABCDE. $h = 536,7$ mm, $b_f = 210,1$ mm, $h_w = (h - 2t_f) = (536,7 - 2 \times 17,4) = 501,9$ mm, $t_f = 17,4$ mm, $t_w = 10,9$ mm, $r = 12,7$ mm.

From Section Tables the $610 \times 229 \times 140$ kg lower support beam at D. $h = 617$ mm, $b_f = 230,1$ mm, $h_w = (h - 2t_f) = (617 - 2 \times 22,1) = 572,8$ mm, $t_f = 22,1$ mm, $t_w = 13,1$ mm, $r = 12,7$ mm.

The transverse shear buckling strength of the upper $533 \times 210 \times 101$ kg load carrying beam (cl 6.2, Eq. (6.1), EN 1993-1-5 (2003)).

$$F_{Rd} = \frac{f_{yw}L_{eff}t_w}{\gamma_{M1}} = \frac{355 \times 501,9 \times 10,9}{1,0 \times 1E3}$$

$$= 1942 > R_D = 525 \text{ kN satisfactory no stiffener required.}$$

The previous calculation for F_{Rd} includes the effective bearing length (L_{eff}) which is calculated as follows:

The stiff bearing length (cl 6.3, Fig. 6.2, EN 1993-1-5 (2003)), provided by the flange and web of the lower $610 \times 229 \times 140$ UB

$$s_s = 2t_f + t_w + (2 - 2^{1/2})r_b$$

$$= 2 \times 22,1 + 13,1 + (2 - 2^{1/2}) \times 12,7 = 64,74 < h_w = 501,9 \text{ mm}$$

Buckling co-efficient for the load application (cl 6.1(4), Type (a), EN 1993-1-5 (2003)) assuming a is large

$$k_F = 6 + 2\left(\frac{h_w}{a}\right)^2 = 6$$

Force (cl 6.4, Eq. (6.5), EN 1993-1-5 (2003))

$$F_{cr} = \frac{0,9\,k_F E t_w^3}{h_w} = \frac{0,9 \times 6 \times 210E3 \times 10,9^3}{501,9 \times 1E3} = 2926 \text{ kN}$$

The dimensionless parameters (cl 6.5, Eq. (6.8), EN 1993-1-5 (2003))

$$m_1 = \frac{f_{yf}b_f}{f_{yw}t_w} = \frac{355 \times 210,1}{355 \times 10,9} = 19,28$$

and (cl 6.5, Eq. (6.9), EN 1993-1-5 (2003))

$$m_2 = 0,02 \left(\frac{h_w}{t_f}\right)^2 = 0,02 \left(\frac{501,9}{17,4}\right)^2 = 16,64$$

Yield length (cl 6.5, Eq. (6.10), EN 1993-1-5 (2003))

$$l_y = s_s + 2t_f[1 + (m_1 + m_2)^{1/2}]$$
$$= 64,74 + 2 \times 17,4[1 + (19,28 + 16,64)^{1/2}] = 308,1 \, \text{mm}$$

Factor (cl 6.4, Eq. (6.4), EN 1993-1-5 (2003))

$$\overline{\lambda}_F = \left(\frac{l_y t_w f_{yw}}{F_{cr}}\right)^{1/2} = \left(\frac{308,1 \times 10,9 \times 355}{2926E3}\right)^{1/2} = 0,638$$

Reduction factor (cl 6.4, Eq. (6.3), EN 1993-1-5 (2003))

$$\chi_F = \frac{0,5}{\overline{\lambda}_F} = \frac{0,5}{0,638} = 0,784 < 1$$

The effective length (cl 6.2, Eq. (6.2), EN 1993-1-5 (2003))

$$L_{eff} = \chi_F l_y = 0,784 \times 308,1 = 241,6 < h_w = 501,9 \, \text{mm}$$

Similar calculations are required for the web buckling strength of the upper $533 \times 210 \times 101 \, \text{kg UB}$ at B, C and E.

Stiffeners are not required for the $533 \times 210 \times 101 \, \text{kg UB}$ but in the past many designers have inserted them where large point loads are applied. Assuming that one is required at D($R_D = 525 \, \text{kN}$) the following calculations are required. Assume a symmetrical stiffener of 10 mm thickness welded to the web and the flanges.

The non-dimensional slenderness ratio of the stiffener (Eq. (6.50), EN 1993-1-1 (2005))

$$\overline{\lambda}_z = \left(\frac{Af_y}{N_{cr}}\right)^{1/2} = \frac{(L_{cr}/i_y) \times 1}{\lambda_1}$$
$$= \frac{(0,75 \times 572,8/15E3) \times 1}{[93,9 \times (235/355)^{1/2}]} = 0,375E\text{-}3$$

assuming a stiffener on both sides of the web, breadth of stiffener

$$b_s = b_f - t_w - 2 \, (\text{weld leg}) = 210,1 - 10,9 - 2 \times 6 = 187,2 \, (\text{use } 180) \, \text{mm}$$
$$i_z = b_s t_s^3/12 = 180 \times 10^3/12 = 15E3 \, \text{mm}^4$$

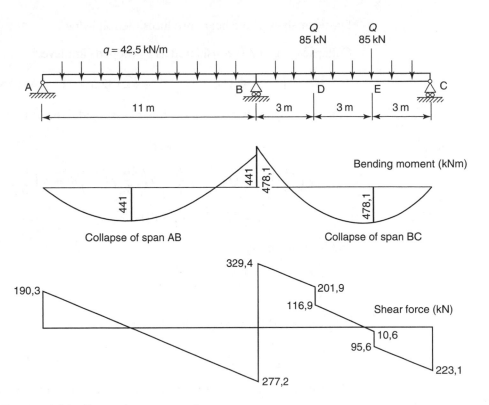

q = 42,5 kN/m

Q 85 kN Q 85 kN

A   B   D   E   C

11 m   3 m   3 m   3 m

441   441   478,1

Bending moment (kNm)

478,1

Collapse of span AB   Collapse of span BC

329,4

190,3   201,9

116,9   Shear force (kN)

10,6

95,6

223,1

277,2

FIGURE 4.29 Example: two span beam

From Fig. 6.4, EN 1993-1-1 (2005) (buckling curve c) with $\bar{\lambda}_z = 0{,}375\text{E-}3$ the reduction factor $\chi_z = 1{,}0$ and the design buckling resistance

$$N_{b,Rd} = \frac{\chi_z A f_y}{\lambda_M} = \frac{1{,}0 \times 180 \times 10 \times 355}{1{,}1 \times 1\text{E}3}$$

$$= 581 > R_D = 525 \text{ kN satisfactory.}$$

The design resistance of four vertical 6 mm fillet welds connecting the stiffener to both sides of the web (cl 4.5.3.3(3), Eq. (4.4), EN 1993-1-8 (2005))

$$F_{w,Rd} = 4h_w\, f_u/(3^{1/2}\beta\gamma_{M2}) \times 0{,}7a$$

$$= 4 \times 572{,}8 \times 510/(3^{1/2} \times 0{,}9 \times 1{,}25) \times 0{,}7 \times 6/1\text{E}3$$

$$= 2519 > R_D = 525 \text{ kN satisfactory.}$$

EXAMPLE 4.13   Two span beam. Determine the size of Universal Beam required to support the design loads at the ultimate limit state as shown in Fig. 4.29. Assume that the compression flange is fully restrained and that lateral torsional buckling does not occur.

Plastic analysis of the beam produces the following.

Collapse of span AB considered as a propped cantilever

$$M_{pl} = \left(1{,}5 - \sqrt{2}\right) qL^2 = \left(1{,}5 - \sqrt{2}\right) 42{,}5 \times 11^2 = 441 \text{ kNm}.$$

Collapse of span BC with plastic hinges assumed to be at E and B (Fig. 4.29)

External work = internal work

$$Q\left(\frac{2L}{3}\right)\theta_1 + Q\left(\frac{2L}{3}\right)\frac{\theta_1}{2} + q\left(\frac{2L}{3}\right)\left(\frac{L}{3}\right)\theta_1 + q\left(\frac{L}{3}\right)\left(\frac{L}{6}\right)\theta_2$$

$$= M_{pl}(\theta_1 + \theta_1 + \theta_2) \tag{i}$$

and from geometry

$$\left(\frac{2L}{3}\right)\theta_1 = \left(\frac{L}{3}\right)\theta_2 \quad \text{hence } \theta_2 = 2\theta_1 \tag{ii}$$

Combining eqs (i) and (ii) and rearranging

$$M_{pl} = \frac{QL}{4} + \frac{qL^2}{12} = \frac{85 \times 9}{4} + \frac{42{,}5 \times 9^2}{12} = 478{,}1 \text{ kNm}$$

This value of M_{pl} is the greater than the value for span AB and therefore is used to determine the size of a section which is continuous for two spans. The assumption that a plastic hinge is at E is not correct, it actually occurs at 3,25 m from C and $M_{pl} = 479{,}1$ kNm. However the error is small and is ignored.

Used Grade S355 steel and assuming a characteristic strength $f_y = 355$ MPa the plastic section modulus required

$$W_{ply} = \frac{M_{max}}{f_y/\gamma_M} = \frac{478{,}1\text{E}6}{355/1{,}0} = 1{,}347\text{E}6 \text{ mm}^3$$

From Section Tables try $457 \times 191 \times 67$ kg UB which is a Class 1 section bending about the y–y axis, $W_{ply} = 1{,}471\text{E}6 \text{ mm}^3$, $t_f = 12{,}7$ mm, $t_w = 8{,}5$ mm, $A_v = 4100 \text{ mm}^2$, $d = 407{,}8$ mm, $b_f = 189{,}9$ mm.

Check deflections at service load.

Using area and area moment methods the deflections due to the imposed load are 20,7 mm for AB and 25 mm for BC.

Deflection limit (Table 4.1)

AB is $\dfrac{L}{250} = \dfrac{11\text{E}3}{250} = 44 > 20{,}7$ mm, satisfactory

BC is $\dfrac{L}{250} = \dfrac{9\text{E}3}{250} = 36 > 25$ mm, satisfactory.

Sketch the shear force diagram (Fig. 4.29) at the ultimate limit state for the collapse of span BC.

For EC

$$E \circlearrowright + 478,1 + 42,5 \times 3 \times 1,5 - 3R_C = 0 \quad \text{hence } R_C = 223,1 \text{ kN}$$

For AB

$$B \circlearrowright + 478,1 - \frac{42,5 \times 11 \times 11}{2} + 11R_A = 0 \quad \text{hence } R_A = 190,3 \text{ kN}$$

For ABC

$$\uparrow + R_A + R_B + R_C - \sum qL - \sum Q = 0$$
$$+190,3 + R_B + 223,1 - 42,5 \times 20 - 2 \times 85 = 0 \quad \text{hence } R_B = 606,6 \text{ kN}$$

Check web shear resistance at B at the ultimate limit state.

Design shear force

$$V_{Ed} = 329,4 \text{ kN (Fig. 4.29)}$$

Design plastic shear resistance

$$V_{pl,Rd} = \frac{A_v \tau_y}{\gamma_M} = \frac{A_v(f_y/3^{1/2})}{\gamma_M} = \frac{4100 \times (355/3^{1/2})}{1,0 \times 1E3}$$
$$= 840 > 329,4 \text{ kN satisfactory.}$$

$V_{Ed}/V_{pl,Rd} = 329,4/840 = 0,392 < 0,5$ therefore no reduction in the plastic section modulus (cl 6.2.8, EN 1993-1-1 (2005)).

Check for buckling of the web (Table 5.2 (sheet 1), EN 1993-1-1 (2005))

$$\frac{c}{t_w} = \frac{407,9}{8,5} = 48,0$$
$$72\varepsilon = 72 \left(\frac{235}{f_y}\right)^{1/2} = 72 \left(\frac{235}{355}\right)^{1/2} = 58,6 > 48,0 \text{ satisfactory.}$$

Check flange buckling (Table 5.2 (sheet 2), EN 1993-1-1 (2005))

$$\frac{c}{t_f} = \frac{[(B - t_w)/2 - r]}{t_f} = \frac{[(189,9 - 8,5)/2 - 10,2]}{12,7} = 6,33$$
$$9\varepsilon = 9 \times \left(\frac{235}{f_y}\right)^{1/2} = 9 \times \left(\frac{235}{355}\right)^{1/2} = 7,32 > 6,33 \text{ satisfactory.}$$

This is a Class 1 section and calculations can be reduced by using Section Tables.

A continuous $457 \times 191 \times 67\,kg$ UB can be used for both spans. Alternatively if a minimum weight design is required then:

(a) A smaller section could be used for span AB but there would have to a splice which would increase fabrication costs.
(b) A smaller section could be used for span AB, with flange plates added to increase the moment of resistance for span BC, but again this would increase fabrication costs.

Calculations, similar to Example 4.12, are required for the web buckling strength of the $457 \times 191 \times 67\,kg$ UB at A, B, and C.

REFERENCES

Astill, A.W., Holmes, M. and Martin, L.H. (1980). *Web buckling of steel 'I' beams*. CIRIA Tech. note 102.

Croxton, P.C.L. and Martin, L.H. (1987 and 1989). *Solving problems in structures Vols. 1 and 2*. Longman Scientific and Technical.

EN 1993-1-1 (2005). *General rules and rules for buildings*. BSI.

EN 1993-1-5 (2003). *Plated structural elements*. BSI.

EN 1993-1-8 (2005). *Design of joints*. BSI.

Horne, M.R. (1971). *Plastic theory of structures*. Nelson.

Horne, M.R. (1958). The full plastic moment of sections subjected to shear force and axial loads, *British Welding Journal*, **5**, 170.

Longbottom, E. and Heyman, J. (1956). *Experimental verification of the strength of plate girders designed in accordance with the revised BS153: tests on full size and on model plate girders*, *Proceedings of the ICE*, **5(III)**, 462.

Megson, T.H.G. (1980). *Strength of materials for civil engineers*. Nelson.

Morris L.J. and Randall A.L. (1979) *Plastic Design*, Constrado.

Moy, S.S. (1981). *Plastic methods for steel and concrete structures*. Macmillan Press.

Zbirohowski-Koscia, K. (1967). *Thin walled beams*. crosby Lockwood.

Chapter 5 / Laterally Unrestrained Beams

In the previous chapter it was assumed that the compression flange of a beam was fully restrained, that is, the flange is unable to move laterally under the effect of loads or actions. In practice, this is rarely the case and although compression flanges may be restrained at discrete points the flange is still capable of buckling between restraints. Such buckling reduces the moment capacity of the member.

5.1 LATERAL TORSIONAL BUCKLING OF ROLLED SECTIONS SYMMETRIC ABOUT BOTH AXES

5.1.1 Basic Theory

A beam under the action of flexure alone due to the application of point moments producing single curvature is considered as the basic case. The supports allow rotation about a vertical axis but do not allow relative displacement of the top and bottom flanges, that is, twisting is not allowed (Fig. 5.1).

The governing equation for flexure about the minor axis is

$$EI_z \frac{d^2 v}{dx^2} = -M\phi \tag{5.1}$$

where EI_z is the flexural rigidity about the minor axis.

The governing equation for torsion is given by

$$GI_t \frac{d\phi}{dx} - EI_w \frac{d^3\phi}{dx^3} = M \frac{dv}{dx} \tag{5.2}$$

where GI_t is the torsional rigidity, EI_w the warping rigidity and the final term is the disturbing torque.

Eliminate the term dv/dx between Eqs (5.1) and (5.2) to give

$$GI_t \frac{d^2\phi}{dx^2} - EI_w \frac{d^4\phi}{dx^4} = -\frac{M^2}{EI_z}\phi \tag{5.3}$$

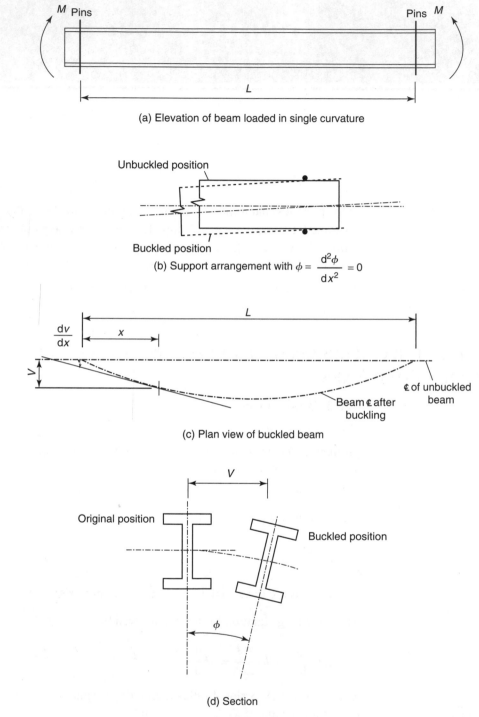

(a) Elevation of beam loaded in single curvature

Unbuckled position

Buckled position

(b) Support arrangement with $\phi = \dfrac{d^2\phi}{dx^2} = 0$

(c) Plan view of buckled beam

Original position

Buckled position

(d) Section

FIGURE 5.1 Lateral torsional buckling of beams

The solution to Eq. (5.3) is given by the first term of a sine series

$$\phi = \phi_0 \sin \frac{\pi x}{L} \tag{5.4}$$

where ϕ_0 is the angle of twist at mid-span.

Substituting Eq. (5.4) into Eq. (5.3) with the boundary condition that $\phi = 0$ at $x = L$ gives the elastic critical moment M_{cr} as

$$M_{cr} = \frac{\pi^2 E I_z}{L^2} \left[\frac{I_w}{I_z} + \frac{L^2 G I_t}{\pi^2 E I_z} \right]^{1/2} \tag{5.5}$$

or

$$M_{cr} = \frac{\pi \sqrt{E I_z G I_t}}{L} \left[1 + \frac{\pi^2 E I_w}{L^2 G I_t} \right]^{1/2} \tag{5.6}$$

The full derivation of the elastic critical moment is given in Kirby and Nethercot (1979) or Trahair and Bradford (1988).

In practice, real beams do not achieve full elastic buckling except at very high slenderness ratios and are also subject to the limit imposed by the section plastic capacity. Reductions in elastic buckling capacity are caused by the existence of residual stresses and any lack of initial straightness. The residual stresses within the section arise in the case of rolled sections due to differential rates of cooling in the web and flanges after hot rolling, and of plate girders due to both the preparation of the web and flange plates and the welding procedure (Nethercot, 1974a). These residual stresses do not affect the plastic capacity (Nethercot, 1974b; Kirby and Nethercot, 1979). The value of the critical moment given by Eq. (5.5) or (5.6) takes no account of major axis flexure. Trahair and Bradford (1988) indicate the critical moment obtained from Eq. (5.5) or (5.6) should be divided by a factor K where K is given by

$$K = \sqrt{\left(1 - \frac{E I_z}{E I_y} \right) \left(1 - \frac{G I_t}{E I_y} \left[1 + \frac{\pi^2 E I_w}{G I_t L^2} \right] \right)} \tag{5.7}$$

In practice for I or H sections $G I_t / E I_y [1 + \pi^2 E I_w / G I_t L^2]$ is negligible compared to unity and thus K may be taken as

$$K = \sqrt{1 - \frac{E I_z}{E I_y}} = \sqrt{1 - \frac{I_z}{I_y}} \tag{5.8}$$

For British Universal Beams the value of K is between 0,94 and 0,97 thus increasing M_{cr} by around 3% to 6%. This enhancement may therefore be neglected for Universal Beams. For Universal Columns, however, K is between 0,80 and 0,83, thus increasing M_{cr} by around 17% to 20%, and should therefore possibly be taken into account.

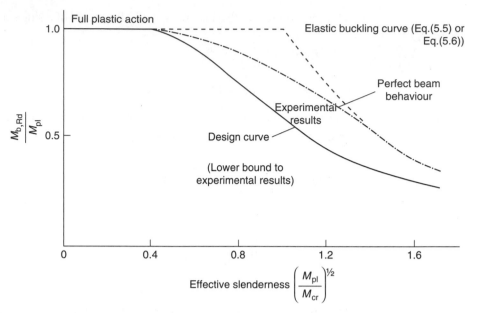

FIGURE 5.2 Beam behaviour with lateral torsional buckling

5.1.2 Interaction between Plastic Moment Capacity, Elastic Critical Moment and Allowable Bending Moment

If the results for beam strength against effective slenderness are plotted in non-dimensional format the results appear as in Fig. 5.2. The performance of beams is reduced below the perfect conditions due to the residual stresses and initial imperfections. A convenient lower bound to test data is given by a Perry–Robertson type approach to modelling the interaction. The imperfection co-efficient makes allowance both for geometric imperfections and residual stress levels. It should be noted that unlike columns there is no theoretical justification for such an approach.

5.1.2.1 Design Interaction between Buckling and Plasticity

For any type of beam, the interaction equation is given by

$$(M_{b,Rd} - M_{cr})(M_{b,Rd} - M_{pl,Rd}) = \eta_{LT}M_{cr}M_{b,Rd} \tag{5.9}$$

where η_{LT} is the imperfection factor for lateral torsional buckling.

Equation (5.9) may be written in a normalized form to give

$$\left(\frac{M_{b,Rd}}{M_{pl,Rd}} - \frac{M_{cr}}{M_{pl,Rd}}\right)\left(\frac{M_{b,Rd}}{M_{pl,Rd}} - 1\right) = \eta_{LT}\frac{M_{cr}}{M_{pl,Rd}}\frac{M_{b,Rd}}{M_{pl,Rd}} \tag{5.10}$$

It is convenient to define two parameters χ_{LT} and $\bar{\lambda}_{LT}$ given as

$$\chi_{LT} = \frac{M_{b,Rd}}{M_{pl,Rd}} \tag{5.11}$$

TABLE 5.1 Values of α_{LT} for the generalized lateral torsional buckling case.

Cross section	Size limits (h/b)	α_{LT}
Rolled I sections	≤ 2	0,21
	>2	0,34
Welded I sections	≤ 2	0,49
	>2	0,76
Other cross-sections		0,76

$$\bar{\lambda}_{LT} = \left[\frac{M_{pl,Rd}}{M_{cr}} \right]^{1/2} \tag{5.12}$$

Equation (5.10) then becomes

$$(\chi_{LT} - 1)(\chi_{LT} - (\bar{\lambda}_{LT})^2) = \eta_{LT} \frac{\chi_{LT}}{(\bar{\lambda}_{LT})^2} \tag{5.13}$$

The solution to which is given by

$$\chi_{LT} = \frac{1}{\Phi_{LT} + [\Phi_{LT}^2 - (\bar{\lambda}_{LT})^2]^{1/2}} \tag{5.14}$$

where Φ_{LT} is given by

$$\Phi_{LT} = 0,5[1 + \eta_{LT} + (\bar{\lambda}_{LT})^2] \tag{5.15}$$

with η_{LT} given as

$$\eta_{LT} = \alpha_{LT}(\bar{\lambda}_{LT} - 0,2) \tag{5.16}$$

The values of α_{LT} are dependant upon the type of section and whether rolled or welded. The appropriate values are given in Table 5.1 (cl 6.3. 2.1. EN 1993-1-1)

In design, full plastic moment capacity may be mobilized when $\bar{\lambda}_{LT}$ is less than $\bar{\lambda}_{LT,0}$ which may be taken as 0,4 or when $M_{Ed}/M_{cr} \leq (\bar{\lambda}_{LT,0})^2$.

5.1.2.2 Rolled Sections and Equivalent Welded Sections (cl 6.3.2.3 EN 1993-1-1)

For rolled sections and welded sections which are by implication symmetric, an alternative formulation of the interaction formula may be used,

$$\chi_{LT} = \frac{1}{\Phi_{LT} + [\Phi_{LT}^2 - \beta(\bar{\lambda}_{LT})^2]^{1/2}} \tag{5.17}$$

where Φ_{LT} is given by

$$\Phi_{LT} = 0,5[1 + \eta_{LT} + \beta(\bar{\lambda}_{LT})^2] \tag{5.18}$$

with η_{LT} as

$$\eta_{LT} = \alpha_{LT}(\bar{\lambda}_{LT} - \bar{\lambda}_{LT,0}) \tag{5.19}$$

TABLE 5.2 Values of α_{LT} for lateral torsional buckling of rolled and welded sections.

Cross section	Size limits (h/b)	α_{LT}
Rolled I sections	≤2	0,34
	>2	0,49
Welded I sections	≤2	0,49
	>2	0,76

The parameter χ_{LT} is subject to the limits $\chi_{LT} \leq 1,0$ and $\leq (\overline{\lambda}_{LT})^2$. The code recommends that the maximum value of $\overline{\lambda}_{LT,0}$ should be 0,4 and the minimum value of β as 0,75. The UK National Annex will adopt these values but will effectively specify the approach should be limited to rolled sections, and that, therefore, the generalized method must be used for welded sections.

Again the values of α_{LT} are dependant upon the type of section and whether rolled or welded. The appropriate values are given in Table 5.2.

Additionally, if this method is used, then the calculation of $M_{b,Rd}$ needs modifying by using a factor $\chi_{LT,mod}$ instead of χ_{LT} where

$$\chi_{LT,mod} = \frac{\chi_{LT}}{f} \tag{5.20}$$

The factor f is given by

$$f = 1 - 0,5(1 - k_c)[1 - 2,0(\overline{\lambda}_{LT} - 0,8)^2] \leq 1,0 \tag{5.21}$$

The factor k_c is determined from the type of loading.

For loading only at points of restraint,

$$k_c = \frac{1}{1,33 - 0,33\psi} \tag{5.22}$$

where ψ is the ratio between the moments at restraint points subject to the condition that $-1 \leq \psi \leq 1$. For loading between restraints values of k_c are given in Table 6.6 of EN 1993-1-1. The additional factor is due to the critical moment being determined elastically but failure will be by generation of plasticity not necessarily at the point of maximum deflection for buckling.

5.1.2.3 Simplified Assessment Methods for Building Structures (cl 6.3.2.4 EN 1993-1-1)

If there are discrete restraints to the compression flange, then lateral torsional buckling will not occur if the length L_c between restraints or the resultant slenderness $\overline{\lambda}_f$ satisfies the following equation

$$\overline{\lambda}_f = \frac{k_c L_c}{i_{fz}\lambda_1} = \overline{\lambda}_{c,0}\frac{M_{c,Rd}}{M_{y,Ed}} \tag{5.23}$$

where $M_{c,Rd}$ is the flexural capacity of the section, k_c is the factor to correct for moment gradient, $i_{f,z}$ is the radius of gyration of the effective compression flange which comprises the actual compression flange together with one-third of the part of the web which is in compression and λ_1 is given by $93,9(235/f_y)^{1/2}$ and $\bar{\lambda}_{C,0}$ is the normalized slenderness ratio of the effective compression flange.

5.1.2.4 Modifications Dependant upon Section Classification (cl 6.3.2.1 (3) EN 1993-1-1)

For sections other than Classes 1 and 2, a further modification is required as for Class 3 sections the capacity is based on the elastic section modulus, and Class 4 an effective section modulus is used. The modification is made by defining the flexural capacity as $W_{el,y}f_y$ for Class 3 and $W_{eff,y}f_y$ for Class 4. Also the buckling capacity $M_{b,Rd}$ is defined as $\chi_{LT}W_{pl,y}f_y/\gamma_{M1}$ for Class 1 or Class 2, $\chi_{LT}W_{el,y}f_y/\gamma_{M1}$ for Class 3, $\chi_{LT}W_{eff,y}f_y/\gamma_{M1}$ for Class 4.

Having established the basic determination of the allowable moment capacity, $M_{b,Rd}$, two additional modifications due to varying support conditions and non-uniform flexural loading need to be examined.

5.1.3 Effect of Support Conditions

If the beam has a rotational end restraint of R (where R takes a value of 0 for no restraint and 1 for full restraint), then the effective length kL is given by Eq. (5.24) (Trahair and Bradford, 1988).

$$\frac{R}{1-R} = -\frac{\pi}{2k}\cot\frac{\pi}{2k} \tag{5.24}$$

With a little loss in accuracy Eq. (5.24) can be written as

$$k = 1 - 0,5R \tag{5.25}$$

Both Eqs (5.24) and (5.25) are plotted in Fig. 5.3. The results from Eq. (5.25) represent ideal conditions therefore in practice the values of kL used are higher at more complete fixity as absolute rigid joints do not exist.

Pillinger (1988) gave some indications of the relationship between practical end conditions and effective lengths. Figure 5.4 gives typical end conditions in terms of connection type and degree of restraint.

For beams the effective length L between restraints can be taken as lying between 0,7 and 1,0 times the actual length with the lower factor implying full rotational restraint to both flanges and the higher factor with both flanges free to rotate in plan. If the rotational restraint parameter R can be assessed, then Eq. (5.25) can be modified to give

$$k = 1 - 0,3R \tag{5.26}$$

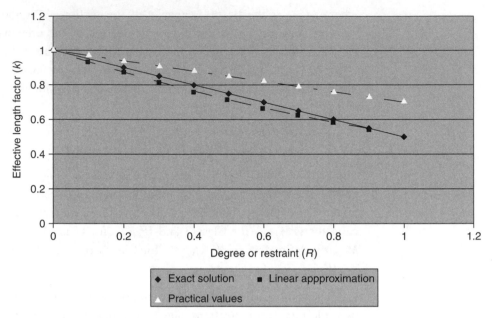

FIGURE 5.3 Values of effective length factor (k)

The values of kL for beams should be increased by 20% for destabilizing loads (BS 5950-1: Part 1 namely 1990 and 2000).

For cantilevers the situation can be more complex as the effective length will depend on both the restraint at the encastré end and at the tip. Table 5.3 (from BS 5950-1 (2000)) gives suitable values. It should be noted that when the loading is destabilizing the effective length factors can be extremely high. Guidance is also given in Nethercot and Lawson (1992).

Although it was suggested that for beams it is generally conservative to take the actual length, this may not be appropriate where the beam is supported solely on its bottom flange with no web or top flange restraint (Fig. 5.5). This type of situation will produce low lateral torsion buckling resistance, and Bradford (1989) suggests that in this case the system length L should be taken as,

$$L = 1 + \alpha \frac{h_s}{6} \left(\frac{t_w}{t_f}\right)^3 \left(\frac{1 + \frac{b}{h_s}}{2}\right) \tag{5.27}$$

where the beam is under a moment gradient, or

$$L = 1 + 10 \frac{h_s}{6} \left(\frac{t_w}{t_f}\right)^{3/2} \left(\frac{1 + \frac{b}{h_s}}{2}\right) \tag{5.28}$$

under a central point load.

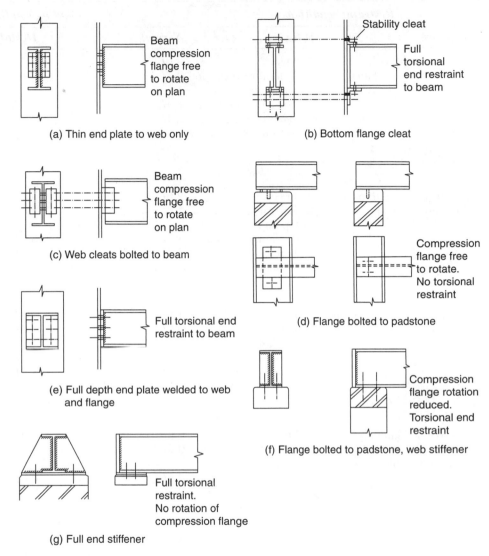

FIGURE 5.4 Typical connection and support detail

In both cases l is the span, h_s the distance between the centroids of the flanges, b the flange width and t_f and t_w the thickness of the flange and web respectively, and α is given by

$$\alpha = 4 + 7\psi + 4\psi^2 \tag{5.29}$$

where ψ is the ratio of the end moments.

5.1.4 Intermediate Restraints

Consider the beam in Fig. 5.6 but with a spring restraints giving a horizontal restraint stiffness of α_t (force/per unit displacement) and rotational restraint stiffness of α_r

TABLE 5.3 Effective length L_E for cantilevers without intermediate restraint.

Restraint conditions		Loading conditions	
At support	*At tip*	*Normal*	*Destabilizing*
(a) Continuous, with lateral restraint to top flange	(1) Free	3.0L	7.5L
	(2) Lateral restraint to top flange	2.7L	7.5L
	(3) Torsional restraint	2.4L	4.5L
	(4) Lateral and torsional restraint	2.1L	3.6L
(b) Continuous, with partial torsional restraint	(1) Free	2.0L	5.0L
	(2) Lateral restraint to top flange	1.8L	5.0L
	(3) Torsional restraint	1.6L	3.0L
	(4) Lateral and torsional restraint	1.4L	2.4L
(c) Continuous, with lateral and torsional restraint	(1) Free	1.0L	2.5L
	(2) Lateral restraint to top flange	0.9L	2.5L
	(3) Torsional restraint	0.8L	1.5L
	(4) Lateral and torsional restraint	0.7L	1.2L
(d) Restrained laterally, torsionally and against rotation on plan	(1) Free	0.8L	1.4L
	(2) Lateral restraint to top flange	0.7L	1.4L
	(3) Torsional restraint	0.6L	0.6L
	(4) Lateral and torsional restraint	0.5L	0.5L

Tip restraint conditions

(1) Free	(2) Lateral restraint to top flange	(3) Torsional restraint	(4) Lateral and torsional restraint
(not braced on plan)	(braced on plan in at least one bay)	(not braced on plan)	(braced on plan in at least one bay)

Source: BS 5950 Part 1:2000

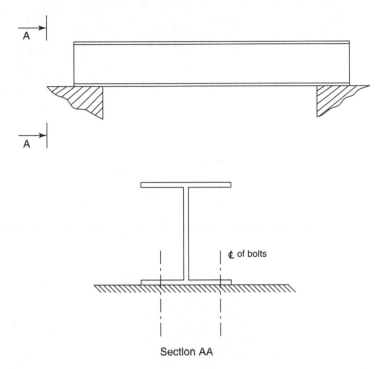

FIGURE 5.5 Beam supported on bottom flange with no restraint to top flange

at mid-span (Fig 5.6(a)) at a height of b_s above the centroidal axis. It can be shown (Mutton and Trahair, 1973) that the elastic critical moment is given by Eq. (5.6) or (5.7) with kL (the effective, or system, length) substituted for L, with k related to α_t and α_r by

$$\frac{\alpha_t L^3}{16EI_z}\left(1 + \frac{\frac{2b_s}{h}}{\frac{2b_0}{h}}\right) = \frac{\left(\frac{\pi}{2k}\right)^3 \cot\frac{\pi}{2k}}{\frac{\pi}{2k}\cot\frac{\pi}{2k} - 1} \qquad (5.30)$$

$$\frac{\frac{\alpha_r L^3}{16EI_w}}{1 - \frac{2b_s}{h}\frac{2b_0}{h}} = \frac{\left(\frac{\pi}{2k}\right)^3 \cot\frac{\pi}{2k}}{\frac{\pi}{2k}\cot\frac{\pi}{2k} - 1} \qquad (5.31)$$

where the distance of the centre of rotation below the shear centre b_0 is given by

$$b_0 = \frac{M_{cr}}{\pi^2 \frac{EI_z}{(kL)^2}} = \frac{h}{2}\sqrt{1 + \left(\frac{k}{K}\right)^2} \qquad (5.32)$$

and K is defined

$$K = \left(\frac{\pi^2 EI_w}{L^2 GI_t}\right)^{1/2} \qquad (5.33)$$

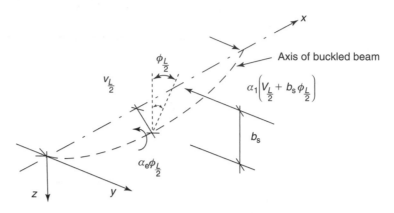

FIGURE 5.6(a) Beam with elastic intermediate restraints

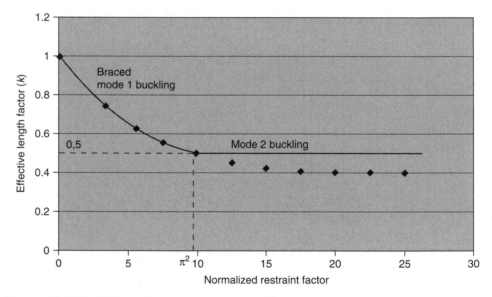

FIGURE 5.6(b) Effect of spring restraint on effective length factor

The relationships from Eqs (5.30) and (5.31) are plotted in Fig 5.6(b). The value of k changes from 1 (when the beam buckles as an entity) to $k = 0,5$ when the beam buckles as two half waves when the right hand sides of Eq. (5.30) or (5.31) reach a value of π^2.

From Eq. (5.32) the minimum value of b_0 is $h/2$. From Eq. (5.30) the maximum value of α_t applied at the top flange ($b_s = h/2$) to force a change into second mode is given by

$$\alpha_t = \frac{4M_{cr}}{Lh} = \frac{4P_f}{L} \tag{5.34}$$

where P_f is the force in the flange.

5.1.5 Loading

The effect of the applied loading on the beam system needs considering under two headings. The first is concerned with how the load is applied whether at restraint points or between restraints. The second is concerned with any possible destabilizing effects of the load.

5.1.5.1 Load Pattern

The background to any amendments to the value of M_{cr} to allow for load pattern is given in Nethercot and Rockey (1971), or Trahair and Bradford (1988).

For loading at points of lateral torsional restraint, the critical moment M_{cr} is modified by a factor C_1 which is dependant solely on the moment gradient within the section of the beam being considered. The moment gradient is defined by the ratio of the applied moments at either end of the beam segment. The small moment within the segment caused by self-weight is not taken into account in this calculation.

From Kirby and Nethercot (1979) C_1 is given by

$$\frac{1}{C_1} = 0{,}57 + 0{,}33\beta_1 + 0{,}1\beta_1^2 \geq 0{,}43 \tag{5.35}$$

where $\beta_1 = 1{,}0$ for single curvature (or -1 for double curvature).

And from Trahair and Bradford (1988) as either,

$$C_1 = 1{,}75 + 1{,}05\beta_2 + 0{,}3\beta_2^2 \leq 2{,}56 \tag{5.36}$$

or,

$$\frac{1}{C_1} = 0{,}6 - 0{,}4\beta_2 \geq 0{,}4 \tag{5.37}$$

where $\beta_2 = -1$ for single curvature (and $+1$ for double curvature).

Rewrite Eqs (5.35) to (5.37) using ψ rather than β_1 or β_2, defining $\psi = 1$ for single curvature to give

$$\frac{1}{C_1} = 0{,}57 + 0{,}33\psi + 0{,}1\psi^2 \leq 0{,}43 \tag{5.38}$$

$$C_1 = 1{,}75 - 1{,}05\psi + 0{,}3\psi^2 \leq 2{,}56 \tag{5.39}$$

$$\frac{1}{C_1} = 0{,}6 + 0{,}4\psi \geq 0{,}4 \tag{5.40}$$

A comparison between the values of C_1 from Eqs (5.38) to (5.39) is given in Table 5.4 and Fig. 5.7 where it is observed that there is little difference between the values, and that this difference is reduced as the value of the normalized slenderness ratio requires the value of $C_1^{1/2}$. For the examples herein Eq. (5.40) will be used.

TABLE 5.4 Comparison of values of C_1.

ψ	Eq. (5.36)	Eq. (5.37)	Eq. (5.38)
1,0	1,00	1,00	1,00
0,75	1,14	1,13	1,11
0,5	1,32	1,30	1,25
0,25	1,52	1,51	1,43
0	1,75	1,75	1,67
−0,25	2,03	2,03	2,00
−0,5	2,33	2,35	2,50
−0,75	2,33	2,56	2,50
−1,0	2,33	2,56	2,50

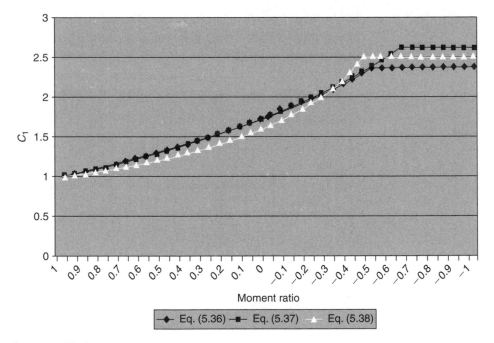

FIGURE 5.7 Values of C_1

For loads between restraints use the n factor method from Tables 15 to 17 of BS 5950-1 (1990) should be used. In this case C_1 is defined by

$$C_1 = \frac{1}{\sqrt{n}} \qquad (5.41)$$

Tables 15 to 17 of BS 5950: Part 1: 1990 are reproduced in Annexe A7.

5.1.5.2 Destabilizing and Stabilizing Loads

A destabilizing load is one which is applied to the compression flange and is free to move as the flange buckles laterally. Such a load has the effect of reducing the elastic critical moment as an additional disturbing torque is introduced (Anderson and Trahair, 1972; Trahair and Bradford, 1988). A stabilizing load, therefore, is one

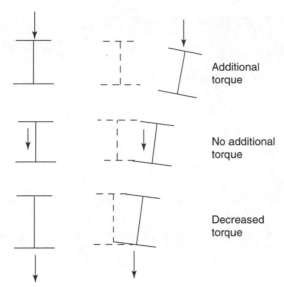

Additional torque

No additional torque

Decreased torque

FIGURE 5.8 Destabilizing loads

applied in such a way that the effect counterbalances the buckling effect and thus enhances the elastic critical moment. Thus for simply supported beams if a load free to move is applied above the shear centre it is destablizing, and below the shear centre it is stabilizing (Fig. 5.8). The reverse is true for a cantilever!

For loading applied at other than the centroid of the section it is only possible to determine the critical moment under a given load pattern.

Timoshenko and Gere (1961) give the value of the critical load for a simply supported I beam under either a uniformly distributed load (UDL) or central point load as

$$P_{cr} = \gamma_2 \frac{\sqrt{EI_z GI_t}}{L^2} \tag{5.42}$$

where γ_2 is dependant upon the position of the load (top flange, centroid or bottom flange) and the factor $L^2 GI_t/EI_w$. Values of γ_2 are plotted in Fig. 5.9 for values of beam stiffness $1/K^2$ (where K is given by Eq. (5.33)).

5.1.5.3 Cantilevers and Beams Cantilevering Over Supports

• Cantilevers

Trahair (1983) demonstrated that for cantilevers built in at the support that the elastic critical moment M_{cr} could be given with little loss of accuracy as

$$M_{cr} = \frac{\sqrt{EI_z GI_t}}{L}(1{,}6 + 0{,}8K) \tag{5.43}$$

where K is given by Eq. (5.33) as $\sqrt{(\pi^2 EI_w/(L^2 GI_t))}$.

It will be noted that this differs from the more usual solution given in Eq. (5.6) where the elastic critical moment is proportional to $(1 + K^2)^{1/2}$.

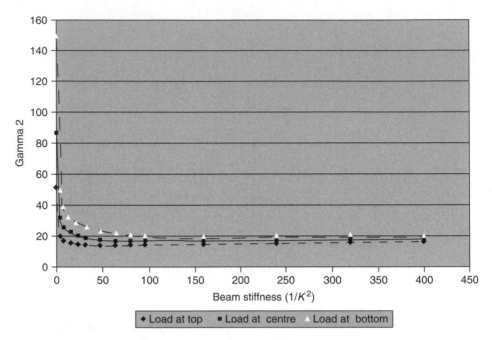

FIGURE 5.9(a) Factor for point load not applied at shear centre

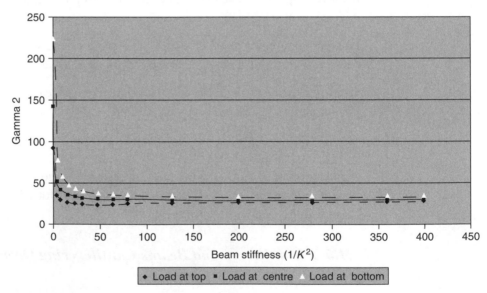

FIGURE 5.9(b) Factor for UDL not applied at shear centre

For a point load P_{cr} at the free end of the cantilever Trahair gives the following expression,

$$P_{cr} = \frac{\sqrt{EI_z GI_t}}{L^2}\left[11\left(1 + \frac{1{,}2\varepsilon}{\sqrt{1 + 1{,}2^2\varepsilon^2}}\right) + 4(K - 2)\left(1 + \frac{1{,}2(\varepsilon - 0{,}1)}{\sqrt{1 + 1{,}2^2(\varepsilon - 0{,}1)^2}}\right)\right]$$

(5.44)

where the parameter ε defines the position relative to the centroidal axis of the point of application of the load, and is given by

$$\varepsilon = \frac{\bar{a}}{L}\sqrt{\frac{EI_z}{GI_t}} = \frac{2\bar{a}}{h}\frac{K}{\pi} \tag{5.45}$$

where $\bar{a}$ is the distance of the point of application of the load

For a UDL of q_{cr} applied to whole length of the cantilever,

$$q_{cr} = \frac{2\sqrt{EI_zGI_t}}{L^3}\left[27\left(1 + \frac{1{,}4(\varepsilon - 0{,}1)}{\sqrt{1 + 1{,}4^2(\varepsilon - 0{,}1)^2}}\right)\right.$$
$$\left. + 10(K - 2)\left(1 + \frac{1{,}3(\varepsilon - 0{,}1)}{\sqrt{1 + 1{,}3^2(\varepsilon - 0{,}1)^2}}\right)\right] \tag{5.46}$$

For loading applied at the centroid, $\varepsilon = 0$, and Eq. (5.44) reduces to

$$P_{cr} = \frac{\sqrt{EI_zGI_t}}{L^2}[3{,}96 + 3{,}52K] \tag{5.47}$$

The moment $M_{cr,P}$ due to this load is given by

$$M_{cr,P} = P_{cr}L = \frac{\sqrt{EI_zGI_t}}{L}[3{,}96 + 3{,}52K] \tag{5.48}$$

Equation (5.46) becomes when loading is applied at the centroid

$$q_{cr} = \frac{2\sqrt{EI_zGI_t}}{L^3}[5{,}83 + 8{,}71K] \tag{5.49}$$

The moment $M_{cr,q}$ due to this load is given by

$$M_{cr,q} = \frac{q_{cr}L^2}{2} = \frac{\sqrt{EI_zGI_t}}{L}[5{,}83 + 8{,}71K] \tag{5.50}$$

Equations (5.43), (5.48) and (5.50) are plotted in Fig. 5.10, where it is observed that as the loading progresses from uniform moment through a point load at the end to a UDL, the equivalent elastic moment increases. This is due to the fact that for the UDL the moment quickly drops off from the support and thus has less of an effect. The moment due to the point load also drops off, but more slowly. It is therefore to be noted that the general UK practice of ignoring the beneficial effects of moment gradients on a cantilever (i.e. $C_1 = 1{,}0$) is extremely conservative.

• Overhanging beams

For the situation shown in Fig. 5.11(a), where a beam overhangs the supports symmetrically and is loaded by end point loads, then the lateral torsional buckling of such a system is controlled by the restraint at the internal supports which may often be of the type shown in Fig. 5.11(b), where such restraints can be considered elastic. Such restraints increase the torsional flexibility of the beam and thereby decrease the buckling moment. The decrease P/P_0 is dependant upon the stiffness

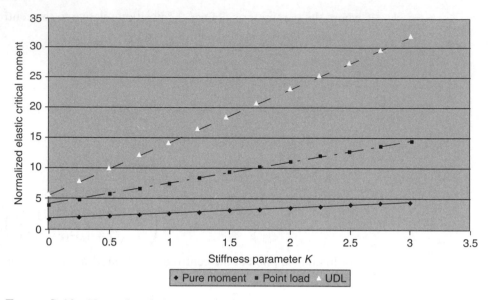

FIGURE 5.10 Normalized elastic critical moment for a cantilever

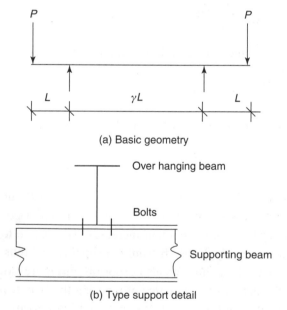

FIGURE 5.11 Overhanging beams

at the support α_R, and Trahair (1983) gives the following equation for beams loaded centroidally,

$$\frac{P}{P_0} = \left(1 - \frac{2\bar{a}}{h}K\beta^2\,(1-\beta)\right)\sqrt{\frac{\frac{\alpha_R L}{GI_t}}{5 + \frac{4K^2}{1+K^2} + \frac{\alpha_R L}{GI_t}}} \tag{5.51}$$

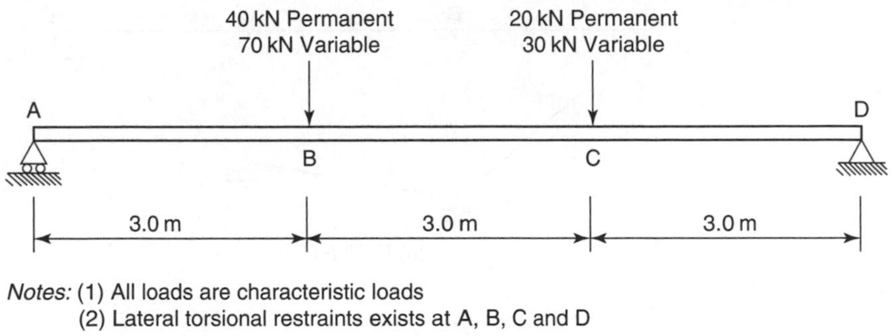
40 kN Permanent
70 kN Variable
20 kN Permanent
30 kN Variable
A
D
B
C
3.0 m
3.0 m
3.0 m
Notes: (1) All loads are characteristic loads
(2) Lateral torsional restraints exists at A, B, C and D

FIGURE 5.12 Design data for Example 5.1

where β is defined by

$$\beta = \frac{\frac{\alpha_R L}{GI_t}}{1 + \frac{\alpha_R L}{GI_t}} \tag{5.52}$$

It should be noted for situations where the internal span is greater than the sum of the overhangs, then the internal span dominates the behaviour.

EXAMPLE 5.1 Beam with loading applied at restraints.

Prepare a design in Grade S355 steel for the beam for which the data are given in Fig. 5.12.

Factored actions at ULS:

at B: $1{,}35 \times 40 + 1{,}5 \times 70 = 159\,\text{kN}$
at C: $1{,}35 \times 20 + 1{,}5 \times 30 = 72\,\text{kN}$
Total load = 231 kN

The BM and SF diagrams are drawn in Fig. 5.13.

The critical section for design is the central section BC as the moment gradient is the least.

Try a 406 × 178 × 74 UKB

$$M_{\text{pl,Rd}} = W_{\text{pl,y}} \frac{f_y}{\gamma_{M0}} = 1501000 \frac{355}{1{,}0} \times 10^{-6} = 533\,\text{kNm}$$

$M_{\text{Sd}} = 390\,\text{kNm}$, beam satisfies the plastic capacity criterion.

Section classification:

Compression flange:

$$c = 0{,}5[b - 2r - t_w] = 0{,}5[179{,}5 - 2 \times 10{,}2 - 9{,}5] = 74{,}8\,\text{mm}$$

$$c/t_f = 74{,}8/16 = 4{,}68$$

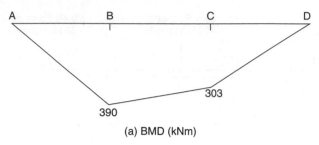

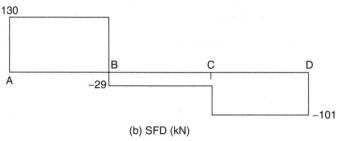

(a) BMD (kNm)

(b) SFD (kN)

FIGURE 5.13   BMD and SFD for Example 5.1

Maximum value for a Class 1 flange is $9\varepsilon = 9(235/355)^{1/2} = 7{,}32$.

Web:

$$c/t = d/t_w = 37{,}9.$$

Maximum value for a Class 1 web is $9\varepsilon = 72(235/355)^{1/2} = 58{,}6$.

Thus a $406 \times 178 \times 74$ UKB Grade S355 is Class 1.

Shear check:

$$V_{pl,Rd} = \frac{1}{\sqrt{3}} A_v \frac{f_y}{\gamma_{M0}} = \frac{1}{\sqrt{3}} 4260 \frac{355}{1{,}0} \times 10^{-3} = 873\,\text{kN}$$

By inspection, the moment capacity is not reduced due to shear.

Calculation of M_{cr}:

$$E = 210\,\text{GPa}; G = 81\,\text{GPa}.$$

The most foolproof method to determine M_{cr} given the use of the cm and dm in section property tables is to work in kN and m.

Use Eq. (5.5) to determine M_{cr}.

System length $L = 3\,\text{m}$.

$$\frac{I_w}{I_z} = \frac{0{,}608 \times 10^{-6}}{1545 \times 10^{-8}} = 0{,}03924\,\text{m}^2$$

$$\frac{\pi^2 E I_z}{L^2} = \frac{\pi^2 \times 210 \times 10^6 \times 1545 \times 10^{-8}}{3^2} = 3558\,\text{kN}$$

$$\frac{L^2 GI_t}{\pi^2 EI_z} = \frac{GI_t}{\frac{\pi^2 EI_z}{L^2}} = \frac{81 \times 10^6 \times 62,8 \times 10^{-8}}{3558} = 0,0143 \, \text{m}^2$$

$$M_{cr} = \frac{\pi^2 EI_z}{L^2} \left[\frac{I_w}{I_z} + \frac{L^2 GI_t}{\pi^2 EI_z} \right]^{1/2} = 3558 \, [0,03924 + 0,0143]^{1/2} = 823 \, \text{kNm}$$

Determine the value of C_1 using Eq. (5.40):

The moment ratio ψ is given by

$$\psi = \frac{M_{Ed,C}}{M_{Ed,B}} = \frac{303}{390} = 0,777$$

$$\frac{1}{C_1} = 0,6 + 0,4\psi = 0,6 + 0,4 \times 0,777 = 0,911$$

or

$$C_1 = \frac{1}{0,911} = 1,098$$

The value of M_{cr} to be used in Eq. (5.12) in the calculation of the normalized slenderness ratio is $C_1 M_{cr}$, thus $\bar{\lambda}_{LT}$ is given as

$$\bar{\lambda}_{LT} = \sqrt{\frac{W_y f_y}{M_{cr}}} = \sqrt{\frac{1501000 \times 355}{1,098 \times 823 \times 10^6}} = 0,768$$

Both methods available to determine the strength reduction factors due to lateral torsional buckling will be used in this example to demonstrate any differences.

(a) General case

The ratio $h/b = 412,8/179,5 = 2,3 > 2$, so from Table 5.1, $\alpha_{LT} = 0,34$ (curve b).

Use Eqs (5.15) and (5.16) to determine Φ_{LT}:

$$\Phi_{LT} = 0,5[1 + \eta_{LT} + (\bar{\lambda}_{LT})^2] = 0,5[1 + 0,34(0,768 - 0,2) + 0,768^2] = 0,891$$

Use Eq. (5.14) to determine χ_{LT}:

$$\chi_{LT} = \frac{1}{\Phi_{LT} + [\Phi_{LT}^2 - (\bar{\lambda}_{LT})^2]^{1/2}} = \frac{1}{0,891 + [0,891^2 - 0,768^2]^{1/2}} = 0,745$$

$$M_{b,Rd} = \chi_{LT} W_{pl,y} \frac{f_y}{\gamma_{M1}} = 0,745 \times 1501000 \frac{355}{1,0} \times 10^{-6} = 397 \, \text{kNm}$$

This is greater than the moment at B (390 kNm).

Clearly the end section AB does not need checking as the system length is the same as BC (therefore the basic value of M_{cr} does not change and since $\psi = 0$, C_1 now becomes 1,75 with the effect of reducing the value of $\bar{\lambda}_{LT}$ and of increasing the value of χ_{LT} and hence $M_{b,Rd}$.

(b) Method for rolled sections

The ratio $h/b = 412,8/179,5 = 2,3 > 2$, so from Table 5.2, $\alpha_{LT} = 0,49$ (curve c).

Determine η_{LT} from Eq. (5.19)

$$\eta_{LT} = \alpha_{LT}(\bar{\lambda}_{LT} - \bar{\lambda}_{LT,0}) = 0,49(0,768 - 0,4) = 0,180$$

Determine Φ_{LT} from Eq. (5.18)

$$\Phi_{LT} = 0,5[1 + \eta_{LT} + \beta(\bar{\lambda}_{LT})^2] = 0,5[1 + 0,180 + 0,75 \times 0,768^2] = 0,811$$

Determine χ_{LT} from Eq. (5.17)

$$\chi_{LT} = \frac{1}{\Phi_{LT} + [\Phi_{LT}^2 - \beta(\bar{\lambda}_{LT})^2]^{1/2}} = \frac{1}{0,811 + [0,811^2 - 0,75 \times 0,768^2]} = 0,784$$

Initially ignore the correction factor f, to give $M_{b,Rd}$ as

$$M_{b,Rd} = \chi_{LT}W_{pl,y}\frac{f_y}{\gamma_{M1}} = 0,784 \times 1501000\frac{355}{1,0} \times 10^{-6} = 418\,kNm$$

Determine f:

The moment ratio $\psi = 303/390 = 0,777$, so from Eq. (5.33),

$$k_c = \frac{1}{1,33 - 0,33\psi} = \frac{1}{1,33 - 0,33 \times 0,777} = 0,931$$

From Eq. (5.32) f is given by

$$f = 1 - 0,5(1 - k_c)[1 - 2,0(\bar{\lambda}_{LT} - 0,8)^2] \leq 1,0$$
$$= 1 - 0,5(1 - 0,931)[1 - 2,0(0,768 - 0,8)^2] = 0,966$$

Using Eq. (5.31) $M_{b,Rd}$ now becomes $418/0,966 = 433\,kNm$. The effect of the factor f is not terribly significant.

Even without the factor, f lateral torsional buckling clauses for rolled sections produces a marginally higher value of $M_{b,Rd}$ by around 20 kNm (or around 5%).

Deflection (under service loading):

$$EI = 210 \times (27430 \times 10^4) \times 10^{-6} = 57603\,kNm^2$$

Mid-span deflection due to an asymmetric point load is given by

$$\delta = \frac{WL^3}{48EI}\left(3\left(\frac{b}{L}\right) - 4\left(\frac{b}{L}\right)^3\right)$$

Variable action deflection:

Load at B: $W = 70\,kN$, $b = 3\,m$, $l = 9\,m$, $\delta = 0,016\,m$

Load at C: $W = 30\,kN$, $b = 3\,m$, $L = 9\,m$, $\delta = 0,007\,m$

Deflection under service variable actions $= 0,023\,m$.

This deflection is equivalent to span/390, and is therefore acceptable.

Deflection under total actions:

Load at B: $W = 110\,\text{kN}$, $b = 3\,\text{m}$, $l = 9\,\text{m}$, $\delta = 0{,}025\,\text{m}$

Load at C: $W = 50\,\text{kN}$, $b = 3\,\text{m}$, $L = 9\,\text{m}$, $\delta = 0{,}011\,\text{m}$

Total deflection $= 0{,}036\,\text{m}$.

This deflection is equivalent to span/250, and is therefore acceptable.

Web check (refer to Section 4.8 and cl 6 EN 1993-1-5):

The web capacity under transverse forces needs checking at A and B.

Note as a rolled section is being used, $f_{yw} = f_{yf} = 355\,\text{MPa}$. Since at either point the length of stiff bearing s_s is not known, set $s_s = 0$ and determine the value required should the check fail

At A: $F_{Sd} = R_A = 130\,\text{kN}$.

For an end support with $c = s_s = 0$, $k_F = 2$ (type c)

Determine m_1:

$$m_1 = \frac{f_{yf}b_f}{f_{yw}t_w} = \frac{b_f}{t_w} = \frac{179{,}5}{9{,}5} = 18{,}9$$

As m_2 is dependant upon $\bar{\lambda}_F$ initially assume $m_2 = 0$.

As s_s and c have been assumed to be zero, then $l_c = 0$, then the least value of l_y is given by

$$l_y = t_f\sqrt{\frac{m_1}{2}} = 16{,}0\sqrt{\frac{18{,}9}{2}} = 49{,}2\,\text{mm}$$

The depth of the web h_w has been taken as d, the depth between fillets.

$$F_{CR} = 0{,}9k_F E\frac{t_w^3}{h_w} = 0{,}9 \times 2 \times 210\frac{9{,}5^3}{360{,}4} = 889\,\text{kN}$$

$$\bar{\lambda}_F = \sqrt{\frac{l_y t_w f_{yw}}{F_{CR}}} = \sqrt{\frac{49{,}2 \times 9{,}5 \times 355}{889 \times 10^3}} = 0{,}432$$

As $\bar{\lambda}_F < 0.5$, $m_2 = 0$.

$$\chi_F = \frac{0{,}5}{\bar{\lambda}_F} = \frac{0{,}5}{0{,}432} = 1{,}16$$

The maximum value of χ_F is 1,0, thus

$$L_{eff} = \chi_F l_y = 1{,}0 \times 49{,}2 = 49{,}2\,\text{mm}$$

$$F_{Rd} = L_{eff}t_w\frac{f_{yw}}{\gamma_{M1}} = 49{,}2 \times 9{,}5\frac{355}{1{,}0} \times 10^{-3} = 166\,\text{kN}$$

As F_{Rd} is greater than F_{Ed} ($=R_A = 130\,\text{kN}$), $\eta_2 < 1{,}0$, therefore the web resistance at A is satisfactory without a stiff bearing.

Check at B:

The applied load is 159 kN, and the applied moment is 390 kNm.

For the situation where the load is applied through the top flange, $k_F = 6$ (type a, with the stiffener spacing a effectively taken as infinity)

Determine m_1:

$$m_1 = \frac{f_{yf}b_f}{f_{yw}t_w} = \frac{b_f}{t_w} = \frac{179,5}{9,5} = 18,9$$

As m_2 is dependant upon $\overline{\lambda}_F$ initially assume $m_2 = 0$.

As s_s has been assumed to be zero, then the value of l_y is given by

$$l_y = 2t_f(1 + \sqrt{m_1}) = 2 \times 16,0(1 + \sqrt{18,9}) = 171 \, \text{mm}$$

The depth of the web h_w has been taken as d, the depth between fillets.

$$F_{CR} = 0,9k_F E \frac{t_w^3}{h_w} = 0,9 \times 6 \times 210 \frac{9,5^3}{360,4} = 2698 \, \text{kN}$$

$$\overline{\lambda}_F = \sqrt{\frac{l_y t_w f_{yw}}{F_{CR}}} = \sqrt{\frac{171 \times 9,5 \times 355}{2698 \times 10^3}} = 0,462$$

As $\overline{\lambda}_F < 0,5$, $m_2 = 0$.

$$\chi_F = \frac{0,5}{\overline{\lambda}_F} = \frac{0,5}{0,462} = 1,08$$

The maximum value of χ_F is 1,0, thus

$$L_{eff} = \chi_F l_y = 1,0 \times 171 = 171 \, \text{mm}$$

$$F_{Rd} = L_{eff} t_w \frac{f_{yw}}{\gamma_{M1}} = 171 \times 9,5 \frac{355}{1,0} \times 10^{-3} = 577 \, \text{kN}$$

$$\eta_2 = \frac{F_{Ed}}{L_{eff} t_w \frac{f_{yw}}{\gamma_{M1}}} = \frac{F_{Ed}}{F_{Rd}} = \frac{159}{577} = 0,276$$

$\eta_2 \leq 1,0$, therefore the web resistance at A is satisfactory without a stiff bearing.

However an interaction equation needs checking owing to the co-existence of shear and bending moment:

$$\eta_2 + 0,8\eta_1 \leq 1,4$$

As there is no axial force and no shift in the neutral axis as the section is Class 1, the equation for η_1 reduces to

$$\eta_1 = \frac{M_{Ed}}{W_{pl} \frac{f_y}{\gamma_{M1}}} = \frac{390 \times 10^6}{1501 \times 10^3 \frac{355}{1,0}} = 0,732$$

$$\eta_2 + 0,8\eta_1 = 0,276 + 0,8 \times 0,732 = 0,862 \leq 1,4$$

The web at B is therefore satisfactory.

EXAMPLE 5.2 Beam design with loads applied between lateral torsional restraints.

Prepare a design in Grade S355 steel for the beam whose data are given in Fig. 5.14

Factored actions:

Factored UDL: $1{,}35 \times 10 + 1{,}5 \times 20 = 43{,}5 \, \text{kN/m}$

Factored point load: $1{,}5 \times 15 = 22{,}5 \, \text{kN}$

Figures 5.15 (b) and (c) show the resultant BM and SF diagrams.

Try a $457 \times 191 \times 98 \, \text{UKB}$ Grade S355

Section classification:

Flanges:

$$c = 0{,}5(b - 2r - t_\text{w}) = 0{,}5(192{,}8 - 2 \times 10{,}2 - 11{,}4) = 80{,}5 \, \text{mm}$$
$$\frac{c}{t_\text{f}} = \frac{80{,}5}{19{,}6} = 4{,}11$$

Class 1 limit:

$$9\varepsilon = 9\sqrt{\frac{235}{355}} = 7{,}32$$

Flanges are Class 1.

Web:

$$\frac{c}{t_\text{w}} = \frac{d}{t_\text{w}} = \frac{407{,}6}{11{,}4} = 35{,}8$$

Class 1 limit:

$$72\varepsilon = 72\sqrt{\frac{235}{355}} = 58{,}56$$

Web is Class 1, therefore the section classification is Class 1.

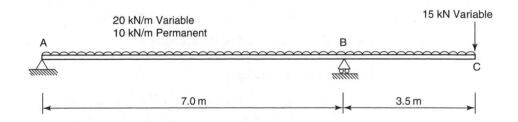

Notes: (1) All loading is characteristic loading
(2) Lateral torsional restraints exist at A and B
(3) The TOP flange is restrained at C

FIGURE 5.14 Design data for Example 5.2

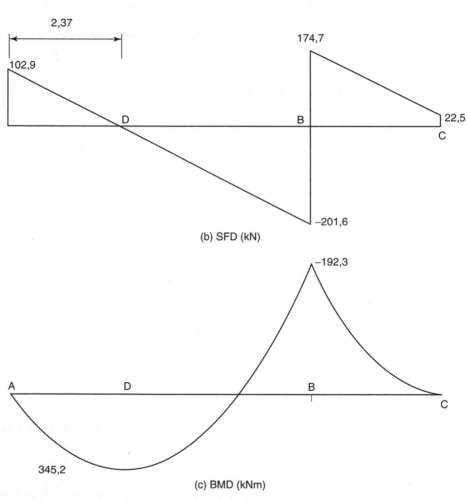

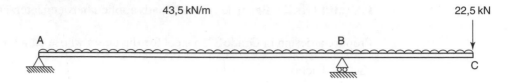

(a) Beam loading at ULS

(b) SFD (kN)

(c) BMD (kNm)

FIGURE 5.15 BMD and SFD for Example 5.2

Plastic moment capacity:

$$M_{pl,Rd} = W_{pl,y} \frac{f_y}{\gamma_{M0}} = 2232 \times 10^3 \frac{355}{1,0} \times 10^{-6} = 793\,\text{kNm}$$

This exceeds the maximum applied moment of 345,2 kNm.

Shear capacity:

$$V_{pl,Rd} = \frac{1}{\sqrt{3}} A_v \frac{f_y}{\gamma_{M0}} = \frac{1}{\sqrt{3}} 5590 \frac{355}{1,0} \times 10^{-3} = 1146\,\text{kN}$$

By inspection, there is no reduction in moment capacity for the effect of shear at the cantilever support.

System length AB:

The system length is taken as the span AB, not the distance between the points of contraflexure calculated from the applied loading as these will not form nodal points in the post-buckled shape of the beam. Such nodal points may only form at points of lateral restraint since at a nodal point the lateral deflection must be zero (Kirby and Nethercot, 1979).

Determination of C_1:

As the loading is between restraints at A and C, the 'n' factor method from BS 5950: Part 1: 1990 must be used (Annexe A7).

From Table 17, the BMD is that of the fifth diagram with $\beta = 0$ (as the BM at A is zero), and γ is negative.

The bending moment M_0 assuming the beam to be simply supported between A and C is given by

$$M_0 = \frac{q_{ult}L^2}{8} = \frac{43{,}5 \times 7^2}{8} = 266{,}4\,\text{kNm}$$

$$\gamma = \frac{M}{M_0} = \frac{-345{,}2}{266{,}4} = -1{,}3$$

From Table 16 with $\beta = 0$ and $\gamma = -1{,}3$, $n = 0{,}53$

$$C_1 = \frac{1}{\sqrt{n}} = \frac{1}{\sqrt{0{,}53}} = 1{,}374$$

Determine M_{cr} from Eq. (5.5):

$$\frac{I_w}{I_z} = \frac{1{,}18 \times 10^{-6}}{2347 \times 10^{-8}} = 0{,}0503\,\text{m}^2$$

$L = 7{,}0\,\text{m}$.

$$\frac{\pi^2 EI_z}{L^2} = \frac{\pi^2 \times 210 \times 10^6 \times 2347 \times 10^{-8}}{7^2} = 993\,\text{kN}$$

$$\frac{L^2 GI_t}{\pi^2 EI_z} = \frac{GI_t}{\frac{\pi^2 EI_z}{L^2}} = \frac{81 \times 10^6 \times 121 \times 10^{-8}}{993} = 0{,}0987\,\text{m}^2$$

$$M_{cr} = \frac{\pi^2 EI_z}{L^2}\left[\frac{I_w}{I_z} + \frac{L^2 GI_t}{\pi^2 EI_z}\right]^{1/2} = 993[0{,}0503 + 0{,}0987]^{1/2} = 383\,\text{kNm}$$

Determine $\bar{\lambda}_{LT}$ using the moment gradient modified value of M_{cr} in Eq. (5.12),

$$\bar{\lambda}_{LT} = \left(\frac{W_{pl,y}f_y}{C_1 M_{cr}}\right)^{1/2} = \left(\frac{2232 \times 10^3 \times 355 \times 10^{-6}}{1{,}374 \times 383}\right)^{1/2} = 1{,}227$$

As in the previous example calculate the strength reduction factor using both methods:

- General method

$h/b = 358{,}0/172{,}2 = 2{,}08 > 2$, thus from Table 5.1, $\alpha_{LT} = 0{,}34$

Determine Φ_{LT} from Eqs (5.15) and (5.16)

$$
\begin{aligned}
\Phi_{LT} &= 0{,}5[1 + \alpha_{LT}(\bar{\lambda}_{LT} - 0{,}2) + (\bar{\lambda}_{LT})^2] \\
&= 0{,}5[1 + 0{,}34(1{,}227 - 0{,}2) + 1{,}227^2] = 1{,}427
\end{aligned}
$$

Determine χ_{LT} from Eq. (5.14)

$$
\chi_{LT} = \frac{1}{\Phi_{LT} + [\Phi_{LT}^2 - (\bar{\lambda}_{LT})^2]^{1/2}} = \frac{1}{1{,}427 + [1{,}427^2 - 1{,}227^2]^{1/2}} = 0{,}464
$$

$$
M_{b,Rd} = \chi_{LT} W_{ply} \frac{f_y}{\gamma_{M1}} = 0{,}464 \times 2232 \times 10^3 \frac{355}{1{,}0} \times 10^{-6} = 368\,\text{kNm}
$$

This exceeds the absolute value of the maximum moment at C of 345,2 kNm.

- Rolled section method:

$h/b = 358{,}0/172{,}2 = 2{,}08 > 2$, thus from Table 5.2, $\alpha_{LT} = 0{,}49$

Determine Φ_{LT} from Eqs (5.18) and (5.19)

$$
\begin{aligned}
\Phi_{LT} &= 0{,}5[1 + \alpha_{LT}(\bar{\lambda}_{LT} - \bar{\lambda}_{LT,0}) + \beta(\bar{\lambda}_{LT})^2] \\
&= 0{,}5[1 + 0{,}49(1{,}227 - 0{,}4) + 0{,}75 \times 1{,}227^2] = 1{,}267
\end{aligned}
$$

Determine χ_{LT} from Eq. (5.17)

$$
\chi_{LT} = \frac{1}{\Phi_{LT} + [\Phi_{LT}^2 - \beta(\bar{\lambda}_{LT})^2]^{1/2}} = \frac{1}{1{,}267 + [1{,}267^2 - 0{,}75 \times 1{,}227^2]^{1/2}} = 0{,}511
$$

Without the correction factor f:

$$
M_{b,Rd} = \chi_{LT} W_{ply} \frac{f_y}{\gamma_{M1}} = 0{,}511 \times 2232 \times 10^3 \frac{355}{1{,}0} \times 10^{-6} = 405\,\text{kNm}
$$

This exceeds the absolute value of the maximum moment at C of 345,2 kNm.

From Table 6.6 of EN 1993-1-1, $k_c = 0{,}91$

Determine f from Eq. (5.32):

$$
\begin{aligned}
f &= 1 - 0{,}5(1 - k_c)[1 - 2{,}0(\bar{\lambda}_{LT} - 0{,}8)^2] \\
&= 1 - 0{,}5(1 - 0{,}91)[1 - 2{,}0(1{,}227 - 0{,}8)^2] = 0{,}971
\end{aligned}
$$

Thus the corrected value of $M_{b,Rd}$ given by Eq. (5.31) is

$$
M_{b,Rd} = \frac{405}{0{,}971} = 417\,\text{kNm}
$$

It is again noted that the rolled section approach even without the factor f gives slightly higher values of $M_{b,Rd}$ than the general case.

System length BC.

Take the conventional approach and adopt $C_1 = 1,0$

Determine M_{cr} from Eq. (5.5):

$$\frac{I_w}{I_z} = \frac{1,18 \times 10^{-6}}{2347 \times 10^{-8}} = 0,0503 \, \text{m}^2$$

$L = 3,5 \, \text{m}$

$$\frac{\pi^2 EI_z}{L^2} = \frac{\pi^2 \times 210 \times 10^6 \times 2347 \times 10^{-8}}{3,5^2} = 3971 \, \text{kN}$$

$$\frac{L^2 GI_t}{\pi^2 EI_z} = \frac{GI_t}{\frac{\pi^2 EI_z}{L^2}} = \frac{81 \times 10^6 \times 121 \times 10^{-8}}{3971} = 0,0247 \, \text{m}^2$$

$$M_{cr} = \frac{\pi^2 EI_z}{L^2} \left[\frac{I_w}{I_z} + \frac{L^2 GI_t}{\pi^2 EI_z} \right]^{1/2} = 3971[0,0503 + 0,0247]^{1/2} = 1088 \, \text{kNm}$$

Determine $\bar{\lambda}_{LT}$ from Eq. (5.12),

$$\bar{\lambda}_{LT} = \left(\frac{W_{pl,y} f_y}{C_1 M_{cr}} \right)^{1/2} = \left(\frac{2232 \times 10^3 \times 355 \times 10^{-6}}{1088} \right)^{1/2} = 0,853$$

As in the previous example calculate the strength reduction factor using both methods:

• General method

$h/b = 358,0/172,2 = 2,08 > 2$, thus from Table 5.1, $\alpha_{LT} = 0,34$

Determine Φ_{LT} from Eqs (5.15) and (5.16)

$$\Phi_{LT} = 0,5[1 + \alpha_{LT}(\bar{\lambda}_{LT} - 0,2) + (\bar{\lambda}_{LT})^2]$$

$$= 0,5[1 + 0,34(0,853 - 0,2) + 0,853^2] = 0,975$$

Determine χ_{LT} from Eq. (5.14)

$$\chi_{LT} = \frac{1}{\Phi_{LT} + [\Phi_{LT}^2 - (\bar{\lambda}_{LT})^2]^{1/2}} = \frac{1}{0,975 + [0,975^2 - 0,853^2]^{1/2}} = 0,691$$

$$M_{b,Rd} = \chi_{LT} W_{pl,y} \frac{f_y}{\gamma_{M1}} = 0,691 \times 2232 \times 10^3 \frac{355}{1,0} \times 10^{-6} = 548 \, \text{kNm}$$

This exceeds the absolute value of the maximum moment at C of 345,2 kNm.

• Rolled section method

$h/b = 358,0/172,2 = 2,08 > 2$, thus from Table 5.2, $\alpha_{LT} = 0,49$

Determine Φ_{LT} from Eqs (5.18) and (5.19)

$$\Phi_{LT} = 0,5[1 + \alpha_{LT}(\bar{\lambda}_{LT} - \bar{\lambda}_{LT,0}) + \beta(\bar{\lambda}_{LT})^2]$$

$$= 0,5[1 + 0,49(0,853 - 0,4) + 0,75 \times 0,853^2] = 0,884$$

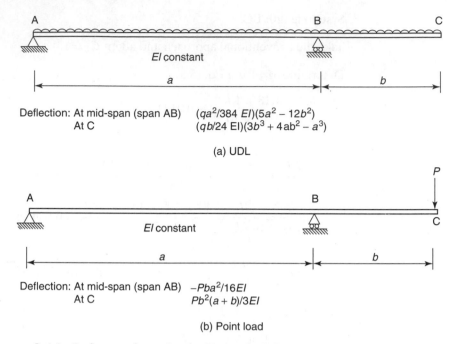

Deflection: At mid-span (span AB) $(qa^2/384 \, EI)(5a^2 - 12b^2)$
At C $(qb/24 \, EI)(3b^3 + 4ab^2 - a^3)$

(a) UDL

Deflection: At mid-span (span AB) $-Pba^2/16EI$
At C $Pb^2(a+b)/3EI$

(b) Point load

FIGURE 5.16 Deflection formulae for Example 5.2

Determine χ_{LT} from Eq. (5.17)

$$\chi_{LT} = \frac{1}{\Phi_{LT} + [\Phi_{LT}^2 - \beta(\overline{\lambda}_{LT})^2]^{1/2}}$$

$$= \frac{1}{0{,}884 + [0{,}884^2 - 0{,}75 \times 0{,}853^2]^{1/2}} = 0{,}730$$

As there appears to be no consideration given to cantilevers in Table 6.6, f will be taken as 1,0.

$$M_{b,Rd} = \chi_{LT} W_{pl,y} \frac{f_y}{\gamma_{M1}} = 0{,}730 \times 2232 \times 10^3 \frac{355}{1{,}0} \times 10^{-6} = 578 \, \text{kNm}$$

This exceeds the absolute value of the maximum moment at C of 345,2 kNm.

It is again noted that the rolled section approach even without the factor f gives slightly higher values of $M_{b,Rd}$ than the general case.

For this particular design case, the span AC is critical.

Deflection check:

$$EI = 210 \times 10^{-6} \times 45730 \times 10^4 = 96033 \, \text{kNm}^2.$$

The relevant formulae are given in Fig. 5.16.

(a) Variable action check:
    UDL: 20 kN/m

Span AB, central deflection

$$\delta = \frac{qa^2}{384EI}(5a^2 - 12b^2) = \frac{20 \times 7^2}{384 \times 96033}(5 \times 7^2 - 12 \times 3,5^2) = 0,0026\,\mathrm{m}$$

Span BC, at C

$$\delta = \frac{qb}{24EI}(3b^3 + 4ab^2 - a^3) = \frac{20 \times 3,5}{24 \times 96033}(3 \times 3,5^3 + 4 \times 7 \times 3,5^2 - 7^3)$$
$$= 0,0039\,\mathrm{m}$$

Point load at C
Span AB, mid-span

$$\delta = -\frac{Pba^2}{16EI} = -\frac{15 \times 3,5 \times 7^2}{16 \times 96033} = -0,0017\,\mathrm{m}$$

Span BC, at C

$$\delta = \frac{P(a+b)b^2}{3EI} = \frac{15(7+3,5)3,5^2}{3 \times 96033} = 0,0067\,\mathrm{m}$$

Net deflections:
    at mid-span $= 0,0026 - 0,0017 = 0,0009\,\mathrm{m}$
    Span deflection ratio is $7/0,0009 = 7780$. This is more than acceptable.
    at $C = 0,0039 + 0,0067 = 0,0106\,\mathrm{m}$
    Span deflection ratio (based on twice the span) is $2 \times 3,5/0,0106 = 660$.
    This is acceptable.
(b) Check under total actions.
    Deflection due to point load as above.
    Total UDL of 30 kN/m:
    Mid-span, $\delta = 0,0039\,\mathrm{m}$; at C, $\delta = 0,0059\,\mathrm{m}$.
    Total deflection at mid-span $= 0,0039 - 0,0017 = 0,0022\,\mathrm{m}$
    Span deflection ratio: $7/0,0022 = 3180$. This is satisfactory.
    Total deflection at $C = 0,0059 + 0,0067 = 0,0126\,\mathrm{m}$
    Span deflection ratio (based on twice the span) is $2 \times 3,5/0,0126 = 556$.
    This is satisfactory.

Web check at B (refer to Section 4.8 and cl 6 EN 1993-1-5):

This is the only point that needs checking, as the other reaction point has a much lower force (even allowing for reduced dispersion length) and no coincident moment.

$R_{Sd} = 124,7\,\mathrm{kN}$ (Reaction at B); $M = 114,5\,\mathrm{kNm}$

Ignore any stiff bearing ($s_s = 0$).

For the situation where the load is applied through the top flange, $k_F = 6$ (type a, with the stiffener spacing a effectively taken as infinity)

Determine m_1:

$$m_1 = \frac{f_{yf}b_f}{f_{yw}t_w} = \frac{b_f}{t_w} = \frac{192,8}{11,4} = 16,9$$

As m_2 is dependant upon $\bar{\lambda}_F$ initially assume $m_2 = 0$.

As s_s has been assumed to be zero, then the value of l_y is given by

$$l_y = 2t_f(1 + \sqrt{m_1}) = 2 \times 19,6(1 + \sqrt{16,9}) = 200\,\text{mm}$$

The depth of the web h_w has been taken as d, the depth between fillets.

$$F_{CR} = 0,9k_FE\frac{t_w^3}{h_w} = 0,9 \times 6 \times 210\frac{11,4^3}{407,6} = 4122\,\text{kN}$$

$$\bar{\lambda}_F = \sqrt{\frac{l_yt_wf_{yw}}{F_{CR}}} = \sqrt{\frac{200 \times 11,4 \times 355}{4122 \times 10^3}} = 0,443$$

As $\bar{\lambda}_F < 0,5$, $m_2 = 0$.

$$\chi_F = \frac{0.5}{\bar{\lambda}_F} = \frac{0,5}{0,443} = 1,13$$

The maximum value of χ_F is 1,0, thus

$$L_{eff} = \chi_F l_y = 1,0 \times 200 = 200\,\text{mm}$$

$$F_{Rd} = L_{eff}t_w\frac{f_{yw}}{\gamma_{M1}} = 200 \times 11,4\frac{355}{1,0} \times 10^{-3} = 809\,\text{kN}$$

$$\eta_2 = \frac{F_{Ed}}{L_{eff}t_w\frac{f_{yw}}{\gamma_{M1}}} = \frac{F_{Ed}}{F_{Rd}} = \frac{124,7}{809} = 0,154$$

$\eta_2 \leq 1,0$, therefore the web resistance at A is satisfactory without a stiff bearing.

However an interaction equation needs checking owing to the co-existence of shear and bending moment:

$$\eta_2 + 0,8\eta_1 \leq 1,4$$

As there is no axial force and no shift in the neutral axis as the section is Class 1, the equation for η_1 reduces to

$$\eta_1 = \frac{M_{Ed}}{W_{pl}\frac{f_y}{\gamma_{M1}}} = \frac{114,5 \times 10^6}{2232 \times 10^3\frac{355}{1,0}} = 0,145$$

$$\eta_2 + 0,8\eta_1 = 0,154 + 0,8 \times 0,145 = 0,27 \leq 1,4$$

The web at B is therefore satisfactory.

5.1.6 Other Section Profiles

A number of special cases need considering, hollow sections as the earlier equations for critical moment are not applicable, rectangular sections as the warping stiffness is zero, and T sections.

5.1.6.1 Rolled Hollow Sections

Rees (1990) indicates that thin wall tubes of circular and triangular cross with uniform thickness cannot warp. For rectangular thin walled tubes only those whose wall thicknesses are in a constant ratio the length of the sides do not warp. These tubes are known as Neuber tubes. If there is no warping then lateral torsional buckling can only be resisted by torsion. For conventional hollow sections where the wall thickness is constant, then lateral torsional buckling is in part resisted by warping. It will be conservative to neglect the warping stiffness, and, therefore as a result, Eq. (5.3) can be reduced to

$$GI_t \frac{d^2\phi}{dx^2} = -\frac{M^2}{EI_z}\phi \tag{5.53}$$

The resultant value of M_{cr} (with no allowance for major axis bending) is given from Eq. (5.6) as

$$M_{cr} = \frac{\pi\sqrt{EI_zGI_t}}{L} \tag{5.54}$$

The factor K from Eq. (5.7) with the warping constant I_w set equal to zero becomes,

$$K = \sqrt{\left(1 - \frac{EI_z}{EI_y}\right)\left(1 - \frac{GI_t}{EI_y}\right)} \tag{5.55}$$

Combining Eqs (5.54) and (5.55) gives the elastic critical moment as

$$M_{cr} = \frac{\pi\sqrt{EI_zGI_t}}{L\sqrt{\left(1 - \frac{EI_z}{EI_y}\right)\left(1 - \frac{GI_t}{EI_y}\right)}} \tag{5.56}$$

or the normalized slenderness ratio, $\bar{\lambda}_{LT}$, is given from Eq. (5.12) with the introduction of the moment gradient factor C_1 as

$$\bar{\lambda}_{LT} = \sqrt{\frac{W_{pl,y}f_y}{C_1M_{cr}}} = \sqrt{\frac{W_{pl,y}f_yL\sqrt{\left(1 - \frac{I_z}{I_y}\right)\left(1 - \frac{GI_t}{EI_y}\right)}}{C_1\pi\sqrt{EI_zGI_t}}} \tag{5.57}$$

The normalized lateral torsional buckling slenderness ratio $\bar{\lambda}_{LT}$ is also given by

$$\bar{\lambda}_{LT} = \frac{\lambda_{LT}}{\lambda_1} \tag{5.58}$$

or

$$\lambda_{LT} = \bar{\lambda}_{LT}\lambda_1 = \pi\sqrt{\frac{E}{f_y}}\bar{\lambda}_{LT} = \sqrt{\frac{\pi E W_{\text{pl},y} L \sqrt{\left(1 - \frac{I_z}{I_y}\right)\left(1 - \frac{GI_t}{EI_y}\right)}}{C_1\sqrt{EI_z GI_t}}} \tag{5.59}$$

Rewrite Eq. (5.59) as

$$\lambda_{LT} = \frac{1}{C_1^{1/2}}\left[\pi\sqrt{\frac{E}{G}}\right]^{1/2}\sqrt{\frac{W_{\text{pl},y} L \sqrt{\left(1 - \frac{I_z}{I_y}\right)\left(1 - \frac{GI_t}{EI_y}\right)}}{\sqrt{I_z I_t}}} \tag{5.60}$$

Define the slenderness ratio λ as

$$\lambda = \frac{L}{i_z} = \frac{L}{\sqrt{\frac{I_z}{A}}} \tag{5.61}$$

when Eq. (5.62) becomes

$$\lambda_{LT} = \frac{1}{C_1^{1/2}}\left[\pi\sqrt{\frac{E}{G}}\right]^{1/2}\sqrt{\frac{W_{\text{pl},y}\lambda\sqrt{\left(1 - \frac{I_z}{I_y}\right)\left(1 - \frac{GI_t}{EI_y}\right)}}{\sqrt{AI_t}}} \tag{5.62}$$

or λ_{LT} is given as

$$\lambda_{LT} = \frac{1}{C_1^{1/2}}\left[\pi\sqrt{\frac{E}{G}}\right]^{1/2}(\phi_b\lambda)^{1/2} \tag{5.63}$$

where ϕ_b is defined as by

$$\phi_b = \left(\frac{W_{\text{pl},y}^2\left[1 - \frac{I_z}{I_y}\right]\left[1 - \frac{GI_t}{EI_y}\right]}{AI_t}\right)^{1/2} \tag{5.64}$$

This is the equation given in Section B2.6.1 of BS 5950-1. The term in rectangular parentheses in Eq. (5.63) has a value of 2,25.

An alternative approach avoiding the calculation of lateral torsion buckling is to determine the critical length l_{crit} (in mm) corresponding to a value of $\bar{\lambda}_{LT}$ equalling 0,4 (below which buckling will not occur) (Rondal, et al., 1992). This value is given by

$$l_{\text{crit}} = \frac{113400(h - t)}{f_y}\frac{\left(\frac{b-t}{h-t}\right)^2}{1 + 3\frac{b-t}{h-t}}\sqrt{\frac{3 + \frac{b-t}{h-t}}{1 + \frac{b-t}{h-t}}} \tag{5.64a}$$

It should be noted that there is an apparent anomaly in Eq. (5.64) in that for a square section ($b = h$), l_{crit} remains finite, whereas Eq. (5.57) indicates ϕ_b (and hence λ_{LT}) = 0.

The reason is that the values from Eq. (5.64) in Rondal, *et al.* are also given in tabular form for discrete values of $(h - t)/(b - t)$, and that interpolation would not be possible if the value for a square section were given as infinity. It should be noted that there is an error in the formula quoted in Rondal, *et al.*, but that the values in Table 15 in the same publication are correct. Eq. (5.168) has been corrected (Kaim, 2006). It should be noted that the values of l_{crit} are extremely safe.

Kaim (2006) suggests the normalized slenderness limit $\bar{\lambda}_{z,lim}$ is given by

$$\bar{\lambda}_{z,lim} = \frac{25}{\frac{h}{b}} \sqrt{\frac{235}{f_y}} \tag{5.64b}$$

5.1.6.2 Rectangular Sections

For rectangular sections of width b and depth h the equation for critical moment M_{cr} given in Eq. (5.6) reduces to

$$M_{cr} = \frac{\pi}{L} \sqrt{EI_z GI_t} \tag{5.65}$$

as the warping constant I_w is zero.

(a) Thin sections

For thin sections, $I_t = hb^3/3$ and $I_z = hb^3/12$, so Eq. (5.65) reduces to

$$M_{cr} = \frac{hb^3}{L} \frac{\pi E}{\sqrt{72(1 + v)}} \tag{5.66}$$

(b) Thick sections

In this case I_t is no longer given by $ht^3/3$. The following approximate formula can be used

$$I_t = \frac{hb^3}{3} \left[1 - 0{,}63\frac{b}{h} \left(1 - \frac{1}{12} \left(\frac{b}{h}\right)^4 \right) \right] \tag{5.67}$$

If $h/b > 2$, Eq. (5.67) can with little loss in accuracy be reduced to

$$I_t = \frac{hb^3}{3} \left[1 - 0{,}63\frac{b}{h} \right] \tag{5.68}$$

However the major axis bending is now important and to ignore it would be too conservative, thus the parameter K from Eq. (5.8) must be introduced to give M_{cr} as

$$M_{cr} = \frac{\pi hb^3}{6L} \sqrt{EG} \sqrt{\frac{1 - 0{,}63\frac{b}{h}}{1 - \left(\frac{b}{h}\right)^2}} \tag{5.69}$$

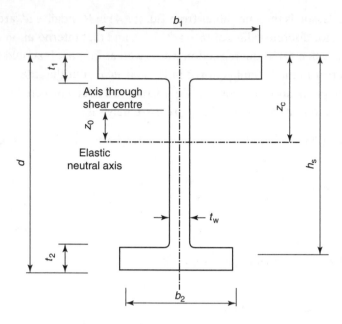

FIGURE 5.17
Calculation of monosymmetry index

5.1.6.3 Monosymmetric Beams

For beams with only one axis of symmetry (usually the minor axis), the centroid and the shear centre do not coincide, thus an additional disturbing torque occurs due to the longitudinal flexural stresses.

From Trahair and Bradford (1988) the elastic critical moment M_{cr} for the monosymmetric beam in Fig. 5.17 is given by

$$M_{cr} = \frac{\pi}{L}\sqrt{EI_z GI_t}\left\{\sqrt{\left[1 + \frac{\pi^2 EI_w}{GI_t L^2} + \left(\frac{\pi\gamma_M}{2}\right)^2\right]} + \frac{\pi\gamma_M}{2}\right\} \tag{5.70}$$

where γ_M is given by

$$\gamma_M = \frac{\beta_y}{L}\sqrt{\frac{EI_z}{GI_t}} \tag{5.71}$$

The monosymmetry parameter for the section β_y is given by

$$\beta_y = \frac{1}{I_y}\left\{\beta_{y1} - \beta_{y2} + \beta_{y3}\right\} - 2z_0 \tag{5.72}$$

where

$$\beta_{y1} = (h_s - z_c)\left[\frac{b_2^3 t_2}{12} + b_2 t_2 (h_s - z_c)\right] \tag{5.73}$$

$$\beta_{y2} = z_c\left[\frac{b_1^3 t_1}{12} + b_1 t_1 z_c^2\right] \tag{5.74}$$

$$\beta_{y3} = \frac{t_w}{4} \left[\left(h_s - z_c - \frac{t_2}{2} \right)^4 - \left(z_c - \frac{t_1}{2} \right)^4 \right]$$ (5.75)

where

$$h_s = d - \frac{t_1 + t_2}{2}$$ (5.76)

$$z_c = \frac{b_2 t_2 h_s + \frac{t_w}{2}(d - t_1 - t_2)(d - t_2)}{b_1 t_1 + b_2 t_2 + (d - t_1 - t_2)t_w}$$ (5.77)

$$z_0 = \alpha h_s - z_c$$ (5.78)

$$\alpha = \frac{1}{1 + \left(\frac{b_1}{b_2} \right)^3 \frac{t_1}{t_2}}$$ (5.79)

The warping constant I_w is now given by

$$I_w = \alpha(1 - \alpha)I_z h_s^2$$ (5.80)

5.1.6.4 'T' Beams

For 'T' beams, it should be first checked that the value of K from Eq. (5.8) is real as a large number of commercial 'T' beams have $I_y > I_z$ in which case lateral torsional buckling cannot occur. Where M_{cr} needs calculating, it should be noted I_w is zero as $\alpha = 0$ (Eq. (5.80)) and $z_0 = -z_c$. The position of the centroid z_c and I_y are tabulated in section property tables.

5.1.6.5 Parallel Flange Channels

The procedure for calculating M_{cr} follows that for 'I' beams except that I_w is calculated as

$$I_w = \frac{t_f b_f^3 h^2}{12} \frac{3b_f t_f + 2h t_w}{6b_f t_f + h t_w}$$ (5.81)

where b_f and t_f are the width and thickness of the flange and h and t_w are the height and thickness of the web. For channels with tapered flanges t_f may be taken as the mean thickness of the flange (Kirby and Nethercot, 1979).

5.2 PURE TORSIONAL BUCKLING

This form of failure can only occur in open sections and is most likely to only where the sections are thin walled. There can then be an interaction between strut buckling or buckling in pure torsion.

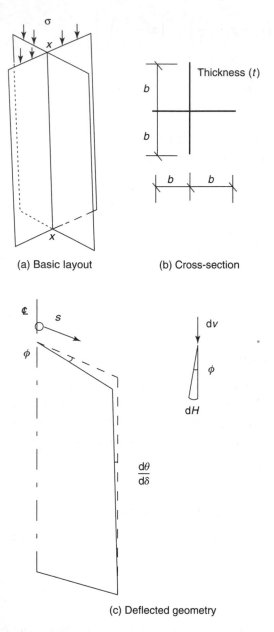

(a) Basic layout

(b) Cross-section

(c) Deflected geometry

FIGURE 5.18   Cruciform strut

5.2.1 Interaction between Torsional Buckling and Strut Buckling

This is best illustrated by considering the case of a cruciform strut loaded by a uniform stress σ over its cross-section. To demonstrate the principles a thin section strut is considered (Fig. 5.18). The load induced by the stress σ remains parallel to the x axis, and therefore induces a lateral force in the yz plane. If the vertical component in Fig 5.18(c) is dV then the horizontal component dH is given by

$$+dH = dV \tan \phi = \phi dV = s\frac{d\theta}{dx}dV \tag{5.82}$$

as ϕ is small, and where $d\theta/dx$ is the angle of twist per unit length and s is a distance measured from the x axis.

The force dV is given by

$$dV = \sigma dA = \sigma t ds \qquad (5.83)$$

where t is the thickness. The incremental torque dT is given by

$$dT = sdH \qquad (5.84)$$

Substitute Eqs (5.81) and (5.83) into Eq. (5.84) to give

$$dT = \sigma \frac{d\theta}{dx} t s^2 ds \qquad (5.85)$$

Integrate Eq. (5.85) over the four arms of the strut to give

$$T = 4\sigma \frac{d\theta}{dx} t \int_0^h s^2 ds = \sigma \frac{d\theta}{dx} \frac{4}{3} b^3 t \qquad (5.86)$$

where $2b$ is the width of the strut.

Equation (5.86) may be rewritten as

$$T = \sigma \frac{d\theta}{dx} I_x \qquad (5.87)$$

where I_x is the polar second moment of area about the x axis, and is given by

$$I_x = \frac{4}{3} b^3 t \qquad (5.88)$$

The second moment of area of the strut about either the z or y axis is given by

$$I_z = I_y = \frac{1}{12}(2b)^3 t = \frac{2}{3} b^3 t \qquad (5.89)$$

as the other arm of the strut has negligible second moment of area ($bt^3/6$) compared with the other direction.

Note that

$$I_x = I_y + I_z = 2\left(\frac{2}{3}b^3 t\right) = \frac{4}{3}b^3 t \qquad (5.90)$$

From torsion theory the torque that may be carried by the section is given by

$$T = GI_t \frac{d\theta}{dx} \qquad (5.91)$$

where G is the shear modulus and T_t is the torsional second moment of area. For the cruciform thin walled section I_t is given by

$$I_t = 4\left(\frac{1}{3}bt^3\right) = \frac{4}{3}bt^3 \qquad (5.92)$$

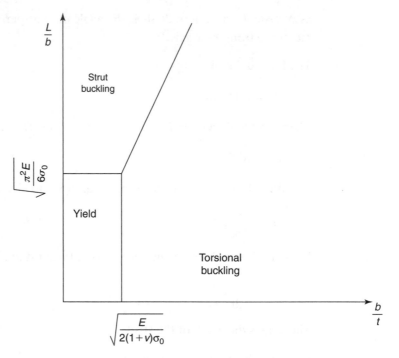

FIGURE 5.19 Interaction diagram

Substitute Eq. (5.92) into Eq. (5.91) to give

$$T = \frac{4}{3}bt^3G\frac{d\theta}{dx} \tag{5.93}$$

For the section to be able to sustain its twisted shape the two values of the torques from Eqs (5.87) and (5.93) must be equal, so

$$\sigma = G\left(\frac{t}{b}\right)^2 \tag{5.94}$$

There also exists the possibility that the strut can undergo normal Euler buckling under a stress σ_{cr} given by

$$\sigma_{cr} = \frac{\pi^2 E I_z}{AL^2} = \frac{\pi^2 E \frac{2}{3}b^3 t}{4btL^2} = \frac{\pi^2 E}{6}\left(\frac{b}{L}\right)^2 \tag{5.95}$$

Assuming no interaction, there are three possibilities of behaviour; the stress produces yield; torsional buckling occurs, or strut buckling occurs (Fig. 5.19). The stress which causes torsional buckling and strut buckling to occur simultaneously is given when the stresses from Eqs (5.94) and (5.95) are equal, or

$$G\left(\frac{t}{b}\right)^2 = \frac{\pi^2 E}{6}\left(\frac{b}{L}\right)^2 \tag{5.96}$$

or

$$\frac{L}{b} = \frac{b}{t}\sqrt{\frac{\pi^2 E}{6G}} = \frac{b}{t}\sqrt{\frac{\pi^2(1+v)}{3}} \tag{5.97}$$

as $G = E/2(1+v)$.

The transition from yield to torsional buckling occurs when the stress in Eq. (5.94) equals the yield stress σ_0, or

$$\frac{b}{t} = \sqrt{\frac{G}{\sigma_0}} = \sqrt{\frac{E}{2(1+v)\sigma_0}} \tag{5.98}$$

and from yield to Euler buckling when the stress in Eq. (5.95) equals σ_0, or

$$\frac{L}{b} = \sqrt{\frac{\pi^2 E}{6\sigma_0}} \tag{5.99}$$

Note, other sections such as thin walled angles may also suffer similar behaviour, but the analysis is more complex as buckling is about the principal axes.

5.2.2 Torsional Buckling Interaction

The critical axial load for torsional buckling N_{cr} is given in Chapman and Buhagiar (1993) as

$$N_{cr,T} = \frac{1}{i_s^2}\left(GI_t + \frac{\pi^2 EI_w}{l_T^2}\right) \tag{5.100}$$

where l_T is the buckling length, and i_s the polar radius of gyration given by

$$i_s^2 = i_z^2 + i_y^2 + y_0^2 + z_0^2 \tag{5.101}$$

where i_z and i_y are the flexural radii of gyration and z_0 and y_0 are distances from the shear centre to the geometric centroid. For a section whose centroid and shear centre co-incide y_0 and z_0 are zero. Alternatively the critical stress $\sigma_{cr,T}$ is given by

$$\sigma_{cr,T} = \frac{1}{I_0}\left(GI_t + \frac{\pi^2 EI_w}{l_T^2}\right) \tag{5.102}$$

where I_0 is the polar moment of area.

Timoshenko and Gere (1961) give the following interaction equation between strut buckling and torsion buckling

$$i_s^2(N - N_{cr,z})(N - N_{cr,y})(N - N_{cr,T}) - N^2 z_0^2(N - N_{cr,y}) - N^2 y_0^2(N - N_{cr,z}) = 0 \tag{5.103}$$

where N is the critical value of the axial load and $N_{cr,z}$ and $N_{cr,y}$ are the Euler buckling loads about the zz and xx axes.

For a section whose centroid and shear centre co-incide, Eq. (5.103) reduces to

$$(N - N_{cr,z})(N - N_{cr,y})(N - N_{cr,T}) = 0 \qquad (5.104)$$

That is, N is therefore the least of $N_{cr,x}$, $N_{cr,y}$ and $N_{cr,T}$.

Chapman and Buhagiar (1993) also indicate that where buckling can occur about both axes (i.e. where the buckling lengths and second moments of area are approximately equal, or where the critical loads are similar), then the imperfection factor η in the standard strut buckling interaction equation should be taken as twice its normal value to allow for torsional buckling.

5.3 PLATE GIRDERS

Plate girders are used either on long spans where a rolled section would need to be spliced and as a result may be inefficient, or to support heavy loads such as on a bridge structure. It is important to note that although plate girders may be lighter than other forms of compound beams, fabrication costs are likely to be much higher. Also as Corus now roll UKB's with a depth of 1016 mm, the use of plate girders in building structures unless spans are extremely high is less likely.

Plate girders are built up from two flange plates and a web plate, generally from the same grade of steel. Continuous automatic electric arc or submerged gas welding is used to form the fillet welds between the flange and web (Fig. 5.20). Such welding is generally performed as a double pass one on either side of the girder. It should be noted that this process may cause very high residual stresses to exist in the flanges and web. The exact magnitudes will depend also on whether the flange plates were sheared or flame cut to size (Nethercot, 1974a).

The stiffeners are then welded in place often manually. Such stiffeners are needed either to help combat the effects of web buckling or to provide support to any concentrated load or reaction. Only straight girders with equal flanges and vertical stiffeners are considered in this text.

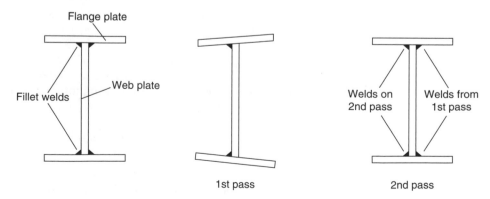

FIGURE 5.20 Plate girder fabrication

5.3.1 Minimum Web Thickness

With no web stiffeners the web should be sized to avoid the flange undergoing local buckling due to the web being unable to support the flange. This is known as flange induced buckling (cl. 8 EN 1993-1-5).

In pure bending the flanges are subjected to equal and opposite forces. The force per unit length is given by

$$\sigma_v t_w = \frac{f_{yf} A_{fc}}{R} \tag{5.105}$$

where σ_v is the vertical stress in the web, A_{fc} is the area of the compression flange and R is the radius of curvature. The curvature is dependant upon the variation of strain $\Delta \varepsilon_f$ occurring at the mid-depth of the flanges. The residual strain due to fabrication ε_f is assumed to have a value of $0,5\varepsilon_y$, and $\Delta \varepsilon_f$ is the sum of the yield strain ε_y and the residual strain ε_f, that is, a total of $1,5\varepsilon_y$. The radius of curvature R is then given by

$$R = \frac{0,5 h_w}{\Delta \varepsilon_f} \tag{5.106}$$

where for simplicity h_w has been taken as the depth between the centroids of the flanges rather than the clear depth of the web. It should be noted any error will be small.

From Eqs (5.105) and (5.106),

$$\sigma_v = 3 \frac{A_{fc}}{A_w} \frac{f_{yf}^2}{E} \tag{5.107}$$

The stress σ_v cannot exceed the elastic critical buckling stress for a simply supported thin plate which is given by (Bulson, 1970) as

$$\sigma_v = \frac{\pi^2 E}{12(1 - v^2)} \left(\frac{t_w}{h_w} \right)^2 \tag{5.108}$$

Equating Eqs (5.107) and (5.108) gives with a slight change in notation with A_c replacing A_{fc}

$$\frac{h_w}{t_w} \leq 0,55 \frac{E}{f_{yf}} \sqrt{\frac{A_w}{A_c}} \tag{5.109}$$

EN 1993-1-5 modifies the co-efficient of 0,55 to allow for situations where higher strains are required. Thus the critical h_w/t_w ratio is given by

$$\frac{h_w}{t_w} \leq k \frac{E}{f_{yf}} \sqrt{\frac{A_w}{A_c}} \tag{5.110}$$

where f_{yf} is the yield strength of the compression flange, A_{fc} is the effective area of the compression flange and A_w is the area of the web. The parameter k takes values of 0,3 where plastic hinge rotation is utilized, 0,4 if the plastic resistance is utilized and 0,55 if the elastic resistance is utilized. Thus for rigid (continuous) design $k = 0,3$ unless the analysis is elastic with no redistribution. For simply supported beams k may be taken as 0,4.

5.3.2 Bending Resistance

The section classification is determined in the same manner as rolled sections.

5.3.2.1 Compression Flange Restrained (i.e. Lateral Torsional Buckling cannot Occur)

There are two methods that can be used for girder design:

(1) The flanges carrying the bending moment and the web the shear force
This is probably best used where the maximum bending moment and maximum shear force are not coincident, as the ability of the flange to contribute towards shear capacity may be utilized. Thus for a restrained beam under a UDL, this method may be advantageous. If the maximum bending moment and maximum shear are coincident in either a simply supported beam under point loading or in a continuous beam at the internal support, then the flange capacity will not be able to be utilized to resist shear, thus probably necessitating a thicker web.
With this method, the moment capacity is only dependant on the section classification of the flanges as the web does not carry compression.
(2) The girder carrying the forces as an entity
This method of design is more complex as the beam is likely to be Class 4 and may not show any resultant economies over the first method, but should be utilized where maximum moment and maximum shear are co-incident.

5.3.2.2 Lateral Buckling May Occur

In this case the second method must be used. The design then follows that for rolled beams, except that the general method for calculating χ_{LT} should be used with the value of α_{LT} appropriate for welded sections. Section properties will need to be calculated from first principles.

5.3.3 Basic Dimensioning

One method of dimensioning is to consider a minimum weight solution. It must be noted that a minimum weight solution is not synonymous with a minimum cost solution (Gibbons, 1995).

Method 1:

Assuming the moment to be resisted by the flanges alone, then

$$M_{Rd} = f_{yd}b_f t_f h_w \tag{5.111}$$

where f_{yd} is the design strength of the flanges, t_f and b_f the thickness and width of the flange plates and h_w the distance between the internal faces of the flanges. Equation (5.111) is slightly conservative for beams of Classes 1 to 3.

The cross-sectional area A is given by

$$A = 2b_f t_f + h_w t \tag{5.112}$$

Eliminate $b_f t_f$ between Eqs (5.111) and (5.112) to give

$$A = \frac{2M_{Rd}}{h_w f_{yd}} + h_w t \tag{5.113}$$

Define the web slenderness ratio h_w/t as λ, then Eq. (5.113) becomes

$$A = \frac{2M_{Rd}}{\lambda t f_{yd}} + \lambda t^2 \tag{5.114}$$

For an optimum solution, $dA/dt = 0$, so Eq. (5.114) becomes,

$$\frac{dA}{dt} = -\frac{2M_{Rd}}{\lambda f_{yd} t^2} + 2\lambda t = 0 \tag{5.115}$$

or,

$$t = \sqrt[3]{\frac{M_{Rd}}{\lambda^2 f_{yd}}} \tag{5.116}$$

and

$$h_w = \sqrt[3]{\frac{\lambda M_{Rd}}{f_{yd}}} \tag{5.117}$$

The area of the web, A_w is then given by

$$A_w = \sqrt[3]{\frac{M_{Rd}^2}{\lambda f_{yd}^2}} \tag{5.118}$$

Using Eq. (5.112), the flange area, A_f is given by

$$A_f = b_f t_f = \frac{M_{Rd}}{f_{yd}\sqrt[3]{\frac{\lambda M_{Rd}}{f_{yd}}}} = \sqrt[3]{\frac{M_{Rd}^2}{\lambda f_{yd}^2}} \tag{5.119}$$

Thus the area of a single flange is equal to that of the web.

Method 2 (Classes 1 and 2):

It is recognized that this is not a likely case but is included for completeness. The moment is resisted by the complete section, when the moment capacity is given by that due to the flanges (Eq. (5.111)) and the additional plastic capacity of the web

$$M_{Rd} = f_{yd} b_f t_f h_w + f_{yd} \frac{t h_w^2}{4} \tag{5.120}$$

where f_{yd} is the design strength of the flanges, t_f and b_f the thickness and width of the flange plates and h_w the distance between the internal faces of the flanges.

The cross-sectional area A is given by Eqs (5.112), thus from Eqs (5.112) and (5.120), $b_f t_f$ is given by

$$b_f t_f = \frac{M_{Rd}}{f_{yd} h_w} - \frac{h_w t}{4} \tag{5.121}$$

Eliminate $b_f t_f$ between Eqs (5.112) and (5.121) to give

$$A = \frac{2M_{Rd}}{h_w f_{yd}} + h_w t - 2\frac{h_w t}{4} = \frac{2M_{Rd}}{h_w f_{yd}} + \frac{h_w t}{2} \tag{5.122}$$

Define the web slenderness ratio h_w/t as λ, then Eq. (5.122) becomes

$$A = \frac{2M_{Rd}}{\lambda t f_{yd}} + \frac{\lambda t^2}{2} \tag{5.123}$$

For an optimum solution, $dA/dt = 0$, so Eq. (5.123) becomes,

$$\frac{dA}{dt} = -\frac{2M_{Rd}}{\lambda f_{yd} t^2} + \lambda t = 0 \tag{5.124}$$

or,

$$t = \sqrt[3]{\frac{2M_{Rd}}{\lambda^2 f_{yd}}} \tag{5.125}$$

and

$$h_w = \sqrt[3]{\frac{2\lambda M_{Rd}}{f_{yd}}} \tag{5.126}$$

The area of the web, A_w is then given by

$$A_w = \sqrt[3]{\frac{4M_{Rd}^2}{\lambda f_{yd}^2}} \tag{5.127}$$

Using Eq. (5.121), the flange area, A_f is given by

$$A_f = b_f t_f = \sqrt[3]{\frac{M_{Rd}^2}{\lambda f_{yd}^2}} \left[\sqrt[3]{\frac{1}{2}} - \sqrt[3]{\frac{1}{16}} \right] \tag{5.128}$$

or,

$$\frac{A_f}{A_w} = \frac{\sqrt[3]{\dfrac{M_{Rd}^2}{\lambda f_{yd}^2}} \left[\sqrt[3]{\dfrac{1}{2}} - \sqrt[3]{\dfrac{1}{16}} \right]}{\sqrt[3]{\dfrac{4M_{Rd}^2}{\lambda f_{yd}^2}}} = \frac{1}{4} \tag{5.129}$$

Thus the area of the web is equal four times that of a single flange.

Method 2 (Class 3):

The moment is resisted by the complete section, when the moment capacity is given by that due to the flanges (Eq. (5.111)) and the additional elastic capacity of the web

$$M_{\mathrm{Rd}} = f_{\mathrm{yd}} b_{\mathrm{f}} t_{\mathrm{f}} h_{\mathrm{w}} + f_{\mathrm{yd}} \frac{t h_{\mathrm{w}}^2}{6} \qquad (5.130)$$

where f_{yd} is the design strength of the flanges, t_{f} and b_{f} the thickness and width of the flange plates and h_{w} the distance between the internal faces of the flanges.

The cross-sectional area A is given by Eq. (5.112), thus from Eqs (5.112) and (5.130), $b_{\mathrm{f}} t_{\mathrm{f}}$ is given by

$$b_{\mathrm{f}} t_{\mathrm{f}} = \frac{M_{\mathrm{Rd}}}{f_{\mathrm{yd}} h_{\mathrm{w}}} - \frac{h_{\mathrm{w}} t}{6} \qquad (5.131)$$

Eliminate $b_{\mathrm{f}} t_{\mathrm{f}}$ between Eqs (5.112) and (5.131) to give

$$A = \frac{2M_{\mathrm{Rd}}}{h_{\mathrm{w}} f_{\mathrm{yd}}} + h_{\mathrm{w}} t - 2\frac{h_{\mathrm{w}} t}{6} = \frac{2M_{\mathrm{Rd}}}{h_{\mathrm{w}} f_{\mathrm{yd}}} + \frac{2h_{\mathrm{w}} t}{3} \qquad (5.132)$$

Define the web slenderness ratio h_{w}/t as λ, then Eq. (5.132) becomes

$$A = \frac{2M_{\mathrm{Rd}}}{\lambda t f_{\mathrm{yd}}} + \frac{2\lambda t^2}{3} \qquad (5.133)$$

For an optimum solution, $\mathrm{d}A/\mathrm{d}t = 0$, so Eq. (5.133) becomes,

$$\frac{\mathrm{d}A}{\mathrm{d}t} = -\frac{2M_{\mathrm{Rd}}}{\lambda f_{\mathrm{yd}} t^2} + \frac{4}{3}\lambda t = 0 \qquad (5.134)$$

or,

$$t = \sqrt[3]{\frac{3M_{\mathrm{Rd}}}{2\lambda^2 f_{\mathrm{yd}}}} \qquad (5.135)$$

and

$$h_{\mathrm{w}} = \sqrt[3]{\frac{3\lambda M_{\mathrm{Rd}}}{2f_{\mathrm{yd}}}} \qquad (5.136)$$

The area of the web, A_{w} is then given by

$$A_{\mathrm{w}} = \sqrt[3]{\frac{9M_{\mathrm{Rd}}^2}{4\lambda f_{\mathrm{yd}}^2}} \qquad (5.137)$$

Using Eq. (5.131), the flange area, A_{f} is given by

$$A_{\mathrm{f}} = b_{\mathrm{f}} t_{\mathrm{f}} = \sqrt[3]{\frac{M_{\mathrm{Rd}}^2}{\lambda f_{\mathrm{yd}}^2}} \left[\sqrt[3]{\frac{2}{3}} - \frac{1}{6}\sqrt[3]{\frac{9}{4}} \right] \qquad (5.138)$$

Thus

$$\frac{A_f}{A_w} = \frac{\sqrt[3]{\frac{M_{Rd}^2}{\lambda f_{yd}^2}}\left[\sqrt[3]{\frac{2}{3}} - \frac{1}{6}\sqrt[3]{\frac{9}{4}}\right]}{\sqrt[3]{\frac{9M_{Rd}^2}{4\lambda f_{yd}}}} = \frac{1}{2} \tag{5.139}$$

Thus the area of the web is equal twice that of a single flange.

It should be noted that a plate girder may well be Class 4 in which case an effective section needs calculating, and thus minimum weight optimization is not directly possible, although it can be simulated by increasing the applied moment (see Example 5.4).

5.3.4 Web Design

Experimental work (Basler, 1961; Porter *et al.*, 1975; Rockey *et al.*, 1978; Davies and Griffith, 1999) showed that after web buckling occurred there was still a reserve of strength in the web. This additional reserve of strength in the web is due to a tension field forming in the central diagonal portion of the web (Fig. 5.21).

The shear capacity in the web is determined using a shear buckling slenderness $\bar{\lambda}_w$ which is dependant upon the critical shear strength τ_{cr} (cl. A.1 EN 1993-1-5)

The critical shear strength τ_{cr} is given by

$$\tau_{cr} = k_\tau \sigma_E \tag{5.140}$$

where k_τ is a shear buckling co-efficient dependant upon the aspect ratio of a web panel and σ_E is the elastic critical stress,

From classical plate buckling theory (Bulson, 1970)

$$\sigma_E = \frac{\pi^2 E}{12(1-\nu^2)}\left(\frac{t}{h_w}\right)^2 = 19000\left(\frac{t}{h_w}\right)^2 \tag{5.141}$$

where t is the thickness and h_w the depth.

The parameter k_τ is given as (Bulson, 1970),

for $a/h_w \geq 1,0$

$$k_\tau = 5,34 + 4,00\left(\frac{h_w}{a}\right)^2 \tag{5.142}$$

for $a/h_w < 1,0$

$$k_\tau = 4,00 + 5,34\left(\frac{h_w}{a}\right)^2 \tag{5.143}$$

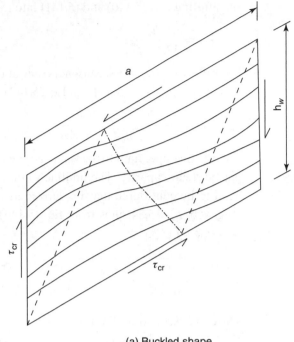

(a) Buckled shape

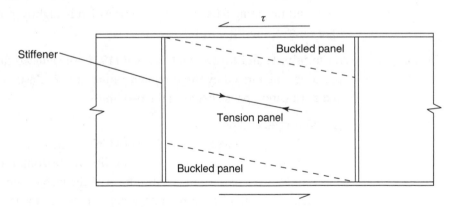

(b) Panel showing tension field

FIGURE 5.21 Shear failure of a plate girder

The non-dimensionalized web slenderness ratio $\bar{\lambda}_w$ is defined as (cl. 5.3 EN 1993-1-5)

$$\bar{\lambda}_w = \left[\frac{\frac{f_{yw}}{\sqrt{3}}}{\tau_{cr}} \right]^{1/2} = 0{,}76 \sqrt{\frac{f_{yw}}{\tau_{cr}}} \tag{5.144}$$

For webs with transverse stiffeners at the supports and either intermediate transverse or longitudinal stiffeners, the normalized web slenderness ratio $\bar{\lambda}_w$ is obtained by substituting Eqs (5.140) and (5.141) into (5.144)

$$\bar{\lambda}_w = \frac{h_w}{37,4\varepsilon t\sqrt{k_\tau}} \tag{5.145}$$

For webs with transverse stiffeners only at the supports, a/h_w is large, hence with little loss in accuracy $k_\tau = 5,34$ from Eq. (5.142). Thus $\bar{\lambda}_w$ is given by

$$\bar{\lambda}_w = \frac{h_w}{37,4\varepsilon t\sqrt{5,34}} = \frac{h_w}{86,4\varepsilon t} \tag{5.146}$$

Clearly the upper limit to the value the shear force that may be carried by the web is the plastic shear capacity. The limiting value of web slenderness beyond which buckling need to be considered may be derived as follows. From Eq. (5.142) for an infinitely long web, $k_I = 5,34$, thus from Eq. (5.140), the limiting value of h_w/t is given when $\tau_{cr} = \tau_{yw} = f_{yk}/\sqrt{3}$, or

$$\frac{h_w}{t} = \sqrt{\frac{5,34\pi^2 E}{235 \times 12(1-v^2)}}\sqrt{\frac{235}{f_{yk}}} = \sqrt{\frac{5,34\pi^2 \times 210 \times 10^3}{12(1-0,3^2) \times 235}}\varepsilon = 65,7\varepsilon \tag{5.147}$$

The code uses a lower limit of $72\varepsilon/\eta$ for unstiffened webs and $31\varepsilon\sqrt{k_\tau}/\eta$ (cl 5.1.(2), EN 1993-1-5). The recommended value of η for steel grades of S460 or lower is 1,2. For steel grades higher than S460, $\eta = 1,0$. With $\eta = 1,2$, the lower limit for steel grades up to and including S460 is $72\varepsilon/1,2 = 60\varepsilon$ which is slightly more conservative than the figure derived in Eq. (5.147).

Basler (1961) and Rockey and Škaloud (1971) recognized that actual web behaviour could be categorized by three regimes: pure shear, elastic buckling at the extremes and a transition phase between the two limits.

(1) Non-rigid end post
    In the case of a non-rigid end post (which cannot generate post-buckling strength as it is incapable of resisting the additional horizontal forces) EN 1993-1-5 only defines two zones for the contribution of the web to shear buckling resistance χ_w where a tension field cannot be generated as the anchorage force is unable to be sustained,

$$\bar{\lambda}_w \leq 0,83/\eta$$
$$\chi_w = \eta \tag{5.148}$$
$$\bar{\lambda}_w > 0,83/\eta$$
$$\chi_w = \frac{0,83}{\bar{\lambda}_w} \tag{5.149}$$

(2) Rigid end post
    Where there is a rigid end stiffener, the anchorage force from tension field theory can be sustained. The original theory behind tension field theory was outlined by Porter et al. (1975) and Rockey et al. (1978). EN 1993-1-5 has adopted Höglund's rotating stress–field theory (Davies and Griffith, 1999) which is easier to apply

than the original tension field theory. Höglund's rotating stress–field theory also mobilizes post-buckling behaviour, but only if there is a rigid end post. In this case the relationships between normalized web slenderness $\bar{\lambda}_w$ and χ_w are given by

$$\bar{\lambda}_w \leq 0,83/\eta$$

$$\chi_w = \eta \tag{5.150}$$

$$0,83/\eta \leq \bar{\lambda}_w < 1,08$$

$$\chi_w = \frac{0,83}{\bar{\lambda}_w} \tag{5.151}$$

$$\bar{\lambda}_w \geq 1,08$$

$$\chi_w = \frac{1,37}{0,7 + \bar{\lambda}_w} \tag{5.152}$$

It will be observed that the difference between the two methods is that the web shear parameter χ_w is enhanced for $\bar{\lambda}_w \geq 1,08$.

The design resistance of a web $V_{b,Rd}$ whether stiffened or unstiffened (cl 5.2 EN 1993-1-5) is given by

$$V_{b,Rd} = \chi_v h_w t \frac{\frac{f_{yw}}{\sqrt{3}}}{\gamma_{M1}} \tag{5.153}$$

where shear co-efficient χ_v is given by

$$\chi_v = \chi_w + \chi_f \leq \eta \tag{5.154}$$

The parameter χ_f represents the contribution to shear resistance from the flanges for $M_{Ed} < M_{f,Rd}$ and is given by

$$\chi_f = \frac{b_f t_f^2 f_{yf} \sqrt{3}}{c t h_w f_{yw}} \left[1 - \left(\frac{M_{Ed}}{M_{f,Rd}} \right)^2 \right] \tag{5.155}$$

where b_f is the width of the flange taken as not greater than $15\varepsilon t_f$ on each side of the web, t_f is the web thickness, $M_{f,Rd}$ is design moment of resistance of the cross-section determined using the effective flanges only and c is the width of the portion of the web between the plastic hinges (see Fig. 5.22) and is given by

$$c = a \left(0,25 + 1,6 \frac{M_{pl,f}}{M_{pl,w}} \right) = a \left(0,25 + \frac{1,6 b_f t_f^2 f_{yf}}{t h_w^2 f_{yw}} \right) \tag{5.156}$$

Equation (5.155) is derived by determining the shear that may be carried by the portion of both flanges of length c with the yield strength f_{yf} reduced by considering the effect of induced axial forces in the flanges.

The background to the simplified method adopted in the Code is given in Davies and Griffith (1999).

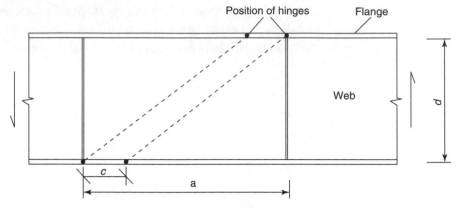

(a) Basic geometry of web panel

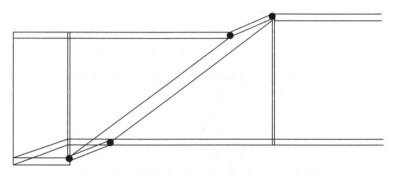

(b) Failure mechanism used to derive post buckling strength of web

FIGURE 5.22   Overall behaviour of web at failure

The calculations for web shear capacity are iterative as both the shear factors and the applied moment are dependant upon the stiffener spacing. The calculations for shear capacity in the examples following were performed on a spreadsheet.

5.3.5 Stiffeners

5.3.5.1 Rigid End Post (cl. 9.3.1 EN 1993-1-5)

This may either be a set of flats welded at the end and above the support with the centroids a distance e apart, or the end post may comprise a rolled section in which case e is the distance between the flange centroids.

The horizontal stress σ_h for large slenderness ratios can be given as

$$\sigma_h = \frac{0.43}{\bar{\lambda}_w} f_y \tag{5.157}$$

Substitute the value of $\bar{\lambda}_w$ from Eq. (5.145) into Eq. (5.157) to give an upper bound for the UDL q_h as

$$q_h = \sigma_h t = 16.1 \frac{f_y t^2 \varepsilon \sqrt{k_\tau}}{h_w} \tag{5.158}$$

Taking the maximum value of ε as 1,0 and the maximum value of k_τ as 9,34, the value of q_h becomes

$$q_h = 49\frac{t^2 f_y}{h_w} \tag{5.159}$$

The distributed load is not uniform over the depth of the girder and the theoretical value of σ_h is high, the co-efficient of 49 in Eq. (5.159) is replaced by 32.

Assuming the end post is simply supported, the maximum moment M_{max} is given by

$$M_{max} = \frac{q_h h_w^2}{8} \tag{5.160}$$

The section modulus, ignoring any contribution from the web, W is given by $A_{min}e$, thus assuming the maximum stress is given by f_y, then

$$\sigma_{max} = f_y = \frac{M_{max}}{W} = \frac{\dfrac{32t^2 f_y}{h_w}\dfrac{h_w^2}{8}}{A_{min}e} \tag{5.161}$$

Thus the minimum cross-sectional area of each pair A_{min} is given by

$$A_{min} = \frac{4h_w^2 t}{e} \tag{5.162}$$

with $e > 0,1h_w$.

The restriction on e would appear to be a detailing requirement.

The end panel may be designed as non-rigid shear panel carrying the whole of the applied shear. This was originally proposed by Basler (1961).

From Eq. (5.153) with $\chi_f = 0$, the required value of χ_w is given by

$$\chi_w = \frac{V_{Ed}\sqrt{3}\gamma_{M1}}{f_{yw}h_w t} \tag{5.163}$$

From Eq. (5.149), the normalized web slenderness $\bar{\lambda}_w$ is given by

$$\bar{\lambda}_w = \frac{0,83}{\chi_w} \tag{5.164}$$

The buckling parameter k_τ is given from Eq. (5.145) as

$$k_\tau = \left(\frac{h_w}{37,4t\varepsilon\bar{\lambda}_w}\right)^2 \tag{5.165}$$

As $a/h_w < 1$, k_τ is given by Eq. (5.143), or the required panel width a is given as

$$a = h_w\sqrt{\frac{5,35}{k_\tau - 4,00}} \tag{5.166}$$

Note, this will give an upper bound to the value of a.

5.3.5.2 Transverse Stiffeners (cl 9.2.1 EN 1993-1-5)

These should be checked as a simply supported beam with an initial sinusoidal imperfection w_0 given by Eq. (5.169) together with any eccentricities. The transverse stiffener should carry the deviation forces from the adjacent panels assuming the adjacent transverse stiffeners are rigid and straight. A second order analysis should be used to determine that the maximum stress does not exceed f_{yd} nor any additional deflection $b/300$. This will be more critical for single sided stiffeners. As double sided plate stiffeners have been used in the ensuing examples, only the stress criterion has been checked.

In the absence of transverse loads or axial forces in the stiffeners, then the strength and deflection criteria are satisfied if they have a second moment of area given by

$$I_{st} = \frac{\sigma_M}{E}\left(\frac{b}{\pi}\right)^4\left(1 + w_0\frac{300}{b}u\right) \tag{5.167}$$

where σ_M is given by

$$\sigma_M = \frac{\sigma_{cr,c}}{\sigma_{cr,p}}\frac{N_{Ed}}{b}\left(\frac{1}{a_1} + \frac{1}{a_2}\right) \tag{5.168}$$

where a_1 and a_2 are the panel lengths either side of the stiffener under consideration, N_{Ed} is the larger compressive force in the adjacent panels, b is the height of the stiffener.

The initial imperfection w_0 is given as

$$w_0 = \frac{1}{300}\text{LEAST}(a_1, a_2, b) \tag{5.169}$$

The parameter u is given by

$$u = \frac{\pi^2 E e_{max}}{\dfrac{300bf_y}{\gamma_{M1}}} \geq 1{,}0 \tag{5.170}$$

The distance e_{max} is taken from the extreme fibre of the stiffener to the centroid of the stiffener.

The critical stress for plate between vertical stiffeners $\sigma_{cr,c}$ is given by

$$\sigma_{cr,c} = \frac{\pi^2 E t^2}{12(1 - v^2)a^2} \tag{5.171}$$

The critical stress $\sigma_{cr,p}$ is given by $k_{\sigma,p}\sigma_E$. The value of σ_E is given by Eq. (5.141). The value for plates with longitudinal stiffeners in Annex A of EN 1993-1-5. For unstiffened plates $\sigma_{cr,p} = \sigma_{cr,c}$

To avoid lateral torsional buckling of the stiffener,

$$\frac{I_T}{I_p} \geq 5{,}3\frac{f_y}{E} \tag{5.172}$$

where I_T and I_p are St. Venant torsional constant for the stiffener alone and I_p is the polar second moment of area about the edge fixed to the plate. Eq. (5.172) can be derived as follows. The critical buckling stress σ_{cr} for open section stiffeners (with negligible warping stiffness) is given by Eq. (5.102) with $I_w = 0$,

$$\sigma_{cr} = G\frac{I_T}{I_p} \tag{5.173}$$

From cl 9.2.1(8) of EN 1993-1-5, the critical stress σ_{cr} is limited by

$$\sigma_{cr} \geq \theta f_y \tag{5.174}$$

For plate stiffeners, θ is taken as 2,0, thus Eq. (5.173) becomes

$$\frac{I_T}{I_p} \geq 5,2\frac{f_y}{E} \tag{5.175}$$

The code replaces the co-efficient 5,2 by 5,3.

5.3.5.3 Intermediate Transverse Stiffeners (cl. 9.3.3 EN 1993-1-5)

The force $N_{s,Rd}$ to be resisted by a stiffener is given by

$$N_{s,Rd} = V_{Ed} - \chi_w h_w t \frac{\frac{f_{yw}}{\sqrt{3}}}{\gamma_{M1}} \tag{5.176}$$

Note, χ_w is calculated for the web panel between adjacent stiffeners assuming the stiffener under consideration is removed. In the case of variable shear, then the check is performed at a distance $0,5h_w$ from the edge of the pane with the larger shear force.

To determine the buckling resistance of the stiffener a portion of the web may taken into account (Rockey et al., 1981). A section of the web in length equal to $15\varepsilon t$ either side of the stiffener may be considered (cl 9.1, EN 1933-1-3) (Fig. 5.23)

For a symmetric stiffener, the effective area A_e is given by

$$A_{equiv} = A_{st} + 30\varepsilon t^2 \tag{5.177}$$

and the effective second moment of area I_{equiv} by

$$I_{equiv} = I_{st} + \frac{1}{12}30\varepsilon t^4 \tag{5.178}$$

where A_{st} is the area of the stiffener and I_{st} is the second moment of area of the stiffener. For end stiffeners the co-efficient of 30 in Eqs (5.177) and (5.178) should be replaced by 15.

The effective length of the stiffener may be taken as $0,75h_w$ and buckling curve 'c' used to determine the strength reduction factor (cl 9.4 EN 1993-1-5).

In order to provide adequate restraint against buckling it was found that the stiffeners need to possess a minimum second moment of area (Rockey et al., 1981).

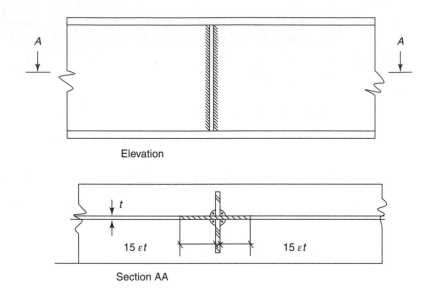

Elevation

Section AA

FIGURE 5.23 Stiffener geometry

The minimum second moment of area I_s is given by

for $a/h_w < \sqrt{2}$

$$I_{st} \geq 1{,}5\frac{h_w^3 t^3}{a^2} \tag{5.179}$$

for $a/h_w \geq \sqrt{2}$

$$I_{st} \geq 0{,}75 h_w t^3 \tag{5.180}$$

It can be demonstrated that for compression buckling a change from 1 to 2 half sine waves occurs at $a/h_w = \sqrt{2}$, and that thereafter the buckling co-efficient is sensibly independent of the aspect ratio of the panel. Thus Eq. (5.180) is determined from Eq. (5.179) by substituting $a = h_w\sqrt{2}$ (cl 9.3.3 (3) EN 1993-1-5).

5.3.5.4 Plate Splices (cl 9.2.3 EN 1993-1-5)

The splice whether in the web or flanges, should ideally occur at a transverse stiffener. If not then the stiffener should be at a distance no greater than $b_0/2$ along the thinner plate where b_0 is the depth of the web (or the least spacing of longitudinal stiffeners).

5.3.5.5 Longitudinal Welds (Web to Flange) (cl 9.3.5 EN 1993-1-5)

The weld between the web and flange(s) should be designed for a shear flow of V_{Ed}/h_w, provided

$$V_{Ed} < \chi_w h_w t \frac{\frac{f_{yw}}{\sqrt{3}}}{\gamma_{M1}} \tag{5.181}$$

If the condition in Eq. (5.181) is not satisfied, the welds should be designed under a shear flow of $\eta t (f_{yw}/\sqrt{3})/\gamma_{M1}$.

EXAMPLE 5.3 Design of a laterally restrained plate girder

Design a plate girder in Grade S355 steel to carry a characteristic variable load of 150 kN/m over a span of 20 m. The compression flange is fully restrained against lateral torsional buckling.

For a plate girder span/depth ratios are generally around 8 or 10 to 1. The higher this ratio is the lower the flange size but at the probable expense of a thicker web to overcome buckling.

To calculate the imposed bending moment, assume the total weight of the beam is 100 kN.

Total loading on the beam $= 1,5 \times 150 + 1,35 \times (100/20) = 232$ kN/m

$M_{Sd} = 232 \times 20^2/8 = 11,6$ MNm

The critical slenderness ratio for the web can be controlled by flange-induced buckling with $k = 0,4$ as plastic rotation is not utilized. If the flanges resist the bending moment, $A_w = A_f$ (from Eq. (5.119)), thus the critical h_w/t ratio is given by Eq. (5.110)

$$\frac{h_w}{t} \le k\frac{E}{f_{yf}}\sqrt{\frac{A_w}{A_c}} = 0,4\frac{210 \times 10^3}{355}\sqrt{1} = 237$$

From Eq. (5.117) calculate h_w

$$h_w = \sqrt[3]{\frac{\lambda M_{Rd}}{f_{yd}}} = \sqrt{\frac{237 \times 11,6 \times 10^9}{\frac{355}{1,0}}} = 1978 \text{ mm}$$

The thickness t is given as

$$t = \frac{h_w}{\lambda} = \frac{1978}{237} = 8,34$$

Use $t = 9$ mm

Maximum ratio flange outstand to flange thickness beyond the weld for a Class 1 section is 9ε where $\varepsilon = (235/355)^{1/2}$ $(=0,814)$.

So maximum flange outstand is $9 \times t_f \times (235/355)^{1/2} = 7,33t_f$

So flange area is $14,66t_f^2$ (ignoring the effect of weld width).

$$M_{pl,Rd} \approx A_f \frac{f_y}{\gamma_{M0}} h_w$$

or

$$A_{\mathrm{f}} = \frac{M_{\mathrm{Ed}}}{h_{\mathrm{w}} \frac{f_y}{\gamma_{\mathrm{M0}}}} = \frac{11{,}6 \times 10^9}{1978 \frac{355}{1{,}0}} = 16{,}520\,\mathrm{mm}^2$$

or

$$14{,}66 t_{\mathrm{f}}^2 = 16{,}520$$

or

Inset $t_f = 33{,}6\,\mathrm{mm}$

Use 35 mm thick plate with a width of 500 mm.

Overall depth, h:

$h = 2 \times 35 + 1978 = 2048\,\mathrm{mm}$. Use $h = 2000\,\mathrm{mm}$.

Check actual h_{w}/t ratio:

Actual web slenderness:

$$\frac{h_{\mathrm{w}}}{t} = \frac{1930}{9} = 214$$

Web is Class 4. However as the webs do not carry any compression, then the section may be treated dependant upon the classification of the flanges (cl 5.5.2 (12) EN 1993-1-1).

Maximum h_{w}/t ratio:

$$\frac{h_{\mathrm{w}}}{t} = k\frac{E}{f_{\mathrm{yf}}}\sqrt{\frac{A_{\mathrm{w}}}{A_{\mathrm{f}}}} = 0{,}4\frac{210 \times 10^3}{355}\sqrt{\frac{1930 \times 9}{500 \times 35}} = 236$$

The actual value is below the allowable, and is therefore satisfactory.

Plastic moment of resistance of the flanges, $M_{\mathrm{pl,Rd}}$:

$$M_{\mathrm{pl,Rd}} = A_{\mathrm{f}}\frac{f_y}{\gamma_{\mathrm{M0}}}(h - t_{\mathrm{f}}) = 35 \times 500\frac{355}{1{,}0}(2000 - 35) = 12{,}2\,\mathrm{MNm}$$

$$V_{\mathrm{pl,Rd}} = 232 \times 10 = 2320\,\mathrm{kN} \text{ (at the support)}$$

$$h_w = h - 2t_{\mathrm{f}} = 2000 - 2 \times 35 = 1930\,\mathrm{mm}$$

The web will be designed both ways, non-rigid and rigid end post:

The determined flange and web contributions take the maximum shear and maximum moment in a web panel, even though they are at opposite ends of the panel. This will be conservative.

Determination of χ_{f}:

To determine χ_{f}, the flange width is limited to $15\varepsilon t_{\mathrm{f}}$ on either side of the web:

$$15\varepsilon t_{\mathrm{f}} = 15 \times 35 \left(\frac{235}{355}\right)^{1/2} = 427\,\mathrm{mm}$$

Actual flange width $= 0,5(500 - 9) = 245$ mm. This is less, therefore use actual width, so $M_{f,Rd} = 12,2$ MNm (from above).

Both the web and flanges have a yield strength of 355 MPa.

Non-rigid end post:

First panel from the support:

Intermediate stiffener is 1,2 m from the support.

Determine c from Eq. (5.156)

$$c = a \left(0,25 + \frac{1,6 b_f t_f^2 f_{yf}}{t h_w^2 f_{yw}} \right) = 1200 \left(0,25 + \frac{1,6 \times 500 \times 35^2 \times 355}{9 \times 1930^2 \times 355} \right) = 335 \text{ mm}$$

$$M_{Ed} = 2320 \times 1,2 - 232\frac{1,2^2}{2} = 2617 \text{ kNm}$$

Determine the flange contribution factor χ_f from Eq. (5.155)

$$\chi_f = \frac{b_f t_f^2 f_{yf} \sqrt{3}}{c t h_w f_{yw}} \left[1 - \left(\frac{M_{Ed}}{M_{f,Rd}} \right)^2 \right] = \frac{500 \times 35^2 \times 355 \times \sqrt{3}}{335 \times 9 \times 1930 \times 355} \left[1 - \left(\frac{2617}{12200} \right)^2 \right]$$

$$= 0,174$$

$$\frac{h_w}{a} = \frac{1930}{1200} = 1,608$$

$$\frac{a}{h_w} = 0,622$$

For $a/h_w < 1,0$, k_τ is given by Eq. (5.143), or

$$k_\tau = 4 + 5,34 \left(\frac{h_w}{a} \right)^2 = 4 + 5,34 \times 1,608^2 = 17,81$$

Determine the normalized web slenderness ratio $\bar{\lambda}_w$ from Eq. (5.145)

$$\bar{\lambda}_w = \frac{h_w}{37,4 \varepsilon t \sqrt{k_\tau}} = \frac{1930}{37,4\sqrt{\frac{235}{355}} \times 9\sqrt{17,81}} = 1,670$$

As $\bar{\lambda}_w > 1,08$, χ_w is given by Eq. (5.149) as

$$\chi_w = \frac{0,83}{\bar{\lambda}_w} = \frac{0,83}{1,670} = 0,497$$

From Eq. (5.154), the total shear co-efficient χ_v is given by

$$\chi_v = \chi_f + \chi_w = 0,174 + 0,497 = 0,671$$

TABLE 5.5 Shear capacity calculations with non-rigid end post (Example 5.3).

Distance from Support (m)	1,2	2,55	4,15	6,25	10
Panel width (m)	1,2	1,35	1,6	2,1	3,75
Moment M_{Ed} (kNm)	2617	5162	7630	9969	11600
Shear V_{Ed} (kN)	2320	2042	1728	1357	870
c (mm)	335	377	447	586	1047
χ_f	0,174	0,133	0,083	0,034	0,006
a/h_w	0,622	0,699	0,829	1,088	1,943
k_τ	17,81	14,91	11,77	8,719	6,400
$\bar{\lambda}_w$	1,670	1,825	2,054	2,387	2,786
χ_w	0,497	0,455	0,404	0,378	0,298
χ_v	0,671	0,588	0,487	0,382	0,304
V_{Rd} (kN)	2389	2093	1735	1361	1081

The shear capacity $V_{b,Rd}$ is determined from Eq. (5.153)

$$V_{b,Rd} = \chi_v h_w t \frac{\frac{f_{yw}}{\sqrt{3}}}{\gamma_{M1}} = \frac{0,671 \times 1930 \times 9}{1000} \frac{\frac{355}{\sqrt{3}}}{1,0} = 2389 \, kN$$

The calculations for subsequent panels are summarized in Table 5.5 (together with the first panel)

Intermediate stiffeners:

Check strength:

First stiffener:

The axial force N_{Ed} is given by Eq. (5.176) as

$$N_{Rd} = V_{Ed} - \chi_w h_w t \frac{\frac{f_{yw}}{\sqrt{3}}}{\gamma_{M1}}$$

As the load is a UDL, V_{Ed} is determined at $0,5 h_w$ from the stiffener in the panel with the higher shear:

$$V_{Ed} = 2320 - 232(1,2 - 0,5 \times 1,930) = 2265 \, kN$$

The web contribution parameter χ_w is calculated assuming the stiffener is removed, thus $a = 1200 + 1350 = 2550 \, mm$

$$\frac{h_w}{a} = \frac{1930}{2550} = 0,757$$

$$\frac{a}{h_w} = 1,321$$

For $a/h_w > 1,0$, k_τ is given by Eq. (5.142), or

$$k_\tau = 5,34 + 4\left(\frac{h_w}{a}\right)^2 = 5,34 + 4 \times 0,757^2 = 7,63$$

Determine the normalized web slenderness ratio $\bar{\lambda}_w$ from Eq. (5.145)

$$\bar{\lambda}_w = \frac{h_w}{37,4\varepsilon t\sqrt{k_\tau}} = \frac{1930}{37,4 \times 9\sqrt{\frac{235}{355}}\sqrt{7,63}} = 2,55$$

As $\bar{\lambda}_w > 1,08$, χ_w is given by Eq. (5.149) as

$$\chi_w = \frac{0,83}{\bar{\lambda}_w} = \frac{0,83}{2,55} = 0,325$$

$$N_{Rd} = V_{Ed} - \chi_w h_w t \frac{\frac{f_{yw}}{\sqrt{3}}}{\gamma_{M1}} = 2265 - 0,325 \times 1930 \times 9 \times 10^{-3}\frac{\frac{355}{\sqrt{3}}}{1,0} = 1108\,\text{kN}$$

Second stiffener:

Axial force N_{Ed} is given by Eq. (5.176) as

$$N_{Rd} = V_{Ed} - \chi_w h_w t \frac{\frac{f_{yw}}{\sqrt{3}}}{\gamma_{M1}}$$

As the load is a UDL, V_{Ed} is determined at $0,5h_w$ from the stiffener in the panel with the higher shear:

$$V_{Ed} = 2320 - 232(2,55 - 0,5 \times 1,930) = 1952\,\text{kN}$$

The web contribution parameter χ_w is calculated assuming the stiffener is removed, thus $a = 1350 + 1600 = 2950\,\text{mm}$

$$\frac{h_w}{a} = \frac{1930}{2950} = 0,654$$

$$\frac{a}{h_w} = 1,528$$

For $a/h_w > 1,0$, k_τ is given by Eq. (5.142), or

$$k_\tau = 5,34 + 4\left(\frac{h_w}{a}\right)^2 = 5,34 + 4 \times 0,654^2 = 7,05$$

Determine the normalized web slenderness ratio $\bar{\lambda}_w$ from Eq. (5.145)

$$\bar{\lambda}_w = \frac{h_w}{37,4\varepsilon t\sqrt{k_\tau}} = \frac{1930}{37,4 \times 9\sqrt{\frac{235}{355}}\sqrt{7,05}} = 2,65$$

As $\bar{\lambda}_w > 1,08$, χ_w is given by Eq. (5.149) as

$$\chi_w = \frac{0,83}{\bar{\lambda}_w} = \frac{0,83}{2,65} = 0,313$$

$$N_{Rd} = V_{Ed} - \chi_w h_w t \frac{\frac{f_{yw}}{\sqrt{3}}}{\gamma_{M1}} = 1952 - 0,313 \times 1930 \times 9 \times 10^{-3}\frac{\frac{355}{\sqrt{3}}}{1,0} = 838\,\text{kN}$$

Third stiffener:

Axial force N_{Ed} is given by Eq. (5.176) as

$$N_{Rd} = V_{Ed} - \chi_w h_w t \frac{\frac{f_{yw}}{\sqrt{3}}}{\gamma_{M1}}$$

As the load is a UDL, V_{Ed} is determined at $0{,}5h_w$ from the stiffener in the panel with the higher shear:

$$V_{Ed} = 2320 - 232(4{,}15 - 0{,}5 \times 1{,}930) = 1581 \text{ kN}$$

The web contribution parameter χ_w is calculated assuming the stiffener is removed, thus $a = 1600 + 2100 = 3700 \text{ mm}$

$$\frac{h_w}{a} = \frac{1930}{3700} = 0{,}522$$

$$\frac{a}{h_w} = 1{,}917$$

For $a/h_w > 1{,}0$, k_τ is given by Eq. (5.142), or

$$k_\tau = 5{,}34 + 4\left(\frac{h_w}{a}\right)^2 = 5{,}34 + 4 \times 0{,}522^2 = 6{,}43$$

Determine the normalized web slenderness ratio $\bar{\lambda}_w$ from Eq. (5.145)

$$\bar{\lambda}_w = \frac{h_w}{37{,}4\varepsilon t \sqrt{k_\tau}} = \frac{1930}{37{,}4 \times 9\sqrt{\frac{235}{355}}\sqrt{6{,}43}} = 2{,}78$$

As $\bar{\lambda}_w > 1{,}08$, χ_w is given by Eq. (5.149) as

$$\chi_w = \frac{0{,}83}{\bar{\lambda}_w} = \frac{0{,}83}{2{,}78} = 0{,}299$$

$$N_{Rd} = V_{Ed} - \chi_w h_w t \frac{\frac{f_{yw}}{\sqrt{3}}}{\gamma_{M1}} = 1581 - 0{,}299 \times 1930 \times 9 \times 10^{-3}\frac{\frac{355}{\sqrt{3}}}{1{,}0} = 517 \text{ kN}$$

Fourth stiffener:

Axial force N_{Ed} is given by Eq. (5.176)

$$N_{Rd} = V_{Ed} - \chi_w h_w t \frac{\frac{f_{yw}}{\sqrt{3}}}{\gamma_{M1}}$$

As the load is a UDL, V_{Ed} is determined at $0{,}5h_w$ from the stiffener in the panel with the higher shear:

$$V_{Ed} = 2320 - 232(6{,}25 - 0{,}5 \times 1{,}930) = 1094 \text{ kN}$$

The web contribution parameter χ_w is calculated assuming the stiffener is removed, thus $a = 2100 + 3750 = 5850$ mm

$$\frac{h_w}{a} = \frac{1930}{5850} = 0{,}330$$

$$\frac{a}{h_w} = 3{,}03$$

For $a/h_w > 1{,}0$, k_τ is given by Eq. (5.142), or

$$k_\tau = 5{,}34 + 4 \left(\frac{h_w}{a}\right)^2 = 5{,}34 + 4 \times 0{,}33^2 = 5{,}78$$

Determine the normalized web slenderness ratio $\overline{\lambda}_w$ from Eq. (5.145)

$$\overline{\lambda}_w = \frac{h_w}{37{,}4\varepsilon t \sqrt{k_\tau}} = \frac{1930}{37{,}4 \times 9\sqrt{\frac{235}{355}}\sqrt{5{,}78}} = 2{,}93$$

As $\overline{\lambda}_w > 1{,}08$, χ_w is given by Eq. (5.145) as

$$\chi_w = \frac{0{,}83}{\overline{\lambda}_w} = \frac{0{,}83}{2{,}93} = 0{,}283$$

$$N_{Rd} = V_{Ed} - \chi_w h_w t \frac{\frac{f_{yw}}{\sqrt{3}}}{\gamma_{M1}} = 1094 - 0{,}283 \times 1930 \times 9 \times 10^{-3} \frac{\frac{355}{\sqrt{3}}}{1{,}0} = 86\,\text{kN}$$

This value is small, thus there is no need to check the centre stiffener.

The minimum stiffness requirement for all but the two panels either side of the central stiffener is given by the case $a < \sqrt{2}h_w$, so design on the least value of a:

$$I_s = 1{,}5\frac{h_w^3 t^3}{a^2} = 1{,}5\frac{1930^3 \times 9^3}{1200^2} = 5{,}46 \times 10^6\,\text{mm}^4$$

Use 9 mm thick plate, then the total breadth of the stiffener b is given by

$$b = \sqrt[3]{\frac{12 \times 5{,}46 \times 10^6}{9}} = 194\,\text{mm}$$

Use $b = 200$ mm. The axial force that can be carried by the stiffener N_{Rd} is given as

$$N_{Rd} = 2 \times 200 \times 9\frac{355}{1{,}0} \times 10^{-3} = 1278\,\text{kN}.$$

Buckling check:

Effective length $= 0{,}75 \times 1930 = 1448$ mm

From Eq. (5.177), the effective area A_{equiv} is given as

$$A_{equiv} = A_{st} + 30\varepsilon t^2 = 200 \times 9 + 30 \times 9\sqrt{\frac{235}{355}} = 2020\,\text{mm}^2$$

From Eq. (5.178), the effective area I_{equiv} is given as

$$I_{equiv} = I_{st} + 30\varepsilon t^4 = \frac{1}{12}200^3 \times 9 + \frac{1}{12}30 \times 9^4\sqrt{\frac{235}{355}} = 6,013 \times 10^6 \, \text{mm}^4$$

$$N_{cr} = \frac{\pi^2 E I_{equiv}}{L^2} = \frac{\pi^2 \times 210 \times 6,013 \times 10^6}{1448^2} = 5944 \, \text{kN}$$

$$\bar{\lambda} = \sqrt{\frac{A_{equiv}f_y}{N_{cr}}} = \sqrt{\frac{2020 \times 355 \times 10^{-3}}{5944}} = 0,121$$

As $\bar{\lambda} \leq 0,2$ strut buckling need not be checked (cl 6.3.1.2 (4) (EN 1993-1-1)).

Thus the minimum stiffener size will be adequate for all the intermediate stiffeners.

End stiffener:

Try a 500 wide by 15 thick plate:

The section must be checked for buckling, but a proportion of the web may be taken into account.

Length of web $= 15\varepsilon t = 15 \times 9\sqrt{(235/355)} = 110 \, \text{mm}$

$$A_{s,eff} = A_s + A_{web} = 500 \times 15 + 110 \times 9 = 8490 \, \text{mm}^2$$

$$I_{s,eff} = I_s + I_{web} = \frac{15 \times 500^3}{12} + \frac{110 \times 9^3}{12} = 0,156 \times 10^9 \, \text{mm}^4$$

$$i = \sqrt{\frac{I_{s,eff}}{A_{s,eff}}} = \sqrt{\frac{0,156 \times 10^9}{8490}} = 135,6 \, \text{mm}$$

Use an effective length of $0,75h_w = 0,75 \times 1930 = 1448 \, \text{mm}$

$$N_{cr} = \frac{\pi^2 EI}{l^2} = \frac{210 \times 0,156 \times 10^9 \pi^2}{1448^2} = 154210 \, \text{kN}$$

$$\bar{\lambda} = \sqrt{\frac{Af_y}{N_{cr}}} = \sqrt{\frac{8490 \times 355}{154210 \times 10^3}} = 0,14$$

From cl 6.3.1.2 (4) (EN 1993-1-1), there is no reduction for strut buckling as $\bar{\lambda} < 0,2$.

So,

$$N_{Rd} = A\frac{f_y}{\gamma_{M0}} = 8490\frac{355}{1.0} \times 10^{-3} = 3014 \, \text{kN}$$

This exceeds the reaction of 2320 kN.

Check cl 9.2.1 (7)

Determine I_p:

The second moment of area about the web centre line, I_y:

$$I_y = \frac{bh^3}{12} = \frac{15 \times 500^3}{12} = 0,156 \times 10^9 \, \text{mm}^4$$

The second moment of area normal to the web centre line about one edge, I_y:

$$I_y = \frac{bh^3}{3} = \frac{500 \times 15^3}{3} = 0{,}563 \times 10^6 \text{ mm}^4$$

$$I_p = I_x + I_y = 0{,}156 \times 10^9 + 0{,}563 \times 10^6 = 0{,}157 \times 10^9 \text{ mm}^4$$

$$I_T = \frac{bh^3}{3} = 0{,}563 \times 10^6 \text{ mm}^4$$

$$\frac{I_T}{I_p} = \frac{0{,}563 \times 10^6}{0{,}157 \times 10^9} = 3{,}59 \times 10^{-3}$$

Limiting value:

$$5{,}3\frac{f_y}{E} = 5{,}3\frac{355}{210 \times 10^3} = 8{,}96 \times 10^{-3}$$

The actual value is less than the limiting value thus the stiffener size must be increased.

A 25 mm thick end plate will satisfy the limiting stiffness criterion ($I_T/I_P = 0{,}01$) (and will clearly satisfy the strength and buckling criteria).

Flange to web welds:

From Table 5.5 $V_{Ed} > h_w t(f_{yw}/\sqrt{3})/\gamma_{M1}$ as the flange contribution χ_f has been mobilized. Thus welds should be designed for a shear flow of

$$\eta t\frac{\frac{f_{yw}}{\sqrt{3}}}{\gamma_{M1}} = 1{,}2 \times 9\frac{\frac{355}{\sqrt{3}}}{1{,}0} = 2214 \text{ N/mm}$$

The final layout of the girder whose self-weight is 91,1 kN is given in Fig 5.24(a).

Rigid end post:

First panel from the support:

Intermediate stiffener is 1,45 m from the support.

Determine c from Eq. (5.156)

$$c = a\left(0{,}25 + \frac{1{,}6b_f t_f^2 f_{yf}}{t_w^2 f_{yw}}\right) = 1450\left(0{,}25 + \frac{1{,}6 \times 500 \times 35^2 \times 355}{9 \times 1930^2 \times 355}\right)$$

$$= 405 \text{ mm}$$

$$M_{Ed} = 2320 \times 1{,}45 - 232\frac{1{,}45^2}{2} = 3120 \text{ kNm}$$

Determine the flange contribution factor χ_f from Eq. (5.155)

$$\chi_f = \frac{b_f t_f^2 f_{yf}\sqrt{3}}{cth_w f_{yw}}\left[1 - \left(\frac{M_{Ed}}{M_{f,Rd}}\right)^2\right] = \frac{500 \times 35^2 \times 355 \times \sqrt{3}}{405 \times 9 \times 1930 \times 355}\left[1 - \left(\frac{3120}{12200}\right)^2\right]$$

$$= 0{,}141$$

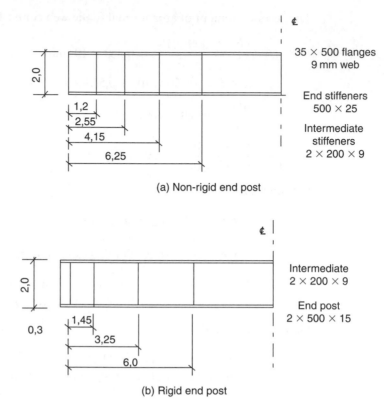

(a) Non-rigid end post

35 × 500 flanges
9 mm web

End stiffeners
500 × 25

Intermediate
stiffeners
2 × 200 × 9

(b) Rigid end post

Intermediate
2 × 200 × 9

End post
2 × 500 × 15

FIGURE 5.24 Final layout for the beam in Example 5.3

$$\frac{h_w}{a} = \frac{1930}{1450} = 1,331$$

$$\frac{a}{h_w} = 0,751$$

For $a/h_w < 1,0$, k_τ is given by Eq. (5.142), or

$$k_\tau = 4 + 5,34 \left(\frac{h_w}{a}\right)^2 = 4 + 5,34 \times 1,331^2 = 13,46$$

Determine the normalized web slenderness ratio $\bar{\lambda}_w$ from Eq. (5.145)

$$\bar{\lambda}_w = \frac{h_w}{37,4\varepsilon t \sqrt{k_\tau}} = \frac{1930}{37,4\sqrt{\frac{235}{355}} \times 9\sqrt{13,46}} = 1,920$$

As $\bar{\lambda}_w > 1,08$, χ_w is given by Eq. (5.152) as

$$\chi_w = \frac{1,37}{0,7 + \bar{\lambda}_w} = \frac{1,37}{0,7 + 1,92} = 0,523$$

From Eq. (5.154), the total shear co-efficient χ_v is given by

$$\chi_v = \chi_f + \chi_w = 0,141 + 0,523 = 0,664$$

The shear capacity $V_{b,Rd}$ is determined from Eq. (5.153)

$$V_{b,Rd} = \chi_v h_w t \frac{\frac{f_{yw}}{\sqrt{3}}}{\gamma_{M1}} = \frac{0,664 \times 1930 \times 9 \frac{355}{\sqrt{3}}}{1000 \quad 1,0} = 2363\,kN$$

The resultant spacing of stiffeners and a summary of the remaining calculations is given in Table 5.6 The final result for the calculation of stiffener forces are also given in Table 5.6.

The minimum stiffness requirement for all but the two panels either side of the central stiffener is given by the case $a < \sqrt{2}h_w$, so design on the least value of a:

$$I_s = 1,5\frac{h_w^3 t^3}{a^2} = 1,5\frac{1930^3 \times 9^3}{1450^2} = 3,74 \times 10^6\,mm^4$$

Use 9 mm thick plate, then the total breadth of the stiffener b is given by

$$b = \sqrt[3]{\frac{12 \times 3,74 \times 10^6}{9}} = 171\,mm$$

Use $b = 200$ mm. The axial force that can be carried by the stiffener N_{Rd} is given as

$$N_{Rd} = 2 \times 200 \times 9\frac{355}{1,0} \times 10^{-3} = 1278\,kN.$$

This exceeds the values of N_{Ed} in the last line of Table 5.6.

Rigid end post:

Use Eqs (5.163)–(5.166) to determine the maximum panel size in accordance with cl 9.3.1(4) (EN 1993-1-5)

Determine χ_w from Eq. (5.163)

$$\chi_w = \frac{V_{Ed}\sqrt{3}\gamma_{M1}}{f_{yw}h_w t} = \frac{2320 \times 10^3 \sqrt{3} \times 1,0}{355 \times 1930 \times 9} = 0,652$$

TABLE 5.6 Shear capacity calculations with rigid end post (Example 5.3).

Distance from support (m)	1,45	3,25	6,0	10
Panel width (m)	1,45	1,80	2,75	4,00
Moment M_{Ed} (kNm)	3120	6315	9744	11 600
Shear V_{Ed} (kN)	2320	1984	1566	928
c (mm)	405	503	768	1117
χ_f	0,141	0,089	0,029	0,005
a/h_w	0,751	0,933	1,425	2,073
k_τ	13,46	10,14	7,310	6,271
$\bar{\lambda}_w$	1,920	2,213	2,607	2,814
χ_w	0,523	0,470	0,414	0,390
χ_v	0,664	0,559	0,443	0,395
V_{Rd} (kN)	2363	1991	1578	1407
Stiffener force N_{Rd} (kN)	1118	758	154	−826

The required normalized web slenderness λ_w from Eq. (5.164) is given by

$$\bar{\lambda}_w = \frac{0,83}{\chi_w} = \frac{0,83}{0,652} = 1,273$$

The buckling parameter k_τ is given by Eq. (5.165) as

$$k_\tau = \left(\frac{h_w}{37,4t\varepsilon\bar{\lambda}_w}\right)^2 = \left(\frac{1930}{37,4 \times 9 \times \sqrt{\frac{235}{355}} \times 1,273}\right)^2 = 30,65$$

As $a/h_w < 1$, k_τ is given by Eq. (5.166), or the required panel width a is given as

$$a = h_w\sqrt{\frac{5,35}{k_\tau - 4,00}} = 1930\sqrt{\frac{5,35}{30,65 - 4}} = 865\,\text{mm}$$

Centre to centre distance of the pair of double stiffeners should be greater than $0,1\,h_w$ (=193 mm). Use a centre to centre distance of 300 mm (The shear resistance will be greater than 2320 kN).

Minimum cross-sectional area of each plate is

$$\frac{4h_w t^2}{e} = \frac{4 \times 1930 \times 9^2}{300} = 2084\,\text{mm}^2$$

As the beam is 500 mm wide use an end plate 500 wide, thus the required thickness is 2084/500 (=4 mm). As for the non-rigid end post, a stiffener 500 mm by 15 mm suffices for resisting the reaction, use the same here.

The layout of the beam whose self-weight is 93 kN is given in Fig 5.24 (b). The self-weight per unit length of the beam is 4,56 kN/m for the non-rigid end post and 4,51 for the rigid end post (i.e. almost identical). It would be possible to further optimize the weight of the girder by decreasing the web thickness towards the centre and the flange width and/or thickness towards the ends of the beam. Note, however, that so doing may also increase the stiffener requirements.

EXAMPLE 5.4 Design of a plate girder with lateral torsional buckling.

Design a plate girder in Grade S355 steel to carry a characteristic variable load of 1500 kN at the centre of a span of 20 m. Lateral torsional restraints exist at the supports and the load.

To calculate the imposed bending moment, assume the total weight of the beam is 150 kN.

$$M_{Ed} = \frac{1,35 \times 150 \times 20}{8} + \frac{1,5 \times 1500 \times 20}{4} = 11760\,\text{kNm}$$

To size the beam allowing for the fact that its classification will be Class 4 (needing the determination of an effective section) and lateral torsion buckling will occur, increase the moment by 50% and use the optimization equations for Class 3.

$$M_{Ed} = \frac{1,5 \times 1500 \times 20}{4} + \frac{1,35 \times 200 \times 20}{8} = 11925 \text{ kNm}$$

Approximate design:

As the moment resistance is calculated on an elastic resistance utilization, the factor k in Eq. (5.110) may be taken as 0,55, with the flange area equal to half the web area for an optimal design under elastic stress distribution (Eq. (5.139)), so

$$\lambda = \frac{h_w}{t} = k\frac{E}{f_{yf}}\sqrt{\frac{A_w}{A_f}} = 0,55\frac{210 \times 10^3}{355}\sqrt{2} = 460$$

To determine the optimal solution increase M_{Ed} by 50% to $1,5 \times 11925$ (=17888 kNm).

From Eq. (5.135), t is given by

$$t = \sqrt[3]{\frac{3M_{Ed}}{2\lambda^2 f_{yd}}} = \sqrt[3]{\frac{3 \times 17888 \times 10^6}{2 \times 460^2 \times 355}} = 7,1 \text{ mm}$$

Use 8 mm plate.

Determine h_w from Eq. (5.136)

$$h_w = \sqrt{\frac{3\lambda M_{Ed}}{2f_{yd}}} = \sqrt{\frac{3 \times 460 \times 17888 \times 10^6}{2 \times 355}} = 3264 \text{ mm}$$

As the plate thickness has been rounded up from 7,1 to 8 mm, then the web height may be rounded down to 3240 mm.

The area of the flange A_f is given by Eq. (5.139)

$$A_f = \frac{A_w}{2} = \frac{8 \times 3240}{2} = 12960 \text{ mm}^2$$

Use the limit of the flange outstand as a Class 1 section (i.e. $7,33t_f$) so the area of the flange is $14,66t_f^2$. Thus

$$14,66t_f^2 = 12960, \text{ or } t_f = 29,7 \text{ mm}$$

Use a flange plate 30 mm thick.

$$b_f \approx 14,66t_f = 14,66 \times 30 = 440 \text{ mm}$$

Use a flange plate 500 mm wide ($A_f = 15000 \text{ mm}^2$).

The reason for keeping the flange classification as low as possible is in order to mobilize as much of the flange capacity as possible to resist shear.

Actual web slenderness

$$\frac{h_w}{t} = \frac{3240}{8} = 405$$

Allowable web slenderness:

$$\frac{h_w}{t} = k\frac{E}{f_{yf}}\sqrt{\frac{A_w}{A_f}} = 0.55\frac{210 \times 10^3}{355}\sqrt{\frac{3240 \times 8}{500 \times 30}} = 427$$

Thus flange-induced buckling will not occur. From the actual web slenderness the web (and therefore the complete section) is Class 4.

Determination of effective section properties (cl 4.4 EN 1993-1-5):

The process is iterative as the amount of web not considered is a function of the stresses at the top and bottom of the web. Note, the calculation takes compressive stresses *positive* rather than the normal stress analysis convention of compressive stresses *negative*.

First Iteration:

Determine the stresses at the top and bottom of the web with no loss of section:

Gross Area, A

$$A = 2b_f t_f + h_w t = 2 \times 500 \times 30 + 3240 \times 8 = 55920\,\text{mm}^2$$

Gross I_y:

$$I_y = \frac{1}{12}[b(h_w + 2t_f)^3 - (b - t)h_w^3] = \frac{1}{12}[500 \times 3300^3 - 492 \times 3240^3]$$

$$= 102,88 \times 10^9\,\text{mm}^4$$

Stress at the top of the web, σ_1:

$$\sigma_1 = \frac{11\,925 \times 10^6 \times 1620}{102,88 \times 10^9} = 188\,\text{MPa}$$

Stress at the top of the web, σ_2:

$$\sigma_2 = -\frac{11\,925 \times 10^6 \times 1620}{102,88 \times 10^9} = -188\,\text{MPa}$$

Determine the stress ratio, ψ:

$$\psi = \frac{\sigma_2}{\sigma_1} = \frac{-188}{188} = -1,0$$

From Table 4.1 of EN 1993-1-5, the buckling factor $k_\sigma = 23,9$ for $\psi = -1,0$.

The normalized slenderness ratio $\bar{\lambda}_p$ is given by

$$\bar{\lambda}_p = \frac{\bar{b}}{28,4t\varepsilon\sqrt{k_\sigma}} = \frac{3240}{28,4 \times 8\sqrt{\frac{235}{355}}\sqrt{23,9}} = 3,585$$

where $\bar{b}$ is the depth of the web.

The reduction factor ρ for an internal compression member is given by

$$\rho = \frac{\bar{\lambda}_{\mathrm{p}} - 0{,}055(3 + \psi)}{(\bar{\lambda}_{\mathrm{p}})^2} = \frac{3{,}585 - 0{,}055(3 + (-1))}{3{,}585^2} = 0{,}271$$

The effective depth b_{eff} is given by

$$b_{\mathrm{eff}} = \frac{\rho h_{\mathrm{w}}}{1 - \psi} = \frac{0{,}271 \times 3240}{1 - (-1)} = 439\,\mathrm{mm}$$

The depth of web left at the top b_{e1}:

$$b_{\mathrm{e1}} = 0{,}4 b_{\mathrm{eff}} = 0{,}4 \times 439 = 176\,\mathrm{mm}$$

The depth of web left at the bottom (above the centroidal axis), b_{e2}:

$$b_{\mathrm{e1}} = 0{,}6 b_{\mathrm{eff}} = 0{,}6 \times 439 = 263\,\mathrm{mm}$$

The ineffective portion of web has a length $l_{\mathrm{w}} = 1620 - 176 - 263 = 1181\,\mathrm{mm}$

The net loss of area of the web, A_{w} is given by

$$A_{\mathrm{w}} = 1181 \times 8 = 9448\,\mathrm{mm}^2$$

Effective area of section, A_{eff}:

$$A_{\mathrm{eff}} = A - A_{\mathrm{w}} = 59920 - 9448 = 50472\,\mathrm{mm}^2$$

Position of effective centroid, z_{eff}:

$$z_{\mathrm{eff}} = \frac{A\frac{h}{2} - A_{\mathrm{w}}\left(z_{\mathrm{eff,prev}} + b_{\mathrm{e2}} + \frac{l_{\mathrm{w}}}{2}\right)}{A_{\mathrm{eff}}}$$

$$= \frac{59920\frac{3300}{2} - 9448\left(\frac{3300}{2} + 263 + \frac{1181}{2}\right)}{50472} = 1490\,\mathrm{mm}$$

Note, $z_{\mathrm{eff,prev}}$ is the neutral axis position at the previous iteration. For the first iteration, $z_{\mathrm{eff,prev}} = h/2$.

Effective second moment of area, $I_{y,\mathrm{eff}}$:

$$I_{y,\mathrm{eff}} = I_y + A\left(\frac{h}{2} - z_{\mathrm{eff}}\right)^2 - \frac{t l_{\mathrm{w}}^3}{12} - A_{\mathrm{w}}\left(z_{\mathrm{eff,prev}} + b_{\mathrm{e2}} + \frac{l_{\mathrm{w}}}{2} - z_{\mathrm{eff}}\right)^2$$

$$= 102{,}88 \times 10^9 + 59920\left(\frac{3300}{2} - 1490\right)^2 - \frac{8 \times 1181^3}{12}$$

$$- 9448\left(\frac{3300}{2} + 263 + \frac{1181}{2} - 1490\right)^2$$

$$= 102{,}88 \times 10^9 + 1{,}53 \times 10^9 - 1{,}10 \times 10^9 - 9{,}70 \times 10^9 = 93{,}61 \times 10^9\,\mathrm{mm}^4$$

Stress at the top of the web, σ_1:

$$\sigma_1 = \frac{11925 \times 10^6 \times (3270 - 1490)}{93{,}61 \times 10^9} = 227\,\text{MPa}$$

Stress at the bottom of the web, σ_2:

$$\sigma_2 = -\frac{11925 \times 10^6 \times (1490 - 30)}{93{,}61 \times 10^9} = -186\,\text{MPa}$$

Determine the stress ratio, ψ:

$$\psi = \frac{\sigma_2}{\sigma_1} = \frac{-186}{227} = -0{,}819$$

From Table 4.1 of EN 1993-1-5, the buckling factor k_σ for $\psi = -0{,}794$ is given by

$$k_\sigma = 7{,}81 - 6{,}29\psi + 9{,}78\psi^2 = 7{,}81 - 6{,}29(-0{,}819) + 9{,}78(-0{,}819)^2 = 19{,}5$$

The normalized slenderness ratio $\bar{\lambda}_p$ is given by

$$\bar{\lambda}_p = \frac{\bar{b}}{28{,}4t\varepsilon\sqrt{k_\sigma}} = \frac{3240}{28{,}4 \times 8\sqrt{\frac{235}{355}}\sqrt{19{,}5}} = 3{,}97$$

where $\bar{b}$ is the depth of the web.

The reduction factor ρ for an internal compression member is given by

$$\rho = \frac{\bar{\lambda}_p - 0{,}055(3 + \psi)}{(\bar{\lambda}_p)^2} = \frac{3{,}97 - 0{,}055(3 + (-0{,}819))}{3{,}97^2} = 0{,}244$$

The effective depth b_{eff} is given by

$$b_{\text{eff}} = \frac{\rho h_w}{1 - \psi} = \frac{0{,}244 \times 3240}{1 - (-0{,}819)} = 435\,\text{mm}$$

The depth of web left at the top b_{e1}:

$$b_{e1} = 0{,}4b_{\text{eff}} = 0{,}4 \times 435 = 174\,\text{mm}$$

The depth of web left at the bottom (above the centroidal axis), b_{e2}:

$$b_{e1} = 0{,}6b_{\text{eff}} = 0{,}6 \times 435 = 261\,\text{mm}$$

The ineffective portion of web has a length $l_w = (3240 - 1490) - 174 - 261 = 1315\,\text{mm}$

The net loss of area of the web, A_w is given by

$$A_w = 1315 \times 8 = 10520\,\text{mm}^2$$

Effective area of section, A_{eff}:

$$A_{\text{eff}} = A - A_w = 59920 - 10520 = 49400\,\text{mm}^2$$

Position of effective centroid, z_{eff}:

$$z_{\text{eff}} = \frac{A\frac{h}{2} - A_{\text{w}}\left(z_{\text{eff,prev}} + b_{\text{e}2} + \frac{l_{\text{w}}}{2}\right)}{A_{\text{eff}}}$$

$$= \frac{59920\frac{3300}{2} - 10520\left(1490 + 261 + \frac{1315}{2}\right)}{49400} = 1488\,\text{mm}$$

Effective second moment of area, $I_{y,\text{eff}}$:

$$I_{y,\text{eff}} = I_y + A\left(\frac{h}{2} - z_{\text{eff}}\right)^2 - \frac{t l_{\text{w}}^3}{12} - A_{\text{w}}\left(z_{\text{eff,prev}} + b_{\text{e}2} + \frac{l_{\text{w}}}{2} - z_{\text{eff}}\right)^2$$

$$= 102{,}88 \times 10^9 + 59920\left(\frac{3300}{2} - 1488\right)^2 - \frac{8 \times 1315^3}{12}$$

$$- 10520\left(1490 + 261 + \frac{1315}{2} - 1488\right)^2$$

$$= 102{,}88 \times 10^9 + 1{,}57 \times 10^9 - 1{,}52 \times 10^9 - 8{,}91 \times 10^9 = 94{,}02 \times 10^9\,\text{mm}^4$$

Stress at the top of the web, σ_1:

$$\sigma_1 = \frac{11925 \times 10^6 \times (3270 - 1488)}{94{,}02 \times 10^9} = 226\,\text{MPa}$$

Stress at the bottom of the web, σ_2:

$$\sigma_2 = -\frac{11925 \times 10^6 \times (1488 - 30)}{94{,}02 \times 10^9} = -185\,\text{MPa}$$

As these stresses are virtually identical to those on the first iteration, there is no need to continue.

Note, that as the changes in $b_{\text{e}1}$ and $b_{\text{e}2}$ were small the iteration could have stopped without further calculations of σ_1 and σ_2 ($\psi = -0{,}819$)

The lesser elastic section modulus $W_{\text{eff},y}$ is given as

$$W_{\text{eff},y} = \frac{I_{\text{eff}}}{z_{\text{eff}} - t_{\text{f}}} = \frac{94{,}02 \times 10^9}{1488 - 30} = 64{,}49 \times 10^6\,\text{mm}^3$$

$$M_{\text{Rd}} = W_{\text{eff},y}\frac{f_y}{\gamma_{\text{M0}}} = 64{,}49 \times 10^6\frac{355}{1{,}0} \times 10^{-6} = 22900\,\text{kNm}$$

Lateral torsional buckling check

Moment gradient factor, C_1 from Eq. (5.40), with $\psi = 0$:

$$\frac{1}{C_1} = 0{,}4\psi + 0{,}6 = 0{,}6$$

The gross section is used to calculate the section properties required for M_{cr}.

$$I_z = \frac{1}{12}[2t_f b_f^3 + h_w t^3] = \frac{1}{12}[2 \times 0,030 \times 0,5^3 + 3,24 \times 0,008^3]$$

$$= 0,625 \times 10^{-3}\,\text{m}^4$$

$$I_w = I_z \frac{h_s^2}{4} = 0,625 \times 10^{-3} \frac{3,270^2}{4} = 1.67 \times 10^{-3}\,\text{m}^6$$

Note: h_s is the distance between the centroids of the flanges.

The torsional second moment of area may be calculated on the thin plate assumption.

$$I_t = \frac{1}{3}[2b_f t_f^3 + h_w t^3] = \frac{1}{3}[2 \times 0,5 \times 0,030^3 + 3,24 \times 0,008^3] = 9,55 \times 10^{-6}\,\text{m}^4$$

Use Eq. (5.5) to determine M_{cr}

$$M_{cr} = \frac{\pi^2 E I_z}{L^2} \left[\frac{I_w}{I_z} + \frac{L^2 G I_t}{\pi^2 E I_z} \right]^{1/2}$$

$L = 10\,\text{m}.$

$$\frac{\pi^2 E I_z}{L^2} = \frac{\pi^2 \times 210 \times 10^6 \times 0,625 \times 10^{-3}}{10^2} = 12954\,\text{kN}$$

$$G I_t = 81 \times 10^6 \times 9,55 \times 10^{-6} = 774\,\text{kNm}^2$$

$$\frac{G I_t}{\frac{\pi^2 E I_z}{L^2}} = \frac{774}{12954} = 0,060\,\text{m}^2$$

$$\frac{I_w}{I_z} = \frac{1,67 \times 10^{-3}}{0,625 \times 10^{-3}} = 2,672\,\text{m}^2$$

$$M_{cr} = 12954[2,672 + 0,060]^{1/2} = 21410\,\text{kNm}$$

From Eq. (5.12), $\bar{\lambda}_{LT}$ is given by

$$\bar{\lambda}_{LT} = \sqrt{\frac{f_y W_{\text{eff},y}}{C_1 M_{cr}}} = \sqrt{0,6 \frac{355 \times 64,49 \times 10^6 \times 10^{-6}}{21410}} = 0,801$$

$h/b > 2$, so $\alpha_{LT} = 0,76$

$$\Phi_{LT} = 0,5[1 + \alpha_{LT}(\bar{\lambda}_{LT} - 0,2) + (\bar{\lambda}_{LT})^2]$$

$$= 0,5[1 + 0,76(0,801 - 0,2) + 0,801^2] = 1,049$$

$$\chi_{LT} = \frac{1}{\Phi_{LT} + (\Phi_{LT}^2 - (\bar{\lambda}_{LT})^2)^{1/2}} = \frac{1}{1,049 + (1,049^2 - 0,801^2)^{1/2}} = 0,579$$

$$M_{b,Rd} = \chi_{LT} W_{\text{eff},y} \frac{f_y}{\gamma_{M1}} = 0,579 \times 64,49 \times 10^6 \frac{355}{1,0} \times 10^{-6} = 13260\,\text{kNm}$$

This exceeds the applied moment of 11925 kN.

The beam is therefore satisfactory for flexure.

Deflection check:

$$I_{\text{gross}} = 102,88 \times 10^9 \text{ mm}^4$$

For a worst case scenario assume that all the variable load contributes to the deflection,

$$\delta = \frac{1}{48}\frac{WL^3}{EI} = \frac{1}{48}\frac{1500 \times 20^3}{210 \times 10^6 \times 102,88 \times 10^{-3}} = 0,012 \text{ m}$$

This is equivalent to span/1667, which is satisfactory.

Web design:

As the web carries compression due to flexure, then the following interaction equation must be satisfied (cl 7.1(1), EN 1993-1-5),

$$\eta_1 + \left(1 - \frac{M_{\text{f,Rd}}}{M_{\text{pl,Rd}}}\right)(2\eta_3 - 1)^2 \le 1,0 \tag{5.182}$$

where $M_{\text{f,Rd}}$ is the plastic moment resistance of the flanges and $M_{\text{pl,Rd}}$ is the plastic moment of resistance of the section (both calculations are irrespective of classification of the section). The shear contribution factor η_3 is defined as (cl. 5.5 (1), EN 1993-1-5)

$$\eta_3 = \frac{V_{\text{Ed}}}{\chi_v h_w t[(f_{\text{yw}}/\sqrt{3})/\gamma_{\text{M1}}]} \le 1,0 \tag{5.183}$$

as there is no axial force the equation for η_1 from cl 4.6 (1) (EN 1993-1-5) reduces to

$$\eta_1 = \frac{M_{\text{ed}}}{\frac{f_y W_{\text{eff}}}{\gamma_{\text{M0}}}} \le 1,0 \tag{5.184}$$

If $\eta_3 < 0,5$ there is no reduction in moment capacity (cl 7.1 (1), EN 1993-1-5)

$$M_{\text{pl,Rd}} = W_{\text{pl}}\frac{f_y}{\gamma_{\text{M0}}} = \frac{500 \times 3300^2 - 492 \times 3240^2}{4}\frac{355}{1,0} \times 10^{-6} = 24870 \text{ kNm}$$

$$M_{\text{f,Rd}} = A_f h_s \frac{f_y}{\gamma_{\text{M0}}} = 500 \times 30 \times (3300 - 30)\frac{355}{1,0} \times 10^{-6} = 17410 \text{ kNm}$$

$$\left(1 - \frac{M_{\text{f,Rd}}}{M_{\text{pl,Rd}}}\right) = 1 - \frac{17410}{24870} = 0,3$$

Flexible end post:

Three intermediate stiffeners are required at 2,8, 5,6 and 7,8 m from the support, together with a load bearing stiffener at the centre.

Stiffener at 2,8 m:

$$M_{\text{Ed}} = 1260 \times 2,8 - 6,75 \times 2,8^2 = 3475 \text{ kNm}$$

$$V_{\text{Ed}} = 1260 \text{ kN (at support)}$$

Determine c from Eq. (5.156)

$$c = a \left(0{,}25 + \frac{1{,}6 b_f t_f^2 f_{yf}}{t h_w^2 f_{yw}} \right) = 2800 \left(0{,}25 + \frac{1{,}6 \times 500 \times 30^2 \times 355}{8 \times 3240^2 \times 355} \right) = 724 \, \text{mm}$$

Determine the flange contribution factor χ_f from Eq. (5.155)

$$\chi_f = \frac{b_f t_f^2 f_{yf} \sqrt{3}}{c t h_w f_{yw}} \left[1 - \left(\frac{M_{Ed}}{M_{f,Rd}} \right)^2 \right]$$

$$= \frac{500 \times 30^2 \times 355 \times \sqrt{3}}{724 \times 8 \times 3240 \times 355} \left[1 - \left(\frac{3475}{17410} \right)^2 \right] = 0{,}040$$

$$\frac{h_w}{a} = \frac{3240}{2800} = 1{,}157$$

$$\frac{a}{h_w} = 0{,}864$$

For $a/h_w < 1{,}0$, k_τ is given by Eq. (5.143), or

$$k_\tau = 4 + 5{,}34 \left(\frac{h_w}{a} \right)^2 = 4 + 5{,}34 \times 1{,}157^2 = 11{,}15$$

Determine the normalized web slenderness ratio $\bar{\lambda}_w$ from Eq. (5.145)

$$\bar{\lambda}_w = \frac{h_w}{37.4 \varepsilon t \sqrt{k_\tau}} = \frac{3240}{37.4 \sqrt{\frac{235}{355}} \times 8 \sqrt{11.15}} = 3.985$$

As $\bar{\lambda}_w > 1{,}08$, χ_w is given by Eq. (5.149) as

$$\chi_w = \frac{0{,}83}{\bar{\lambda}_w} = \frac{0{,}83}{3{,}985} = 0{,}208$$

From Eq. (5.154), the total shear co-efficient χ_v is given by

$$\chi_v = \chi_f + \chi_w = 0{,}040 + 0{,}208 = 0{,}248$$

The shear capacity $V_{b,Rd}$ is determined from Eq. (5.153)

$$V_{b,Rd} = \chi_v h_w t \frac{\frac{f_{yw}}{\sqrt{3}}}{\gamma_{M1}} = \frac{0{,}248 \times 3240 \times 8 \frac{355}{\sqrt{3}}}{1000} \frac{}{1{,}0} = 1318 \, \text{kN}$$

$$\eta_3 = \frac{V_{Ed}}{\chi_v h_w t [(f_{yw}/\sqrt{3})/\gamma_{M1}]} = \frac{1260}{1318} = 0{,}956$$

As $\eta_3 > 0{,}5$ the interaction equation between moment and shear must be considered (Eq. (5.152))

$$\eta_1 = \frac{M_{ed}}{\frac{f_y W_{eff}}{\gamma_{M0}}} = \frac{3475}{22900} = 0{,}152$$

TABLE 5.7 Shear capacity calculations with non-rigid end post (Example 5.4).

Distance from support (m)	2,8	5,6	7,8	10
Panel width (m)	2,8	2,8	2,2	2,2
Moment M_{Ed} (kNm)	3475	6844	9417	11 925
Shear V_{Ed} (kN)	1260	1241	1222	1207
c (mm)	724	724	568	568
χ_f	0,040	0,035	0,038	0,028
a/h_w	0,864	0,864	0,679	0,679
k_τ	11,15	11,15	15,58	15,58
$\overline{\lambda}_w$	3,990	3,990	3,371	3,371
χ_w	0,208	0,208	0,246	0,246
χ_v	0,248	0,243	0,284	0,274
V_{Rd} (kN)	1318	1293	1506	1457
η_1	0,152	0,209	0,411	0,521
η_3	0,956	0,960	0,811	0,829
Interaction equation	0,401	0,553	0,527	0,650
Stiffener force (kN)	1079	210	−134	−932

$$\eta_1 + \left(1 - \frac{M_{f,Rd}}{M_{pl,Rd}}\right)(2\eta_3 - 1)^2 = 0,152 + 0,3(2 \times 0,956 - 1)^2 = 0,402 \leq 1,0$$

The calculations for the remaining stiffeners are carried out in Table 5.7, together with the calculation of the forces on the stiffeners.

Intermediate stiffeners:

For all the panels, $a \leq \sqrt{2}h_w$ (=$3,24\sqrt{2}=4,58$ m), thus based on the lesser value of a (=2,2 m) the stiffener requirement, I_{st} is given by

$$I_{st} = 1,5\frac{h_w^3 t^3}{a^2} = 1,5\frac{3240^3 \times 8^3}{2200^2} = 5,4 \times 10^6 \text{ mm}^4$$

Use 8 mm plate, so

$$b = \sqrt[3]{\frac{12 \times 5,4 \times 10^6}{8}} = 200 \text{ mm}$$

Use double intermediate stiffeners of 200×8 mm plates either side of the web.

Load capacity of stiffeners N_{Rd} (with no buckling):

$$N_{Rd} = A_{st}\frac{f_y}{\gamma_{M0}} = 2 \times 200 \times 8\frac{355}{1,0} \times 10^{-3} = 1136 \text{ kN}$$

This exceeds the maximum stiffener force of 1079 kN.

As in earlier examples strut buckling is not critical, it will not be checked in this example.

End post:

$$V_{Ed} = 1260 \text{ kN}$$

Use a 500 wide by 10 mm thick plate.

The section must be checked for buckling, but a proportion of the web may be taken into account.

Length of web $= 15\varepsilon t = 15 \times 8\sqrt{(235/355)} = 98\,\text{mm}$

From Eq. (5.177) A_{equiv} is given by

$$A_{\text{equiv}} = A_{\text{st}} + 15\varepsilon t^2 = 500 \times 10 + 98 \times 8 = 5784\,\text{mm}^2$$

and I_{equiv} by

$$I_{\text{equiv}} = I_{\text{st}} + \frac{1}{12}15\varepsilon t^4 = \frac{10 \times 500^3}{12} + \frac{98 \times 8^3}{12} = 0{,}104 \times 10^9\,\text{mm}^4$$

$$i = \sqrt{\frac{I_{\text{equiv}}}{A_{\text{equiv}}}} = \sqrt{\frac{0{,}104 \times 10^9}{5784}} = 134{,}1\,\text{mm}$$

Use an effective length of $0{,}75h_{\text{w}} = 0{,}75 \times 3240 = 2430\,\text{mm}$

$$N_{\text{cr}} = \frac{\pi^2 EI}{l^2} = \frac{210 \times 0{,}104 \times 10^9 \pi^2}{2430^2} = 36500\,\text{kN}$$

$$\overline{\lambda} = \sqrt{\frac{Af_y}{N_{\text{cr}}}} = \sqrt{\frac{5784 \times 355}{36500 \times 10^3}} = 0{,}237$$

Use buckling curve 'c' ($\alpha = 0{,}49$)

$$\Phi = 0{,}5[1 + \alpha(\overline{\lambda} - 0{,}2) + (\overline{\lambda})^2] = 0{,}5[1 + 0{,}49(0{,}237 - 0{,}2) + 0{,}237^2] = 0{,}537$$

$$\chi = \frac{1}{\Phi + \sqrt{\Phi^2 - (\overline{\lambda})^2}} = \frac{1}{0{,}537 + \sqrt{0{,}537^2 - 0{,}237^2}} = 0{,}981$$

$$N_{\text{Rd}} = \chi A \frac{f_y}{\gamma_{\text{M0}}} = 0{,}981 \times 5784 \frac{355}{1{,}0} \times 10^{-3} = 2014\,\text{kN}$$

This exceeds the reaction of 1260 kN.

Check cl 9.2.1 (7)

Determine I_{p}:

The second moment of area about the web centre line, I_y:

$$I_y = \frac{bh^3}{12} = \frac{10 \times 500^3}{12} = 0{,}104 \times 10^9\,\text{mm}^4$$

The second moment of area normal to the web centre line about one edge, I_y:

$$I_y = \frac{bh^3}{3} = \frac{500 \times 10^3}{3} = 0{,}042 \times 10^6\,\text{mm}^4$$

$$I_{\text{p}} = I_x + I_y = 0{,}104 \times 10^9 + 0{,}042 \times 10^6 = 0{,}104 \times 10^9\,\text{mm}^4$$

$$I_{\text{T}} = \frac{bh^3}{3} = 0{,}042 \times 10^6\,\text{mm}^4$$

$$\frac{I_{\text{T}}}{I_{\text{p}}} = \frac{0{,}042 \times 10^6}{0{,}104 \times 10^9} = 0{,}4 \times 10^{-3}$$

Limiting value:

$$5{,}3\frac{f_y}{E} = 5{,}3\frac{355}{210 \times 10^3} = 8{,}96 \times 10^{-3}$$

The actual value is less than the limiting value thus the stiffener size must be increased.

A 25 mm thick end plate will satisfy the limiting stiffness criterion ($I_T/I_P = 0{,}01$) (and will clearly satisfy the strength and buckling criteria).

Central stiffener:

$$N_{Ed} = 1{,}5 \times 1500 = 2250 \,\text{kN}$$

Use two 240 wide by 15 mm thick plates.

The section must be checked for buckling, but a proportion of the web may be taken into account.

Length of web $= 30\varepsilon t = 30 \times 8\sqrt{(235/355)} = 195\,\text{mm}$

$$A_{equiv} = A_{st} + 30\varepsilon t^2 = 2 \times 240 \times 15 + 195 \times 8 = 8760\,\text{mm}^2$$

$$I_{sequiv} = I_{st} + 30\varepsilon t^4 = \frac{15 \times (240 + 8 + 240)^3}{12} + \frac{195 \times 8^3}{12} = 0{,}145 \times 10^9\,\text{mm}^4$$

$$i = \sqrt{\frac{I_{equiv}}{A_{equiv}}} = \sqrt{\frac{0{,}145 \times 10^9}{8760}} = 128{,}7\,\text{mm}$$

Use an effective length of $0{,}75\,h_w = 0{,}75 \times 3240 = 2430\,\text{mm}$

$$N_{cr} = \frac{\pi^2 EI}{l^2} = \frac{210 \times 0{,}145 \times 10^9 \pi^2}{2430^2} = 50895\,\text{kN}$$

$$\bar{\lambda} = \sqrt{\frac{Af_y}{N_{cr}}} = \sqrt{\frac{8760 \times 355}{50895 \times 10^3}} = 0{,}247$$

Use buckling curve 'c' ($\alpha = 0{,}49$):

$$\Phi = 0{,}5[1 + \alpha(\bar{\lambda} - 0{,}2) + (\bar{\lambda})^2] = 0{,}5[1 + 0{,}49(0{,}247 - 0{,}2) + 0{,}247^2] = 0{,}542$$

$$\chi = \frac{1}{\Phi + \sqrt{\Phi^2 - (\bar{\lambda})^2}} = \frac{1}{0{,}542 + \sqrt{0{,}542^2 - 0{,}247^2}} = 0{,}976$$

$$N_{Rd} = \chi A \frac{f_y}{\gamma_{M0}} = 0{,}976 \times 8760 \frac{355}{1{,}0} \times 10^{-3} = 3035\,\text{kN}$$

This exceeds the applied load of 2250 kN.

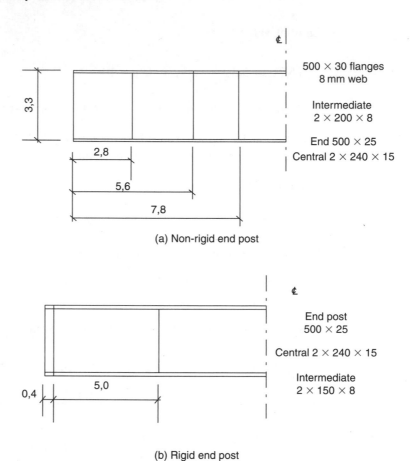

(a) Non-rigid end post

(b) Rigid end post

FIGURE 5.25 Final layout for the girder of Example 5.4

Flange to web welds:

From Table 5.7, $V_{Ed} > h_w t [(f_{yw}/\sqrt{3})/\gamma_{M1}]$ as the flange contribution χ_f has been mobilized. Thus welds should be designed for a shear flow of

$$\eta t \frac{\frac{f_{yw}}{\sqrt{3}}}{\gamma_{M1}} = 1.2 \times 8 \frac{\frac{355}{\sqrt{3}}}{1.0} = 1968\,\text{N/mm}$$

The final layout is given in Fig 5.25 (a). The self-weight of the beam is 101 kN (or 5,05 kN/m).

(b) Rigid end post

First panel from the support:

Intermediate stiffener is 5 m from the support.

$$M_{Ed} = 1260 \times 5 - 6,75 \times 5^2 = 6131\,\text{kNm}$$

$$V_{Ed} = 1260\,\text{kN (at the support)}$$

Determine c from Eq. (5.156)

$$c = a\left(0{,}25 + \frac{1{,}6 b_f t_f^2 f_{yf}}{t h_w^2 f_{yw}}\right) = 5000\left(0{,}25 + \frac{1{,}6 \times 500 \times 30^2 \times 355}{8 \times 3240^2 \times 355}\right) = 1293 \text{ mm}$$

Determine the flange contribution factor χ_f from Eq. (5.155)

$$\chi_f = \frac{b_f t_f^2 f_{yf} \sqrt{3}}{c t h_w f_{yw}}\left[1 - \left(\frac{M_{Ed}}{M_{f,Rd}}\right)^2\right]$$

$$= \frac{500 \times 30^2 \times 355 \times \sqrt{3}}{1293 \times 8 \times 3240 \times 355}\left[1 - \left(\frac{6131}{17410}\right)^2\right] = 0{,}020$$

$$\frac{h_w}{a} = \frac{3240}{5000} = 0{,}648$$

$$\frac{a}{h_w} = 1{,}543$$

For $a/h_w > 1{,}0$, k_τ is given by Eq. (5.142), or

$$k_\tau = 5{,}34 + 4\left(\frac{h_w}{a}\right)^2 = 5{,}34 + 4 \times 0{,}648^2 = 7{,}02$$

Determine the normalized web slenderness ratio $\bar{\lambda}_w$ from Eq. (5.145)

$$\bar{\lambda}_w = \frac{h_w}{37{,}4 \varepsilon t \sqrt{k_\tau}} = \frac{3240}{37{,}4\sqrt{\frac{235}{355}} \times 8\sqrt{7{,}02}} = 5{,}02$$

As $\bar{\lambda}_w > 1{,}08$, χ_w is given by Eq. (5.149) as

$$\chi_w = \frac{1{,}37}{0{,}7 + \bar{\lambda}_w} = \frac{1{,}37}{0{,}7 + 5{,}02} = 0{,}240$$

From Eq. (5.154), the total shear co-efficient χ_v is given by

$$\chi_v = \chi_f + \chi_w = 0{,}020 + 0{,}240 = 0{,}260$$

The shear capacity $V_{b,Rd}$ is determined from Eq. (5.153)

$$V_{b,Rd} = \chi_v h_w t \frac{\frac{f_{yw}}{\sqrt{3}}}{\gamma_{M1}} = \frac{0{,}260 \times 3240 \times 8 \frac{355}{\sqrt{3}}}{1000 \quad 1{,}0} = 1380 \text{ kN}$$

$$\eta_3 = \frac{V_{Ed}}{\chi_v h_w t [(f_{yw}/\sqrt{3})/\gamma_{M1}]} = \frac{1260}{1380} = 0{,}913$$

As $\eta_3 > 0,5$ an interaction equation between moment and shear must be considered,

$$\eta_1 + \left(1 - \frac{M_{\text{f,Rd}}}{M_{\text{pl,Rd}}}\right)(2\eta_3 - 1)^2 \leq 1,0$$

$$\eta_1 = \frac{M_{\text{ed}}}{\dfrac{f_y W_{\text{eff}}}{\gamma_{\text{M0}}}} = \frac{6131}{22900} = 0,268$$

$$\eta_1 + \left(1 - \frac{M_{\text{f,Rd}}}{M_{\text{pl,Rd}}}\right)(2\eta_3 - 1)^2 = 0,268 + 0,3(2 \times 0,913 - 1)^2 = 0,473 \leq 1,0$$

The resultant spacing of stiffeners and a summary of the remaining calculations is given in Table 5.8. The final result for the calculation of stiffener forces are also given in Table 5.8.

Note, it is possible to eliminate the intermediate stiffener by increasing the web thickness to 9 mm. This may be more economic when considering the overall materials and fabrication costs.

The minimum stiffness requirement for both panels either side of the central stiffener is given by the case $a > \sqrt{2}h_{\text{w}}$, so design on the least value of a:

$$I_{\text{st}} = 0,75h_{\text{w}}t^2 = 0,75 \times 3240 \times 9^2 = 0,197 \times 10^6 \text{ mm}^4$$

Use 8 mm thick plate, then the total breadth of the stiffener b is given by

$$b = \sqrt[3]{\frac{12 \times 0,197 \times 10^6}{8}} = 67 \text{ mm}$$

Minimum area to carry a stiffener force of 757 kN is given as

$$A = \frac{757 \times 10^3}{355} = 2132 \text{ mm}^2$$

TABLE 5.8 Shear capacity calculations with rigid end post (Example 5,4).

Distance from support (m)	5	10
Panel width (m)	5	5
Moment M_{Ed} (kNm)	6131	11 925
Shear V_{Ed} (kN)	1260	1226
c (mm)	1293	1293
χ_{f}	0,020	0,012
a/h_{w}	1,543	1,543
k_τ	7,02	7,02
$\bar{\lambda}_{\text{w}}$	5,02	5,02
χ_{w}	0,239	0,239
χ_{v}	0,259	0,261
V_{Rd} (kN)	1380	1337
η_1	0,268	0,521
η_3	0,913	0,917
Interaction equation	0,473	0,729
Stiffener force N_{Rd} (kN)	757	−419

This would require a total width of $2132/8 = 267\,\text{mm}$. Use two plates $150\,\text{mm}$ by $8\,\text{mm}$ as intermediate stiffeners.

Design of rigid end post.

Use Eqs (5.163) to (5.166) to determine the maximum panel size in accordance with cl 9.3.1(4) (EN 1993-1-5)

Determine χ_w from Eq. (5.163)

$$\chi_w = \frac{V_{Ed}\sqrt{3}\gamma_{M1}}{f_{yw}h_w t} = \frac{1260 \times 10^3 \sqrt{3} \times 1.0}{355 \times 3240 \times 8} = 0{,}237$$

The required normalized web slenderness $\bar{\lambda}_w$ from Eq. (5.164) is given by

$$\bar{\lambda}_w = \frac{0{,}83}{\chi_w} = \frac{0{,}83}{0{,}237} = 3{,}502$$

The buckling parameter k_τ is given by Eq. (5.165) as

$$k_\tau = \left(\frac{h_w}{37{,}4t\varepsilon\bar{\lambda}_w}\right)^2 = \left(\frac{3240}{37{,}4 \times 8 \times \sqrt{\frac{235}{355}} \times 3{,}502}\right)^2 = 11{,}44$$

As $a/h_w < 1$, k_τ is given by Eq. (5.143), or the required panel width a is given as

$$a = h_w\sqrt{\frac{5{,}35}{k_\tau - 4{,}00}} = 3240\sqrt{\frac{5{,}35}{11{,}44 - 4}} = 2750\,\text{mm}$$

Centre to centre distance of the pair of double stiffeners should be greater than $0.1\,h_w$ ($=324\,\text{mm}$). Use a centre to centre distance of $400\,\text{mm}$ (the shear resistance will be greater than $1260\,\text{kN}$).

Minimum cross-sectional area of each plate is

$$\frac{4h_w t^2}{e} = \frac{4 \times 3240 \times 8^2}{400} = 2074\,\text{mm}^2$$

As the beam is $500\,\text{mm}$ wide use an end plate 500 wide, thus the required thickness is $2074/500$ ($=4\,\text{mm}$).

As for the non-rigid end post, a stiffener $500\,\text{mm}$ by $25\,\text{mm}$ is needed.

Stiffener under the point load:

The calculations are the same as those for the non-rigid end post.

The final layout is given in Fig 5.25 (b). The self-weight is $105\,\text{kN}$ ($5{,}06\,\text{kN/m}$)

REFERENCES

Anderson, J.M. and Trahair, N.S. (1972). Stability of mono-symmetric beams and cantilevers, *Journal of the Structural Division, Proceedings of the American Society of Civil Engineers*, **98**, 269–286.

Basler, K. (1961). Strength of plate girders in shear, *Journal of the Structural Division, Proceedings of the American Society of Civil Engineers*, **87**, 151–180.

Bradford, M.A. (1989). Buckling of beams supported on seats, *Structural Engineer*, **69(23)**, 411–414.

Bulson, P.S. (1970). *The stability of flat plates*. Chatto and Windus.

Chapman, J.C. and Buhagiar, D. (1993). Application of Young's buckling equation to design against torsional buckling. *Proceedings of the Institution of Civil Engineers, Structures and Buildings*, **99**, 359–369.

Davies, A.W. and Griffiths, D.S.C. (1999). Shear strength of steel plate girders, *Proceedings of the Institution of Civil Engineers*, **134**, 147–157.

Gibbons, C. (1995). Economic steelwork design, *Structural Engineer*, **73(15)**, 250–253.

Kaim, P. (2006). Buckling of members with rectangular hollow sections, In *Tubular structures XI* (eds Packer and Willibald). Taylor and Francis Group, 443–449.

Kirby, P.A. and Nethercot, D.A. (1979). *Design for structural stability*. Granada.

Mutton, B.R. and Trahair, N.S. (1973). Stiffness requirements for lateral bracing, *Journal of the Structural Division, ASCE*, **99**, 2167.

Nethercot, D.A. (1974a). Residual stresses and their influence upon the lateral buckling of rolled steel beams, *Structural Engineer*, **52(3)**, 89–96.

Nethercot, D.A. (1974b). Buckling of welded beams and girders, *IABSE*, **34**, 163–182.

Nethercot, D.A. and Lawson, R.M. (1992). *Lateral stability of steel beams and columns – common cases of restraint*, Publication 093. Steel Construction Institute.

Nethercot, D.A. and Rockey, K.C. (1971). A unified approach to the elastic critical buckling of beams, *Structural Engineer*, **49(7)**, 321–330.

Pillinger, A.H. (1988). Structural steelwork: a flexible approach to the design of joints in simple construction, *Structural Engineer*, **66(19)**, 316–321.

Porter, D.M., Rockey, K.C. and Evans, H.R. (1975). The collapse of plate girders loaded in shear, *Structural Engineer*, **53(8)**, 313–325.

Rees, D.W.A. (1990). *Mechanics of solids and structures*. McGraw Hill.

Rockey, K.C. and Škaloud, M. (1971). The ultimate load behaviour of plate girders loaded in shear, *IABSE Colloquium – design of plate girders for ultimate strength*. London, 1–19.

Rockey, K.C., Evans, H.R. and Porter, D.M. (1978). A design method for predicting the collapse behaviour of plate girders, *Proceedings of The Institution of Civil Engineers*, **65**, 85–112.

Rockey, K.C., Valtinat, G. and Tang, K.H. (1981). The design of transverse stiffeners on webs loaded in shear – an ultimate load approach, *Proceedings of The Institution of Civil Engineers*, **71(2)**, 1069–1099.

Rondal, J., Würker, K-G., Dutta, D., Wardenier, J. and Yeomans, N. (1992). *Structural stability of hollow sections*. Verlag TÜV Rheinland.

Timoshenko, S.T. and Gere, J.M. (1961). *Theory of elastic stability*. McGraw Hill.

Trahair, N.S. (1983). Lateral buckling of overhanging beams, In *Instability and plastic collapse of steel structures* (ed L.J. Morris). Granada, 503–518.

Trahair, N.S. and Bradford, M.A. (1988). *The behaviour and design of steel structures* (2nd Edition). Chapman and Hall.

BS 5950-1: *Structural use of steelwork in building – Part 1: Code of practice for design – Rolled and welded sections*. BSI.

EN 1993-1-1. *Eurocode 3: Design of steel structures, Part 1-1: General rules and rules for buildings*. CEN/BSI.

EN 1993-1-5. *Eurocode 3: Design of steel structures, Part 1-5: Plated structural elements*. CEN/BSI.

Chapter 6 / Axially Loaded Members

6.1 AXIALLY LOADED TENSION MEMBERS

A member subject to axial tension extends and tends to remain straight or, if there is a small initial curvature, to straighten out as the axial load is increased. Tension members (ties) occur in trusses, bracing and hangers for floor beams. A flat can be used as a tie, but this is generally impractical because it buckles if it goes into compression. Tie sections are therefore angles and tees for small loads and 'I' sections for larger loads. In situations where the load is not applied axially then the member is designed to resist an axial force plus a bending moment.

A tension member extends when subject to an axial load and is deemed to have failed when the yield or ultimate stress is reached. The failure load is independent of the length of the member which is in contrast to an axially loaded compression member which fails by buckling.

A member which is purely in tension does not buckle locally or overall and is therefore not affected by the classification of sections. The characteristic stress is not reduced, for design purposes, except by the material factor.

6.1.1 Angles as Tension Members (cl 4.13, EN 1993-1-8 (2005))

Generally angles are connected by one leg at the end of the member and this introduces an eccentric load. Where an angle is connected to one side of a gusset plate (as in a truss) bending moments are introduced in addition to the direct axial force. For a tension member these moments produce lateral deflections which reduce the eccentricity of the load near the middle of the member. Thus under increasing load the bending stresses become concentrated more towards the ends of the member. For angles connected by one leg the principal sectional axes are inclined to the plane containing the bending moment. Secondary deflections therefore occur normal to the plane of bending and, because of the restraints provided by the gusset plates, twisting also takes place.

Generally eccentrically loaded members are designed to resist an axial load and bending moment. However angle and tee experiments (Nelson (1953); Regan and Salter

(1984)) demonstrated that the above effects could be compensated for in design by reducing the cross-sectional area of the member. If there are holes then these also reduce the area of the cross-section.

Angles may be treated as axially loaded members provided that the net area is reduced to the effective area (cl 4.13, EN 1993-1-8 (2005)). For an equal angle, or an unequal angle connected by the larger leg the effective area is the gross area. For an unequal angle connected by the smaller leg the effective area is twice that of the smaller leg.

6.1.2 Design Value of a Tension Member (cl 6.2.3, EN 1993-1-1 (2005))

The design value of the tensile force N_{Ed} at each cross-section should satisfy (Eq. (6.5), EN 1993-1-1 (2005))

$$\frac{N_{\text{Ed}}}{N_{\text{t,Rd}}} \leq 1 \tag{6.1}$$

where $N_{\text{t,Rd}}$ is the design tension resistance taken as the smaller of:

(a) The design plastic resistance of the gross cross-section (Eq. (6.6), EN 1993-1-1 (2005))

$$N_{\text{pl,Rd}} = \frac{A f_y}{\gamma_{\text{M0}}} \tag{6.2a}$$

(b) The design ultimate resistance of the net cross-section at holes for fasteners (Eq. (6.7), EN 1993-1-1 (2005))

$$N_{\text{u,Rd}} = 0.9 \frac{A_{\text{net}} f_u}{\gamma_{\text{M2}}} \tag{6.2b}$$

(c) The design ultimate resistance of the net cross-section at holes for fasteners which are preloaded or non-preloaded (Eq. (6.8), EN 1993-1-1 (2005) and cl 3.4.2.1(1), EN 1993-1-8 (2005))

$$N_{\text{net,Rd}} = \frac{A_{\text{net}} f_y}{\gamma_{\text{M0}}} \tag{6.2c}$$

EXAMPLE 6.1 An angle in tension connected by one leg. A $100 \times 65 \times 6$ mm single angle tie is connected through the smaller leg by two 20 mm diameter bolts in line with a pitch of $2.5\,d_0$. Determine the design ultimate resistance of the angle assuming S275 steel and material factors of $\gamma_{\text{M2}} = 1.25$ and $\gamma_{\text{M0}} = 1$.

Net area of angle connected by the smaller leg and allowing for holes

$$A_{\text{net}} = (b - d_o)t + bt = (65 - 22) \times 6 + 65 \times 6 = 648\,\text{mm}^2$$

Compare this value with the gross area $= 1120\,\text{mm}^2$ (Section Tables).

Design ultimate tensile resistance of the net cross-section for a two bolt angle connection (Eq. (6.7), EN 1993-1-1 (2005))

$$N_{u,Rd} = 0,9 \frac{A_{net} f_u}{\gamma_{M2}} = \frac{0,9 \times 648 \times 430}{1,25 \times 1E3} = 200,6\,kN.$$

or

$$N_{pl,Rd} = \frac{A f_y}{\gamma_{M0}} = \frac{(2 \times 65 \times 6) \times 275}{1,0 \times 1E3} = 214,5 > 200,6\,kN.$$

A check must be made for the strength of the bolts as shown in Chapter 7.

6.2 COMBINED BENDING AND AXIAL FORCE – EXCLUDING BUCKLING (cl 6.2.9, EN 1993-1-1 (2005))

Combining an axial force (tension or compression) and bending moment induces stresses which vary across a ductile steel section. At a certain point the stresses may combine to produce a yield stress but this does not produce collapse of the member because collapse only occurs if the entire section is at yield stress (i.e. the section is plastic). If the axial force is compressive it is assumed that buckling does not occur.

6.2.1 Rectangular Sections (cl 6.2.9.1(3), EN 1993-1-1 (2005))

A rectangular section subject to an axial force and bending moment produces a stress diagram as shown in Fig. 6.1. The neutral axis is displaced from the equal area position and the stress diagram can be represented in two parts, one for the axial load and one for the reduced plastic modulus.

Plastic moment of resistance about y–y axis

$$M_y = f_y b h_1 (h - h_1) \tag{6.3}$$

Axial load

$$N = f_y b h (h - 2h_1) \tag{6.4}$$

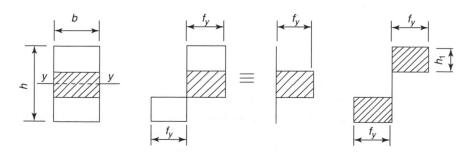

FIGURE 6.1 Effect of axial force on the plastic moment of resistance of a rectangular section

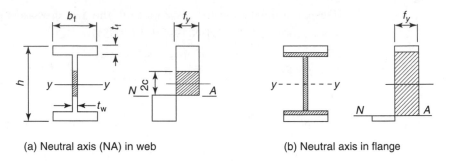

(a) Neutral axis (NA) in web   (b) Neutral axis in flange

FIGURE 6.2   Effect of axial force on the plastic moment of resistance of an 'I' section

Combining Eqs (6.3) and (6.4) and eliminating h_1

$$M_y/(bh^2f_y/4) + [N/(bhf_y)]^2 = 1 \qquad (6.5a)$$

This relationship is plotted on Fig. 6.3. The term $(bh^2f_y/4)$ is the plastic moment of resistance for a rectangular section. This form of the equation is given in Eq. (6.32), EN 1993-1-1 (2005) and is applicable to rectangular solid sections without holes.

$$M_{\text{N,Rd}} = M_{\text{pl,Rd}}\left[1 - \left(\frac{N_{\text{Ed}}}{N_{\text{pl,Rd}}}\right)^2\right] \qquad (6.5b)$$

6.2.2 'I' Sections

The relationship between bending moment and axial force for an 'I' section is more complicated. An 'I' section subject to an axial load and bending moment produces a stress diagram as shown in Fig. 6.2. The neutral axis is displaced from the equal area position. This stress diagram can be represented in two parts, one for the axial load and one for the reduced plastic modulus.

A convenient ratio to determine the reduced plastic section modulus is

$$n = \frac{\sigma}{f_y}$$

where

σ is the mean axial stress

f_y is the specified minimum yield strength of steel.

If the axial force is small the neutral axis is in the web. Alternatively if the axial force is large the neutral axis is in the flange as shown in Fig. 6.2. For bending about the y–y axis, the neutral axis moves from the web into the flange when $n > n_c$ where

$$n_c = t_w\left(\frac{h - 2t_f}{A}\right)$$

A is the total area of the section.

Subtracting the plastic modulus of the hatched area from the whole section.

$$W_{yr} = W_y - W_{(\text{shaded area})}$$

$$W_{(\text{shaded area})} = \frac{t_w(2c)^2}{4}$$

From equilibrium

$$\sigma A = (2c)t_w f_y$$

and since by definition $n = \sigma/f_y$

When $n \leq n_c$

$$W_{yr} = W_y - an^2$$

When $n > n_c$

$$W_{yr} = b(1-n)(c+n)$$

where

$$a = \frac{A^2}{4t_w}$$

$$b = \frac{A^2}{4b_f}$$

$$c = \frac{2b_f h}{A} - 1$$

For bending about the y–y axis the change point for n is

$$n_c = \frac{t_w h}{A}$$

and the values of

$$a = \frac{A^2}{4h}$$

$$b = \frac{A^2}{8t_f}$$

$$c = \frac{4t_f b_f}{A} - 1$$

Values of a, b, c and n_c for 'I' sections are given in Section Tables.

The plastic section modulus determined in this way is applicable in tension and for short columns. For longer columns instability effects due to deflection of the column reduce this value. The relationship between N/N_p and M/M_p for a typical 'I' section is plotted in Fig. 6.3.

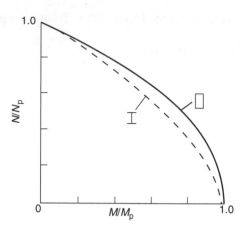

FIGURE 6.3 Relationship between N/N_p and M/M_p at collapse

The above theory for 'I' and 'H' sections is simplified in the European Code by the following recommendations (cl 6.2.9.1(4), EN 1993-1-1 (2005)).

(a) For Class 1 and Class 2 cross-sections the plastic moment of resistance is not reduced if axial loads are limited.

Bending about the y–y axis (Eqs (6.33) and (6.44), EN 1993-1-1 (2005))

$$N_{Ed} \leq 0,25 N_{pl,Rd}$$

$$N_{Ed} \leq 0,5 \frac{h_w t_w f_y}{\gamma_{M0}}$$

Bending about the z–z axis

$$N_{Ed} \leq \frac{h_w t_w f_y}{\gamma_{M0}}$$

(b) For Class 3 and Class 4 cross-sections the maximum longitudinal stress is limited to f_y/γ_{M0}. In the case of Class 4 sections the effective area of the section is used.

6.3 BUCKLING OF AXIALLY LOADED COMPRESSION MEMBERS

Compression members are present in many structures, for example, trusses, bracing and columns. They are generally greater in cross-sectional area than tension members because they may fail in buckling (Fig. 6.4). If buckling is likely to occur then sections must be capable of resisting bending moments.

6.3.1 Compression Members (cl 6.3.1, EN 1993-1-1 (2005))

An efficient cross-section for a strut is a hot finished tube because residual stresses are a minimum and the buckling resistance is the same for all axes of bending. However the use of a tube is not always practical because of the difficulties of making connections.

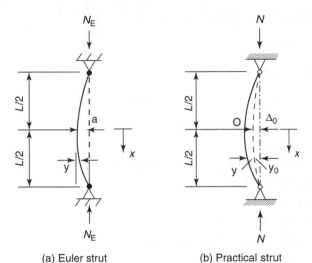

(a) Euler strut   (b) Practical strut

FIGURE 6.4 Buckling behaviour based on Euler strut and practical strut

Connections to Universal Column 'I' sections are simpler but the section is not as efficient because of the weaker z–z axis of bending. Where loads are relatively small angles and 'T' sections are used as struts (e.g. in roof trusses).

The strength of a compression member is not reduced significantly by welding or holes with fasteners but they must be arranged sensibly.

Short steel compression members fail by squashing at the yield stress, while long, or more accurately slender members, fail by buckling. Buckling may occur at an axial stress which is less than the yield stress and is related to slenderness ratio, lack of straightness and non-axial loads. As buckling progresses the load becomes progressively more eccentric to the longitudinal axis of the member and a bending moment is introduced as shown in Fig. 6.4(a).

6.3.2 Buckling Theory

The Euler theory was the first attempt to produce a rational explanation of buckling behaviour of a strut. It was based on the differential equation, related to Fig. 6.4(a), which shows the final deflected form of a pin-ended strut

$$EI\frac{d^2y}{dx^2} = M = N_E(a - y) \tag{6.6}$$

The solution of Eq. (6.6) shows that an axially loaded pin-ended strut becomes elastically unstable and buckles at the Euler critical stress

$$f_E = \frac{\pi^2 E}{\lambda^2}$$

where

E Young's elastic modulus

L length of a pin-ended strut

i radius of gyration

λ slenderness ratio $= L/i$

The Euler buckling theory makes no allowance for:

(a) homogeneity of column material,
(b) isotropy of column material,
(c) variation of E value and elastic–plastic behaviour of column material,
(d) loading not axial,
(e) residual stresses,
(f) lack of straightness of a column,
(g) cross-section of a column not rectangular,
(h) local buckling,
(i) alternative end conditions to a column.

These factors are present in practice and reduce the Euler buckling load.

The Euler differential equation can be modified in a number of ways (Bleich (1952)) to take account of these factors. One method produces the Perry–Robertson formula which is used in the European Code and is related to the Euler buckling stress.

Small unavoidable eccentricities of loading and lack of initial straightness can be simulated mathematically by assuming an initial curvature which produces a small central deflection Δ_0 (Fig. 6.4(b)).

When a load N is applied the deflection at x is increased by y and the differential equation of bending similar to Eq. (6.6) is

$$EI\frac{d^2y}{dx^2} = M = -N[y + y_0] \tag{6.7}$$

Adopting a sinusoidal function for the initial curvature

$y_0 = \Delta_0 \cos(\pi x/L)$ and putting $\mu^2 = N/EI$ then

$$\frac{d^2y}{dx^2} + \mu^2\left[y + \Delta_0 \cos\left(\frac{\pi x}{L}\right)\right] = 0$$

The solution to this equation is

$$y = A\sin\mu x + B\cos\mu x + \frac{\mu^2\Delta_0\cos(\pi x/L)}{\pi^2/L^2 - \mu^2}$$

when $x = \pm L/2$, $y = 0$ and hence $A = B = 0$ and

$$y = \frac{\mu^2\Delta_0\cos(\pi x/L)}{\pi^2/L^2 - \mu^2}$$

If the Euler buckling load $N_E = \pi^2 EI/L^2$, then

$$y = \frac{N\Delta_0\cos(\pi x/L)}{N_E - N}$$

If $N/A = f$ and $N_E/A = f_E$ then the deflection

$$y = \frac{f}{f_E - f} \Delta_0 \cos\left(\frac{\pi x}{L}\right)$$

and the total deflection at any point

$$y_0 + y = \left[\frac{f_E}{(f_E - f)} + 1\right] \Delta_0 \cos\left(\frac{\pi x}{L}\right)$$

The maximum deflection at $x = 0$

$$y_{max} = \left(\frac{f_E}{f_E - f}\right) \Delta_0$$

and the maximum bending moment is

$$M_{max} = \left(\frac{f_E}{f_E - f}\right) \Delta_0 N$$

If d_{exf} is the distance of the extreme fibre from the neutral axis then the maximum compressive stress

$$f_{exf} = \left(\frac{\Delta_0 d_{exf} N}{I}\right) \left(\frac{f}{f_E - f}\right) + f$$

If $(\Delta_0 d_{exf} N/I) = (\Delta_0 d_{exf} Af/I) = (\Delta_0 d_{exf}/i^2)f = \eta f$ where i is the radius of gyration of the column section then

$$f_{exf} = f\left[\frac{\eta f_E}{(f_E - f)} + 1\right]$$

Assuming that the critical buckling load is reached when yielding commences in the extreme fibres of the strut, that is, when $f_{exf} = f_y$ and $f = f_{PR}$ then rearranging the Perry–Robertson buckling stress

$$f_{PR} = 0{,}5[f_y + (\eta + 1)f_E] - \{0{,}5^2[f_y + (\eta + 1)f_E]^2 - f_y f_E\}^{1/2} \tag{6.8}$$

where the Euler critical buckling stress $f_E = \pi^2 E/\lambda^2$.

Equation (6.8) is known as the Perry–Robertson formula and its adoption is explained by Dwight (1975). The value of the function η has varied over the years and the value originally obtained experimentally by Robertson in 1925 related to the slenderness ratio (λ) for circular sections was

$$\eta = 0{,}003\lambda \tag{6.9}$$

The value suggested later by Godfrey (1962) to give more economical designs, based on experimental work by Duthiel in France, was

$$\eta = 0{,}3\left(\frac{\lambda}{100}\right)^2 \tag{6.10}$$

Equation (6.8) has been rearranged in the European Code (Eq. (6.49), EN 1993-1-1 (2005)) to express the buckling stress (f_{PR}) in terms of the stress ratio ($f_E/f_y = \bar{\lambda}$) and a reduction factor (χ) related to column imperfections. If

$$\varsigma = 0{,}5[f_y + (\eta + 1)f_E] \text{ then}$$

$$f_{PR} = \varsigma - (\varsigma^2 - f_y f_E)^{1/2} = \frac{f_y f_E}{[\varsigma + (\varsigma^2 - f_y f_E)^{1/2}]}$$

$$= \frac{f_y}{\left[\frac{\varsigma}{f_E} + \left\{\left(\frac{\varsigma}{f_E}\right)^2 - \left(\frac{f_y}{f_E}\right)\right\}^{1/2}\right]} = \chi f_y \qquad (6.11)$$

In the European Code $\varsigma/f_E = \Phi, f_y/f_E = \bar{\lambda}$ and the reduction factor (χ) is then expressed in terms of $\bar{\lambda}$

$$\Phi = \frac{\varsigma}{f_E} = \frac{[f_y/f_E + (\eta + 1)]}{2}$$

$\eta = 0{,}001a\ (\lambda - \lambda_0)$ but not less than zero, where a varies from 2 to 8 depending on the shape of the section and the limiting slenderness ratio

$$\lambda_0 = 0{,}2\sqrt{\left(\frac{\pi^2 E}{f_y}\right)} > 0$$

Combining these equations (cl 6.3.1.2, EN 1993-1-1 (2005))

$$\Phi = 0{,}5[1 + \alpha(\bar{\lambda} - 0{,}2) + \bar{\lambda}^2]$$

where $\alpha = 0{,}001a\ (\pi^2 E/f_y)^{1/2}$ is an imperfection factor.

For uniform members the buckling stress reduction factor (Eq. (6.49), EN 1993-1-1 (2005))

$$\chi = 1/[\Phi + (\Phi^2 - \bar{\lambda}^2)^{1/2}] \leq 1 \qquad (6.12)$$

The buckling curves are related to the shape of the sections, axis of buckling, and thickness of material as shown in Tables 6.1 and 6.2, and Fig. 6.4, EN 1993-1-1 (2005). The curves are based on experimental results for real columns as described in the ECCS (1976) report, and expressed theoretically by Beer and Schultz (1970).

As the lowest value of the central deflection (Δ) is related to the initial curvature, bending moments are generated as soon as the axial load is applied and the buckling process starts immediately. Therefore there is no condition of elastic instability as defined by Euler (1759) and the average compressive stress may never reach the Euler critical stress for a strut of finite length. Nevertheless the failure of a strut is sudden when compared with the ductile failure of a tension member.

For a fuller development of the buckling theory see Trahair and Bradford (1988).

6.3.3 Local Buckling (cl 5.5.2(2) and Table 5.2, EN 1993-1-1 (2005))

Slender elements of a section (e.g. flanges), which are primarily in compression, may buckle locally before overall buckling of the member occurs. This is likely to occur with Class 4 sections for high values of the width (c) to thickness (t) ratios of elements of a cross-section. To allow for local buckling the European Code reduces the cross-sectional area to an effective area and consequently the member supports less load.

6.3.4 Column Cross-section

Studies (Bleich, 1952) over the year have shown that the strength of a column is influenced by the cross-section. Initial theoretical investigations showed that material concentrated at the centre of gravity was more effective in resisting buckling.

Later research identified other factors (Table 6.2, EN 1993-1-1 (2005)) which affect buckling, for example, hot finished or cold formed, yield strength, buckling axis, welding and shape of cross-section. For design purposes these factors are related to an appropriate buckling curve (Fig. 6.4, EN 1993-1-1 (2005)).

6.3.5 Buckling Length of a Column

The buckling theory, developed previously, is based on the assumption that the ends of the column are pinned, that is, frictionless joints which prevent linear movement but which can rotate freely about any axis of the section. Pin-ended columns are rare in practice but the pin-ended condition is a useful theoretical concept which can be shown to relate to other end conditions.

Alternative end support conditions can be simulated by replacing the actual length of the column by an effective length. Consider the theoretical end conditions shown in Fig. 6.5 which vary from complete end fixity to complete freedom at the end of a cantilever. The theoretical effective length (l) is expressed as a proportion of the actual length (L) and is the distance between real or theoretical pins (i.e. points of contraflexure). The effective length is important in design calculations because the Euler buckling load is inversely proportional to the square of the effective length.

In practice the full rigidity of a fixed built-in end is never achieved and some rotation occurs due to the flexibility of the connection or the support. Only a small rotation is necessary to transform a built-in end to a pinned end and thus reduce the buckling resistance of a column. In comparison small translational movements at supports are not so critical and may be limited by supporting members.

Practical end conditions therefore allow for some rotation and translation at the ends of a real column. For the column in Fig. 6.6 the theoretical effective length is $0,7L$ but in practice the built-in end can rotate and the real effective length is $0,85L$, which

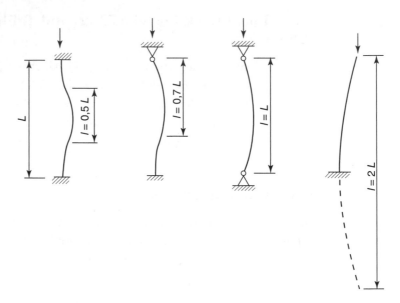

FIGURE 6.5 Theoretical effective lengths

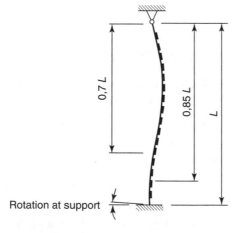

Rotation at support

FIGURE 6.6 Rotation of the end of a strut

reduces the axial load at failure. Because perfect rigidity and completely free ends do not occur in practice the range of common practical values is between $0,7L$ and $1,5L$.

Methods exist (Williams and Sharp, 1990; Wood, 1974) for determining the effective lengths for columns in rigid frames which are based on the relative total column stiffness at a joint to the total stiffness of all members at the joint. For situations where sway of the column does not occur, for example, where cross bracing is present in simple construction, the effective length is less than the real length ($0,5L$ to $1,0L$). Where sway occurs, for example, an unpropped portal frame, the effective length is greater than the real length ($1,0L$ to $2,0L$).

For single angles used as struts eccentricity of axial load may generally be ignored. As a rough guide the effective length is approximately $0,85L$ for two fasteners (or weld) and $1,0L$ for one fastener. Buckling failure about the minor axis must be considered for which a single equal angle, is at 45° to the major axis of bending. An alternative method of determining a modified slenderness ratio for angles is given in cl BB 1.2, EN 1993-1-1 (2005). For double angle struts the effective lengths are similar to those for a single angle but failure about the 45° axis may not be possible because of restraints.

6.3.6 Maximum Slenderness Ratios

At high values of the slenderness ratio struts become so flexible that deflections under their own weight are sufficient to introduce stresses in excess of the Perry–Robertson buckling formula. The following empirical limits have been used in the past.

The slenderness ratio (λ) should not generally exceed the following:

(a) for members resisting loads other than wind loads   180
(b) for members resisting self-weight and wind loads only   250
(c) for any member normally acting as a tie but subject to reversal of
     stress resulting from the action of wind   350

Members whose slenderness exceeds 180 should be checked for self-weight deflection. If this exceeds (length/1000) the effect of the bending should be taken into account in design.

6.3.7 Intermediate Restraints

A member that provides an intermediate restraint and prevents buckling of a strut, reduces the effective length and increases the strength of the strut. The restraint need not be rigid and may be elastic provided that its stiffness exceeds a certain value (Trahair and Bradford, 1988). Often restraining members associated with built-up members are required to resist not less than 1% of the axial force in the restrained member.

6.3.8 Combined Bending, Shear and Axial Force (cl 6.2.10, EN 1993-1-1 (2005))

Th effect of a shear force on the Euler buckling load for solid sections is small (Bleich, 1952). However as explained in Chapter 4 the plastic moment of resistance is reduced when the design shear force is greater than 50% of the plastic shear resistance. Also for large values of shear shear buckling may reduce the resistance of the section (cl 5, EN 1993-1-5 (2005)).

6.3.9 Design Buckling Resistance (cl 6.3.1, EN 1993-1-1 (2005))

Equation (6.8) expresses failure as a buckling stress but the European Code expresses failure as a design load where local buckling and the buckling stress factor (χ) reduce the buckling load.

The design buckling resistance of a compression member for Classes 1, 2 and 3 cross-sections

$$N_{b,Rd} = \chi A \frac{f_y}{\gamma_{M1}} \qquad (6.13a)$$

and for Class 4 sections

$$N_{b,Rd} = \chi A_{eff} \frac{f_y}{\gamma_{M1}} \qquad (6.13b)$$

where

χ is the reduction factor for the relevant buckling mode which is generally 'flexural buckling'. In other cases 'torsional' or 'flexural–torsional' modes may govern.

$$\chi = 1[\Phi + (\Phi^2 - \bar{\lambda}^2)^{1/2}] \le 1$$

$$\Phi = 0{,}5[1 + \alpha(\bar{\lambda} - 0{,}2) + \bar{\lambda}^2]$$

$$\bar{\lambda} = \left(A\frac{f_y}{N_{cr}}\right)^{1/2} = \frac{(L_{cr}/i)}{\lambda_1}$$

$$\lambda_1 = \pi \left(\frac{E}{f_y}\right)^{1/2} = 93{,}9\varepsilon$$

$$\varepsilon = \left(\frac{235}{f_y}\right)^{1/2}$$

The imperfection factor (α) corresponding to the appropriate buckling curve is obtained from Table 6.1, EN 1993-1-1 (2005).

The value of the cross sectional area is A for Classes 1, 2 and 3 sections with no reductions for local buckling provided that the maximum width-to-thickness ratios are within the limits of Table 5.2, EN 1993-1-1 (2005). For Class 4 sections the effective area, reduced because of local buckling, is calculated from cl 4, EN 1993-1-5 (2003).

EXAMPLE 6.2   Angle strut in a roof truss. A steel roof truss is composed of angles and tee sections and has been analysed assuming pin joints. A particular member is subject to the following axial forces. Permanent action ($F_G = -7{,}2$ kN), variable snow load ($F_s = -10{,}7$ kN), variable wind downwards ($F_w = -2{,}0$ kN) and variable wind upwards ($F_w = +19{,}1$ kN). If an angle, welded at the ends, is chosen to resist the loads, determine the size if the actual length is 2,1 m.

Design load cases:

(a) no wind

$$1{,}35F_G + 1{,}5\,F_s = 1{,}35(-7{,}2) + 1{,}5(-10{,}7) = -25{,}77\,\text{kN (compression)}$$

(b) wind down

$$1,35\,F_G + 1,5(F_s + F_w) = 1,35(-7,2) + 1,5(-10,7 - 2)$$
$$= -28,77\,\text{kN (compression)}$$

(c) wind up

$$1,0\,F_G + 1,5\,F_w = -7,2 + 1,5(+19,1) = +21,45\,\text{kN (tension)}$$

Design value $N_{b,Ed} = 28,77\,\text{kN}$ assuming uniform compressive stress across the section. Try a $65 \times 50 \times 6\,\text{mm}$ angle grade S275 steel, long leg attached ($A = 659\,\text{mm}^2$, $i_v = 10,7\,\text{mm}$ from Section Tables).

Design as an axially loaded Class 3 compression member, ignoring eccentric loads (cl BB 1.2, EN 1993-1-1 (2005)). The effective area need not be reduced because of end holes (cl 6.3.1.1(4), EN 1993-1-1 (2005)).

Check maximum width-to-thickness ratios (Table 5.2, EN 1993-1-1 2005)

$$\frac{h}{t} = \frac{65}{6} = 10,8 < \left[15 \times \left(\frac{235}{275} \right)^{1/2} = 13,9 \right] \text{ satisfactory}$$

$$\frac{b+h}{2t} = \frac{50+65}{2 \times 6} = 9,58 < \left[11,5 \times \left(\frac{235}{275} \right)^{1/2} = 10,6 \right]$$

satisfactory, no reduction of area for local buckling.

Non-dimensional slenderness about the v–v axis (Eq. (6.50), EN 1993-1-1 (2005))

$$\bar{\lambda}_v = \left(\frac{Af_y}{N_{cr}} \right)^{1/2} = \frac{(L_{cr}/i_v)}{\lambda_1} = \frac{(L_{cr}/i_v)}{93,9\varepsilon}$$

$$= \frac{2100/10,7}{[93,9 \times (235/275)^{1/2}]} = 2,26$$

Largest effective slenderness ratio is about the v–v axis for an angle (cl BB 1.2, EN 1993-1-1 (2005))

$$\bar{\lambda}_{eff,v} = 0,35 + 0,7\bar{\lambda}_v = 0,35 + 0,7 \times 2,26 = 1,93$$

For $\bar{\lambda}_{eff,v} = 1,93$ and curve b (Table 6.2, EN 1993-1-1 (2005)) the buckling reduction factor (Fig. 6.4, EN 1993-1-1 (2005))

$$\chi = 0,23$$

Or by calculation (Eq. (6.49), EN 1993-1-1 (2005)) the buckling reduction factor

$$\chi = \frac{1}{[\Phi + (\Phi^2 - \bar{\lambda}_{eff,v}^2)^{1/2}]} = \frac{1}{[2,66 + (2,66^2 - 1,93^2)^{1/2}]}$$

$$= 0,223 < \text{(graph value 0,23)}$$

where

$$\Phi = 0{,}5[1 + \alpha(\overline{\lambda}_{\text{eff,v}} - 0{,}2) + \overline{\lambda}_{\text{eff,v}}^2]$$
$$= 0{,}5 \times [1 + 0{,}34 \times (1{,}93 - 0{,}2) + 1{,}93^2] = 2{,}66$$

which includes the imperfection factor for buckling curve b (Table 6.1, EN 1993-1-1 (2005))

$$\alpha = 0{,}34$$

Design buckling resistance (Eq. (6.47), EN 1993-1-1 (2005))

$$N_{\text{b,Rd}} = \chi A \frac{f_y}{\gamma_{\text{M1}}} = 0{,}223 \times 659 \times \frac{275}{1{,}1 \times 1\text{E}3}$$
$$= 36{,}7 > (N_{\text{b,Fd}} = 28{,}77) \text{ kN satisfactory.}$$

6.4 COMBINED BENDING AND AXIAL FORCE – WITH BUCKLING (cl 6.3.3, EN 1993-1-1 (2005))

6.4.1 Introduction

In practical situations axial forces in columns are accompanied by bending moments acting about the major and minor axes of bending. The axial force and bending moment vary along the length of the member and an exact analysis is complicated (Culver, 1966). In some situations overall buckling, lateral torsional buckling and local buckling can occur together. This is more likely to be a problem with Class 4 sections where outstand/thickness ratios are large.

Exact theoretical solutions for buckling of columns are not available and in any case would be too complicated for design. Alternatively design interaction equations (Eqs (6.61) and (6.62), EN 1993-1-1 (2005)) are used based on elastic and plastic limits. A steel member subject to bending moments and axial forces fails when a stress is greater than first yield but less than full plasticity of the section. First yield theories are conservative while full plasticity theories are unsafe.

Joint rotation reduces the buckling load of a column and this can now be incorporated into the analysis of frames (cl 2, EN 1993-1-8 (2005)). This can be important because even small rotations can significantly reduce load capacity.

In design it is often assumed that parts of frames can be analysed separately and the compression members can be isolated and analysed accordingly. For traditional simple design methods beam and column structures are assumed to be connected together with pin joints and braced to prevent sidesway collapse. The pin joints are assumed eccentric to the column axis and thus introduce bending moments to the column which reduces the failure load.

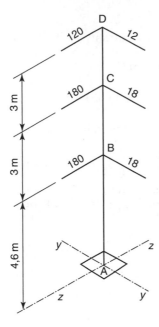

FIGURE 6.7 Example: three storey corner column

EXAMPLE 6.3 Bending moments in a three storey corner column. The first three storeys of a corner stanchion are sketched in Fig. 6.7. The beam support reactions are indicated in kN on each beam. Assuming simple design, eccentric pin joints and cross bracing to reduce sway, determine the bending moments in the columns AB and BC.

Try the following column sizes from Section Tables.

Columns AB and BC: $203 \times 203 \times 60$ UC

$\quad h = 209{,}6$ mm, $I_y = 6103$ cm^4, $I_z = 2047$ cm^4.

Column CD: $203 \times 203 \times 46$ UC

$\quad h = 203{,}2$ mm, $I_y = 4565$ cm^4, $I_z = 1539$ cm^4.

Consider a corner column buckling with bending about the major and minor axes.

Bending about the major y–y axis

Stiffness of columns (cm units)

$\quad$ Column AB, $K_{AB} = I_y/L_{AB} = 6103/460 = 13{,}27$

$\quad$ Column BC, $K_{BC} = I_y/L_{BC} = 6103/300 = 20{,}34$

$\quad$ Column CD, $K_{CD} = I_y/L_{CD} = 4564/300 = 15{,}21$

(a) First floor level (at joint B):

Moment about the y–y axis from eccentricity of the joint assuming

$$M_{By} = \text{(beam reaction)}(100 + h/2)$$

$$= 180 \times \frac{(100 + 209{,}6/2)}{1E3} = 36{,}86 \text{ kNm}.$$

For simple design moments are distributed in proportion to stiffness

$$M_{BA} = \frac{M_{By}K_{AB}}{K_{AB} + K_{BC}} = \frac{36{,}86 \times 13{,}27}{13{,}27 + 20{,}34} = 14{,}55 \text{ kNm}$$

$$M_{BC} = \frac{M_{By}K_{BC}}{K_{AB} + K_{BC}} = \frac{36{,}86 \times 20{,}34}{13{,}27 + 20{,}34} = 22{,}31 \text{ kNm}$$

(b) Second floor level (at joint C):

Assuming that the splice between columns BC and CD lies above first floor level, then the design moment is the same at the first floor ($M_{Cy} = 36{,}86$ kNm). If moments are distributed equally for simple multi-storey construction

$$M_{CB} = M_{CD} = \frac{M_{Cy}}{2} = \frac{36{,}86}{2} = 18{,}43 \text{ kNm}.$$

It should be appreciated that the simple method does not incorporate joint stiffness which reduces moments at mid-span for the beams. This is beneficial, but moments in the column are inaccurate. However, the method has been used extensively in the past.

EXAMPLE 6.4 Two storey corner column. Determine the size of a column section required for a two storey corner column shown in Fig. 6.8. The design loads (kN) shown are the end reactions from the beams. The beams are connected to the column using cleats and bolts and pin joints are assumed. The columns and beams are encased in concrete.

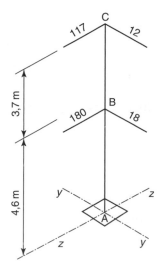

FIGURE 6.8 Example: two storey corner column

Axial design loads on the column:

(a) First floor | kN
   Floor beam (self-weight included) | 117
   Wall beam (self-weight included) | 12
   Self-weight of column BC (estimated) | 10
   Total | 139

(b) Ground floor | kN
   Floor beam (self-weight included) | 180
   Wall beam (self-weight included) | 18
   Upper storey | 139
   Total (excluding self-weight of column AB) | 337

Try $152 \times 152 \times 37\,$kg UC grade S275 steel for both storeys. From Section Tables $h = 161{,}8$ mm, $b = 154{,}4$ mm, $t_w = 8{,}1$ mm $t_f = 11{,}5$ mm, $A_{gross} = 4730$ mm^2, $i_z = 38{,}7$ mm, $i_y = 68{,}4$ mm, $W_{el,y} = 274$E3 mm^3, $W_{el,z} = 91{,}5$E3 mm^3, $W_{pl,y} = 309$E3 mm^3, $W_{pl,z} = 140$E3 mm^3, $I_z = 706$E4 mm^4, $I_y = 2213$E4 mm^4, $I_w = 0{,}0399$ dm^6, $I_t = 19{,}3$E4 mm^4.

This is a Class 1 section, that is, c/t outstand ratios are within limits (Table 5.2, EN 1993-1-1 (2005)) and local buckling does not occur. However, the section must be checked for overall buckling and lateral torsional buckling.

Assuming simple nominal bending moments from eccentricity of the beam reactions (R) at join B:

$$M_{By} = \frac{R(100 + h/2)}{1\text{E3}} = \frac{180(100 + 161{,}8/2)}{1\text{E3}} = 32{,}56\,\text{kNm}$$

$$M_{Bz} = \frac{R(100 + t_w/2)}{1\text{E3}} = \frac{18(100 + 8{,}1/2)}{1\text{E3}} = 1{,}87\,\text{kNm}$$

Design bending moments applied to the column (same section throughout)

$$M_{y,Ed} = \frac{M_{By}(I_{BC}/L_{BC})}{[(I_{BC}/L_{BC}) + (I_{BA}/L_{BA})]}$$

$$= \frac{32{,}56(1/3{,}7)}{[(1/3{,}7) + (1/4{,}6)]} = 32{,}56 \times 0{,}554 = 18{,}03\,\text{kNm}$$

$$M_{z,Ed} = M_{Bz} \times 0{,}554 = 1{,}87 \times 0{,}554 = 1{,}036\,\text{kNm}$$

Check combined axial load, bending and lateral torsional buckling about the stronger y–y axis using the interaction formula (Eq. (6.61), EN 1993-1-1 (2005))

$$\frac{N_{Ed}}{\chi_y N_{Rk}/\gamma_{M1}} + \frac{k_{yy}(M_{y,Ed} + \Delta M_{y,Ed})}{\chi_{LT} M_{y,Rk}/\gamma_{M1}} + \frac{k_{yz}(M_{z,Ed} + \Delta M_{z,Ed})}{M_{z,Rk}/\gamma_{M1}} \leq 1$$

$$= \frac{337}{946} + \frac{0{,}465 \times (18{,}03 + 0)}{0{,}751 \times 77{,}3} + \frac{0{,}453 \times (1{,}036 + 0)}{35}$$

$$= 0{,}356 + 0{,}144 + 0{,}013 = 0{,}513 < 1 \text{ satisfactory.}$$

Check combined axial load, bending and lateral torsional buckling about the weaker z–z axis using the interaction formula (Eq. (6.62), EN 1993-1-1 (2005))

$$\frac{N_{Ed}}{\chi_z N_{Rk}/\gamma_{M1}} + \frac{k_{zy}(M_{y,Ed} + \Delta M_{y,Ed})}{\chi_{LT} M_{y,Rk}/\gamma_{M1}} + \frac{k_{zz}(M_{z,Ed} + \Delta M_{z,Ed})}{M_{z,Rk}/\gamma_{M1}} \leq 1$$

$$= \frac{337}{532} + \frac{0,453 \times (18,03 + 0)}{0,751 \times 77,3} + \frac{0,755 \times (1,036 + 0)}{35}$$

$$= 0,633 + 0,141 + 0,022 = 0,796 < 1 \text{ satisfactory.}$$

The two previous equations include the following values.

For buckling load calculations, consider column AB just below B, assuming the buckling length about the y–y and z–z axes, $L_{cr} = 0,85 L$. Use buckling curve b for buckling about the y–y axis and curve c for the z–z axis (Table 6.2, EN 1993-1-1 (2005)) based on $h/b < 1,2$; $t_f < 100$ mm and steel grade S275.

The non-dimensional slenderness ratio (Eq. (6.50), EN 1993-1-1 (2005))

$$\bar{\lambda}_y = \left(\frac{A f_y}{N_{cr}}\right)^{1/2} = \left(\frac{L_{cr}}{i_y}\right) \times \frac{1}{\lambda_1}$$

$$= \frac{(0,85 \times 4600)}{68,4} \times \frac{1}{[93,9 \times (235/275)^{1/2}]} = 0,659$$

From Fig. 6.4, EN 1993-1-1 (2005) (buckling curve b), $\bar{\lambda}_y = 0,659$ and a reduction factor $\chi_y = 0,80$ the design buckling resistance for a Class 1 section (Eq. (6.47), EN 1993-1-1 (2005))

$$N_{b,Rd} = \chi_y A \frac{f_y}{\gamma_{M1}} = 0,80 \times 4730 \times \frac{275}{1,1 \times 1E3} = 946 \text{ kN.}$$

$$M_{y,Rd} = W_{pl,y} \frac{f_y}{\gamma_{M1}} = 309E3 \times \frac{275}{1,1 \times 1E6} = 77,3 \text{ kNm.}$$

$$\bar{\lambda}_z = \left(\frac{A f_y}{N_{cr}}\right)^{1/2} = \left(\frac{L_{cr}}{i_z}\right) \times \frac{1}{\lambda_1}$$

$$= \left(\frac{0,85 \times 4600}{38,7}\right) \times \frac{1}{[93,9 \times (235/275)^{1/2}]} = 1,16$$

From Fig. 6.4, EN 1993-1-1 (2005) (buckling curve c) and $\bar{\lambda}_z = 1,16$ the reduction factor $\chi_z = 0,45$ and the design buckling resistance for a Class 1 section (Eq. (6.48), EN 1993-1-1 (2005))

$$N_{b,Rd} = \chi_z A \frac{f_y}{\gamma_{M1}} = 0,45 \times 4730 \times \frac{275}{1,1 \times 1E3} = 532 \text{ kN.}$$

$$M_{z,Rd} = W_{pl,z} \frac{f_y}{\gamma_{M1}} = 140E3 \times \frac{275}{1,1 \times 1E6} = 35 \text{ kNm.}$$

For a Class 1 section (Table 6.7, EN 1993-1-1 (2005))

$$\Delta M_{y,\mathrm{Ed}} = \Delta M_{z,\mathrm{Ed}} = 0$$

Interaction factor for a Class 1 section (Table B1, EN 1993-1-1 (2005)). The lesser value of

$$k_{yy} = c_{\mathrm{my}} \left[1 + \frac{(\bar{\lambda}_y - 0,2)N_{\mathrm{Ed}}}{\chi_y N_{\mathrm{Rk}}/\gamma_{\mathrm{M1}}} \right]$$

$$= 0,4 \times \left[1 + \frac{(0,659 - 0,2) \times 337\mathrm{E}3}{0,80 \times 4730 \times 275/1,1} \right] = 0,465$$

or

$$k_{yy} = c_{\mathrm{my}} \left[1 + \frac{0,8 N_{\mathrm{Ed}}}{\chi_y N_{\mathrm{RK}}/\gamma_{\mathrm{M1}}} \right]$$

$$= 0,4 \times \left[1 + \frac{0,8 \times 337\mathrm{E}3}{0,80 \times 4730 \times 275/1,1} \right]$$

$$= 0,514 > 0,465 \text{ use } 0,465$$

where (Table B3, EN 1993-1-1 (2005))

$$c_{\mathrm{my}} = 0,6 + 0,4\varphi = 0,6 + 0,4 \times (-0,5) = 0,4$$

Interaction factors for a Class 1 section (Table B1, EN 1993-1-1 (2005)). The lesser value of

$$k_{zz} = c_{\mathrm{mz}} \left[1 + (2\bar{\lambda}_z - 0,6)\frac{N_{\mathrm{Ed}}}{\chi_z N_{\mathrm{Rk}}/\gamma_{\mathrm{M1}}} \right]$$

$$= 0,4 \times \left[1 + (2 \times 1,16 - 0,6) \times \frac{337\mathrm{E}3}{0,45 \times 4730 \times 275/1,1} \right] = 0,836$$

or

$$k_{zz} = c_{\mathrm{mz}} \left[1 + \frac{1,4 N_{\mathrm{Ed}}}{\chi_z N_{\mathrm{Rk}}/\gamma_{\mathrm{M1}}} \right]$$

$$= 0,4 \times \left[1 + \frac{1,4 \times 337\mathrm{E}3}{0,45 \times 4730 \times 275/1,1} \right]$$

$$= 0,755 < 0,836 \text{ use } 0,755$$

where (Table B3, EN 1993-1-1 (2005))

$$c_{\mathrm{mz}} = 0,6 + 0,4\varphi = 0,6 + 0,4 \times (-0,5) = 0,4$$

and

$$k_{yz} = 0,6 k_{zz} = 0,6 \times 0,755 = 0,453$$

Reduction factor for lateral torsional buckling (Eq. (6.57), EN 1993-1-1 (2005))

$$\chi_{LT} = \frac{1}{[\Phi_{LT} + (\Phi_{LT}^2 - \beta\bar{\lambda}_{LT}^2)^{1/2}]}$$

$$= \frac{1}{[0,881 + (0,881^2 - 0,75 \times 0,905^2)^{1/2}]} = 0,751 < 1$$

or

$$\frac{1}{\bar{\lambda}_{LT}^2} = \frac{1}{0,905^2} = 1,22 > 1, \text{ use } 0,751$$

which includes the factor

$$\Phi_{LT} = 0,5[1 + \alpha_{LT}(\bar{\lambda}_{LT} - \bar{\lambda}_{LT,0}) + \beta\bar{\lambda}_{LT}^2]$$

$$= 0,5 \times [1 + 0,21 \times (0,905 - 0,2) + 0,75 \times 0,905^2] = 0,881$$

where the imperfection factor $\alpha_{LT} = 0,21$ (Table 6.3, EN 1993-1-1 (2005)).

$$\bar{\lambda}_{LT} = \left(\frac{W_{ply}f_y}{M_{cr}}\right)^{1/2} = \left(\frac{309E3 \times 275}{103,85E6}\right)^{1/2} = 0,905$$

The elastic critical bending moment Eq. (5.5)

$$M_{cr} = \left(\frac{\pi^2 EI_z}{L^2}\right)\left[\frac{I_w}{I_z} + \frac{L^2 GI_t}{\pi^2 EI_z}\right]^{1/2}$$

$$= \left(\frac{\pi^2 \times 1483}{4,6^2}\right)\left[\frac{0,0399E\text{-}6}{706E\text{-}8} + \frac{4,6^2 \times 15,59}{\pi^2 \times 1483}\right]^{1/2}$$

$$= 103,85 \text{ kNm}$$

where

$$L = 4,6 \text{ m}$$

$$EI_z = 210E6 \times 706E\text{-}8 = 1483 \text{ kNm}^2$$

$$GI_t = 80,77E6 \times 19,3E\text{-}8 = 15,59 \text{ kNm}^2$$

$$EI_w = 210E6 \times 0,0399E\text{-}6 = 8,379 \text{ kNm}^4$$

For further information on elastic critical bending moments see Chapter 5.

Check the self-weight of column BC. Minimum overall dimensions of cased column

$$H = 161,8 + 100 = 261,8, \text{ say } 270 \text{ mm}$$

$$B = 154,4 + 100 = 254,4, \text{ say } 260 \text{ mm}$$

$$A_c = A_g - A_s = 270 \times 260 - 4740 = 65460 \text{ mm}^2.$$

Total weight = steel column + concrete casing

$$= L\rho g + LA_c \rho g$$

$$= 3{,}7 \times 37 \times 9{,}81/1E3 + 3{,}7 \times 65460 \times 2400 \times 9{,}81/1E9$$

$$= 7\,\text{kN} < 10\,\text{kN (assumed) satisfactory.}$$

These calculations show that a $152 \times 152 \times 37\,\text{kg}$ UC section grade 275 steel is satisfactory, however a lesser weight of $152 \times 152 \times 30\,\text{kg}$ UC might be suitable.

Calculations are more extensive for Class 4 sections which are reduced in area to allow for local buckling (cl 5.5.2(2) EN 1993-1-1; and cl 4.3 EN 1993-1-5). Section properties are based on the effective cross-sections.

REFERENCES

Beer, H. and Schultz, G. (1970). *Theoretical basis for the European column curves*. Construction Metallique, No. 3.

Bleich, F. (1952). *Buckling strength of metal structures*. McGraw-Hill.

Culver, G.C. (1966). *Exact solution for biaxial bending equations*. *American Society of Civil Engineers*, Str. Div., 92(ST2).

Dwight, J.B. (1975). *Adaption of the Perry formula to represent the new European steel column curves, Steel Construction, AISC*, **9(1)**.

The background to British Standards for structural steel work. Imperial College, London, and Constrado.

ECCS (1976). *Manual on stability, introductory report, second international colloquium on stability*. European Convention for Structural Steelwork. Liege.

EN 1993-1-1 (2005). *General rules and rules for buildings*. BSI.

EN 1993-1-5 (2003). *Plated structural elements*. BSI.

EN 1993-1-8 (2005). *Design of joints*. BSI.

Euler, L. (1759). *Sur la force de collones*. Memoires de l'Acadamie de Berlin.

Godfrey, G.B. (1962). The allowable stress in axially loaded struts, *Structural Engineer*, March 1962.

Nelson, H.M. (1953). *Angles in tension*. British Constructional Steelwork Association publication No. 7.

Regan, P.E. and Salter, P.R. (1984). Tests on welded angle tension members, *Structural Engineer*, **62B(2)**.

Robertson, A. (1925). *The strength of struts, Selected Engineering Paper No. 28*, Institution of Civil Engineers.

Trahair, N.S. and Bradford, M.S. (1988). *The Behaviour and Design of Steel Structures*. Chapman and Hall.

Williams, F.W. and Sharp, G. (1990). Simple elastic critical load and effective length calculations for multi-storey rigid sway frames. *Proceedings of the ICE*, 90.

Wood, R.H. (1974). Effective lengths for columns in multi-storey buildings. *Structural Engineer*, **52**.

Chapter 7 / Structural Joints (EN 1993-1-8, 2005)

7.1 INTRODUCTION

Structural steel connections, referred to as joints in the Code, are required to ensure continuity at the intersection members and foundations. They are also used to form splices and to construct brackets to support loads. Generally structural steel joints are composed of plates, or parts of sections, shop welded in controlled conditions and bolted together on site. Welding can be carried out on site but it needs to be carefully supervised and is limited because of the expense. The physical appearance of some joints is shown in Figs 7.22 and 7.23.

Structural joints transmit internal forces and moments in a structure and strength is of major importance. However, the rigidity of joints also needs to be considered. All joints are semi-rigid with associated small linear and larger rotational movements. The linear movements at a joint are generally small and generally need not be considered, but the rotational movements affect the distribution of forces and moments which must be taken into account in structural analysis.

For theoretical purposes in the analysis of structures, joints are classified (Table 5.1, EN 1993-1-8 (2005)) by strength and rigidity as:

(a) Pinned – low moment of resistance.
(b) Rigid – full strength and all deformations are insignificant.
(c) Semi-rigid – characteristics of the connection lie between (a) and (b).

These theoretical and practical descriptions are important to recognize when analyzing a structure to determine the distribution of forces and moments using global analysis (cl 5.1, EN 1993-1-8 (2005)). There are three methods of global analysis:

(1) Elastic – joints are classified by rotational stiffness.
(2) Rigid-plastic – joints are classified by strength.
(3) Elastic plastic – joints are classified by stiffness and strength.

7.2 THE IDEAL STRUCTURAL JOINT

The types of joints in common use have been developed and modified to suit the manufacturing and assembly processes and the ideal requirements are:

(a) Simple to manufacture and assemble.
(b) Standardized for situations where the dimensions and loads are similar thus avoiding a multiplicity of dimensions, plate thicknesses, weld sizes and bolts.
(c) Manufactured from materials and components that are readily available.
(d) Designed and detailed so that work is from the top of the joint not from below where the workman's arms will be above his head. There should also be sufficient room to locate a spanner, or space to weld if required.
(e) Designed so that the welding is confined to the workshops to ensure a good quality and to reduce costs.
(f) Detailed to allow sufficient clearance and adjustment to accommodate the lack of accuracy in site dimensions.
(g) Designed to withstand normal working loads and also erection forces.
(h) Designed to avoid the use of temporary supports during erection.
(i) Designed to develop the required load–deformation characteristics at service and ultimate loads.
(j) Detailed to resist corrosion and to be of reasonable appearance.
(k) Low in cost and cheap to maintain.

7.3 WELDED JOINTS

Welding is a method of connecting components by heating the materials to a suitable temperature so that fusion occurs. The most common method for heating steelwork is by means of an electric arc between a coated wire electrode and the materials being joined. The electrical circuit is shown in Fig. 7.1(a). During the process, which is illustrated in Fig. 7.1(b), the coated electrode is consumed, the wire becomes the filler material and the coating is converted partly into a shielding gas, partly into slag, and some part is absorbed by the weld metal. This method, known as the manual metal arc welding process, is still the most common for structural joints because of low capital cost and flexibility. However, for long continuous welds automatic processes are preferred because of consistency.

Generally the electrode is stronger than the parent metal. For manual metal arc welding the electrodes should be compatible with the steel being welded (BSEN 499, 1995; Gourd, 1980). The main reason for the flux covering to the electrode in the manual metal arc welding process is to provide an inert gas which shields the molten metal from atmospheric contamination. In addition the flux forms a slag to protect the weld until it is cooled to room temperature, when the slag should be easily detachable. Other functions of the flux include: arc stabilization, control of surface profile, control of weld metal composition, alloying and deoxidization. However, it should be noted that

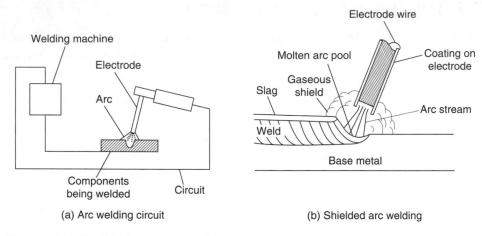

(a) Arc welding circuit

(b) Shielded arc welding

FIGURE 7.1 Shielded metal arc welding

the flux can be a source of hydrogen contamination from absorbed and chemically combined moisture. The absorbed moisture can be removed by drying.

The particular advantage of welding is that it forms a rigid joint, however the manufacture of welded joints requires more skill and supervision than bolted joints. Most welded structural joints are affected using the manual metal arc process but long continuous welds, which occur in built-up girders, are laid down with automatic welding equipment. The automatic processes achieve the exclusion of atmospheric pollution by gas shielding, flux core or submerged arc.

Types of welds used in structural engineering and allowed in the European Code are fillet, slot, butt, plug and flare groove. Some common types are shown in Fig. 7.2.

The two types in most common use are butt and fillet welds. Butt welds, often used to lengthen plates in the end on position, may be considered as strong as the parent plate as long as full penetration for the weld is achieved. For thin plates penetration is achieved without preparing the plate, but on thicker plates V or double J preparation is required. Butt welds are also used to connect plates at right angles but the plates require edge preparation. Partial penetration butt welds are not favoured in design and should not be used intermittently or in fatigue situations (BSEN 1011-1 (1998)).

Fillet welds are generally formed with equal leg lengths. They do not require special edge preparation of the plates and are therefore cheaper than butt welds.

Generally in connections, plates intersect at right angles but intersection angles of between 60° and 120° can be used provided that the correct throat size is used in design calculations. In order to accommodate lack of fit the minimum leg length of fillet weld in structural engineering is 5 mm although 6 mm is often preferred. The maximum size of fillet weld from a single run metal arc process is 8 mm, but 6 mm is

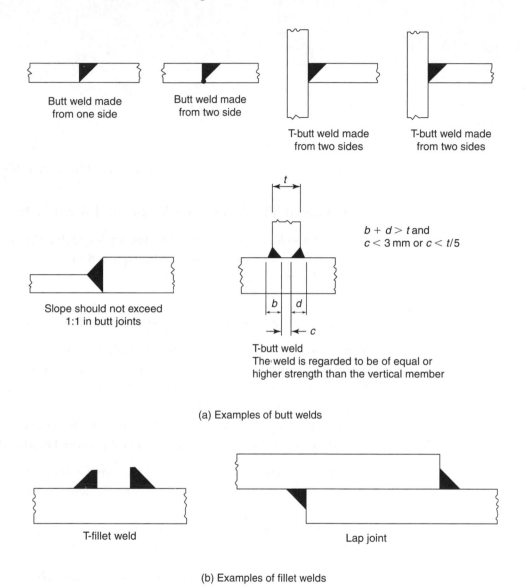

Butt weld made from one side

Butt weld made from two side

T-butt weld made from two sides

T-butt weld made from two sides

Slope should not exceed 1:1 in butt joints

$b + d > t$ and $c < 3$ mm or $c < t/5$

T-butt weld
The weld is regarded to be of equal or higher strength than the vertical member

(a) Examples of butt welds

T-fillet weld

Lap joint

(b) Examples of fillet welds

FIGURE 7.2 Types of welds in structural joints

preferred to guarantee quality. When larger fillet welds are required they are formed from multiple runs.

The use of intermittent butt and fillet welds is permitted (cl 4.3.2.2, EN 1993-1-8 (2005)). Intermittent welds are not favoured in structural engineering because they introduce stress discontinuities, act as stress raisers, may introduce fatigue cracks, may act as corrosion pockets and are difficult to produce with an automatic welding machine. The spacing of intermittent fillet welds is shown in Fig. 4.1, EN 1993-1-8 (2005).

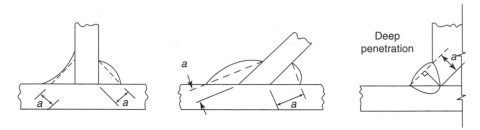

FIGURE 7.3 Throat thickness of fillet welds (Figs 4.3 and 4.4, EN 1993-1-8)

7.3.1 Throat Thickness of a Weld (cl 4.5.2, EN 1993-1-8 (2005))

The size of a weld is often described by the leg length but the strength is calculated using the effective throat thickness (a) as defined in Fig. 7.3. The effective throat thickness should not be less than 3 mm.

7.3.2 Effective Length of a Weld (cl 4.5.1, EN 1993-1-8 (2005))

The effective length of a fillet weld should be taken as the length over which the fillet is full size. The minimum length allowed to transmit loading is six times the throat thickness, and not less than 30 mm. In practice, fillet welds terminating at the ends, or sides, of parts are returned continuously round the corners for a distance of not less than twice the leg length, unless impracticable. The continuation round the corner is to reduce stress concentrations and its strength is generally ignored in strength calculations.

The effective length of a weld is reduced if a component distorts under load in situations similar to that shown In Fig. 7.4, where the deformations in the weld adjacent to the web are greater than those at the end of the flange. The larger deformations at the web initiate failure in the weld at this point with consequent loss of strength for the total length of the weld (Elzen, 1966; Rolloos, 1969).

For design the effective breadth (b_{eff}) of a weld (cl 4.10, EN 1993-1-8 (2005)) is: For a rolled 'I' or 'H' section (Eqs (4.6a) and (4.6b), EN 1993-1-8 (2005))

$$b_{\text{eff}} = t_{\text{w}} + 2r + 7kt_{\text{f}} \leq t_{\text{w}} + 2r + 7\left(\frac{t_{\text{f}}}{t_{\text{p}}}\right)\left(\frac{f_{y,\text{f}}}{f_{y,\text{p}}}\right)t_{\text{f}} \tag{7.1a}$$

For box or channel sections where the widths of the connected plate and the flange are similar

$$b_{\text{eff}} = 2t_{\text{w}} + 5\,t_{\text{f}} \leq 2t_{\text{w}} + 5\left(\frac{t_{\text{f}}}{t_{\text{p}}}\right)\left(\frac{f_{y,\text{f}}}{f_{y,\text{p}}}\right)t_{\text{f}} \tag{7.1b}$$

7.3.3 Long Welded Joints (cl 4.11, EN 1993-1-8 (2005))

The stress distribution along the length of a long lap joint is not uniform being greatest at the ends. To allow for this the length of the weld is reduced. For joints longer

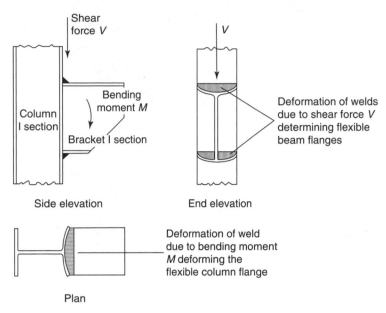

FIGURE 7.4 Reduction in strength of welds associated with flange deformations

than 150 times the throat thickness (a), the reduction factor (Eq. (4.9), EN 1993-1-8 (2005))

$$\beta_{LW,1} = 1,2 - 0,2L_j/(150a) \leq 1 \tag{7.2}$$

where

L_j is the overall length of the lap in the direction of the force transfer.

7.3.4 Design Resistance of Fillet Welds (cl 4.5.3, EN 1993-1-8 (2005))

The real external forces acting on a 90° fillet weld are probably those shown in Fig. 7.5 (a) (Clarke, 1971). Experiments (Biggs *et al.*, 1981) on 90° fillet welds of equal leg length loaded to failure show that the fracture plane varies between 10° and 80° depending on the combination of external forces. The actual distribution of stress on the failure plane is uncertain but a theoretical distribution (Kato and Morita, 1974) shows peak stresses at the root of the weld which reduce towards the face of the weld. This distribution appears to be confirmed by experimental observations of cracks initiating at the root. The situation is complicated further by residual stresses and variables such as the type of electrode type of steel, ratio of the size of weld to the plate thickness, the quality of weld and whether the loading is static or dynamic. If stresses on the failure plane are assumed to be uniform then the relationship between the average shear stress and tensile stress on the failure plane has been shown (Biggs *et al.*, 1981) to approximate to an ellipse. An ellipse of failure stresses combined with a variable

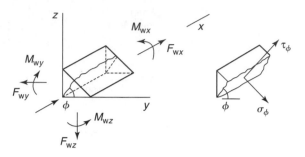

(a) Complex system of forces

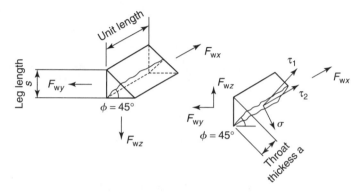

(b) Simple system of forces

FIGURE 7.5   Forces acting on a 90° fillet weld

fracture angle can be used theoretically to predict the magnitude of the external forces, but the method is unnecessarily complicated for design purposes.

For design purposes a complex system of external forces acting on a fillet weld is reduced forces acting in three perpendicular directions on a unit length of weld as shown in Fig. 7.5 (b). The vector sum of all the design forces should not exceed the design resistance of a fillet weld (cl 4.5.3.3, EN 1993-1-8 (2005)) and may be expressed as

$$F_{wx}^2 + F_{wy}^2 + F_{wz}^2 \leq F_{w,Rd}^2 \tag{7.3}$$

The term $F_{w,Rd} = a\, f_{vw,d} = a(f_u/3^{1/2})/(\beta_w\, \gamma_{M2})$ is the design strength of a fillet weld per unit length.

f_u = nominal ultimate tensile strength of the weaker part joined (Table 3.1, EN 1993-1-1 (2005))

β_w = correlation factor dependent on steel grade (Table 4.1, EN 1993-1-8 (2005))

γ_{M2} = partial safety factors for joints, recommended value 1,25 (Table 2.1, EN 1993-1-8 (2005)).

The design strength of a fillet weld has been shown (Ligtenberg, 1968) to be related to the strength of the parent material. The correct type and strength of electrode must be used for each grade of steel.

An alternative directional method of design (cl 4.5.3.2, EN 1993-1-8 (2005)) involves calculating the normal and shear stresses on the throat section of the weld and combining them using the yield criteria developed in Chapter 2. This method is more laborious and introduces the possibility of further errors when resolving the forces onto the critical plane. The relationship between the forces F_x, F_y and F_z for this method can be expressed I the same form as Eq. (7.3) for a 45° plane (Holmes and Martin, 1983). The size of the fillet weld obtained using this method is slightly less than using the vector addition method.

7.3.5 Load–Deformation Relationships for Fillet Welds

The strength of the weld in a connection is of primary importance but the load–deformation characteristics of the weld should also be considered. The deformation at the maximum load varies from approximately 0.6 to 1.4 mm depending on the orientation of the weld in relation to the applied load (Clarke, 1970) as shown in Fig. 7.6. The maximum deformation for the side fillet weld which is parallel to the applied load and the minimum is for an end fillet weld. To allow for this effect design stresses are based on the weaker side fillet welds. At the stress the disparity in deformations for end and side fillet welds is less than at failure.

7.3.6 Conditions Affecting the Strength of Welded Joints

The following conditions affect the strength of welded connections:

(a) Use of an incorrect steel (BS 7668, 1994).
(b) Use of an incorrect electrode (BSEN 499, 1995).
(c) Cavities and slag inclusions in the weld. These may be detected by non-destructive testing.

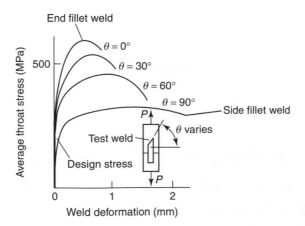

FIGURE 7.6 Load-deformation relationships for an 8 mm fillet weld (Clarke 1970)

(d) Excessive lack of fit between components.

(e) Stress concentrations combined with oscillating loads producing fatigue.

(f) Residual stresses introduced from differential heating during welding.

(g) Hydrogen cracks associated with welding occur when the cooling rate is too rapid (Fig. 7.7(a)). Excessive hardening occurs in the heat affected zone which cracks under the action of residual stresses if sufficient hydrogen is present in the weld. This defect can be avoided by controlling the cooling and the hydrogen input to the weld (BSEN 1011-1 (1998)).

(h) Lamellar tearing may occur when welding plate connections of the type shown in Fig. 7.7 (b). the cracks are produced by a combination of low ductility in the plate in the transverse direction and high point restraint in the weld which induces tensile forces adjacent to the connection. The low ductility in the plate is produced by inclusions of non-metallic substances formed in the steel making process. When the ingot is rolled to make steel these inclusions form as plates parallel to the direction of rolling. Only a small percentage of plates are susceptible to lamellar tearing, and where it occurs joint details can be changed to reduce the chances of if affecting the strength of the connection (Farrar and Dolby, 1972).

(i) Brittle fracture.

(j) Corrosion which reduces the size of components or causes pitting which may initiate fatigue cracks.

(k) Insufficient penetration of the parent metal which leads to a reduction in strength of the weld. The welder uses a voltage and arc length which produces a stable arc and a satisfactory weld profile. The current then becomes the main factor in controlling penetration. Another important factor in depth of penetration is edge preparation. Plates of 6 mm with square edges can be butt welded from one side, but the edges of thicker material must be bevelled to provide access for the arc.

(l) Lack of side wall fusion occurs if there is poor bond between the parent and weld metal. Good bonding can only occur when the surface of the parent metal has been melted before the weld metal is allowed to flow into the joint.

Further information on faults in welds can be found in Gourd (1980).

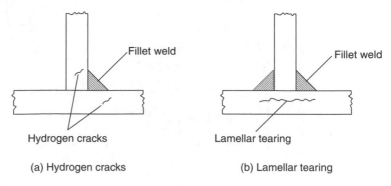

(a) Hydrogen cracks   (b) Lamellar tearing

FIGURE 7.7 Faults associated with welding

7.3.7 Design Strength of Fillet Welds (cl 4.5.3.3 and Table 4.1, EN 1993-1-8 (2005))

The design strength of a unit length of fillet weld (Eq. (4.4), EN 1993-1-8 (2005))

$$F_{w,Rd} = \text{(throat thickness)} \times \text{(unit length)} \times \text{(design shear stress)/factors}$$

$$= \frac{a\left(\frac{f_u}{3^{1/2}}\right)}{\beta_w \gamma_{M2}} \tag{7.4}$$

The factor (β_w) is related to steel grade, and varies between 0,8 and 1,0 (Table 4.1, EN 1993-1-8 (2005)). The ultimate stress (f_u) is that of the weaker material joined (Table 3.1, EN 1993-1-1 (2005)). Typical values are $f_u = 430$ MPa for S275 grade steel, and $f_u = 510$ MPa for S355 grade steel. The value $(f_u/3^{1/2})$ is from the shear distortion strain energy theory as explained in Chapter 2. It should be noted that although the strength of the weld is calculated using the throat thickness (a) (Figs 4.3 and 4.4, EN 1993-1-8 (2005)) the weld is often specified by the leg length. A table of the strength of fillet welds is given in Annex Al.

7.4 BOLTED JOINTS (cl 3, EN 1993-1-8 (2005))

The advantage of bolted joints is that they require less supervision than welding, and therefore are ideal for site conditions. Other advantages are that the connection can be fastened quickly, supports load as soon as the bolts are tightened and accommodates minor discrepancies in dimensions.

Disadvantages of bolted connections are that for large forces the space required for the joint is extensive, and the connection is not as rigid as a welded connection even when friction grip bolts are used.

Steel bolts are identified by their gross diameter, strength and use. The preferred sizes of bolts in general use are 16, 20, 24, 30 and 36 mm diameter. The most common size use in structural connections is 20 mm.

The types of bolt in common use (Table 3.1, EN 1993-1-8 (2005)) are:

(a) Ordinary bolts Classes 4.6–6.8 and includes foundation bolts.
(b) Pre-loaded bolts Classes 8.8 and 10.9.

A Class 4.6 bolt is low in cost, can be installed with the use of simple tools and requires little supervision during the erection. At fracture the bolt has a relatively large extension of 25%, a property which is preferred at plastic collapse.

Where forces are large, or where space for the connection is limited, or where erection costs can be reduced by using fewer bolts, then the higher grade bolts are used. The percentage elongation of 12% at failure is less, but is still acceptable for design purposes.

Where a more rigid bolted connection is required, for example in plastic methods of design, pre-loaded bolts are used. The strength of these bolts is greater with an increased cost for the additional site supervision which is necessary to ensure that the bolts are axially pre-loaded in tension to the design values. The object of the pre-load is to ensure that the friction between the 'faying' surfaces prevents slip when subject to external shear forces and thus produce a more rigid joint.

The nuts of pre-loaded bolts are tightened with a torque wrench which is calibrated in relation to the required axial pre-load. A simpler method of measuring the axial force in the bolts is to use a load indicating washer, under the head of the nut, which reduces in thickness to a specified value for a specified pre-load. The washer is less accurate than the torque spanner (Bahia and Martin, 1981). A further alternative method of ensuring that the bolt is pre-loaded is to specify 'turns of the nut'. Investigations (Fisher and Struck, 1974) showed that in general the pre-load produced by the torque wrench and 'turns of the nut' method on site exceeded the specified value.

Close tolerance turned bolts are used only where accurate alignment of components or structural elements is required. The shank of the bolt is at least 2 mm greater in diameter than the threaded portion of the bolt and the hole is only 0,15 mm greater than the shank diameter. This small tolerance necessitates the use of special methods to ensure that the holes align correctly.

Foundation bolts, or holding down bolts, are used for connection structural elements to concrete pads or concrete foundations. Generally the bolts are cast into the concrete before erection of the steel work and thus require accurate setting out. Where uplift forces occur the bolts must be anchored by a washer plate. Most bolts used are Class 4.6 but higher strengths are available. Sometimes bolts are grouted in the holes during erection using epoxy resin.

Rivets were used extensively in the past in the fabrication shop and on site. They were difficult and expensive to place but they resulted in a rigid connection because the hot rivet, after driving, expanded to fill the hole. Rivets have now been superseded by welding and bolts.

There are five categories of bolted joints (Table 3.2, EN 1993-1-8 (2005)) related to the type of connection or bolt and pre-loading.

Shear connections

Category A: *bearing type* where there is no pre-loading nor special provision for contact surfaces. Design for shear and bearing resistance. This is the cheapest type of connection where complete rigidity and plasticity are not important.

Category B: *slip resistant at serviceability limit state*. Design for slip resistance at the serviceability limit state and shear and bearing resistance at the ultimate limit state. Connection used to provide full rigidity in the elastic stage of behaviour when deflections are critical.

Category C: *slip resistant at ultimate limit state*. Design for slip resistance and bearing resistance at the ultimate limit state. Connection used to limit movements at the ultimate plastic limit state.

Tension connections

Category D: No pre-loading of bolts.

Category E: Pre-loaded bolts.

7.4.1 Washers

In British practice most bolts have steel washers under the head and under the nut. The washer distributes the bolt force and prevents the nut, or bolt head, from damaging the component or member. However washers are not essential in all cases (ECSS, 1981), and they are now being omitted in British practice.

Ordinary washers (BSEN 4320, (1998)) are in general use but hardened washers (BSEN 14399-1 to 5 (2005)) are used for pre-loaded bolts. The outside diameter of a washer is an important dimension when detailing, for example to avoid overlapping an adjacent weld.

7.4.2 Bolt Holes

Bolt holes are usually drilled, but may be punched full size, or punched under size and reamed. Holes should never be formed by gas cutting because of the inaccuracy and the effect on the local properties of the steel.

Punched holes are preferred by steel fabricators because it saves time and reduces cost. However research (Owens *et al.*, 1981) shows that distortion in the vicinity of a hole reduces toughness and ductility and can lead to brittle fracture. Punched holes should not be used in locations where plastic tensile straining can occur.

Bolt holes are made larger than the bolt diameter to facilitate erection and to allow for inaccuracies. The clearance is 2 mm for bolts not exceeding 24 mm diameter and 3 mm for bolts exceeding 24 mm diameter. Oversize and slotted holes are allowed but not often used. Slotted holes are sometimes used for pre-loaded bolts to facilitate erection with unusual shaped structures, or alternatively they can be used to accommodate movement in a structure. The clearance for a close tolerance turned bolt is 0,15 mm.

Bolt holes reduce the gross cross-sectional area of a plate to the net cross-sectional area. The net value is used for calculations where the structural element, or parts of an element, are in tension. Bolt holes also produce stress concentrations, but it is argued that these are offset by the fact that at yield the highly stressed cross-section will work harder before fracture and yield will by then have occurred at adjacent cross-sections. The gross cross-section of a member is used in compression because at yield the bolt hole deforms and the shank of the bolt resists part of the load in bearing. Consideration should be given to corrosion and local buckling when deciding the position of holes.

7.4.3 Spacing of Bolt Holes (Table 3.3, EN 1993-1-8 (2005))

The longitudinal spacing between the centre line of bolts in the direction of the axial stress in a member is called the pitch. The minimum spacing in the direction of the load of $2,2 \times$ (diameter of the hole) is specified to prevent excessive reduction in the cross-sectional area of a member, to provide sufficient space to tighten the bolts and to prevent overlapping of the washers. Other critical distances are given in Table 3.3 and Fig. 3.1, EN 1993-1-8 (2005). These values are specified to prevent buckling of plates in compression between bolts, to ensure that bolts act together as a group to resist forces, and to minimize corrosion.

7.4.4 Edge and End Distances for Bolt Holes (Table 3.3, EN 1993-1-8 (2005))

Edge and end distances are specified to resist the load, to prevent local buckling, to limit corrosion and to provide space for the bolt head, washer and nut.

The edge distance is from the centre of a hole to the nearest edge measured at right angles to the direction of the load. The minimum edge distance specified is $1,2 \times$ (diameter of the hole) and the maximum should not exceed $4t + 40$ mm.

The end distance is from the centre of a hole to the adjacent edge in the direction of the load transfer. The minimum end distance in the direction of the load is $1,2 \times$ (diameter of the hole) and the maximum should not exceed $4t + 40$ mm. The end distance should also be sufficient for bearing capacity.

There are recommended positions, spacing and diameter of holes in Section Tables. These distances are based on providing sufficent clearances to the web and adequate edge distances (Annexes A4 to A6).

7.4.5 Deductions for Holes in Tension Members (cl 3.10, EN 1993-1-8 (2005))

Holes are drilled in tension members to accommodate fasteners at connections. A hole reduces the gross cross-sectional area and weakens a tension member because the fastener in the hole does not transmit the axial force. In contrast a hole in a compression member has little effect on the buckling strength because as the member compresses the axial force is transmitted by bearing on the shank of the bolt.

When designing a tie the net cross-sectional area is used in calculations to determine the design axial force. The net cross-sectional area is the gross area reduced by the maximum sum of the sectional areas of the holes. These holes may be in line at right

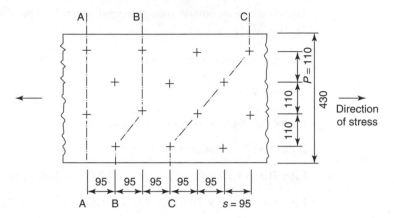

FIGURE 7.8 Example: net area of a plate

angles to the axial stress in the member (line AA in Fig. 7.8), or staggered (lines BB and CC in Fig. 7.8).

Typical areas to be deducted for bolt holes are:

Bolt diameter + 2 mm clearance for bolts not exceeding 24 mm in diameter or Bolt diameter + 3 mm clearance for bolts exceeding 24 mm in diameter.

For staggered fastener holes the area to be deducted shall be the greater of:

(a) the deduction for non-staggered holes;
(b) the sum of the sectional areas of all holes in any diagonal or zig-zag line extending progressively across the member, or part of the member, less $s^2 t/4p$ for each gauge space in the chain of holes.

where (Fig. 7.8)

s is the staggered pitch

p is the spacing of the holes

t is the thickness of the holed material.

For sections such as angles with holes in both legs the spacing is approximately the sum of the back marks to each hole, less the leg thickness. The arrangement and spacing of holes in a member should not significantly weaken a member at a section.

EXAMPLE 7.1 Net area of a plate with holes. Calculate the net cross-sectional area for the plate shown in Fig. 7.8 which is subject to a tensile force. The plate is 20 mm thick and contains four lines of staggered holes drilled for 24 mm diameter bolts.

From Fig. 7.8, $s = 95$ mm and $p = 110$ mm.

Diameter of hole $d_0 = d + 2 = 24 + 2 = 26$ mm.

Gross cross-sectional area perpendicular to the direction of stress $= 20 \times 430 = 8600\,\text{mm}^2$

Areas to be deducted at possible failure lines are:

$$n_\text{h}t\, d_\text{o} - n_\text{gs}\frac{s^2 t}{4p}$$

where n_gs is the number of gauge spaces in the chain of holes.

Line AA: $2 \times 20 \times 26 = 1040\,\text{mm}^2$

Line BB: $3 \times 20 \times 26 - 1 \times 95^2 \times 20/(4 \times 110) = 1150\,\text{mm}^2$

Line CC: $4 \times 20 \times 26 - 3 \times 95^2 \times 20/(4 \times 110) = 849\,\text{mm}^2$.

Minimum net area for line BB $= 8600 - 1150 = 7450\,\text{mm}^2$.

7.4.6 Design Resistance of Single Bolts (Tables 3.1 and 3.4, EN 1993-1-8 (2005))

Bolted connections consist of two or more bolts and each bolt may be subject to any combination of tension, shear or bearing forces. The design resistance of a single bolt when subject to these forces is now considered in detail.

A bolt in tension fails at the smallest cross-section, that is, the root of the threads where the net area is approximately 80% of the gross area. Where the bolt fails across the reduced cross-section at the root of the thread the tension resistance

$$F_\text{t,Rd} = k_2 f_\text{ub}\frac{A_\text{s}}{\gamma_\text{M2}} \tag{7.5}$$

where the ultimate tensile strength of the bolt (f_ub) is obtained from Table 3.1. EN 1993-1-8 (2005) and $k_2 = 0{,}9$ except for a countersunk bolt where $k_2 = 0{,}63$.

Where the bolt assembly fails by the bolt head, or nut, shear punching through the plate then the tension resistance of the plate (Table 3.4, EN 1993-1-8 (2005))

$$B_\text{p,Rd} = 0{,}6\,\pi\, d_\text{m}t_\text{p}\frac{f_\text{u}}{\gamma_\text{M2}} \tag{7.6}$$

where

f_u is the ultimate tensile stress of the plate.

t_p is the thickness of the plate under the head of the bolt.

d_m is the mean of across points and across flats dimensions of the bolt head or the nut, whichever is smaller.

The shear resistance per shear plane for a single bolt (Table 3.4, EN 1993-1-8 (2005))

$$F_\text{v,Rd} = \alpha_\text{v} f_\text{ub}\frac{A}{\gamma_\text{M2}} \tag{7.7}$$

Where the shear plane passes through the unthreaded portion of the bolt $\alpha_v = 0{,}6$ and A is the gross area.

Where the shear plane passes through the threads $\alpha_v = 0{,}5$ or $0{,}6$, depending on the grade of the bolt, and $A = A_s$ the reduced area.

The values of the reduced bolt areas used in this chapter (BS 3692, 2001; BS 4190, 2001) are:

Bolt diameter (mm)	12	16	20	22	24	27	30	36
Reduced area (mm^2)	84,3	157	245	303	353	459	561	817

Values may also be obtained from Section Tables.

Shearing of a bolt occurs on the shank, that is, the gross area of the bolt, if the thread length on a bolt is carefully specified, but it is safer to assume that it occurs on the reduced area. It also simplifies calculations and avoids confusion. Experiments (Bahia and Martin, 1980) and other investigators have found shear values that vary between $0{,}62f_u$ and $0{,}71f_u$. The value from the Huber–Von–Mises–Hencky shear distortion strain energy theory is $f_u/3^{1/2}$ as shown in Chapter 2.

Ordinary bolts deform when subject to shear stresses but it is important to realize that the shear deformation of the connection is increased by the bearing stresses on the plate. The higher the bearing stresses the greater the deformation as shown in Fig. 7.9.

7.4.7 Design Resistance of a Bolt Subject to Shear and Tension Forces (Table 3.4, EN 1993-1-8 (2005))

Shear and tensile resistances are related by the linear interaction formula

$$\frac{F_{v,Ed}}{F_{v,Rd}} + \frac{F_{t,Ed}}{1{,}4F_{t,Rd}} \leq 1{,}0 \tag{7.8}$$

Generally it is assumed that the failure plane passes through the threaded portion of the bolt.

This equation is to be compared with a non-linear experimental relationship (Chesson *et al.*, 1965) based on the net cross-sectional area of the bolt (Fig. 7.10). Alternative elliptical relationships are given in ECSS (1981) and BS 5400 (2000).

7.4.8 Design Bearing Resistance for a Bolt (Table 3.4, EN 1993-1-8 (2005))

A bolt subject to a shear force, such as shown in Fig. 7.11, comes in contact with the plate when the shear load is applied and slip occurs. The bearing stresses between the bolt and plate need to be controlled to limit deformations.

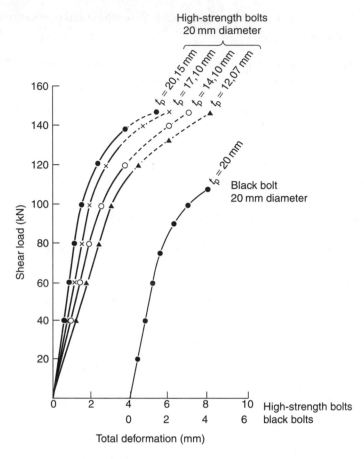

FIGURE 7.9 Relation between shear load and deformation for single bolt tests (Bahia and Martin, 1980)

Notes: Permanent deformation of holes shown as a broken line; bolts failed in single shear across the threads; t_p is the plate thickness.

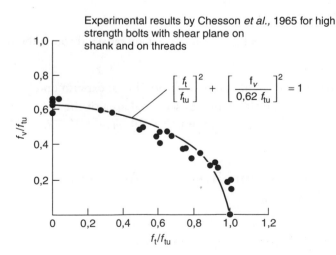

Experimental results by Chesson *et al.,* 1965 for high strength bolts with shear plane on shank and on threads

$$\left[\frac{f_t}{f_{tu}}\right]^2 + \left[\frac{f_v}{0{,}62\,f_{tu}}\right]^2 = 1$$

FIGURE 7.10 Relationship between shear and tensile stresses for bolts

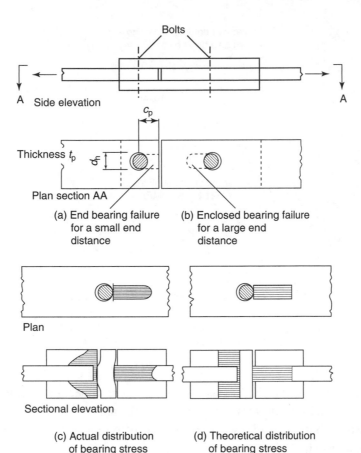

Bolts

A   Side elevation   A

Thickness t_p

Plan section AA

(a) End bearing failure
    for a small end
    distance

(b) Enclosed bearing failure
    for a large end
    distance

Plan

Sectional elevation

(c) Actual distribution
    of bearing stress

(d) Theoretical distribution
    of bearing stress

FIGURE 7.11 Bearing
stresses for bolts

The bolt or plate may deform because of high local bearing stresses between bolt and
plate, or a bolt may shear through the end of the plate. The bearing resistance

$$F_{b,Rd} = k_1 \alpha_b f_u \frac{dt}{\gamma_{M2}}$$ (7.9)

Values of k_1 control end distances and values of α_b control edge distances (Table 3.4,
EN 1993-1-8 (2005)). The value of f_u is the weaker of the bolt grade or the adjacent
plate or section.

The values of the ultimate tensile strength for bolts (Table 3.1 EN 1993-1-8 (2005))
used in this chapter are:

Bolt class	4.6	4.8	5.6	5.8	6.8	8.8	10.9
f_{ub} MPa	400	400	500	500	600	800	1000
α_v	0,6	0,5	0,6	0,5	0,5	0,6	0,5

Values of steel grades used in examples in this chapter are S275 ($f_u = 430$ MPa) and
S355 ($f_u = 510$ MPa).

Equation (7.9) assumes uniform bearing stresses as shown in Fig. 7.11(d) whereas in
reality they are closer to those shown in Fig. 7.11(c). Design bearing stresses are high

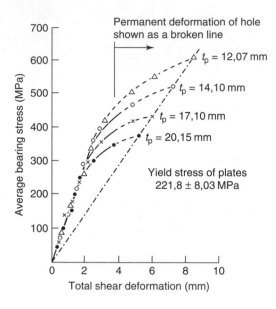

FIGURE 7.12 Relationships between bearing stress and deformation for an M20 high stress single bolt (Bahia and Martin, 1980)

Note: Bolts failed on the threads in single shear at a shear stress of 600 MPa

in relation to the yield stress because material subject to bearing stresses is generally confined by other parts which restricts deformation. High bearing stresses are not disastrous but lead to excessive deformation of a connection as shown in the experimental results (Fig. 7.12).

Equations (7.5)–(7.9) express the design tensile and shear strengths of a bolt and can be presented in the form of tables to reduce calculations (Annex A2).

7.4.9 Bolts Through Packings (cl 3.6.1(12), EN 1993-1-8 (2005))

Where the total thickness of the packing (t_p) is greater than three times the nominal diameter (d) of the bolts the design shear resistance is multiplied by a reduction factor

$$\beta_p = \frac{9d}{8d + 3t_p} \leq 1 \tag{7.10}$$

The reason for the reduction in strength is because as the grip length increases the bolt is subject to greater bending moments from shear forces which move further apart.

7.4.10 Long Bolt Joints (cl 3.8, EN 1993-1-8(2005))

For long joints the load is not shared equally by the bolts or rivets. The fasteners on the end resist the greatest force and the resistance gradually reduces to the centre line of the joint. The reduction is due to friction, errors in marking out and deformations

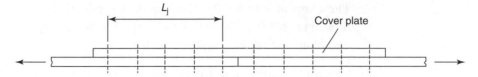

FIGURE 7.13 Length (L_j) for long bolted joints

in the materials. Where the length of the joint $L_j > 15d$ (Fig. 7.13) measured in the direction off the transfer force then the design shear resistance reduction factor

$$\beta_{Lf} = 1 - \frac{L_j - 15d}{200d} \qquad (7.11)$$

but $0,75 \leq \beta_{Lf} \leq 1,0$

7.4.11 Design of Slip Resistant Joints (cl 3.9, EN 1993-1-8 (2005))

High strength bolts can be pre-loaded when installed and designed to be slip resistant at working load or ultimate load. The design slip resistance of a pre-loaded bolt when subject to external tensile and shear forces is now considered in detail.

The design slip resistance of pre-loaded high strength bolt of Classes 8.8 or 10.9 (cl 3.9.1, EN 1993-1-8 (2005)) is

$$F_{s,Rd} = \frac{k_s \, n \, \mu F_{p,C}}{\gamma_{M3}} \qquad (7.12)$$

where

$F_{p,C} = 0,7 f_{ub} \, A_s$ is the design preloading force.

$\mu =$ slip factor as listed in Table 3.7, EN 1993-1-8 (2005).

$k_s = 1$ for a bolt in a clearance hole and reduced for slotted holes (Table 3.6, EN 1993-1-8 (2005))

$n =$ number of friction interfaces.

Typical Surface treatment	Slip factor (μ)
A – blasted with shot or grit, loose rust removed, no pitting	0,50
B – ditto, painted with zinc	0,40
C – cleaned and loose rust removed	0,30
D – surfaces not treated	0,20

Table 7.2 Slip factors(μ) for pre-loaded bolts (Table 3.7, EN 1993-1-8 (2005)).

The effect of a tensile force acting on a pre-loaded bolt is to reduce the frictional resistance (cl 3.9.2, EN 1993-1-8 (2005)). The design slip resistance per bolt subject to a combination of shear and tension forces is

Category B connection (Eq. (3.8a), EN 1993-1-8 (2005)) at service load

$$F_{s,Rd,ser} = \frac{k_s n \mu (F_{p,C} - 0.8 F_{t,Ed,ser})}{\gamma_{M3,serv}}$$ (7.13)

Category C connection (Eq. (3.8b), EN 1993-1-8 (2005)) at ultimate load

$$F_{s,Rd} = \frac{k_s n \mu (F_{p,C} - 0.8 F_{t,Ed})}{\gamma_{M3}}$$ (7.14)

The factor of 0,8 is introduced to allow for the fact that the minimum shank tension may not be achieved.

The design tensile, shear and bearing strength of a parallel shank pre-loaded bolt and can be presented in the form of a table to reduce calculations (Annex A3).

7.5 PLATE THICKNESSES FOR JOINT COMPONENTS

Plates, or parts of sections acting as flange plates, often form part of structural connections. The length and breadth of plates are generally determined from the geometry of the connection but the thickness is calculated from the elastic or plastic theory of bending.

Backing plates are used to strengthen flanges of columns as shown in Fig. 6.3, EN 1993-1-8 (2005). In particular, they are used to strengthen T-stubs and methods are given in Table 6.2, EN 1993-1-8 (2005). The limits on dimensions of the plates are given in cl 6.2.4.3, EN 1993-1-8 (2005).

7.5.1 Plastic Methods for Plates (cl 6.2.4, EN 1993-1-8 (2005))

Parts of connections may be idealized as T-stubs as shown in Fig. 7.14 where the external force (F_t) is balanced by bolt force (B_t) and prying force (Q). A prying force is an additional axial tensile force that is induced in a bolt due to the flexing and reaction of components.

There are three possible conditions of equilibrium for a T-stub at the ultimate limit state (Fig. 7.15).

Case 1: Bolt failure as shown in Fig. 7.15(a)

Resolving forces vertically

$$F_t + \Sigma B_t = 0$$ (7.15)

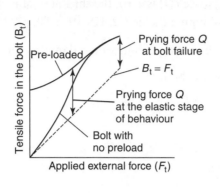

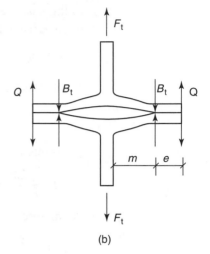

(a) Relationship between external force and bolt force for a T-stub

(b)

FIGURE 7.14   Prying forces for T-stubs

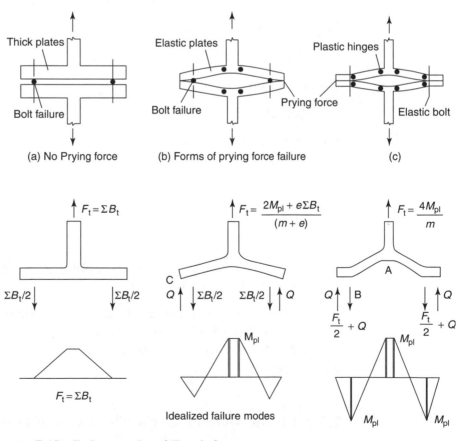

FIGURE 7.15   Failure modes of T-stub flanges

In this case the prying force (Q) is zero, the thickness of plate is a maximum and the force in the bolt is a minimum (Table 6.2, EN 1993-1-8 (2005), mode 3).

Case 2: Bolt failure with partial yielding of the flange as shown in Fig. 7.15(b)

Taking moments of forces about C

$$-M_{pl} + \frac{F_t}{2(m+e)} + \frac{\Sigma B_t}{2e} = 0$$

Rearranging, the design tension resistance

$$F_t = \frac{2M_{pl} + \Sigma B_t e}{m+e} \tag{7.16}$$

Prying force

$$Q = \frac{\Sigma B_t / 2m - M_{pl}}{m+e} \tag{7.17}$$

This is the case that is most likely to occur in practice because plate thicknesses are limited to those available (Table 6.2, EN 1993-1-8 (2005), mode 2).

Case 3: Complete yielding of the flange as shown in Fig. 7.15(c)

Taking moments of forces about A

$$-M_{pl} + (F_t/2 + Q)m - Q(m+e) = 0 \tag{i}$$

Taking moments about B

$$M_{pl} - Qe = 0 \tag{ii}$$

Combining (i) and (ii) to eliminate Q

$$F_t = \frac{4M_{pl}}{m} \tag{7.18}$$

They prying force

$$Q = \frac{M_{pl}}{e} \tag{7.19}$$

This case results in the smallest thickness of plate but the largest bolt force (Table 6.2, EN 1993-1-8 (2005), mode 1).

If needed the prying force can be calculated more accurately (Holmes and Martin, 1983) and the theory has been shown to agree with the experimental results (Bahia *et al.*, 1981). Prying forces can be avoided by using non-flexible components or by the use of stiffeners.

7.5.2 Plastic Method for the Thickness of Flange Plates

When considering T-stubs the length of the flange is known, but when analysing column flanges in similar situations (Fig. 7.16(a)) an effective length needs to be determined.

The method of determining the effective length is illustrated by a simple example for the plastic yield lines shown in Fig. 7.16(b).

For a single bolt force then from virtual work

$$\frac{F_t m \Delta}{m+e} = \left(\frac{t_p^2 f_{yp}}{4}\right) \Delta \left[\frac{4(m+e)}{x} + \frac{2x}{m+e}\right] \tag{i}$$

Differentiating with respect to x to determine the value of x for which F_t is a minimum

$$\frac{\partial F_t}{\partial x} = -4\frac{(m+e)}{x^2} + \frac{2}{m+e} = 0 \quad \text{hence } x = (m+e)\sqrt{2} \tag{ii}$$

Combining Eqs (i) and (ii) and rearranging, the thickness of the plate

$$t_p = \sqrt{\left(\frac{F_t m f_{yp}}{\sqrt{2}(m+e)}\right)} \tag{7.20}$$

From Eq. (i) the effective length

$$l_{eff,b} = \left[\frac{4m+e}{x} + \frac{2x}{m+e}\right](m+e) = 4\sqrt{2}(m+e)$$

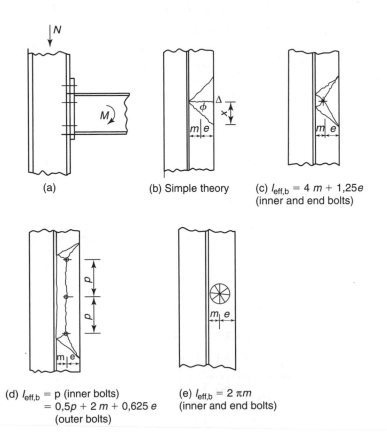

(a)   (b) Simple theory   (c) $l_{eff,b} = 4\,m + 1{,}25e$ (inner and end bolts)

(d) $l_{eff,b} = p$ (inner bolts) $= 0{,}5p + 2\,m + 0{,}625\,e$ (outer bolts)   (e) $l_{eff,b} = 2\,\pi m$ (inner and end bolts)

FIGURE 7.16 Effective lengths for column flanges

The work Eq. (i) can therefore be written as an equilibrium equation

$$F_t m = \left(\frac{t_p^2 f_{yp}}{4} \right) l_{eff,b} \tag{7.21}$$

The above theory assumes simple idealized conditions and ignores, welds, washers, limited rotation at the plastic hinge and strain hardening. However the simple example shows the basic theoretical approach. Other solutions are related to research work (Stark and Bijlaard, 1988).

In practice the yield lines are more complicated (Figs 7.16(c), (d) and (e)) and more difficult to analyse. Analysis using yield lines is avoided in the European Code by giving the effective length of plate ($l_{eff,b}$) (Fig. 7.16) and applying the T-stub equations (Table 6.4, EN 1993-1-8 (2005)).

7.5.3 Elastic Methods for Plates

(a) *Base plates* are used to distribute the load from column sections as shown by the area enclosed by the dotted line in Fig. 7.17. Elastic stress distribution with the maximum stress at the yield strength is used at the ultimate limit state to ensure that large displacements do not occur. If the load is axial the pressure (f_j) beneath the base plate is uniform and the projection (c) of the steel plate beyond the edge of the column, then for a cantilever from the simple theory of elastic bending at first yield per unit width

$$M = f_y W$$
$$\frac{f_j c^2}{2} = \frac{f_y t^2}{6}$$

rearranging, the projection of the plate (Eq. (6.5), EN 1993-1-8 (2005))

$$c = t \left[\frac{f_y}{(3 f_j \gamma_{M0})} \right]^{1/2} \tag{7.22}$$

Justification for this theory is given in Holmes and Martin (1983).

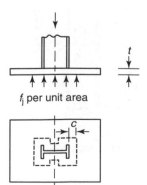

f_j per unit area

FIGURE 7.17 Bending strength of a column base plate

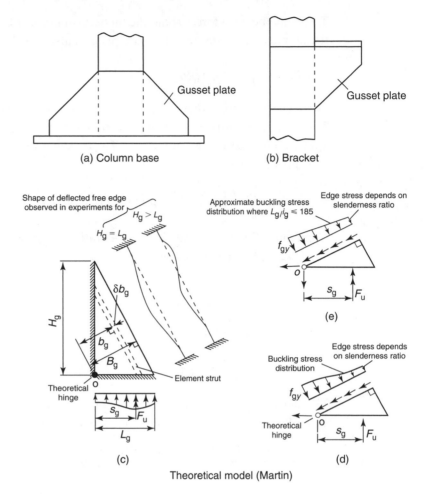

FIGURE 7.18 Gusset plates

(b) *Gusset plates* are used to stiffen base plates and brackets as shown in Fig. 7.18. The following theory for the buckling of a gusset plate is based on experimental work (Martin, 1979; Martin and Robinson, 1981). No advice is given in the European Code.

The basic structural unit is a triangular plate with loading applied to one edge as shown in Fig. 7.18(c). For theoretical purposes the plate is assumed to be composed of a series of fixed ended struts parallel to the free edge. The distribution of direct stress across the width b_g is shown on an element of the gusset plate in Fig. 7.18(d). The buckling stress varies depending on the slenderness ratio of the elemental strut. At the hinge the stress is for a very short strut (i.e. yield). At the free edge the value is for the slenderness ratio of the elemental strut at the free edge.

For simplicity the buckling stress distribution shown in Fig. 7.18(d) can be replaced by a linear distribution as shown in Fig. 7.18(e), provided that the slenderness ratio of the free edge $l_g/i_g < 185$. This restraint is acceptable because slenderness ratios of gusset plates in structural engineering do not often exceed this value.

Taking moments of forces about the theoretical hinge at O (Fig. 7.18(e)) and ignoring the moment of resistance of the base plate as justified (Martin and Robinson, 1981).

$$F_u s_g = \int_0^{B_g} (b_g f_g t_g) \delta b_g \tag{7.23}$$

For each strip the buckling stress (f_g) is linearly related to the slenderness ratio (l_g/i_g). The effective length $l_g = b_g$ when $L_g = H_g$, and from experiments (Martin, 1979) this is approximately correct when $L_g = H_g$. The buckling stress for each strip can therefore be expressed as

$$f_g = f_{gy} \left[1 - \left(\frac{b_g}{185 i_g} \right) \right] \quad \text{provided that } l_g/i_g < 185 \tag{7.24}$$

Combining Eqs (7.23) and (7.24), integrating, expressing the radius of gyration as $i_g = t_g/(2 \times 3^{1/2})$, and rearranging

$$t_g = \frac{2 F_u s_g}{f_{gy} B_g^2} + \frac{B_g}{80} \tag{7.25}$$

where from the geometry of the plate

$$B_g = \frac{L_g}{[(L_g/H_g)^2 + 1]^{1/2}} \tag{7.26}$$

The slenderness ratio of the gusset plate may be defined as the slenderness ratio of a strip of unit width parallel to the free edge. From this definition and Eq. (7.26)

$$\frac{l_g}{i_g} = \frac{(2 \times 3^{1/2}) B_g}{t_g} = \frac{2 \times 3^{1/2} (L_g/t_g)}{[(L_g/H_g)^2 + 1]^{1/2}} \tag{7.27}$$

This theory is for non-slender gusset plates, that is, for $l_g/i_g < 185$. The theory for slender gusset plates is given elsewhere (Martin, 1979).

7.6 JOINTS SUBJECT TO SHEAR FORCES

Two simple connections subject to shear forces are shown in Fig. 7.19. The forces in the members are assumed to be axial and to act through the centroidal axes of the members. This is correct in some situations, for example the bolted joint shown in Fig. 7.19(a). However it is not correct for the welded lap connection shown in Fig. 7.19(b) because the eccentricity of the force produces a moment which results in distortion at ultimate load. It is not correct for a roof truss joint as shown in Fig. 7.24 because although the centroidal axes intersect and there are axial forces in the members there are also secondary moments.

A further assumption for simple joints is that the external forces are distributed evenly to the bolts or welds. This is not correct for long bolted and welded joints and allowance must be made for this.

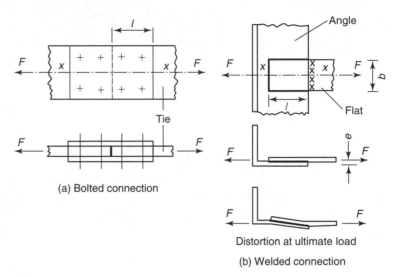

FIGURE 7.19   Joints subject to shear forces

The overlap distance (l) is important for simple joints. For bolted joints the minimum of two bolts and the required end distances generally ensure that the lap is sufficient. However for welds (Fig. 7.19) the greater strength may indicate that the lap distance can be small, but it must be appreciated that there must be room for stop and start lengths and that stress concentrations can occur. The minimum lap length is generally not be less than four times the thickness of the thinner part joined where the weld is continuous. The minimum length of weld is 40 mm or six times the throat thickness.

For a joint with side welds only the lap is generally not less than the width of the member and there should be end returns of twice the leg length of the weld (cl 4.3.2.1(4), EN 1993-1-8 (2005)) to reduce stress concentrations.

7.7 JOINTS SUBJECT TO ECCENTRIC SHEAR FORCES

Joints, such as shown in Fig. 7.20(a), are subject to eccentric shear forces which tend to rotate the joint. This produces a resultant shear force on a fastener (bolt, or unit length of weld) from the direct shear force and the moment.

The forces acting on a group of fasteners can be idealized as shown in Fig. 7.20(b). The bolt group rotates about the theoretical instantaneous centre of rotation which varies in position depending on the magnitudes of the external forces V and H and the eccentricity e. In the linear elastic stage of behaviour it is reasonable to assume that the force acting on a fastener is proportional to the distance from the centre of rotation. At ultimate load this assumption is not strictly correct but the error involved is not great. For a rigorous solution to the theory and accuracy at ultimate load see Bahia and Martin (1980).

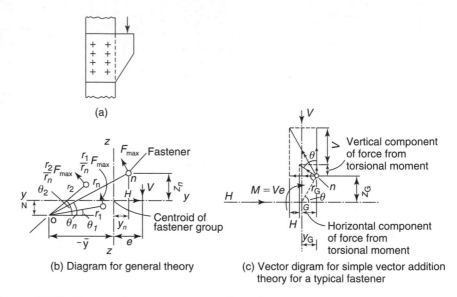

(a)

(b) Diagram for general theory

(c) Vector digram for simple vector addition theory for a typical fastener

FIGURE 7.20 Joints subject to eccentric shear forces

Although this is the correct approach to the theory there is a simpler more practical method in common use which gives the same values. It is assumed that rotation occurs about the centroid of the fastener group and for convenience the forces acting on a fastener are parallel to the z–z and y–y axes as shown in Fig. 7.20(c) There forces are combined vectorially and the resultant force on a fastener furthest from the centre of rotation is

$$F_R = [F_y^2 + F_z^2]^{1/2} = \left[\left(\frac{V}{n} + \frac{My_G}{I_x} \right)^2 + \left(\frac{H}{n} + \frac{Mz_G}{I_x} \right)^2 \right]^{1/2} \tag{7.28}$$

where

n is the number of fasteners in the group

y_G and z_G are coordinates of a fastener related to the centroid of the fastener group

$$I_X = I_Y + I_Z$$

I_X, I_Y and I_Z are second moments of area of unit size fasteners about the X–X, Y–Y and Z–Z axes

The method is used in practice for ordinary bolts, high strength friction grip bolts and welds.

7.8 JOINTS WITH END BEARING

Some joints involve end bearing between components (Fig. 7.21). End bearing can occur in beam-to-column (Fig. 6.15, EN 1993-1-8 (2005)), bracket-to-column, beam-to-beam and column-to-base joints.

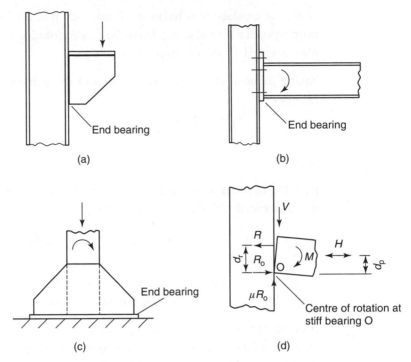

FIGURE 7.21 Joints with end bearing

Where end bearing occurs, rotation takes place about a stiff axis of rotation, axis O–O shown in Fig. 7.21(d). The reaction force R_o is generally large and the bearing may have to be reinforced if it is not to distort under the load. The balancing tensile force R is resisted by bolts or welds. If there is slip at the stiff bearing a frictional force μR_o develops parallel to the stiff bearing surface.

Consider an end bearing joint subject to external forces V, H and moment M as shown in Fig. 7.21(d). If slip occurs resolving forces vertically

$$V - \mu R_o < 0 \tag{7.29}$$

Taking moments of forces about force R

$$M + -H(d_r - d_p) - R_o d_r = 0 \tag{7.30}$$

where R is the resultant force of the fasteners acting at a distance d_r from the axis of rotation O–O.

Combining Eqs (7.29) and (7.30) to eliminate R_o, then slip will not occur if

$$\frac{\mu[M \pm H(d_r - d_p)]}{Vd_r} > 1 \tag{7.31}$$

Most joints with end bearing do not slip and therefore the fasteners are not subject to the external shear force. One exception is a bracket supporting a load with a small eccentricity.

In the elastic stage of behaviour it is assumed that the forces acting on a fastener are proportional to the distance from the axis of rotation O–O. If, conservatively, this is also assumed to occur at ultimate load then:

Taking moment of forces about axis O–O the maximum tensile force resisted by a fastener

$$F_{t(max)} = \frac{[M \pm Hd_p]z_{max}}{I_O} \qquad (7.32)$$

where

$I_O = \Sigma z^2$ is the second moment of area of unit size fasteners about axis O–O Resolving forces vertically the shear force on a fastener

$$F_s = \frac{(V - \mu R_o)}{n} = \frac{\left\{V - \mu\left[\frac{M}{d_r} \pm H(1 - d_p/d_r)\right]\right\}}{n} \qquad (7.33)$$

This equation is formed assuming that bolts of the same size and design strength resist equal shear forces and also that slip has taken place.

If the fasteners are welds then $F_{t(max)}$ and F_s are combined vectorially. If the fasteners are ordinary bolts the forces are combined using Eq. (7.8) and related to design strengths. If friction grip bolts are used the forces are combined using Eqs (7.13) and (7.14) and related to design strengths.

Traditional design methods ignore the existence of the frictional force which errs on the side of safety. However, research (Bahia *et al.*, 1981) has shown that the frictional force does resist part of the shear force. If pre-loaded bolts are used then slip may occur at service load or at ultimate load.

7.9 'PINNED' JOINTS (CL 3.13, EN 1993-1-8 (2005))

Some simple joints (e.g. a tie bar) are connected by real pins as shown in Fig. 7.22(a). Provided that the pins are not corroded, or blocked with debris, they will act as pin joints, that is, they will resist forces but not moments. Tie bars are rarely used now because of the cost of manufacture, risk of seizure from corrosion or debris, and because safety depends on a single pin.

Other connections shown in Fig. 7.22 are designated as 'pins' because the rotational restraint is small. In the past these joints have been designed assuming that the rotational resistance is zero and the connection resists direct forces only. The general approach to design of these 'pinned' connections follows.

7.9.1 Pinned Beam-to-Column Joints (Figs 7.22(b), (c), (e) and (f))

In design calculations for the joint shown in Fig. 7.22 (b) it is assumed that the shear force is resisted by the four bolts connecting the bottom cleat to the column flange.

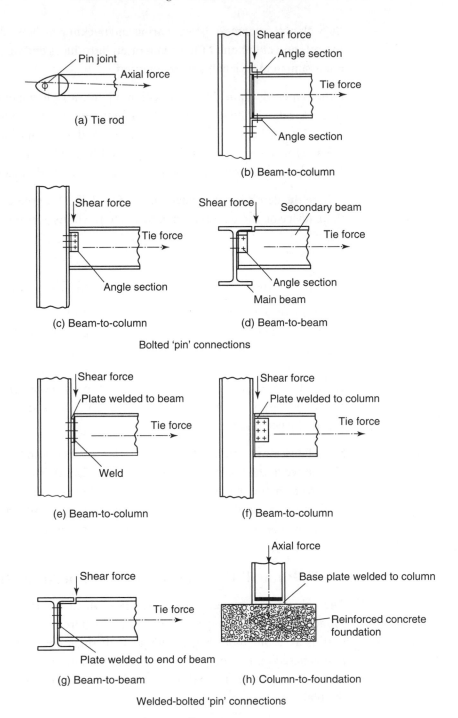

FIGURE 7.22 Examples of 'pinned' joints

During erection the bottom cleat, which is bolted or welded to the column, is used as a marker by the crane operator when the beam is placed. The top cleat is assumed to resist no vertical load but it does provide torsional resistance which is important for lateral stability. The top and bottom cleats also resist the tie force. The resistance of the web of the beam to shear, bearing and buckling must be checked. At ultimate load the rotation at the end of the beam often introduces end moments. In practice, these are assumed to be small and are ignored.

Other types of 'pinned' beam-to-column joints are shown in Figs. 7.22(c), (e) and (f) where the depth of the joint is kept to a minimum. For example, the end plate depth for Fig. 7.22(e) is kept to a minimum to reduce end moments and the empirical thickness is 8 mm for UB sizes up to 457×191 kg, and 10 mm for sizes greater than 533×210 kg. Examples of 'pin' joints used in practice are given by Pillinger (1988).

Where the depth of the connection for Fig. 7.22(e) is greater and is also slip resistant then advice on the distribution of forces to the bolts is given in Fig. 6.15, EN 1993-1-8 (2005).

7.9.2 'Pinned' Beam-to-Beam Joints (Figs 7.22 (d) and (g))

The transverse secondary beam is connected to the main beam through angle cleats as shown in Fig. 7.22(d). It is assumed in design that the shear force is transferred to the main beam via the bolts in the web of the main beam. These bolts are therefore designed for single shear and bearing on the web of the main beam and on the angle cleats. It follows therefore that the shear force is eccentric to the bolts in the web of the secondary beam. These bolts are double shear and bearing on the web of the secondary beam and the angle cleats.

Steel fabricators and erectors often prefer a welded and plate (Fig. 7.22 (g)) as an alternative to the angle cleats shown in Fig. 7.22 (d). This results in a more rigid joint and an end moment is introduced to the end of the secondary beam which is dependant on the torsional stiffness of the main beam. If there are secondary beams on both sides of the main beam the secondary moment can be large.

7.9.3 'Pinned' Column-to-Foundation Joints (Fig. 7.22 (h))

The column is fastened to the base plate which is connected to the foundation by foundation bolts. This type of joint is used where the predominant force in the column is axial but there is generally a small shear force. The size of the foundation bolt is based on resisting the forces, but with a minimum size of M16. The thickness of the base plate is related to pressure beneath the base and the cantilever effect the plate.

7.10 'RIGID' JOINTS

'Rigid' (or 'fixed') joints (Fig. 7.23) exhibit small rotational displacements in the elastic stage of behaviour. They are now more common as connections to members are welded in the workshops and bolted on site. They are useful to limit deflections of members,

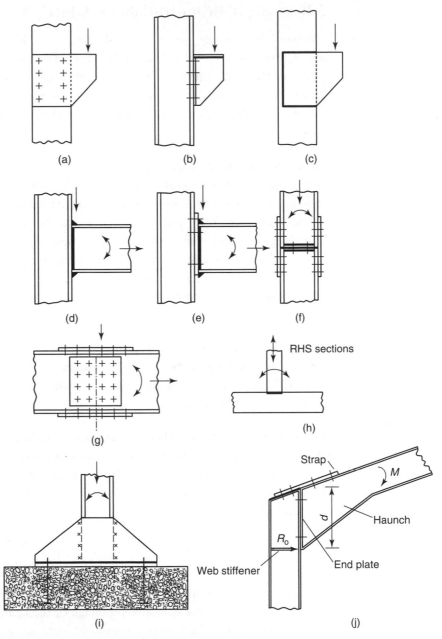

FIGURE 7.23 Examples of 'rigid' joints

resist fatigue and resist impact loading. However, generally, the design procedure for the frame requires global analysis and the components are more highly stressed.

7.10.1 'Rigid' Column Bracket Joints (Figs 7.23 (a), (b) and (c))

For the brackets shown in Figs 7.23 (a) and (c) no end bearing is involved whereas the bracket shown in Fig. 7.23 (b) involves end bearing.

7.10.2 'Rigid' Beam-to-Column Joints (Figs 7.23 (d) and (e))

The most rigid type of joint is where the beam is welded directly to the column on site, as shown in Fig. 7.23(d), but this is expensive and it is difficult to control the quality of the weld. Alternatively stub cantilever beams can be welded to the column in the workshops and a suspended beam site bolted between the ends of the cantilever. The bolted connection is positioned as close to the point of contraflexure as possible.

The method commonly used in Britain is to weld end plates to the beam in the workshops and to bolt these to the columns on site as shown in Fig. 7.23(e). The numbers of bolts is usually six, as shown, because of the limited depth available. If the moment of resistance needs to be increased then it is necessary to increase the lever arm by haunching the beam at the end. This connection is less rigid than welding the beam directly to the column but it is easier to manufacture and erect. The amount of rotation depends on the thickness of the end plate, thickness of the column flanges and the extensibility of the bolts. Advice on details is given in cl 6.2.7.2, EN 1993-1-8 (2005).

If the connection is close to a plastic hinge then it must be decided whether the plastic hinge should forming the beam, or the column, or the connection. Recent research favours the formation of the hinge in the beam and therefore the column and the connection must be overdesigned.

End bearing occurs between beam and column and the first step in design is to check whether slip occurs using Eq. (7.31). Generally, because the bending moment is large, slip does not occur and the size of the top four bolts required can be determined approximately by taking moments of forces about the compression flange of the beam. The tensile force in a bolt

$$F_t = \frac{M}{4(h - t_f)} \tag{7.34}$$

The tensile force in a bolt is increased by a prying force. Equation (7.34) assumes that the two bolts close to the compression flange of the beam do not resist any part of the bending moment. Lever arm recommendations are given in Fig. 6.15 (EN 1993-1-8 (2005)), The thicknesses of the end plate is determined by assuming equivalent T-stubs and associated yield lines. This method can be compared with a survey of existing literature on end plates by Mann and Morris (1979) who recommended

$$\left[\frac{M}{d_{bf}f_y \left(\frac{4w_p}{s_v} + \frac{d_{bf}}{s_h} \right)} \right]^{1/2} < t_p < \left[\frac{Mb_p}{2w_p d_{bf}f_y} \right]^{1/2} \tag{7.35}$$

provided that $B_p = 9d_b$, $s_h = 6d_b$, $s_v = 6d_b$ and $e_b > 2{,}5d_b$.

The thickness of the column flange can also be determined from equivalent T-stubs with yield line patterns. For comparison a further survey by Mann and Morris (1979) recommended:

For unstiffened column flanges the thickness of the flange

$$0{,}28\left[\frac{M}{d_{bt}f_y}\right]^{1/2} < t_{cf} < 0{,}39\left[\frac{M}{d_{bt}f_y}\right]^{1/2} \tag{7.36}$$

For stiffened column flanges the thickness of the flange

$$0{,}23\left[\frac{M}{d_{bt}f_y}\right]^{1/2} < t_{cf} < 0{,}32\left[\frac{M}{d_{bt}f_y}\right]^{1/2} \tag{7.37}$$

Equations (7.36) and (7.37) are valid provided that $b_c = 2{,}5d_b$ and $c_c = t_{st} + 5d_b$. Where the column flange thickness is inadequate backing plates can be used.

The force R_o at the axis of rotation may produce failure by bearing or buckling of the web of the column. If failure is likely to occur then stiffeners can be welded into the web of the column. However, stiffeners increase costs and reduce the room available for bolts. The tensile force balancing R_o (Fig. 7.21) which acts on the web of the column is often not critical but the strength of the column web must be checked.

The size of the fillet weld connecting the end plate to the beam is determined by assuming rotation about the axis O–O at the bottom flange of the beam, but an alternative more conservative method is to assume rotation about the centroidal axis of the weld.

7.10.3 'Rigid' Beam Splices (Fig. 7.23(g))

Beam splices are introduced to extend standard bar lengths, or to facilitate construction and transport. Splices are generally located at sections where the forces are minimum, and to avoid local geometrical deformations of the structure pre-loaded bolts are used. The connection is usually made on site and therefore bolts are used.

A simple design method is to assume that the entire shear force is resisted by the web splice and the flanges resist the entire bending moment. These assumptions are not correct but it simplifies the design and the errors are generally not large. Alternatively part of the moment is assumed to be resisted by the web. The bolts in the web, in double shear, are subject to an eccentric shear load. The bolts in the flange are also in double shear for two flange plates and the force on a bolt is obtained a simple moment equation $F = M/(nd)$.

7.10.4 'Rigid' Column-to-Column Joints (Fig. 7.23(f))

Column splices are used to extend standard bar lengths, to facilitate erection and transport, and for economy by reducing the section size. Where bending moments are large then, as with beam splices, pre-loaded bolts are used to maintain the axial line of the column. Where sections change size steel packings and an end plate are required to ensure that a good fit is obtained.

Generally the ends of the columns are in contact, or in contact with the end plate, and therefore end bearing occurs. Any shear force will be resisted by the friction at the end bearing and the web plates (or cleats). Where forces are small sizes of components are decided from experience, practicality and corrosion resistance.

7.10.5 'Rigid' Hollow Section Joints (Fig. 7.23(h))

This is a typical 'T' joint for a Vierendeel girder where members of different width intersect. If forces and moments are not too large then the connection can be made without using stiffeners. Rectangular hollow sections are also used in braced triangulated trusses but it is difficult to arrange for the centre lines of the members to intersect at a point. The offsets introduce moments which should be taken into account in the analysis of the structure (Purkiss and Croxton, 1981). Further information on analysis and design of connections is given by Davies (1981).

Circular hollow sections are also used for trusses but the geometry of the connection is more difficult and connections are more expensive to form. A review of methods of analysis of the strength of these connections is given by Stamenkovic and Sparrow (1981).

The failure modes for hollow section joints are (cl 7.2.2, EN 1993-1-8 (2005)):

(a) chord face failure,
(b) chord side wall failure,
(c) chord shear failure,
(d) punching shear failure,
(e) brace failure,
(f) local buckling.

The design resistance of numerous types of joints are shown in Tables 7.2–7.24, EN 1993-1-8 (2005). These take into account in plane and out of plane forces.

Guidance on strength of welds for connecting members is given in cl 7.5.1, EN 1993-1-8 (2005). These are only applicable if they meet the joint parameters of Table 7.8, EN 1993-1-8 (2005).

7.10.6 'Rigid' Column-to-Foundation Joints (Fig. 7.23(i))

Where bending moments are small a simple slab base is satisfactory with the bolts in line with the column axis. As the bending moment increases the bolts are off-set from the column axis. A built-up base is used where moments are large. The basic design method is based on 'T' stub theory (cl 6.2.8, EN 1993-1-8 (2005)).

For a gussetted slab base subject to an axial load and bending moment it is assumed that at ultimate load the distribution of stresses beneath the steel base plate are as shown in Fig. 7.34(c). The depth of the compression zone is determined approximately by taking moments of forces about the tensile bolts

$$x = \frac{\left[\frac{M}{d} + \frac{N}{2} \right]}{b_c f_{ij}} \tag{7.38}$$

where b_c is based on the cantilever length (c) and f_{ij} is the allowable bearing pressure.

Alternatively it can be assumed that the centroid of the compression zone is located under the column flange (Table 6.7, EN 1993-1-8 (2005)).

From taking moments of forces about the compressive force the tensile force in a bolt is

$$F_t = \frac{\left[\frac{M}{d} - \frac{N}{2} \right]}{n} \tag{7.39}$$

The thickness of the base plate is found based on elastic bending of the cantilever length (c).

For the built-up base the base plate thickness is determined by the same method. The gusset plate size is determined by the method shown in Section 7.5.3. The size of the welds connecting the gusset plate to the base and to the column can be determined assuming end bearing. Alternatively end bearing can be ignored and a larger size determined by assuming rotation about the centroid of the weld group.

7.10.7 'Rigid' Knee Joint for a Portal Frame (Fig. 7.23(j))

This type of joint is similar to the beam-to-column connection shown in Fig. 7.23(e). However because portal frames are often designed using plastic analysis the magnitude of the reaction $R_o = M/d$ is generally high and consequently the shear stresses in the web are close to the limit. The magnitude of R_o is reduced by haunching the beam and consequently increasing the distance d, but web stiffeners are generally required for the column. To reduce the shear deformation in the column web stiffeners can be introduced as described and tested by Morris and Newsome (1981).

The tensile force which balances R_o can be resisted by a strap, or by a group of bolts through the flange of the column. The strap may interfere with the placing of purlins in the roof and the alternative group of bolts may cause distortion of the column flanges if the column flange thickness is small. The distortion can be controlled by the use of flange stiffeners, or increase the size of the column section, but both increase the cost.

EXAMPLE 7.2   Design of a 'pin' joint for a roof truss (Fig. 7.24). The forces, size of angles and tees have been obtained from an analysis at ultimate load assuming pin joints and axial forces in the members (cl 5.1.5(2), EN 1993-1-8 (2005)).

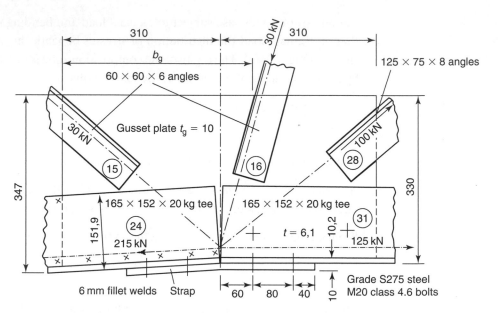

FIGURE 7.24 'Pinned' roof truss joint

The centroidal axes of the members intersect so there is no eccentricity to be taken into account (cls 2.7 and 3.10.3, EN 1993-1-8 (2005)). The thickness of the gusset plate is at least 6 mm to resist corrosion, and at least equal to the minimum thickness of the angle or tee (6,1 mm). Use a 10 mm thick plate grade S275 steel. A rectangular plate is simple to mark and cut, and low in fabrication cost. Alternatively a more complicated shape can be used which is aesthetically more acceptable but the fabrication cost is greater.

Member 24, structural tee cut from a UB ($165 \times 152 \times 20$ kg) welded to a gusset plate, design force $N_{Ed} = 215$ kN.

Tensile resistance of the tee Grade S275 steel (Eqs (6.6) and (6.7), EN 1993-1-1 (2005))

$$N_{pl,Rd} = \frac{A f_y}{\gamma_{M0}} = \frac{2580 \times 275}{1,00 \times 1E3}$$

$$= 709,5 > 215 \text{ kN satisfactory.}$$

$$N_{u,Rd} = 0,9 A_{net} \frac{f_u}{\gamma_{M2}} = 0,9 \times (2580 - 2 \times 22) \times \frac{430}{(1,25 \times 1E3)}$$

$$= 766,9 > 215 \text{ kN satisfactory.}$$

Assuming a 6 mm fillet weld for Grade S275 steel (Eq. (4.4), EN 1993-1-8 (2005))

$$F_{w,Rd} = \frac{f_u a}{(3^{1/2} \beta_w \gamma_{M2})}$$

$$= \frac{430 \times 0,7 \times \frac{6}{1E3}}{(3^{1/2} \times 0,85 \times 1,25)} = 0,981 \text{ kN/mm}$$

Effective length of weld required to resist the tensile force using the simplified method (cl 4.5.3.3, EN 1993-1-8 (2005))

$$\frac{N_{Ed}}{F_{w,Rd}} = \frac{215}{0,981} = 219,2 \, \text{mm}$$

Two side fillet welds of 110 mm length would be satisfactory but in practice the lengths would probably be the full overlap (i.e. $2 \times 310 = 620$ mm).

Member 31, structural tee cut from a UB ($165 \times 152 \times 20$ kg) bolted to the gusset plate and strap, design force $N_{Ed} = 125$ kN.

Resistance of 4-M20 class 4.6 bolts in single shear (Table 3.4, EN 1993-1-8 (2005))

$$\Sigma F_{v,Rd} = n_b \left(0,6 A_s \frac{f_{ub}}{\gamma_{M2}} \right)$$

$$= 4 \times \left[0,6 \times 245 \times \frac{400}{1,25 \times 1E3} \right] = 188 > (N_{Ed} = 125) \, \text{kN satisfactory.}$$

Resistance of 2-M20 class 4.6 bolts in bearing on the web of the tee section ($t = 6,1$ mm) (Table 3.4, EN 1993-1-8 (2005))

$$\Sigma F_{b,Rd} = n_b \left(\frac{k_1 \alpha_b f_{ub} d t}{\gamma_{M2}} \right)$$

$$= 2 \times \left[\frac{2,5 \times 0,909 \times 400 \times 20 \times 6,1}{1,25 \times 1E3} \right]$$

$$= 177,4 > (N_{Ed} = 125) \, \text{kN satisfactory}$$

where

for an end bolt $\alpha_b = e_1/(3 d_o) = 60/(3 \times 22) = 0,909$

for an edge bolt $k_1 = 2,8 e_2/d_o - 1,7 = 2,8 \times 55/22 - 1,7 = 5,3 > 2,5$

The strap increases the out of plane stiffness of the truss. Connections for other members can be designed by the same method.

EXAMPLE 7.3   'Rigid' column bracket. Determine the size of the components required to connect the bracket to the column shown in Fig. 7.25 using Grade S355 steel. The forces shown are applied to one gusset plate at ultimate load.

For the 10 bolts (Fig. 7.25(a)) of unit cross-sectional area the properties of the bolt group are:

Second moment of area of the bolt group about the centroidal y–y axis

$$I_y = \Sigma(\partial A) z^2 = 4(80^2 + 160^2) = 128E3 \, \text{mm}^4$$

Second moment of area of the bolt group about the centroidal z–z axis

$$I_z = \Sigma(\partial A) y^2 = 10(70)^2 = 49E3 \, \text{mm}^4$$

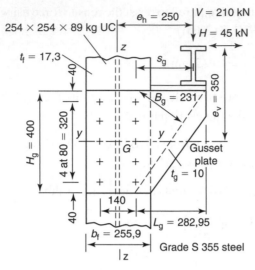

(a) Bracket bolted to a UC   (b) Bracket welded to a compound column

FIGURE 7.25 'Rigid' column brackets

Second moment of area of the bolt group about the centroidal polar x–x axis

$$I_x = I_y + I_z = (128 + 49)\text{E3} = 177\text{E3 mm}^4$$

From Eq. (7.28) the maximum vector force in the direction of the z–z axis on a bolt furthest from the centroid of the bolt group

$$F_z = \frac{V}{n_b} + \frac{(Ve_h + He_v)y_n}{I_x}$$

$$= \frac{210}{10} + \frac{(210 \times 250 + 45 \times 350)70}{177\text{E3}} = 48 \text{ kN}$$

The maximum vector force in the direction of the y–y axis on the same bolt

$$F_y = \frac{H}{n_b} + \frac{(Ve_h + He_v)z_n}{I_x}$$

$$= \frac{45}{10} + \frac{(210 \times 250 + 45 \times 350)160}{177\text{E3}} = 66,2 \text{ kN}$$

Resultant vector design force on this bolt

$$F_r = (F_z^2 + F_y^2)^{1/2} = (48^2 + 66,2^2)^{1/2} = 81,77 \text{ kN}$$

Solution (a) using class 4.6 bolts. Shear resistance of a M30 class 4.6 bolt in single shear (Table 3.4, EN 1993-1-8 (2005))

$$F_{v,Rd} = 0{,}6A_s \frac{f_{ub}}{\gamma_{M2}}$$

$$= 0{,}6 \times 561 \times \frac{400}{1{,}25 \times 1\text{E3}} = 107{,}7 > (F_{r,Ed} = 81{,}77) \text{ kN satisfactory.}$$

However the recommended maximum bolt diameter for a column flange width of 254 mm is 24 mm (Annex A4). Use a higher class of bolt.

Solution (b) using M20 class 8.8 bolts not pre-loaded. Shear resistance of an M20 class 8.8 bolt in single shear (Table 3.4, EN 1993-1-8 (2005))

$$F_{v,Rd} = 0,6A_s \frac{f_{ub}}{\gamma_{M2}}$$

$$= 0,6 \times 245 \times \frac{800}{1,25 \times 1E3} = 94,1 > (F_{r,Ed} = 81,77)\,kN \text{ satisfactory.}$$

M20 class 8.8 bolt in bearing on the gusset plate ($t = 10$ mm) (Table 3.4, EN 1993-1-8 (2005)).

Bearing strength

$$F_{b,Rd} = \frac{k_1 \alpha_b f_{up} dt}{\gamma_{M2}}$$

$$= \frac{2,5 \times 0,606 \times 510 \times 20 \times 10}{1,25 \times 1E3} = 123,6 > (F_{r,Ed} = 81,77)\,kN$$

where

for an end bolt $\alpha_b = e_1/(3d_o) = 40/(3 \times 22) = 0,606$

for an edge bolt $k_1 = 2,8e_2/d_o - 1,7 = 2,8 \times 58/22 - 1,7 = 5,68 > 2,5$

Solution (c) using pre-loaded M22 class 10.9 bolts (cl 3.9, Eqs (3.6) and (3.7), EN 1993-1-8 (2005))

$$F_{s,Rd} = \frac{k_s n \mu_b F_{p,C}}{\gamma_{M3}}$$

$$= \frac{1,0 \times 1,0 \times 0,5 \times 0,7 \times 1E3 \times \frac{303}{1E3}}{1,25} = 84,8 > (F_{r,Ed} = 81,77)\,kN$$

To determine the thickness of the gusset plate for the bolted joint Fig. 7.25(a)

$$L_g = 225 + \frac{(255,9 - 140)}{2} = 282,95\,mm$$

$$s_g = 150 + \frac{(255,9 - 140)}{2} = 207,95\,mm$$

Width of the gusset plate perpendicular to the free edge (Eq. (7.26))

$$B_g = \frac{L_g}{\left[\left(\frac{L_g}{H_g}\right)^2 + 1\right]^{1/2}} = \frac{282,95}{\left[\left(\frac{282,95}{400}\right)^2 + 1\right]^{1/2}} = 231,0\,mm$$

From Eq. (7.25), replacing the term $(P_u s_g)$ with $(V s_g + H h_g)$, the thickness of the gusset plate Grade S355 steel

$$t_g = \frac{2(V S_g + H h_g)}{\left(\frac{f_{gy} B_g^2}{\gamma_{M1}}\right)} + \frac{B_g}{80}$$

$$= \frac{2 \times (210 \times 207{,}95 + 45 \times 150)\text{E3}}{\left(\frac{355 \times 231^2}{1{,}0}\right)} + \frac{231}{80}$$

$$= 8{,}21 \text{ mm; use a 10 mm thick plate of Grade S355 steel}$$

Check the slenderness ratio of the gusset plate (Eq. (7.27)).

$$\frac{l_g}{i_g} = 2 \times 3^{1/2} \frac{B_g}{t_g} = 2 \times 3^{1/2} \times \frac{231}{10}$$

$$= 80{,}02 < 185 \text{ the limit of the slenderness ratio for the application of the}$$
$$\text{theory, satisfactory.}$$

Solution (d) using welds (Fig. 7.25(b)).

Where it is not possible to bolt to a column, for example the compound channel column shown in Fig. 7.25(b), then welds are used. The connection is rigid and for welds of unit size the properties of the weld group are:

Total length of weld

$$L_w = 2(d_w + b_w) = 2(400 + 200) = 1200 \text{ mm}$$

Second moment of area of the weld group about the centroidal y–y axis

$$I_y = \Sigma\,(\partial A)\,z^2 = 2\left[\frac{d_w^3}{12} + b_w\left(\frac{d_w}{2}\right)^2\right]$$

$$= 2\left[\frac{400^3}{12} + 200\left(\frac{400}{2}\right)^2\right] = 26{,}67\text{E6 mm}^4$$

Second moment of area of the weld group about the centroidal z–z axis

$$I_z = \Sigma(\partial A)y^2 = 2\left[\frac{b_w^3}{12} + d_w\left(\frac{b_w}{2}\right)^2\right]$$

$$= 2\left[\frac{200^3}{12} + 400\left(\frac{200}{2}\right)^2\right] = 9{,}33\text{E6 mm}^4$$

Second moment of area of the weld group about the centroidal polar x–x axis

$$I_x = I_y + I_z = (26{,}67 + 9{,}33)\text{E6} = 36\text{E6 mm}^4$$

Maximum vector force in the direction of the z–z axis on a weld element furthest from the centroid of the weld group from Eq. (7.28).

$$F_z = \frac{V}{L_w} + \frac{(Ve_h + He_v)y_n}{I_x}$$

$$= \frac{210}{1200} + \frac{(210 \times 250 + 45 \times 350)100}{36E6} = 0{,}365 \text{ kN/mm}$$

Maximum vector force in the direction of the y–y axis on the same weld element

$$F_y = \frac{H}{L_w} + \frac{(Ve_h + He_v)z_n}{I_x}$$

$$= \frac{45}{1200} + \frac{(210 \times 250 + 45 \times 350)200}{36E6} = 0{,}417 \text{ kN/mm}$$

Resultant vector design force on this weld element

$$F_{r,Ed} = (F_z^2 + F_y^2)^{1/2} = (0{,}365^2 + 0{,}417^2)^{1/2} = 0{,}554 \text{ kN/mm}$$

Assuming a 6 mm fillet weld for Grade S275 steel (Eq. (4.4), EN 1993-1-8 (2005))

$$F_{w,Rd} = \frac{f_u a}{(3^{\frac{1}{2}} \beta_w \gamma_{M2})} = \frac{430 \times 0{,}7 \times \frac{6}{1E3}}{(3^{1/2} \times 0{,}85 \times 1{,}25)}$$

$$= 0{,}981 < (F_{r,Ed} = 0{,}554) \text{ kN/mm satisfactory.}$$

To determine the thickness of the gusset plate for the welded joint Fig. 7.25(b).

From Eq. (7.26) the width of the gusset plate perpendicular to the free edge

$$B_g = \frac{L_g}{\left[\left(\frac{L_g}{H_g}\right)^2 + 1\right]^{1/2}} = \frac{225}{\left[\left(\frac{225}{400}\right)^2 + 1\right]^{1/2}} = 196{,}1 \text{ mm}$$

From Eq. (7.25) replacing the term $P_u s_g$ by $(Vs_g + Hh_g)$

$$t_g = \frac{2(Vs_g + Hh_g)}{\left(\frac{f_{gy}B_g^2}{\gamma_{M1}}\right)} + \frac{B_g}{80}$$

$$= \frac{2 \times (210 \times 150 + 45 \times 150)E3}{\left(\frac{355}{1{,}0} \times 196{,}1^2\right)} + \frac{196{,}1}{80}$$

$$= 8{,}05 \text{ mm, use an 10 mm thick plate of Grade S355 steel.}$$

Check the slenderness ratio of the gusset plate from Eq. (7.27)

$$\frac{l_g}{i_g} = \frac{2 \times 3^{1/2}B_g}{t_g} = \frac{2 \times 3^{1/2} \times 196{,}1}{10}$$

$$= 67{,}93 < 185 \text{ the limit of the slenderness ratio for the application of this theory.}$$

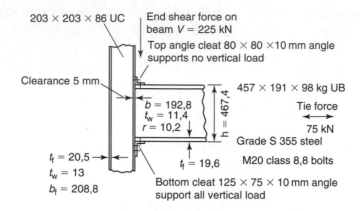

203 × 203 × 86 UC

End shear force on beam $V = 225$ kN

Top angle cleat 80 × 80 ×10 mm angle supports no vertical load

Clearance 5 mm

$b = 192,8$
$t_w = 11,4$
$r = 10,2$

$h = 467,4$

457 × 191 × 98 kg UB

Tie force

75 kN

Grade S 355 steel

M20 class 8,8 bolts

$t_f = 20,5$
$t_w = 13$
$b_f = 208,8$

$t_f = 19,6$

Bottom cleat 125 × 75 × 10 mm angle support all vertical load

FIGURE 7.26 'Pinned' beam-to-column joint

EXAMPLE 7.4 'Pinned' beam-to-column connection. Check the size of components for the connection shown in Fig. 7.26 at ultimate load.

(a) If the design shear force of $F_{v,Ed} = 225$ kN is resisted by 4-M20 grade 8.8 bolts in the bottom cleat (125 × 75 × 10 mm angle), then the design shear force per bolt

$$F_{v,Ed} = \frac{225}{4} = 56,25 \text{ kN}$$

Shear resistance of an M20 grade 8.8 bolt (Table 3.4, EN 1993-1-8 (2005))

$$F_{v,Rd} = 0,6A_s \frac{f_{ub}}{\gamma_{M2}}$$

$$= 0,6 \times 245 \times \frac{800}{1,25 \times 1E3} = 94,1 > (F_{v,Ed} = 56,25) \text{ kN satisfactory.}$$

Bearing resistance of an M20 class 8.8 bolt bearing on the leg of the angle ($t_a = 10$ mm) (Table 3.4, EN 1993-1-8 (2005)).

$$F_{b,Rd} = \frac{k_1 \alpha_b f_{ua} d t_a}{\gamma_{M2}} = \frac{2,12 \times 0,682 \times 510 \times 20 \times 10}{1,25 \times 1E3}$$

$$= 118 > (F_{v,Ed} = 56,25) \text{ kN satisfactory.}$$

where for an end bolt $\alpha_b = e_1/(3d_o) = 45/(3 \times 22) = 0,682$
and for an edge bolt $k_1 = 2,8 \ (e_2/d_o) - 1,7 = 2,8 \times (30/22) - 1,7 = 2,12 < 2,5$

Assume the 4-M20 class 8.8 bolts connecting the bottom cleat to the column which resist the shear force of 225 kN also resist the tensile force of 75 kN. Design tensile force per bolt from the 75 kN tie force is

$$F_{t,Ed} = \frac{75}{4} = 18,75 \text{ kN}$$

Tensile resistance of an M-20 class 8.8 bolt

$$F_{t,Rd} = 0,9A_s \frac{f_u}{\gamma_{M2}} = 0,9 \times 245 \times \frac{800}{1,25 \times 1E3} = 141,1 \text{ kN}$$

Combined shear and tension for a bolt (Table 3.4, EN 1993-1-8 (2005))

$$\frac{F_{v,Ed}}{F_{v,Rd}} + \frac{F_{t,Ed}}{(1,4F_{t,Rd})}$$

$$= \frac{56,25}{94,1} + \frac{18,75}{(1,4 \times 141,1)} = 0,693 < 1 \text{ satisfactory.}$$

(b) Alternatively the four bolts in the vertical leg of the bottom angle could be replaced by fillet welds along the two vertical edges of the angle.

Resistance of two 6 mm fillet welds for Grade S355 steel (Eq. (4.4), EN 1993-1-8 (2005))

$$2l_w F_{w,Rd} = \frac{2l_w f_u a}{(3^{1/2}\beta_w \gamma_{M2})}$$

$$= \frac{2 \times 125 \times 510 \times 0,7 \times \frac{6}{1E3}}{(3^{1/2} \times 0,9 \times 1,25)} = 275 > (V_{Ed} = 225) \text{ kN.}$$

The top cleat is used to provide torsional resistance against lateral buckling of the beam and to resist tie and erection forces. The angle must be at least 6 mm thick to resist corrosion and the leg of sufficient length to accommodate M20 bolts. From Section Tables a 80 × 80 × 10 mm angle is chosen.

For resistance to transverse shear forces for the beam see Example 4.12

EXAMPLE 7.5 'Pinned' beam-to-beam connection. Determine the size of the components required for the connection shown in Fig. 7.27. The beam sizes have been determined from bending calculations at ultimate load.

Assuming that the M20 class 4.6 bolts through the web of the main beam B (Grade S275 steel) are subject to single shear forces.

Shear resistance of a M20 class 4.6 bolt in single shear (Table 3.4, EN 1993-1-8 (2005))

$$F_{v,Rd} = 0,6A_s \frac{f_{ub}}{\gamma_{M2}} = 0,6 \times 245 \times \frac{400}{1,25 \times 1E3} = 47,0 \text{ kN}$$

M20 class 4.6 bolt in bearing on the web of the transverse beam A ($t_w = 6,9$ mm) (Table 3.4 EN 1993-1-8 (2005)).

For an end bolt $\alpha_b = e_1/(3d_o) = 40/(3 \times 22) = 0,606$

For an edge bolt $k_1 = 2,8e_2/d_o - 1,7 = 2,8 \times 30/22 - 1,7 = 2,12 < 2,5$

$$F_{b,Rd} = \frac{k_1\alpha_b f_{ub}dt_w}{\gamma_{M2}}$$

$$= \frac{2,12 \times 0,606 \times 400 \times 20 \times 6,9}{1,25 \times 1E3} = 56,7 > (F_{v,Rd} = 47,0) \text{ kN.}$$

$$n_b = \frac{V_{Ed}}{F_{v,Rd}} = \frac{150}{47,0} = 3,19 \text{ use 4-M20 class 4.6 bolts.}$$

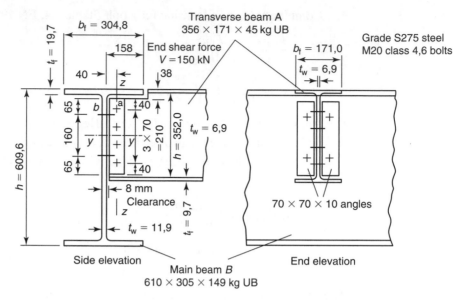

FIGURE 7.27 'Pinned' beam-to-beam joint

Assuming that the bolts connecting the angle cleats to the web of the transverse beam A are in double shear and subject to an eccentric load.

Second moments of area of the bolt group about the centroidal axis for bolts of unit area are

$$I_y = \Sigma(\partial A)z^2 = 2(35^2 + 105^2) = 24{,}5\text{E}3\,\text{mm}^4$$

$$I_z = 0$$

$$I_x = I_y + I_z = 24{,}5\text{E}3\,\text{mm}^4$$

Maximum shear force on a bolt in the y direction from Eq. (7.28)

$$F_y = \frac{V_{Ed}ez_{max}}{I_x} = \frac{150 \times 40 \times 105}{24{,}5\text{E}3} = 25{,}71\,\text{kN}$$

Average shear force on a bolt in the z direction

$$F_z = \frac{V_{Ed}}{n_b} = \frac{150}{4} = 37{,}5\,\text{kN}$$

Maximum resultant design shear force on a bolt

$$F_{r,Ed} = (F_y^2 + F_z^2)^{1/2} = (25{,}71^2 + 37{,}5^2)^{1/2} = 45{,}47\,\text{kN}$$

Double shear strength of an M20 class 4.6 bolt

$$2F_{v,Rd} = 2 \times 47{,}0 = 94{,}0 > (F_{r,Ed} = 45{,}47)\,\text{kN, satisfactory.}$$

Check 'block shear tearing' (cl 3.10.2, EN 1993-1-8 (2005)) for line of holes in end of transverse beam A

$$V_{\text{eff,2,Rd}} = 0{,}5f_u \frac{A_{\text{nt}}}{\gamma_{\text{M2}}} + \frac{1}{3^{1/2}} f_y \frac{A_{\text{nv}}}{\gamma_{\text{MO}}}$$

$$= 0 + \frac{1}{3^{1/2}} \times 275 \times \frac{1194}{1{,}0 \times 1\text{E}3}$$

$$= 189{,}6 > (V_{\text{Ed}} = 150) \text{ kN satisfactory.}$$

which includes

$$A_{\text{nv}} = (3 \times \text{hole spacing} + \text{end distance} - 3{,}5 \times \text{hole diameter})t_w$$

$$= (3 \times 70 + 40 - 3{,}5 \times 22) \times 6{,}9 = 1194 \text{ mm}^2$$

EXAMPLE 7.6 'Pinned' column-to-foundation connection. Determine the size of the components for the axially loaded base shown in Fig. 7.28 at the ultimate limit state. Concrete cylinder crushing strength $f_{\text{ck}} = 20$ MPa.

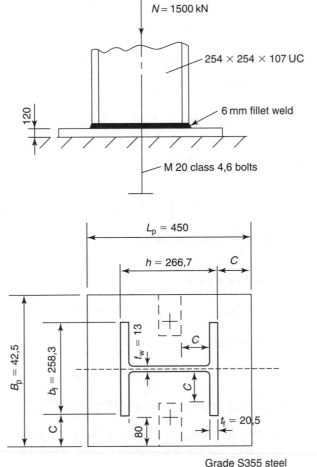

FIGURE 7.28 'Pinned' column to foundation joint

Assuming that the bearing area is bounded by the dotted line shown previously in Fig. 7.17 and the pressure beneath base plate (cl 6.2.5(7), EN 1993-1-8 (2005)),

$$\frac{N}{A} = f_j = \frac{2}{3}f_{ck}$$

Rearranging and inserting known numerical values, bearing area required

$$A = \frac{N_{Ed}}{f_j} = \frac{N_{Ed}}{\left(\frac{2}{3}f_{ck}\right)} = \frac{1500E3}{\left(\frac{2\times20}{3}\right)} = 112{,}5E3\,\text{mm}^2 \tag{i}$$

To determine the minimum thickness of the steel base plate the bearing area enclosed by the dotted line (Fig. 7.17)

$$\begin{aligned}
A &= (b_f + 2c)(h + 2c) - (h - 2t_f - 2c)(b_f - t_w) \\
&= (258{,}3 + 2c)(266{,}7 + 2c) - (266{,}7 - 2\times20{,}5 - 2c)(258{,}3 - 13) \\
&= 13524{,}4 + 1540{,}6c + 4c^2 \tag{ii}
\end{aligned}$$

Equating Eqs (i) and (ii) the projection of the dotted area

$$c = (192{,}57^2 + 24744)^{1/2} - 192{,}57 = 56{,}1\,\text{mm}$$

Check if areas overlap along the bolt line between flanges

$$\frac{(h - 2t_f)}{2} = \frac{(266{,}7 - 2\times20{,}5)}{2}$$
$$= 112{,}85 > (c = 56{,}1)\,\text{mm satisfactory.}$$

Thickness of base plate Grade S355 steel (cl 6.2.5(4), EN 1993-1-8(2005))

$$t_p = c\left(\frac{3f_j\gamma_{M0}}{f_y}\right)^{1/2} = 56{,}1\left(\frac{3\times\frac{2}{3}\times20\times1}{355}\right)^{1/2}$$
$$= 18{,}8\,\text{mm, use 20 mm thick base plate.}$$

Minimum length of base plate

$$D_p = h + 2c = 266{,}7 + 2\times56{,}1 = 322{,}8\,\text{mm}$$

Minimum breadth of base plate

$$B_p = b_f + 2c = 258{,}3 + 2\times56{,}1 = 370{,}5\,\text{mm}$$

Use $450 \times 425 \times 20$ mm base plate Grade S355 steel.

If the end of the column is machined then the load is assumed to be transferred directly to the base plate and a minimum size of fillet weld of 6 mm is used to connect the base plate to the column.

Alternatively if the end of the column is not machined then the force per unit length of weld is approximately

$$F_{w,Ed} = \frac{N_{Ed}}{(4b_f + 2h)} = \frac{1500}{(4 \times 258,3 + 2 \times 266,7)} = 0,957 \text{ kN/mm}$$

Assuming a 6 mm fillet weld for Grade S355 steel (Eq. (4.4), EN 1993-1-8(2005))

$$F_{w,Rd} = \frac{f_u a}{(3^{1/2} \beta_w \gamma_{M2})} = \frac{510 \times 0,7 \times 6/1E3}{3^{1/2} \times 0,9 \times 1,25)}$$
$$= 1,1 > (F_{w,Ed} = 0,957) \text{ kN/mm satisfactor.}$$

The base plate is subject to a compressive force which is not transferred to the holding down bolts. The bolts are therefore subject only to erection forces and if these are not known then experience has shown that a bolt size approximately equal to the plate thickness is suitable. Use 2M20 class 4.6 holding down bolts.

If there is a bending moment applied to the column and hence to the base plate then the bearing area is located beneath the column flange as shown in Fig. 6.4, EN 1993-1-8 (2005).

EXAMPLE 7.7 'Rigid' column bracket. Determine the size of fillet welds for the bracket shown in Fig. 7.29 at the ultimate limit-state.

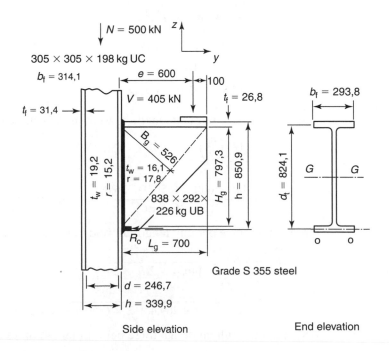

FIGURE 7.29 'Rigid' column bracket

There are two possible solutions based on failure mechanisms (a) assuming rotation about axis G–G which is the simple traditional conservative method and (b) assuming rotation about axis O–O which is more correct but the calculations are more extensive.

(a) *Rotation about axis G–G*

The fillet weld is continuous round the bracket section as shown in Fig. 7.29. If there are no stiffeners in the web of the column then the strength of the weld around the flanges of the bracket is reduced because of the flexibility of the column flange.

Effective length of the column flange weld (Eq. (4.6a), EN 1993-1-8(2005))

$$b_{eff} = t_w + 2r + 7kt_f = 19,2 + 2 \times 15,2 + 7 \times 1 \times 31,4 = 269,4 \, \text{mm}$$

which includes

$$k = \left(\frac{t_f}{t_p}\right)\left(\frac{f_{y,f}}{f_{y,p}}\right) = \left(\frac{31,4}{26,8}\right)\left(\frac{355}{355}\right) = 1,17 > 1 \text{ use } 1.$$

Check if stiffeners required for column web (Eq. (4.7), EN 1993-1-8(2005))

$$b_{eff} = \left(\frac{f_{yp}}{f_{up}}\right)b_p = \frac{355}{510} \times 293,8$$

$$= 204,5 < (b_{eff} = 269,4) \text{ mm therefore no stiffeners required.}$$

Rotation about axis G–G

The effective second moment of area of the weld group about axis G–G

$$I_G = \frac{2d_w^3}{12} + 4b_{eff}(d_f/2)^2$$

$$= \frac{2(850,9 - 2 \times 26,8)^3}{12} + 4 \times 269,4 \times \left[\frac{(850,9 - 26,8)}{2}\right]^2$$

$$= (84,5 + 183,0)\text{E6} = 267,5\text{E6} \, \text{mm}^4$$

Maximum force per unit length on weld in the y direction (Eq. (7.28))

$$F_y = \frac{(Ve)\left(\frac{d_f}{2}\right)}{I_G}$$

$$= \frac{405 \times 600 \times (850,9 - 26,8)}{(2 \times 267,5\text{E6})} = 0,374 \, \text{kN/mm}$$

Maximum force per unit length of weld in the z direction

$$F_z = V/L_w = \frac{V}{[4b_{eff} + 2(h - 2t_f)]}$$

$$= \frac{405}{4 \times 269,4 + 2 \times (850,9 - 2 \times 26,8)} = 0,152 \, \text{kN/mm}$$

Maximum resultant design force per unit length of weld

$$F_{r,Ed} = (F_y^2 + F_z^2)^{1/2} = (0,374^2 + 0,152^2)^{1/2} = 0,404 \, \text{kN/mm}$$

For a 6 mm fillet weld for Grade S355 steel (Eq. (4.4), EN 1993-1-8(2005))

$$F_{w,Rd} = \frac{f_u a}{(3^{1/2}\beta_w \gamma_{M2})} = \frac{510 \times 0,7 \times 6/1E3}{(3^{1/2} \times 0,9 \times 1,25)}$$

$$= 1,10 > (F_{r,Ed} = 0,404) \text{ kN/mm satisfactory.}$$

Rotation about axis O–O

The second moment of area of the weld group about axis O–O

$$I_o = \left(\frac{2}{3}\right) d_w^3 + 2b_{eff}d_f^2$$

$$= \left(\frac{2}{3}\right)(850,9 - 2 \times 26,8)^3 + 2 \times 269,4 \times (850,9 - 26,8)^2$$

$$= (337,9 + 365,9)E6 = 703,8E6 \text{ mm}^4$$

Maximum force per unit length of weld in the y direction

$$F_y = (Ve)d_f/I_o = (405 \times 600) \times (850,9 - 26,8)/703,8E6 = 0,285 \text{ kN/mm}$$

Effective length of weld resisting shear

$$L_{eff} = 4b_{eff} + 2d_w = 4 \times 269,4 + 2(850,9 - 2 \times 26,8) = 2672 \text{ mm}$$

Distance (d_r) from the axis O–O to the resultant force in the weld is determined from equating the moments of the forces in the weld group about the axis O–O

moment of the parts = moments of the whole

$$(2b_{eff} + 0,5 \times 2d_f)F_x d_r = F_x I_o/d_f$$

Rearranging and putting $I_o = (2/3)d_f^3 + 2b_{eff}d_f^2$

$$\frac{d_r}{d_f} = \frac{\left(\frac{2}{3} + \frac{2b_{eff}}{d_f}\right)}{\left(1 + \frac{2b_{eff}}{d_f}\right)}$$

$$= \frac{\left[\frac{2}{3} + \frac{2 \times 269,4}{850,9 - 26,8}\right]}{\left[1 + \frac{2 \times 269,4}{850,9 - 26,8}\right]} = 0,798$$

Check whether slip occurs by substituting in Eq. (7.31)

$$\frac{\mu_s M}{(Vd_r)} = \frac{\mu_s e}{d_r} = \frac{0,45 \times 600}{[0,798(850,9 - 26,8)]}$$

$$= 0,411 < 1, \text{ therefore slip occurs.}$$

Maximum force per unit length of weld in the z direction

$$F_z = \frac{V}{L_{eff}} - \frac{\mu_s R}{L_{eff}} = \frac{V}{L_{eff}} - \frac{\mu_s(Ve/d_r)}{L_{eff}}$$

$$= \frac{405}{2672} - \frac{0,45\left[\frac{405 \times 600}{0,798(850,9 - 26,8)}\right]}{2672} = 0,089 \text{ kN/mm}$$

Maximum resultant design force per unit length of weld

$$F_{r,Ed} = (F_y^2 + F_z^2)^{1/2} = (0.285^2 + 0.089^2)^{1/2} = 0.299 \, \text{kN/mm}$$

For a 6 mm fillet weld and Grade S355 steel (Eq. (4.4), EN 1993-1-8(2005))

$$F_{w,Rd} = \frac{f_u a}{(3^{1/2} \beta_w \gamma_{M2})} = \frac{510 \times 0.7 \times 6/1E3}{(3^{1/2} \times 0.9 \times 1.25)}$$

$$= 1.10 > (F_{r,Ed} = 0.299) \, \text{kN/mm satisfactory.}$$

An alternative method related to the European Code:

Assume the applied vertical shear force is resisted by the two 6 mm web welds

$$V_{R,Ed} = 2 d_f F_{w,Rd} = 2 \times (850.9 - 2 \times 26.8) \times 1.1$$

$$= 1450 > (V_{Ed} = 405) \, \text{kN satisfactory,}$$

and the applied bending moment is resisted by two 6 mm effective flange welds with rotation about axis O–O (cl 6.2.7.1(4), EN 1993-1-8(2005))

$$M_{R,Ed} = 2 b_{eff} d_f F_{w,Rd} = 2 \times 269.4 \times (850.9 - 26.8) \times 1.1/1E3$$

$$= 488.4 > (M_{Ed} = 243) \, \text{kNm satisfactory.}$$

Gusset plate design

Check the thickness of the web of the 838 × 292 × 226 kg UB acting as a gusset plate. From Eq. (7.26)

$$B_g = \frac{L_g}{[(L_g/H_g)^2 + 1]^{1/2}} = \frac{700}{[(700/797.3)^2 + 1]^{1/2}} = 526.0 \, \text{mm}$$

Required thickness of the web of the UB acting as a gusset plate (Eq. (7.25))

$$t_g = \frac{2 P_u s_g}{\left(\frac{f_{gy} B_g^2}{\gamma_{M1}}\right)} + \frac{B_g}{80}$$

$$= \frac{2 \times 405E3 \times 600}{(355/1.0 \times 526)^2} + \frac{526}{80}$$

$$= 11.52 < 16.1 \, \text{mm (thickness of web of UB), satisfaactory.}$$

Check the slenderness ratio of the web of UB acting as a gusset plate (Eq. (7.27))

$$\frac{l_g}{i_g} = 2\sqrt{3}\frac{B_g}{t_g} = 2\sqrt{3} \times \frac{526}{16.1}$$

$$= 113.2 < 185 \, \text{limit of application of the theory, acceptable.}$$

Column web in transverse compression

Reaction R_o may buckle or crush the web of the 305 × 305 × 198 kg UC.

Reaction (Eq. (7.30))

$$R_{o,Ed} = \frac{Ve}{d_r} = \frac{405 \times 600}{[0,798(850,9 - 26,8)]} = 369,5 \, kN$$

The design resistance of the unstiffened column web (Eq. (6.9), EN 1993-1-8(2005))

$$F_{c,wc,Rd} = \frac{\omega k_{wc} b_{eff,c,wc} t_{we} f_{ywe}}{\gamma_{M0}}$$

$$= \frac{0,764 \times 1 \times 271,7 \times 19,2 \times 355}{1,0 \times 1E3}$$

$$= 1415 > (R_{o,Ed} = 369,5) \, kN \text{ satisfactory.}$$

or

$$F_{c,wc,Rd} = \frac{\omega k_{wc} \rho b_{eff,c,wc} t_{we} f_{ywe}}{\gamma_{M1}}$$

$$= \frac{0,764 \times 1 \times 1 \times 271,7 \times 19,2 \times 355}{1,0 \times 1E3}$$

$$= 1415 > (R_{o,Ed} = 369,5) \, kN \text{ satisfactory.}$$

From (Eq. (6.10), EN 1993-1-8(2005)) for a welded connection

$$b_{eff,c,wc} = t_{fb} + 2 \times 2^{1/2} a_b + 5(t_{fc} + s)$$

$$= 26,8 + 2 \times 2^{1/2} \times 0,7 \times 6 + 5 \times (31,4 + 15,2) = 271,7 \, mm$$

The maximum longitudinal stress in the flange of the column from the axial and eccentric loads (Eq. (6.14), EN 1993-1-8(2005))

$$\sigma_{com,Ed} = \frac{(N + V)}{A} + \frac{V(e + h/2)}{W_{el}}$$

$$= \frac{(500 + 405)E3}{252E2} + \frac{405E3(600 + 339,9/2)}{2993E3}$$

$$= 140,1 < (0,7f_{y,wc} = 0,7 \times 355 = 248,5) \, MPa \text{ hence } k_{wc} = 1$$

and (Table (6.3), EN 1993-1-8(2005))

$$\beta = 1(\text{Table 5.4, EN 1993-1-8(2005)})$$

$$\omega = \omega_1 = \frac{1}{\left[1 + 1,3\left(\frac{b_{eff,c,wc} t_{wc}}{A_{vc}}\right)^2\right]^{1/2}}$$

$$= \frac{1}{\left[1 + 1,3 \times \left(271,7 \times \frac{19,2}{7032}\right)^2\right]^{1/2}} = 0,764$$

where the shear area of the column (cl 6.2.6(3), EN 1993-1-1(2005)) or from Section Tables

$$A_{vc} = A - 2bt_f + (t_w + 2r)t_f$$
$$= 252\text{E}2 - 2 \times 314,1 \times 31,4 + (19,2 + 2 \times 15,2) \times 31,4 = 7032 \text{ mm}^2$$
$$\eta h_w t_w = 1,0 \times (339,9 - 2 \times 31,4) \times 19,2 = 5320 < (A_{vc} = 7032) \text{ mm}^2$$

Plate slenderness (cl 6.2.6.2(1), EN 1993-1-8(2005))

$$\bar{\lambda}_p = 0,932 \left[\frac{b_{eff,c,wc} d_{wc} f_{y,wc}}{(Et_{wc}^2)} \right]^{1/2}$$
$$= 0,932 \times \left[\frac{271,7 \times 246,7 \times 355}{(210\text{E}3 \times 19,2^2)} \right]^{1/2}$$
$$= 0,516 < 0,72 \text{ use } \rho = 1,0$$

Shear strength of the column web (cl 6.2.6.1, EN 1993-1-8(2005))

$$V_{wp,Rd} = 0,9 \frac{A_{vc} f_{y,wc}}{(3^{1/2} \gamma_{M0})} = 0,9 \times \frac{7032 \times 355}{3^{1/2} \times 1 \times 1\text{E}3}$$
$$= 1297 > (R_{o,Ed} = 369,5) \text{ kN satisfactory.}$$

EXAMPLE 7.8 'Rigid' beam-to-column connection. Determine the size of the components for the connection shown in Fig. 7.30 at the ultimate limit state.

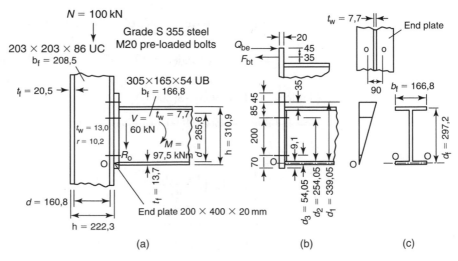

FIGURE 7.30 'Rigid' beam-to-column joint

Check for slip assuming rotation about axis O–O (Eq. (7.31))

$$\frac{\mu M}{[V(h - t_{fb})]} = \frac{0,45 \times 97,5E6}{[60E3 \times (310,9 - 13,7)]} = 2,46 > 1$$

therefore rotation about the compression flange of the beam at O without slip.

Assuming rotation about axis O–O the design tensile force acting on a single bolt

$$F_{t,Ed} = \frac{M}{[4(h - t_{fb})]} = \frac{97,5E6}{[4 \times (310,9 - 13,7) \times 1E3]} = 82\,kN$$

Design tensile resistance of an M20 pre-loaded class 8.8 bolt (Table 3.4, EN 1993-1-8 (2005)) assuming not subject to a shear force

$$F_{t,Rd} = k_2 A_s \frac{f_{ub}}{\gamma_{M2}} = 0,9 \times 245 \times \frac{800}{1,25 \times 1E3}$$
$$= 141 > (F_{t,Ed} = 82)\,kN\ satisfactory.$$

Shear resistance of an M20 class 8.8 pre-loaded bolt in single shear and subject to a tensile force of 82 kN (Eq. (3.8a), EN 1993-1-8 (2005))

$$F_{s,Rd} = \frac{k_s n \mu}{\gamma_{M3}}(F_{p,C} - 0,8F_{t,Ed})$$
$$= \frac{1,0 \times 1,0 \times 0,5}{1,25} \times \frac{(0,7 \times 800 \times 245 - 0,8 \times 82E3)}{1E3}$$
$$= 28,6 > \left(F_{s,Ed} = \frac{60}{6} = 10\right)\,kN\ satisfactory.$$

M20 class 8.8 bolt in bearing on end plate ($t = 20\,mm$) (Table 3.4, EN 1993-1-8 (2005)).

$$F_{b,Rd} = \frac{k_1 \alpha_b f_{up} dt}{\gamma_{M2}}$$
$$= \frac{2,5 \times 0,682 \times 510 \times 20 \times 20}{1,25 \times 1E3} = 278,2 > (F_{s,Ed} = 10)\,kN$$

for an end bolt $\alpha_b = e_1/(3d_o) = 45/(3 \times 22) = 0,682$

for an edge bold $k_1 = 2,8\,(e_2/d_o) - 1,7 = 2,8 \times (55/22) - 1,7 = 5,3 > 2,5$

Thickness of end plate related to a single bolt (Table 6.2, EN 1993-1-8 (2005), Method 1)

$$t_f = \left[\frac{4F_{t,Ed}m}{\left(\frac{l_{eff}f_y}{\gamma_{M0}}\right)}\right]^{1/2}$$
$$= \left[\frac{4 \times 82E3 \times 35}{100 \times 355/1,0}\right]^{1/2} = 18,0\,mm;\ use\ 20\,mm\ plate.$$

Design resistance of the unstiffened column web at O (Eq. (6.9), EN 1993-1-8 (2005))

$$F_{c,wc,Rd} = \frac{\omega k_{wc} b_{eff,c,wc} t_{we} f_{ywe}}{\gamma_{M0}}$$

$$= \frac{0{,}685 \times 1 \times 224{,}1 \times 13 \times 355}{1{,}0 \times 1E3} = 708{,}4 > 328\,\text{kN satisfactory.}$$

or

$$F_{c,wc,Rd} = \frac{\omega k_{wc} \rho b_{eff,c,wc} t_{we} f_{ywe}}{\gamma_{M1}}$$

$$= \frac{0{,}685 \times 1 \times 1 \times 224{,}1 \times 13 \times 355}{1{,}0 \times 1E3}$$

$$= 708{,}4 > 328\,\text{kN satisfactory.}$$

where for a bolted end plate connection (Eq. (6.12), EN 1993-1-8 (2005))

$$b_{eff,c,wc} = t_{fb} + 2 \times 2^{1/2} a_p + 5(t_{fc} + s) + s_p$$

$$= 13{,}7 + 2 \times 2^{1/2} \times 0{,}7 \times 6 + 5 \times (20{,}5 + 15{,}2) + 20 = 224{,}1\,\text{mm}$$

The maximum longitudinal stress in the flange of the column from the axial load and eccentric load (Eq. (6.14), EN 1993-1-8 (2005))

$$\sigma_{com,Ed} = \frac{(N+V)}{A} + \frac{(Vh/2)}{W_{el}} = \frac{(100+60)E3}{110E2} + \frac{\frac{60E3 \times 223{,}2}{2}}{851E3}$$

$$= 22{,}4 < (0{,}7 f_{y,wc} = 0{,}7 \times 355 = 248{,}5)\,\text{MPa hence } k_{wc} = 1$$

Table 5.4, EN 1993-1-8 (2005)

$$\beta = 1$$

Table 6.3, EN 1993-1-8 (2005)

$$\omega = \omega_1 = \frac{1}{\left[1 + 1{,}3 \left(\frac{b_{eff,c,wc} t_{wc}}{A_{vc}}\right)^2\right]^{1/2}}$$

$$= \frac{1}{\left[1 + 1{,}3 \times \left(224{,}1 \times \frac{13}{3124}\right)^2\right]^{1/2}} = 0{,}685$$

where (cl 6.2.6(3), EN 1993-1-1 (2005)) or from Section Tables

$$A_{vc} = A - 2bt_f + (t_w + 2r)t_f$$

$$= 110E2 - 2 \times 208{,}8 \times 20{,}5 + (13 + 2 \times 10{,}2) \times 20{,}5 = 3124\,\text{mm}^2$$

$$\eta h_w t_w = 1{,}0 \times (222{,}3 - 2 \times 20{,}5) \times 13 = 2357 < (A_{vc} = 3124)\,\text{mm}^2$$

Plate slenderness (cl 6.2.6.2(1), EN 1993-1-8 (2005))

$$\bar{\lambda}_p = 0{,}932\left[\frac{b_{\text{eff,c,wc}}d_{\text{wc}}f_{y,\text{wc}}}{(Et_{\text{wc}}^2)}\right]^{1/2}$$

$$= 0{,}932\left[\frac{224{,}1 \times 161{,}8 \times 355}{210\text{E3} \times 13^2}\right]^{1/2} = 0{,}561 < 0{,}72 \text{ use } \rho = 1{,}0$$

Reaction $R_o = 4F_t = 4 \times 82 = 328$ kN

Shear strength of the column web (cl 6.2.6.1, EN 1993-1-8 (2005))

$$V_{\text{wp,Rd}} = 0{,}9\frac{A_{\text{vc}}f_{y,\text{wc}}}{(3^{1/2}\gamma_{\text{M0}})} = \frac{0{,}9 \times 3124 \times 355}{(3^{1/2} \times 1 \times 1\text{E3})}$$

$$= 576 > (R_o = 328) \text{ kN satisfactory.}$$

Alternative calculations for comparison

Previous calculations assume the top four bolts resist all of the applied moment with a single lever arm. Alternatively assume a linear variation of forces from axis O–O to the bolts furthest from the axis.

Maximum tensile force acting on the bolt furthest from axis O–O

$$F_t = \frac{M}{[2\,(d_1 + d_2 + d_3)]}$$

$$= \frac{97{,}5\text{E3}}{[2(339{,}05 + 254{,}05 + 54{,}05)]} = 75{,}3 \text{ kN}$$

Prying force for each bolt assuming $\partial_b = 0$ for a pre-loaded bolt and elastic behavior (Holmes and Martin, 1983).

$$Q_{\text{be}} = \frac{\left[F_{\text{be}} - \frac{2EI\partial_b}{a_p b_p^2}\right]}{\left[\frac{2a_p}{b_p} + \left(\frac{1}{3}\right)\left(\frac{a_p}{b_p}\right)^2\right]}$$

$$= \frac{[90{,}62 - 0]}{\left[\frac{2 \times 45}{35} + \left(\frac{1}{3}\right) \times \left(\frac{45}{35}\right)^2\right]} = 29{,}02 \text{ kN}$$

and the maximum tensile force on a bolt

$$F_{\text{bt}} + Q_{\text{be}} = 75{,}3 + 29{,}0 = 104{,}3 < (F_{\text{t,Rd}} = 141) \text{ kN satisfactory for no slip.}$$

For the welded connection between the end plate and the beam assume the applied vertical shear force is resisted by two 6 mm web welds

$$V_{\text{R,Ed}} = 2d_f F_{\text{w,Rd}} = 2 \times (310{,}9 - 2 \times 13{,}7) \times 1{,}1$$

$$= 623{,}7 > (V_{\text{Ed}} = 60) \text{ kN satisfactory.}$$

Assume that the applied bending moment is resisted by two 10 mm effective flange welds with rotation about axis O–O (cl 6.2.7.1(4), EN 1993-1-8 (2005))

$$M_{R,Ed} = 2b_{eff}d_fF_{w,Rd} = 2 \times 91,2 \times (310,9 - 13,7) \times \frac{1,83}{1E3}$$

$$= 99,2 > (M_{Ed} = 97,5) \text{ kNm satisfactory (cl 6.2.3(4), EN 1993-1-8 (2005)).}$$

For strength of welds ($F_{w,Rd}$) see Annex 1 and for b_{eff} see Eq. (4.6a), EN 1993-1-8 (2005).

EXAMPLE 7.9   'Rigid' beam-to-beam joint. Determine the size of the components for the rigid beam-to-beam joint shown in Fig. 7.31 at the ultimate limit state.

Assuming rotation about axis O–O and the applied moment of 97,5 kNm is resisted entirely by the cover plate then the design tensile force in the flange cover plate

$$F_f = \frac{M}{h} = \frac{97,5E6}{303,8 \times 1E3} = 320,9 \text{ kN.}$$

Thickness of flange connection plate, $b_p = 165$ and 20 mm pre-loaded bolts

$$t_p = \frac{F_f}{\left[(b_p - 2d_h)\left(\frac{f_y}{\gamma_{M1}}\right)\right]} = \frac{320,9E3}{\left[(165 - 2 \times 22) \times \frac{355}{1,0}\right]}$$

$$= 7,47 \text{ mm, use 8 mm thick Grade S355 steel plate.}$$

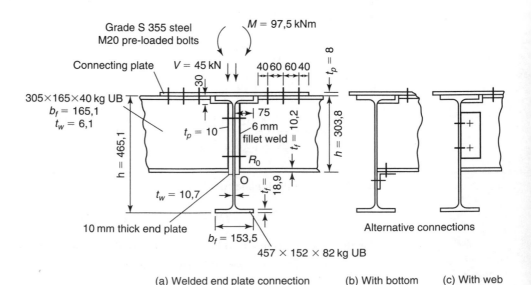

(a) Welded end plate connection   (b) With bottom cleats   (c) With web cleats

FIGURE 7.31   'Rigid' beam-to-beam joint

Single shear resistance of M20 class 8.8 pre-loaded bolt (Table 3.4, EN 1993-1-8 (2005))

$$F_{s,Rd} = \frac{k_s n \mu 0,7 F_{p,C} A_s}{\gamma_{M2}}$$

$$= \frac{1,0 \times 1,0 \times 0,5 \times 0,7 \times 800 \times \frac{245}{1E3}}{1,25} = 54,9 \, \text{kN}$$

M20 class 8.8 bolt in bearing on plate ($t = 8 \, \text{mm}$) (Table 3.4, EN 1993-1-8 (2005)).

$$F_{b,Rd} = \frac{k_1 \alpha_b f_{up} dt}{\gamma_{M2}}$$

$$= \frac{2,34 \times 0,606 \times 510 \times 20 \times 8}{1,25 \times 1E3} = 92,6 > (F_{s,Rd} = 54,9) \, \text{kN}$$

for an end bold $\alpha_b = e_1/(3d_o) = 40/(3 \times 22) = 0,606$

for an edge bolt $k_1 = 2,8(e_2/d_o) - 1,7 = 2,8 \times 31,75/22 - 1,7 = 2,34 < 2,5$

Number of bolts required for the connecting plate

$$n_b = \frac{F_{s,Ed}}{F_{s,Rd}} = \frac{320,9}{54,9} = 5,85 \text{ use 6-M20 class 8.8 pre-loaded bolts.}$$

Reaction at the hinge is equal to the force in the flange, $R_o = F_f = 320,9 \, \text{kN}$ and the frictional resistance at the hinge

$$\mu R_o = 0,45 \times 320,9$$

$$= 144,4 > (V_{Ed} = 45) \, \text{kN therefore no slip occurs.}$$

Use a 10 mm thick end plate welded to the end of the $305 \times 165 \times 40 \, \text{kg}$ UB and bolted to the $457 \times 152 \times 82 \, \text{kg}$ UB as shown in Fig. 7.31(a).

Shear resistance of 4-M20 pre-loaded bolts in double shear in the end plate

$$= 2 \times 4 \times 54,9 = 439,2 > (V_{Ed} = 90) \, \text{kN satisfactory.}$$

Shear resistance of two 6 mm fillet welds connecting the end plate to the $305 \times 165 \times 40 \, \text{kg}$ UB using grade S355 steel

$$F_{w,Rd} = \frac{f_u a}{(3^{1/2} \beta_w \gamma_{M2})}$$

$$= \frac{510 \times 0,7 \times \frac{6}{1E3}}{(3^{1/2} \times 0,9 \times 1,25)} = 1,10 \, \text{kN/mm}$$

$$2 l_w F_{w,Rd} = 2 \times (303,8 - 30 - 10,2) \times 1,10$$

$$= 580 > (V_{Ed} = 45) \, \text{kN satisfactory.}$$

EXAMPLE 7.10 'Rigid' beam splice. Determine the size of components for the beam splice shown in Fig. 7.32 at the ultimate limit state.

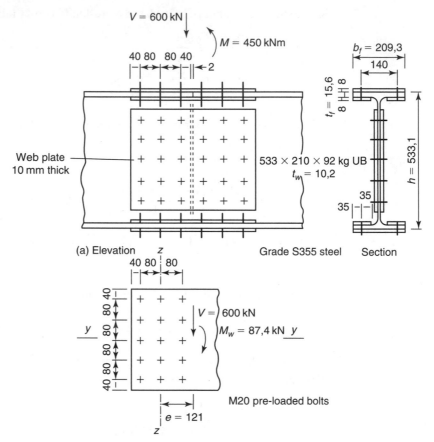

(a) Elevation

Web plate 10 mm thick

533 × 210 × 92 kg UB $t_w = 10,2$

Grade S355 steel   Section

(b) Web connection resisting shear force and web bending moment

FIGURE 7.32 'Rigid' beam splice

Check if the beam is in the elastic stage of behaviour

$$f = \frac{M}{W_e} = \frac{450E6}{2076E3} = 216,8 < (f_y = 355) \text{ MPa, therefore elastic behaviour.}$$

Second moment of area of the web of the beam

$$I_{web} = \frac{t_w(h - 2t_f)^3}{12} = \frac{10,2 \times (533,1 - 2 \times 15,6)^3}{12} = 107,5E6 \text{ mm}^4$$

From Section Tables the gross second moment of area of the beam section

$$I_{gross} = 553,5E6 \text{ mm}^4$$

Assumed proportion of the applied bending moment taken by the web

$$M_{web} = \left(\frac{I_{web}}{I_{gross}}\right) M = \left(\frac{107,5E6}{553,5E6}\right) \times 450 = 87,39 \text{ kNm}$$

Check the strength of the arrangement of bolts in shear in the web plate (Fig. 7.32(b))

Second moment of area of bolts of unit area about the centroidal y–y axis

$$I_y = \Sigma(\partial A)z^2 = 6 \times (80^2 + 160^2) = 192\text{E3 mm}^4$$

Second moment of area of bolts of unit area about the centroidal z–z axis

$$I_z = \Sigma(\partial A)y^2 = 10 \times 80^2 = 64\text{E3 mm}^4$$

Second moment of area of bolts of unit area about the centroidal polar x–x axis

$$I_x = I_y + I_z = (192 + 64)\text{E3} = 256\text{E3 mm}^4$$

Eccentricity of the applied shear force relative to the centroid of half the bolt group is 121 mm (Fig. 7.32(b)). This eccentricity produces a moment which is increased by the bending moment resisted by the web. The equivalent eccentricity

$$e' = e + \frac{M_{\text{web}}}{V} = 121 + \frac{87,4\text{E3}}{600} = 266,6\,\text{mm}$$

Maximum vector shear force in the y–y direction acting on a bolt furthest from the centroid of the web bolt group

$$F_y = Ve'\frac{z_n}{I_x} = 600 \times 266,6 \times \frac{160}{256\text{E3}} = 99,99\,\text{kN}$$

Maximum vector shear force in the z–z direction acting on the same bolt

$$F_z = \frac{V}{n} + Ve'\frac{y_n}{I_x} = \frac{600}{15} + 600 \times 266,6 \times \frac{80}{256\text{E3}} = 90,0\,\text{kN}$$

Resultant maximum vector force acting on the same bolt

$$F_r = [F_y^2 + F_z^2]^{1/2} = [99,99^2 + 90,0^2]^{1/2} = 134,5\,\text{kN}$$

Double shear strength of an M20 pre-loaded bolt class 10,9 ($f_{pu} = 1000\,\text{MPa}$, $\mu = 0,5$) in the web (cl 3.9.1, EN 1993-1-8(2005))

$$F_{v,Rd} = \frac{k_s n \mu F_{p,C}}{\gamma_{M3}} = \frac{1,0 \times 2,0 \times 0,5 \times 245 \times 0,7 \times \frac{1000}{1\text{E3}}}{1,25}$$
$$= 137,2 > (F_r = 134,5)\,\text{kN, satisfactory.}$$

M20 class 8.8 bolt in bearing on the web ($t = 10,2\,\text{mm}$) (Table 3.4, EN 1993-1-8(2005)).

$$F_{b,Rd} = \frac{k_1 \alpha_b f_{up} d t_w}{\gamma_{M2}}$$
$$= \frac{2,5 \times 0,606 \times 510 \times 20 \times 10,2}{1,25 \times 1\text{E3}} = 126,1 < (F_{v,Rd} = 137,2)\,\text{kN}$$

for an end bolt $\alpha_b = e_1/(3d_o) = 40/(3 \times 22) = 0,606$

for an edge bolt $k_1 = 2,8e_2/d_o - 1,7 = 2,8 \times 106,5/22 - 1,7 = 11,9 > 2,5$

Required number of M20 pre-loaded bolts in bearing for the flange splice

$$n_b = \frac{F_f}{F_{b,Rd}} = \frac{\left[\frac{(M-M_{web})}{(h-t_f)}\right]}{F_{b,Rd}}$$

$$= \frac{\left[\frac{(450-87,4)E3}{(533,1-15,6)}\right]}{126,1} = 5,56 \text{ use 6 bolts.}$$

Reduction factor for length of lap (cl 3.8, EN 1993-1-8 (2005))

$$\beta_{Lf} = 1 - \frac{(L_j - 15d)}{200d} = 1 - \frac{(2 \times 80 - 15 \times 20)}{200 \times 20}$$

$$= 1,035 \text{ use } 1,0$$

Thickness of the outer and inner flange cover plates

$$t_p = \frac{F_f}{\frac{(b_f - 2d_h + 2w_p - 2d_h)f_y}{\gamma_{M1}}}$$

$$= \frac{700,7E3}{(209,3 - 2 \times 22 + 2 \times 70 - 2 \times 22) \times \frac{355}{1,0}} = 7,55 \text{ mm}$$

Use 8 mm plates Grade S355 steel as shown in Fig. 7.32.

EXAMPLE 7.11 'Rigid' column splice. Determine the size of the components for the rigid column splice shown in Fig. 7.33 at the ultimate limit state.

Where column sections are of the same serial size it is possible to connect them directly with web and flange plates. The ends of the column are machined and will be in contact. Rotation will take place about an axis near the outer edge of the flange of the upper column.

Thickness of the flange plate, from moments of forces about the axis of rotation

$$t_p = \frac{\left(M - \frac{Nh_u}{2}\right)}{\left[(b_p - 2d_h)\left(\frac{f_y}{\gamma_{M1}}\right)h_u\right]}$$

$$= \frac{\left[480E6 - \frac{712,5E3 \times 355,6}{2}\right]}{\left[(365 - 2 \times 24)\left(\frac{355}{1,0}\right) \times 355,6\right]}$$

$$= 8,83 \text{ mm, use 10 mm Grade S355 steel plate.}$$

Single shear strength of an M22 pre-loaded bolt class 10,9 ($f_{pu} = 1000$ MPa, $\mu = 0,5$) in the flange (cl 3.9.1, EN 1993-1-8 (2005))

$$F_{v,Rd} = \frac{k_s n \mu F_{p,C}}{\gamma_{M2}}$$

$$= \frac{1,0 \times 1,0 \times 0,5 \times 303 \times 0,7 \times \frac{1000}{1E3}}{1,25} = 84,8 \text{ kN.}$$

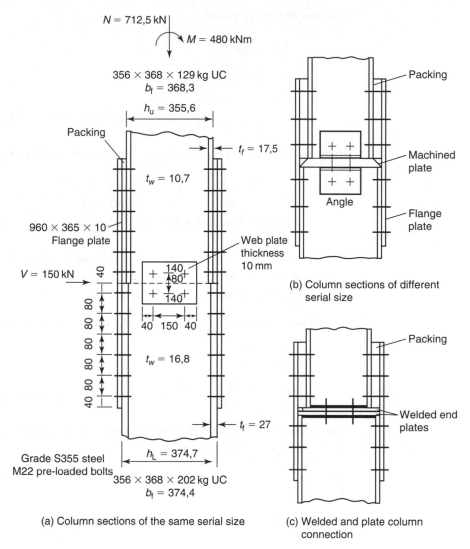

$N = 712,5$ kN

$M = 480$ kNm

$356 \times 368 \times 129$ kg UC
$b_f = 368,3$

$h_u = 355,6$

Packing

$t_f = 17,5$

$t_w = 10,7$

$960 \times 365 \times 10$
Flange plate

$V = 150$ kN

40

80
80
80
80
40

140
80
140

40 150 40

$t_w = 16,8$

$t_w = 16,8$

Web plate
thickness
10 mm

$t_f = 27$

Grade S355 steel
M22 pre-loaded bolts

$h_L = 374,7$

$356 \times 368 \times 202$ kg UC
$b_f = 374,4$

(a) Column sections of the same serial size

Packing

Machined
plate

Angle

Flange
plate

(b) Column sections of different
serial size

Packing

Welded end
plates

(c) Welded and plate column
connection

FIGURE 7.33 'Rigid' splices in steel columns

M22 class 10,9 bolt in bearing on plate ($t = 10$ mm) (Table 3.4, EN 1993-1-8 (2005)).

$$F_{b,Rd} = \frac{k_1 \alpha_b f_{up} d t_w}{\gamma_{M2}}$$

$$= \frac{2,5 \times 0,555 \times 510 \times 22 \times 10}{1,25 \times 1E3} = 124,5 > (F_{v,Rd} = 84,8)\text{kN}$$

for an end bolt $\alpha_b = e_1/(3d_o) = 40/(3 \times 24) = 0,555$

for an edge bolt $k_1 = 2,8e_2/d_o - 1,7 = 2,8 \times 108/24 - 1,7 = 10,9 > 2,5$

Number of bolts required

$$n_b = \frac{\left[M - \frac{Nh_u}{2}\right]}{(F_{v,Rd}h_u)}$$

$$= \frac{\left[480E6 - \frac{712,5E3 \times 355,6}{2}\right]}{84,8E3 \times 355,6} = 11,7 \text{ use 12-M22 bolts.}$$

Reduction factor for length of lap (cl 3.8, EN 1993-1-8 (2005))

$$\beta_{Lf} = 1 - \frac{(L_j - 15d)}{200d} = 1 - \frac{(5 \times 80 - 15 \times 22)}{200 \times 22}$$
$$= 0,984. \text{ This factor does not affect the number of bolts required.}$$

Where the ends of the column are machined and in contact the horizontal shear force on the column is resisted by the friction force, in part or whole, at the point of contact, that is, at the axis of rotation.

Assuming machined surfaces $\mu = 0,15$ the frictional resistance

$$= \mu \left(\frac{M}{h_u} + \frac{N}{2}\right)$$
$$= 0,15 \times \left(\frac{480E3}{355,6} + \frac{712,5}{2}\right) = 255,9 > 150 \text{ kN (applied shear force).}$$

Theoretically no shear connection required but in practice a web plate is generally provided to align the webs.

If the frictional resistance is ignored then the web splice is designed to resist the entire shear force as follows:

Second moments of area of two bolts of unit area on one side of the web connection about the centroidal axes are:

$$I_y = 0$$
$$I_z = 2 \times 75^2 = 11,25E3 \text{ mm}^4$$
$$I_x = I_y + I_z = 11,25E3 \text{ mm}^4$$

Vector force on a bolt in the y–y direction
$$F_x = \frac{V}{n_b} = \frac{150}{2} = 75 \text{ kN}$$

Vector force on a bolt furthest from the centre of rotation in the z–z direction
$$F_z = (Ve)\frac{y_n}{I_x} = 150 \times 40 \times \frac{75}{11,25E3} = 40 \text{ kN}$$

Maximum vector shear force on the same bolt
$$F_r = (F_y^2 + F_z^2)^{1/2} = (75^2 + 40^2)^{1/2} = 85 < (F_{b,Rd} = 124,5) \text{ kN, acceptable.}$$

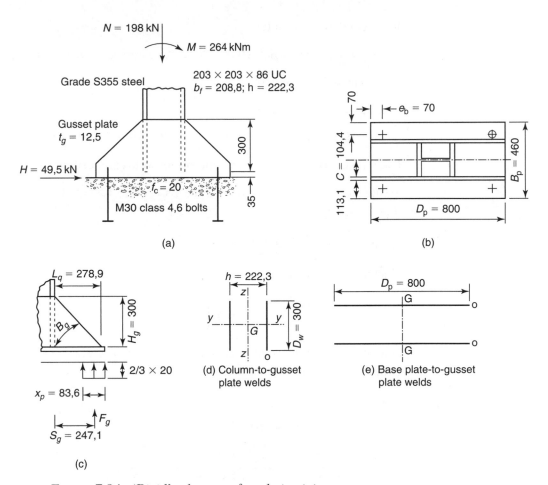

FIGURE 7.34 'Rigid' column-to-foundation joint

EXAMPLE 7.12 'Rigid' built-up column base connection. Determine the size of the components for the connection shown in Fig. 7.34 at the ultimate limit state assuming $f_c = 20$ MPa.

Tensile strength of an M30 class 4.6 holding down bolt (Table 3.4, EN 1993-1-8 (2005))

$$F_{t,Rd} = \frac{0.9A_s f_{ub}}{\gamma_{M2}} = \frac{0.9 \times 561 \times 400}{1.25 \times 1E3} = 161.6 \, \text{kN}$$

Distance required between holding down bolts (Eq. (7.39))

$$d_p = \frac{M}{\left[\left(\frac{N}{2}\right) + nF_{t,Rd}\right]} = \frac{264E3}{\left(\frac{198}{2} + 2 \times 161.6\right)} = 625.3 \, \text{mm}$$

The size of the base plate is determined as follows:

Assume a bolt edge distance of approximately $2d = 2 \times 30 = 60$ mm use 70 mm.

Total length of base plate

$$D_p = d_p + 2e_b = 625,3 + 2 \times 70 = 765,3 \, \text{mm, use length of 800 mm}$$

Minimum width of base plate

$$B_p = b_f + 2t_g + \text{washer} + 2 \times \text{welds} + 2e_b$$

$$= 208,8 + 2 \times 12,5 + 66 + 2 \times 10 + 2 \times 70 = 459,8 \, \text{mm. Use width of 460 mm.}$$

The thickness of gusset plate(t_g) and size of welds are assumed at this stage.

Assume the projection length for the base plate is half the width of the column (Fig. 7.34(b))

$$c = \frac{b_f}{2} = \frac{208,8}{2} = 104,4 \, \text{mm}$$

Thickness of base plate Grade S355 steel (Eq. (6.5), EN 1993-1-8 (2005))

$$t_p = c \left(\frac{3 f_{jd} \gamma_{M0}}{f_y} \right)^{1/2} = 104,4 \left(\frac{3 \times \frac{2}{3} \times 20 \times 1,0}{355} \right)^{1/2} = 35 \, \text{mm}$$

Length of concrete compression zone beneath the steel baseplate (Eq. (7.38)) assuming lever arm $l_a = 660 \, \text{mm}$

$$x_p = \frac{\left(\frac{M}{l_a} + \frac{N}{2} \right)}{(b_f + 2t_g + 2c) f_{jd}}$$

$$= \frac{\left(\frac{264E6}{660} + \frac{198E3}{2} \right)}{[(208,8 + 2 \times 15 + 2 \times 104,4) \times \frac{2}{3} \times 20]}$$

$$= 83,6 \, \text{mm}$$

Lever arm for resistance of concrete in bending at ultimate limit state

$$l_a = D_p - e_b - \frac{x_p}{2} = 800 - 70 - \frac{83,6}{2} = 688,2 \, \text{mm}$$

Tensile force in a holding down bolt (Eq. (7.39))

$$F_{bt} = \frac{M - N \left(\frac{D_p}{2} - \frac{x_p}{2} \right)}{n \frac{l_a}{b}}$$

$$= \frac{264E3 - 198 \times \left(\frac{800}{2} - \frac{83,6}{2} \right)}{2 \times 688,2} = 140,3 < (F_{t,Rd} = 161,6) \, \text{kN}$$

Use $800 \times 460 \times 35 \, \text{mm}$ base plate Grade S355 steel.

Force from bearing pressure applied to each gusset plate (Fig. 7.34(c))

$$F_g = f_{jd} \left(\frac{B_j}{2} \right) x_p = \frac{2}{3} \times 20 \times \left(\frac{460}{2} \right) \times \frac{83,6}{1E3} = 256,4 \, \text{kN}$$

Length of gusset plate allowing for 10 mm for weld

$$L_g = \frac{(D_p - 2 \times 10 - h)}{2} = \frac{(800 - 20 - 222,3)}{2} = 278,9 \text{ mm}$$

Assume height of gusset plate $H_g = 300$ mm

Eccentricity of force F_g in relation to the inner corner of the gusset plate (Fig. 7.34(c))

$$s_g = \frac{(D_p - h)}{2} - x_p/2 = \frac{(800 - 222,3)}{2} - 83,6/2 = 247,1 \text{ mm}$$

Width of gusset plate (Eq. (7.26))

$$B_g = \frac{L_g}{\left[1 + \left(\frac{L_g}{H_g}\right)^2\right]^{1/2}} = \frac{278,9}{\left[1 + \left(\frac{278,9}{300}\right)^2\right]^{1/2}} = 204,2 \text{ mm}$$

Thickness of gusset plate Grade S355 steel (Eq. (7.25))

$$t_g = \frac{2F_g s_g}{\left(\frac{f_{yg}}{\gamma M1} B_g^2\right)} + \frac{B_g}{80}$$

$$= \frac{2 \times 256,4E3 \times 247,1}{\left(\frac{355}{1,0} \times 204,2^2\right)} + \frac{204,2}{80} = 11,1 \text{ mm}$$

Use 12,5 mm thick Grade S355 steel gusset plate.

Check slenderness ratio of gusset plate (Eq. (7.27))

$$\frac{l_g}{i_g} = \frac{2\sqrt{3}B_g}{t_g} = \frac{2\sqrt{3} \times 204,2}{12,5} = 56,6 < 185, \text{ satisfactory.}$$

Minimum length of foundation bolt (Holmes and Martin, 1983) (cl 6.2.6.12, EN 1993-1-8 (2005))

$$L_b = \sqrt{\frac{F_{bt}}{\pi f_{tc}}} = \sqrt{\frac{161,6E3}{\pi \times 0,3 \times \frac{20^{2/3}}{1,5}}} = 186,8 \text{ mm}$$

Use 4-M30 class 4.6 holding down bolts, 300 mm long anchored by washer plates.

The size of the fillet weld connecting the base plate to the gusset assuming no friction is obtained as follows:

Length of weld (Fig. 7.34(e))

$$L_w = 2D_p = 2 \times 800 = 1,6E3 \text{ mm}$$

Second moment of area about centroid of the weld group for unit size weld

$$I_{wG} = \frac{2D_p^3}{12} = \frac{2 \times 800^3}{12} = 85,33E6 \text{ mm}^4$$

Maximum vertical force compressive per unit length weld in the z direction

$$F_{wz} = \frac{N}{L_w} + \frac{M\left(\frac{D_p}{2}\right)}{I_{wg}}$$

$$= \frac{198}{1,6E3} + \frac{264E3 \times \left(\frac{800}{2}\right)}{85,33E6} = 1,36 \,\text{kN/mm}$$

Horizontal force per unit length of weld in the x direction

$$F_{wx} = \frac{H}{L_w} = \frac{49,5}{1,6E3} = 0,0309 \,\text{KN/mm}$$

Resultant vector force per unit length of weld

$$F_{wr} = (F_{wz}^2 + F_{wx}^2)^{1/2} = (1,36^2 + 0,0309^2)^{1/2} = 1,362 \,\text{kN/mm}$$

Assuming a 8 mm fillet weld for Grade S355 steel (Eq. (4.4), EN 1993-1-8 (2005))

$$F_{w,Rd} = \frac{f_u a}{(3^{1/2}\beta_w \gamma_{M2})} = \frac{510 \times 0,7 \times \frac{8}{1E3}}{(3^{1/2} \times 0,9 \times 1,25)}$$

$$= 1,47 > (F_{wr} = 1,362) \,\text{kN/mm}$$

Alternatively if surface between the base plate and edge of the gusset plate are machined and bearing is assumed at the axis of rotation O–O (Fig. 7.34(e)).

From Eq. (7.30)

$$R_o = \left(\frac{\frac{ND_p}{6} + M}{\frac{2D_p}{3}}\right) = \left(\frac{\frac{198 \times 800}{6} + 264E3}{\frac{2 \times 800}{3}}\right)$$

$$= 544,5 \,\text{kN}$$

Frictional resistance at R_o

$$= \mu R_o = 0,15 \times 544,5 = 81,65 > (H = 49,5) \,\text{kN, satisfactory.}$$

Second moment of area about axis O–O for unit size weld (Fig. 7.34(e))

$$I_{wO} = \frac{2D_p^3}{3} = \frac{2 \times 800^3}{3} = 341,3E6 \,\text{mm}^4$$

Maximum vertical force per unit length of weld in the z direction

$$F_{wz} = \frac{M - \frac{ND_w}{2}}{I_{wO}} = \frac{264E3 - \frac{198 \times 800}{2}}{341,3E3} = 0,542 \,\text{kN/mm}$$

Assuming a 6 mm fillet weld for Grade S355 steel (Eq. (4.4), EN 1993-1-8 (2005))

$$F_{w,Rd} = \frac{f_u a}{(3^{1/2}\beta_w \gamma_{M2})} = \frac{510 \times 0,7 \times \frac{6}{1E3}}{(3^{1/2} \times 0,9 \times 1,25)}$$

$$= 1,10 > (F_{wz} = 0,542) \,\text{kN/mm}$$

If the end of the column is not machined then rotation is assumed to be about axis G–G and the size of the weld connecting the column to the gusset plate is obtained as follows.

Length of weld (Fig. 7.34(d))

$$L_w = 4D_w = 4 \times 300 = 1,2E3 \, \text{mm}$$

Second moments of area about the centroid of the weld group (axis G–G) fo unit size welds

$$I_{wGy} = \frac{4D_w^3}{12} = \frac{4 \times 300^3}{12} = 9E6 \, \text{mm}^4$$

$$I_{wGz} = 4D_w \left(\frac{h}{2}\right)^2 = 4 \times 300 \left(\frac{222,3}{2}\right)^2 = 14,83E6 \, \text{mm}^4$$

Polar second moment of area

$$I_{wGx} = I_{wGy} + I_{wGz} = (9,0 + 14,83)E6 = 23,83E6 \, \text{mm}^4$$

Maximum vertical force per unit length of weld in the z direction on an element furthest from the axis of rotation

$$F_{wz} = \frac{N}{L_w} + \left(M - \frac{HD_w}{2}\right)\frac{\left(\frac{h}{2}\right)}{I_{wGx}}$$

$$= \frac{198}{1,2E3} + \left(264E3 - \frac{49,5 \times 300}{2}\right) \times \frac{\left(\frac{222,3}{2}\right)}{23,83E6}$$

$$= 1,362 \, \text{kN/mm}$$

Horizontal force per unit length of weld in the y direction on the same element

$$F_{wy} = \frac{H}{L_w} + \left(M - \frac{HD_w}{2}\right)\frac{\left(\frac{D_w}{2}\right)}{I_{wGx}}$$

$$= \frac{49,5}{1,2E3} + \left(264E3 - \frac{49,5 \times 300}{2}\right) \times \frac{\left(\frac{300}{2}\right)}{23,83E6}$$

$$= 1,656 \, \text{kN/mm}$$

Resultant vector force per unit length of weld

$$F_{wr} = (F_{wy}^2 + F_{wz}^2)^{1/2} = (1,656^2 + 1,362^2)^{1/2} = 2,144 \, \text{kN/mm}$$

Assuming a 12 mm fillet weld for Grade S355 steel (Eq. (4.4), EN 1993-1-8 (2005))

$$F_{w,Rd} = \frac{f_u a}{(30^{1/2}\beta_w \gamma_{M2})} = \frac{510 \times 0,7 \times \frac{12}{1E3}}{(3^{1/2} \times 0,9 \times 1,25)}$$

$$= 2,20 > (F_{wr} = 2,144) \, \text{kN/mm}$$

Alternatively if the contact between the base plate and the end of the column is machined then the axis of rotation is O–O (Fig. 7.34(d)).

Second moments of area about centroid of the weld group for unit size weld about axis O–O

$$I_{wOy} = \frac{4D_w^3}{3} = \frac{4 \times 300^3}{3} = 36E6 \text{ mm}^4$$

$$I_{wbz} = 2D_w h^2 = 2 \times 300 \times 222{,}3^2 = 29{,}65E6 \text{ mm}^4$$

Polar second moment of area

$$I_{wOx} = I_{wOy} + I_{wOz} = (36 + 29{,}65)E6 = 65{,}65E6 \text{ mm}^4$$

Force per unit length of weld in the z direction on an element furthest from the axis of rotation

$$F_{wz} = \frac{\left(M - \frac{Nh}{2}\right)h}{I_{wOx}}$$

$$= \left(264E3 - 198 \times \frac{222{,}3}{2}\right) \times \frac{222{,}3}{65{,}65E6} = 0{,}819 \text{ kN/mm}$$

Force per unit length of weld in the y direction on the same element

$$F_{wy} = \frac{\left(M - \frac{Nh}{2}\right)D_w}{I_{wOx}}$$

$$= \left(264E3 - 198 \times \frac{222{,}3}{2}\right) \times \frac{300}{65{,}65E6} = 1{,}106 \text{ kN/mm}$$

Resultant vector force per unit length of weld

$$F_{wr} = (F_{wy}^2 + F_{wz}^2)^{1/2} = (1{,}106^2 + 0{,}819^2)^{1/2} = 1{,}376 \text{ kN/mm}$$

Assuming a 8 mm fillet weld for Grade S355 steel (Eq. (4.4), EN 1993-1-8 (2005))

$$F_{w,Rd} = \frac{f_u a}{(3^{1/2} \beta_w \gamma_{M2})} = \frac{510 \times 0{,}7 \times \frac{8}{1E3}}{(3^{1/2} \times 0{,}9 \times 1{,}25)}$$

$$= 1{,}47 > (F_{wr} = 1{,}376) \text{ kN/mm}$$

Check shear resistance beneath the base plate (cl 6.2.2(8), Eq. (6.3), EN 1993-1-8(2005))

$$F_{1,v,Rd} = 107{,}7 \text{ kN (see Annex A2)}$$

$$\alpha_b = 0{,}44 - 0{,}0003 f_{yb} = 044 - 0{,}0003 \times 240 = 0{,}368$$

$$F_{2,v,Rd} = \alpha_b f_{ub} \frac{A_s}{\gamma_{Mb}} = 0{,}368 \times 400 \times \frac{561}{1{,}25} = 66{,}1 < 107{,}7 \text{ kN}$$

$$F_{v,Rd} = 0{,}2N + n_b F_{1,v,Rd} = 0{,}2 \times 198 + 4 \times 66{,}1$$

$$= 304 > (H = 49{,}5) \text{ kN satisfactory.}$$

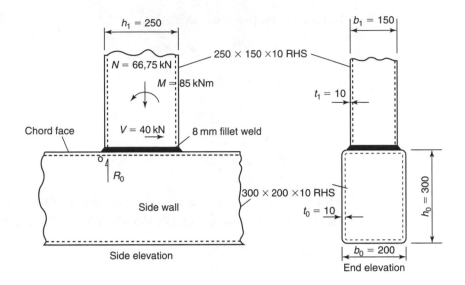

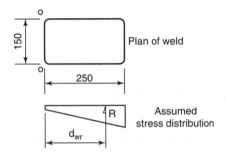

FIGURE 7.35 'Rigid' RHS joint

EXAMPLE 7.13 'Rigid' RHS connection. Check the strength of the rigid rectangular hollow section connection at the ultimate limit state assuming Grade S355 steel (Fig. 7.35).

Check validity of the joint (Table 7.8, EN 1993-1-8 (2005))

$$\frac{b_1}{t_1} = \frac{150}{10} = 15 < 35 \text{ satisfactory.}$$

$$\frac{h_1}{t_1} = \frac{250}{10} = 25 < 35 \text{ satisfactory.}$$

Check for chord face failure of the horizontal member (Table 7.14, EN 1993-1-8 (2005)).

$$M_{\text{ip,1,Rd}} = k_n f_{yo} t_o^2 h_1 \frac{\left[\frac{1}{(2\eta)} + \frac{2}{(1-\beta)^{1/2}} + \frac{\eta}{(1-\beta)}\right]}{\gamma_{M5}}$$

$$= 1{,}0 \times 355 \times 10^2 \times 250 \times \frac{\left[\frac{1}{(2\times 5/3)} + \frac{2}{(1-0{,}75)^{1/2}} + \frac{5/3}{(1-0{,}75)}\right]}{(1{,}0 \times 1\text{E}6)}$$

$$= 97{,}3 > (M_{\text{Ed}} = 85) \text{ kNm satisfactory.}$$

Included in the previous calculations

$$\beta = \frac{b_1}{b_0} = \frac{150}{200} = 0{,}75 < 0{,}85$$

$$k_n = 1{,}3 - 0{,}4n/\beta$$

$$= 1{,}3 - 0{,}4 \times \frac{0{,}282}{0{,}75}$$

$$= 1{,}15 > 1 \text{ use } 1$$

$$\eta = \frac{h_1}{b_0} = \frac{250}{150} = \frac{5}{3}$$

If the weld group is assumed to rotate about the axis O–O (Fig. 7.35) then the effective width of the weld (Table 7.13, EN 1993-1-8 (2005)) furthest from the axis O–O

$$b_{eff} = 10 \frac{f_{y0}t_0 b_1}{\left[\left(\frac{b_0}{t_0}\right)f_{y1}t_1\right]}$$

$$= 10 \times \frac{355 \times 10 \times 150}{\left[\left(\frac{200}{10}\right) \times 355 \times 10\right]} = 75\,\text{mm}$$

$$\frac{b_{eff}}{b_1} = \frac{75}{150} = 0{,}5$$

Distance(d_{wr}) from the axis O–O from the resultant force in the weld is obtained as follows

moment of the parts = moment of the whole

$$\left(b_{eff} + \frac{1}{2} \times 2d_w\right)F_z d_{wr} = \frac{F_z I_{oe}}{d_{wr}}$$

rearranging and substituting $I_{oe} = 2d_w^3/3 + b_{eff}d_w^2$

$$\frac{d_{wr}}{d_w} = \frac{\left(\frac{2}{3} + \frac{b_{eff}}{d_w}\right)}{\left(1 + \frac{b_{eff}}{d_w}\right)}$$

$$= \frac{\left(\frac{2}{3} + \frac{75}{250}\right)}{\left(1 + \frac{75}{250}\right)} = 0{,}744$$

Resultant reaction from Eq. (7.30)

$$R_o = \frac{M + N(d_{wr} - d_w/2)}{d_{wr}}$$

$$= \frac{\left[85E3 + 66{,}75 \times \left(0{,}744 \times 250 - \frac{250}{2}\right)\right]}{0{,}744 \times 250} = 478{,}9\,\text{kN}$$

Frictional force at the stiff bearing if the end of the vertical RHS is machined

$$\mu R_o = 0{,}15 \times 478{,}9 = 71{,}8 > 40\,\text{kN (applied shear force)}.$$

The weld group is subject to the actions from N and M acting about axis O–O.

Second moment of area about axis O–O of the weld group (Fig. 7.35) for unit size welds

$$I_O = \frac{2d_w^3}{3} + b_{\text{eff}}d_w^2 = \frac{2 \times 250^3}{3} + 75 \times 250^2 = 15{,}1\text{E6 mm}^4$$

Moment applied about axis O–O

$$M' = M - \frac{Nd_w}{2} = 85 - 66{,}75 \times \frac{250}{2 \times 1\text{E3}} = 76{,}65 \text{ kNm}$$

Maximum tensile force per unit length of weld furthest from the axis of rotation

$$F_w = \frac{M'd}{I_O} = \frac{76{,}65\text{E3} \times 250}{15{,}1\text{E6}} = 1{,}27 \text{ kN/mm}$$

Assuming a 8 mm fillet weld for Grade S355 steel (Eq. (4.4), EN 1993-1-8 (2005))

$$F_{w,Rd} = \frac{f_u a}{(3^{1/2}\beta_w \gamma_{M2})} = \frac{510 \times 0{,}7 \times \frac{8}{1\text{E3}}}{(3^{1/2} \times 0{,}9 \times 1{,}25)}$$
$$= 1{,}47 > (F_{wy} = 1{,}27) \text{ kN/mm}$$

An alternative more conservative calculation to determine the size of weld is to assume rotation about the centroid of the weld group. This results in a larger weld size.

For interest check for side wall crushing in horizontal member at O–O (Table 7.14, EN 1993-1-8 (2005))

$$M_{ip,1,Rd} = \frac{0{,}5f_{yk}t_0(h_1 + 5t_0)^2}{\gamma_{M5}}$$
$$= \frac{0{,}5 \times 355 \times 10 \times (300 + 5 \times 10)^2}{(1{,}0 \times 1\text{E6})} = 217{,}4 > M_{ip,1,Rd} = 155 \text{ kNm}.$$

This calculation is not necessary because the ratio $\beta = b_1/b_0 = 150/200 = 0{,}75 < 0{,}85$ indicates that it is not critical.

EXAMPLE **7.14**   'Rigid' knee connection for a portal frame. Check the strength of the knee joint components for the frame (Fig. 7.36) at the ultimate limit state.

Moment acting about axis O–O

$$M_O = 250{,}2 - 116{,}8 \times 0{,}5379 = 187{,}4 \text{ kNm}$$

Thickness of strap assuming 190 mm wide and 22 mm diameter hole

$$t_{st} = \frac{M_O}{\left(\frac{l_a b_{st} f_{y,st}}{\gamma_{M1}}\right)}$$
$$= \frac{187{,}4\text{E6}}{\left[717{,}2 \times \cos 10° \times (190 - 2 \times 22) \times \frac{355}{1{,}0}\right]}$$
$$= 5{,}12 \text{ mm. Use a 8 mm thick strap.}$$

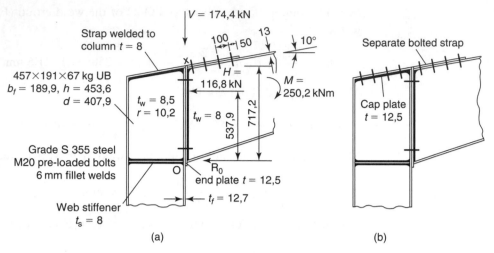

FIGURE 7.36 'Rigid' knee joint for a portal frame

Strap welded to the top of the column and bolted to the rafter (Fig. 7.36(a)).

For a single M20 pre-loaded bolt slip resistance (Eq. (3.6), EN 1993-1-8 (2005))

$$F_{s,Rd} = k_s n \; \mu f_{ub} \frac{A_s}{\gamma_{M3}}$$

$$= 1,0 \times 1,0 \times 0,4 \times 0,7 \times 800 \times \frac{245}{1,25 \times 1E3} = 43,9 \text{ kN}$$

M20 class 8.8 bolt in bearing on the plate ($t = 8$ mm) (Table 3.4, EN 1993-1-8 (2005)).

$$F_{b,Rd} = \frac{k_1 \alpha_b f_{up} d t_w}{\gamma_{M2}}$$

$$= \frac{2,5 \times 0,758 \times 510 \times 20 \times 8}{1,25 \times 1E3} = 123,7 > (F_{s,Rd} = 43,9) \text{ kN}$$

for an end bolt $\alpha_b = e_1/(3d_o) = 50/(3 \times 22) = 0,758$

for an edge bolt $k_1 = 2,8e_2/d_o - 1,7 = 2,8 \times 50/22 - 1,7 = 4,66 > 2,5$

Number of M20 pre-loaded bolts in single shear for the strap

$$n_b = \frac{M_O}{(l_a F_{s,Rd})} = \frac{187,4E3}{717,2 \times \cos 10° \times 43,9} = 6,04 \text{ use 6 bolts.}$$

Force per unit length acting on the two welds connecting strap to head of column

$$F_{w,Ed} = \frac{M_O}{l_a n_w l_w}$$

$$= \frac{187,4E3}{717,2 \times \cos 10° \times 2 \times \frac{407,9}{\cos 10°}} = 0,320 \text{ kN/mm}$$

Shear resistance of a 6 mm fillet weld (cl 4.5.3.3, EN 1993-1-8 (2005))

$$F_{w,Rd} = \frac{f_u a}{(3^{1/2} \beta_w \gamma_{M2})}$$

$$= \frac{510 \times 0,7 \times 6}{(3^{1/2} \times 0,9 \times 1,25 \times 1E3)} = 1,1 > (F_{w,Ed} = 0,320)\,kN/mm$$

These calculations assume, conservatively, that the force is resisted only by the two welds along the web of the column, but the strap is also welded to the flanges of the column which increases the strength.

Moments of forces about X to determine the reaction R_o at O

$$-717, 2R_o + 250, 2E3 + 116, 8 \times (717, 2 - 537, 9) = 0; \quad \text{hence } R_o = 378, 1\,kN$$

Vertical frictional force at O

$$= \mu R_o = 0,4 \times 378,1 = 151,2 < (V_{Ed} = 174,4)\,kN, \text{ therefore slip occurs.}$$

Bolt end plate ($t_{ep} = 12{,}5$ mm) to the flange of the column using M20 pre-loaded bolts ($f_{pu} = 800$ MPa, $\mu = 0{,}4$). For a single bolt slip resistance (Eq. (3.6), EN 1993-1-8 (2005)). $F_{s,Rd} = 43{,}9$ kN as calculated previously.

Number of bolts required

$$n_b = \frac{V_{Ed}}{F_{v,Rd}} = \frac{174,4}{43,9} = 3,97 \text{ use 4 bolts.}$$

Shear strength of the web of the column (cl 6.2.6.1 (2), EN 1993-1-8 (2005))

$$V_{pl,Rd} = 0,9 A_{vc} \frac{\left(\frac{f_{y,wc}}{\gamma_{M0}}\right)}{3^{1/2}} = 0,9 \times 4094 \times \frac{\left(\frac{355}{1,0}\right)}{(3^{1/2} \times 1E3)}$$

$$= 755,2 > (R_o = 378,1)\,kN, \text{ satisfactory.}$$

Design resistance of the unstiffened column web at O (Eq. (6.9), EN 1993-1-8(2005))

$$F_{c,wc,Rd} = \frac{\omega k_{wc} b_{eff,c,wc} t_{wc} f_{ywe}}{\gamma_{M0}}$$

$$= \frac{0,932 \times 1 \times 0,731 \times 164,4 \times 8,5 \times 355}{1,0 \times 1E3}$$

$$= 338,0 < (R_o = 378,1)\,kN \text{ not satisfactory.}$$

or

$$F_{c,wc,Rd} = \frac{\omega k_{wc} \rho b_{eff,c,wc} t_{wc} f_{ywe}}{\gamma_{M1}}$$

$$= \frac{0,932 \times 1 \times 0,731 \times 164,4 \times 8,5 \times 355}{1,00 \times 1E3}$$

$$= 338,0 < (R_o = 378,1)\,kN \text{ not satisfactory stiffeners required.}$$

Previous calculations include the effective width (Eq. (6.11), EN 1993-1-8 (2005))

$$b_{\text{eff,c,wc}} = t_{\text{fb}} + 2 \times 2^{1/2} a_{\text{p}} + 5(t_{\text{fc}} + s) + s_{\text{p}}$$

$$= 13 + 2 \times 2^{1/2} \times 0,7 \times 6 + 5 \times (12,7 + 10,2) + 2 \times 12,5 = 164,4 \, \text{mm}$$

Maximum longitudinal stress in the flange of the column from the axial load and eccentric load (cl 6.2.6.2(2), EN 1993-1-8 (2005))

$$\sigma_{\text{com,Ed}} = \frac{V}{A} + \frac{\left(M + \frac{Vh}{2}\right)}{W_{\text{el}}}$$

$$= \frac{174,4\text{E}3}{85,5\text{E}2} + \frac{\left(250,2\text{E}6 + 174,4\text{E}3 \times \frac{453,6}{2}\right)}{1297\text{E}3}$$

$$= 244 < (0,7 f_{\text{y,wc}} = 0,7 \times 355 = 248,5) \, \text{MPa hence } k_{\text{wc}} = 1$$

$\beta = 1$ (Table 5.4, EN 1993-1-8 (2005)) and from Table 6.3, EN 1993-1-8 (2005)

$$\omega = \omega_1 = \frac{1}{\left[1 + 1,3 \left(\frac{b_{\text{eff,c,wc}} t_{\text{wc}}}{A_{\text{vc}}}\right)^2\right]^{1/2}}$$

$$= \frac{1}{\left[1 + 1,3 \left(\frac{164,4 \times 8,5}{4094}\right)^2\right]^{1/2}} = 0,932$$

which includes (cl 6.2.6(3), EN 1993-1-1 (2005)) or from Section Tables

$$A_{\text{vc}} = A - 2bt_{\text{f}} + (t_{\text{w}} + 2r)t_{\text{f}}$$

$$= 85,5\text{E}2 - 2 \times 189,9 \times 12,7 + (8,5 + 2 \times 10,2) \times 12,7 = 4094 \, \text{mm}^2$$

$$\eta h_{\text{wc}} t_{\text{wc}} = 1,0 \times (453,6 - 2 \times 12,7) \times 8,5$$

$$= 3640 < (A_{\text{vc}} = 4094) \, \text{mm}^2 \text{ satisfactory.}$$

Plate slenderness (cl 6.2.6.2(1), EN 1993-1-8 (2005))

$$\overline{\lambda}_{\text{p}} = 0,932 \left[\frac{b_{\text{eff,c,wc}} d_{\text{we}} f_{\text{y,wc}}}{(E \, t_{\text{wc}}^2)}\right]^{1/2}$$

$$= 0,932 \left[\frac{164,4 \times 407,9 \times 355}{(210\text{E}3 \times 8,5^2)}\right]^{1/2} = 1,167 > 0,72 \text{ therefore}$$

$$\rho = \frac{\overline{\lambda}_{\text{p}} - 0,2}{\overline{\lambda}_{\text{p}}^2} = \frac{1,196 - 0,2}{1,196^2} = 0,731$$

Thickness of the load bearing web stiffeners in compression in the column (cl 6.2.6.2(4), EN 1993-1-8 (2005))

$$t_s = \frac{R_o}{\left[(b_{cf} - t_{cw} - 2r_c)\frac{f_{yw}}{\gamma_{M1}}\right]}$$

$$= \frac{378,1E3}{\left[(189,9 - 8,5 - 2 \times 10,2)\frac{355}{1,0}\right]}$$

$= 6{,}62$ mm. Use 8 mm thick stiffeners either side of the web of the column.

Force per unit length of weld connecting the web stiffener to the column web

$$F_w = \frac{R_o}{(n_w l_w)} = \frac{378,1}{2 \times 407,9} = 0{,}463 \text{ kN/mm}$$

As previously the shear resistance of a 6 mm fillet weld (cl 4.5.3.3, EN 1993-1-8 (2005))

$$F_{w,Rd} = 1{,}1 > (F_w = 0{,}463) \text{ kN/mm}$$

It is assumed conservatively that the force is resisted by the welds along the web.

To avoid damage to the strap in transit an alternative arrangement is shown in Fig. 7.36(b).

7.11 JOINT ROTATIONAL STIFFNESS (CL 6.3, EN 1993-1-8 (2005))

Previous sections in this chapter show calculations for joint strength. Another aspect of joint behaviour that needs consideration is the relationship between joint rotational stiffness and member stiffness. This relationship is important in the analysis of structures (cl 5, EN 1993-1-8 (2005)).

A simple example of joint rotational stiffness (Fig. 7.37(a)) is as follows. The theoretical rotational stiffness related to the elastic and plastic behaviour of the top cleat can be developed for the joint. Initially assuming elastic behaviour of the steel top cleat with deformations as shown

$$S_j = \frac{M}{\phi} = \frac{F_t z}{\left(\frac{\Delta}{z}\right)} = \frac{2\left(6\frac{E\,l_a\Delta}{m^3}\right)z^2}{\Delta} = 12E\left(\frac{b_a t_a^3}{12m^3}\right)z^2$$

$$= \frac{E\,z^2}{\left[\frac{1}{(b_a t_a^3/m^3)}\right]} \tag{7.40}$$

This equation considers only deformation of the top cleat and can be compared with the theoretical expression (Eq. (6.27), EN 1993-1-8 (2005)) which replaces b_a with l_{eff} and includes factors for other deformations (Table 6.11, EN 1993-1-8 (2005)).

$$S_j = \frac{E\,z^2}{\left[\mu\Sigma\left(\frac{1}{k_1}\right)\right]} \tag{7.41}$$

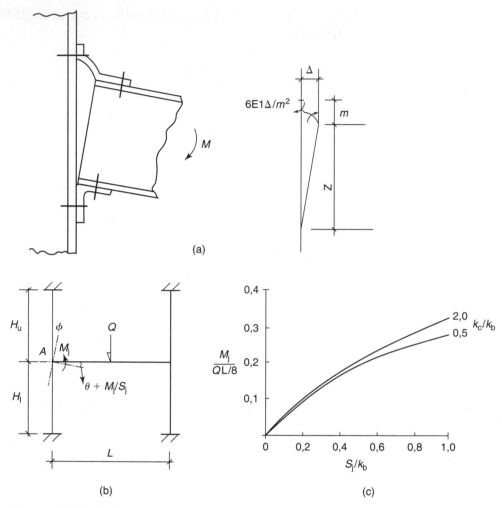

FIGURE 7.37 Stiffness of joints

Other factors that affect the rotational stiffness of a joint are: extension and shear deformation of bolts, and deformation of column flanges, end plates and column webs. Each factor (Tables 6.10 and 6.11, EN 1993-1-8 (2005)) produces values of $1/k_1$ which are added and inserted in Eq. (7.41) as shown in Example 7.15.

Experimental results (Maxwell *et al.*, 1981) which show that the rotational stiffness of a joint is non-linear as ultimate resistance approaches. The stiffness of the joint reduces and to allow for this the value of μ increases (cl 6.3.1(6), EN 1993-1-8 (2005)).

It is also necessary to check for other forms of failure. For this joint (Fig. 7.37(a)) the ultimate moment of resistance based on yielding of the top angle

$$M_{\text{jya,Rd}} = 2\left(\frac{b\,t_a^2}{4}\right)\left(\frac{f_y}{mz}\right)$$

(7.42a)

Alternatively for this joint if the bolts reach ultimate load

$$M_{\text{jub,Rd}} = n_b A_b f_u z \tag{7.42b}$$

7.12 FRAME-TO-JOINT STIFFNESS

Ideally the analysis of a structure should incorporate the stiffness of the joints and the members. An example of the theoretical relationship between the stiffness of joint and members for a simple frame follows.

Consider the elastic behaviour of the simple frame in Fig. 7.37(b). From the application of the area moment method (Croxton and Martin, 1987 and 1989), the moment of resistance of the joint (M_j) for a point load at the mid-span of the beam

For the beam at A

$$EI_b\left(\theta + \frac{M_j}{S_j}\right) = \frac{1}{2} \times \frac{L_b}{2} \times \frac{QL_b}{4} - \frac{M_j L_b}{2} \tag{i}$$

For the upper column at A

$$EI_{cu}\theta = \frac{1}{4} \times M_{ju} H_u \tag{ii}$$

For the lower column at A

$$EI_{cl}\theta = \frac{1}{4} \times M_{jl} H_l \tag{iii}$$

and

$$M_j = M_{ju} + M_{jl} \tag{iv}$$

Combining (i) to (iv) to eliminate θ and expressing stiffness as $k = El/L$

$$M_j = \frac{\left(\frac{QL}{8}\right)}{1 + \frac{1}{2(k_{cU}/k_b + k_{cL}/k_b)} + \frac{2k_b}{S_j}} \tag{7.43}$$

The stiffness of the joint (S_j), beam (k_b) and columns (k_c) can be seen to affect the moment of resistance at the joint (M_j).

EXAMPLE 7.15 Stiffness of a 'pin' joint. Determine the stiffners of the 'Pin' shown in Fig. 7.26.

Joint stiffness (Eq. (6.27), EN 1993-1-8(2005)) assuming $\mu = 1$

$$S_j = \frac{E z^2}{\left[\mu \Sigma\left(\frac{1}{k}\right)\right]}$$

$$= \frac{210E3 \times 517{,}4^2}{[(1 \times 2{,}48 \times 1E6)]} = 22668 \text{ kNm/radian}$$

where (Fig. 6.15(b), EN 1993-1-8 (2005))

$$z = h + e_a + \frac{t_a}{2} = 467,4 + 45 + \frac{10}{2} = 517,4\,\text{mm}$$

and for a column web panel in shear (Table 6.11, EN 1993-1-8 (2005))

$$k_1 = \frac{0,38A_{vc}}{\beta z} = \frac{0,38 \times 3130}{1 \times 517,4} = 2,3$$

and for a column web panel in compression (Table 6.11, EN 1993-1-8 (2005))

$$k_2 = \frac{0,7b_{eff,c,wc}t_{wc}}{d_c} = \frac{0,7 \times 179,5 \times 13}{160,8} = 10,2$$

where (Table 6.12, EN 1993-1-8 (2005))

$$b_{eff,c,we} = 2t_a + 0,6r_a + 5(t_{fc} + s)$$
$$= 2 \times 10 + 0,6 \times 10 + 5 \times (20,5 + 10,2) = 179,5\,\text{mm}$$

and for a column web in tension (Table 6.11, EN 1993-1-8 (2005))

$$k_3 = \frac{0,7b_{eff,c,wc}t_{fc}}{d_c} = \frac{0,7 \times 210 \times 13}{160,8} = 11,9$$

where (Table 6.4, EN 1993-1-8 (2005)) $b_{eff,c,wc}$ is the lesser value of

$$2\pi m = 2\pi \times 45 = 283\,\text{mm and}$$
$$\pi m + 2e_1 = \pi \times 45 + 2 \times \frac{(208,8 - 140)}{2} = 210\,\text{mm}$$

and for a column flange in bending (Table 6.11, EN 1993-1-8 (2005))

$$k_4 = \frac{0,9l_{eff}t_{fe}^3}{m^3} = \frac{0,9 \times 146,5 \times 20,5^3}{45^3} = 12,5$$

where (Table 6.4, EN 1993-1-8 (2005)) l_{eff} is the lesser value of

$$l_{eff} = 4m + 1,25e = 4 \times 45 + 1,25 \times \frac{(208,8 - 140)}{2} = 223\,\text{mm or}$$
$$l_{eff} = 2m + 0,625e + e_1$$
$$= 2 \times 45 + 0,625 + \frac{(208,8 - 140)}{2} + 35 = 146,5\,\text{mm}$$

and for a cleat in bending (Table 6.11, EN 1993-1-8 (2005))

$$k_6 = \frac{0,9l_{eff}t^3}{m^3} = \frac{0,9(208,8)}{2} \times \frac{10^3}{(45 - 10/2)^3} = 1,03$$

and for bolts in tension (Table 6.11, EN 1993-1-8 (2005))

$$k_{10} = \frac{1,6A_s}{L_b} = \frac{1,6 \times 245}{\left(20,5 + 10 + 3,7 + \frac{13}{2} + 16\right)} = 6,91$$

and for bolts in shear (Table 6.11, EN 1993-1-8 (2005))

$$k_{11} = \frac{16 n_b d^2 f_{ub}}{E \, d_{M16}} = \frac{16 \times 2 \times 20^2 \times 800}{210E3 \times 14,1} = 3,45$$

and for bolts in bearing (Table 6.11, EN 1993-1-8 (2005))

$$k_{12} = \frac{24 n_b k_b k_t d f_u}{E} = \frac{24 \times 2 \times 1,06 \times 1,06 \times 20 \times 510}{210E3} = 2,62$$

where

$$k_b = \frac{0,25 e_b}{d} + 0,5 = \frac{0,25 \times 45}{20} + 0,5 = 1,06$$

$$k_t = \frac{1,5 t_j}{d_{M16}} = \frac{1,5 \times 10}{14,1} = 1,06$$

The total value of

$$\mu \Sigma \left(\frac{1}{k} \right) = 1 \times \left(\frac{1}{2,3} + \frac{1}{10,2} + \frac{1}{11,9} + \frac{1}{12,5} + \frac{1}{1,03} + \frac{1}{6,91} + \frac{1}{3,45} + \frac{1}{2,62} \right) = 2,48.$$

EXAMPLE 7.16   Effect of 'pin' joint stiffness on a simple frame shown in Fig. 7.37(b). Assume $Q = 52$ kN, $L_b = 10$ m, $k_b = 9612$ kNm, $k_b/k_c = 1$ and $S_j = 22668$ kNm obtained in Example 7.15.

The joint design moment of resistance at A (Eq. (7.43))

$$M_{j,Rd} = \frac{\left(\frac{QL}{8} \right)}{\left(1 + \frac{1}{2\left(\frac{k_{cu}}{k_b} + \frac{k_{cl}}{k_b} \right)} + \frac{2 k_b}{S_j} \right)}$$

$$= \frac{\frac{52E3 \times 10E3}{(8 \times 1E6)}}{\left(1 + \frac{1}{2(1+1)} + \frac{2 \times 9612}{22668} \right)}$$

$$= 31 \text{ kNm}$$

The effect of introducing the joint stiffness to the above equation is to reduce the end moment on the beam ($M_{j,Rd}$) from 51,8 to 31 kNm.

If $M_{j,Ed} = 20$ kNm then

$M_{j,Ed}/M_{j,Rd} = 20/31 = 0,65 < 2/3$ and the adoption of $\mu = 1$ is satisfactory (cl 6.3.1(6), EN 1993-1-8 (2005)).

Check the ultimate moment of resistance of the joint based on yielding of the top angle (Eq. (7.42a))

$$M_{jya,Rd} = 2 \left(\frac{b \, t_a^2}{4} \right) \left(\frac{f_y}{m \, z} \right)$$

$$= 2 \times \left(208,8 \times \frac{10^2}{4} \right) \times \frac{\left(\frac{355}{40} \times 517,4 \right)}{1E6} = 47,94 > (M_{j,Rd} = 31) \text{ kNm}$$

which is the maximum moment of resistance for the joint.

Check the ultimate moment of resistance of the two top bolts (2M-20 class 8.8) failing in tension (Eq. (7.42b))

$$M_{\text{jub,Rd}} = 2A_{\text{b}}f_{\text{u}}z = 2 \times 245 \times 800 \times \frac{517{,}4}{1\text{E}6}$$

$$= 202{,}8 > (M_{\text{jya,Rd}} = 47{,}94) \text{ kNm}$$

The effect of varying joint to beam stiffness can be shown for the simple structure (Fig. 7.37(b)) using Eq. (7.43). The ratio $(M_{\text{j}}/(QL/8))$, which includes the moment of resistance at the end of the beam, varies with the ratios of joint-to-beam stiffness $(S_{\text{j}}/k_{\text{b}})$ and column-to-beam stiffness$(k_{\text{c}}/k_{\text{b}})$ as shown in Fig. 7.37(c). This relationship is related to the structure in Fig. 7.37(b) for a beam stiffness of 4E3 kNm. Elastic behaviour is assumed but plastic failure may limit the value of the joint resistance.

For example the joint moment (M_{j}) reduces the moment at mid-span for the beam, which is beneficial, but introduces a moment to the columns which if slender can reduce their capacity. However for more complicated structures an increase in connection stiffness reduces slenderness ratios and column deflections, which increase the load capacity of a column. Numerical investigations (Jones *et al.*, 1981) show that including joint stiffnesses in the analysis of frames can reduce the weight of steel.

REFERENCES

Bahia, C.S. and Martin, L.H. (1980). Bolt groups subject to torsion and shear, *Proceedings of the I.C.E.* Pt2, V69.

Bahia, C.S. and Martin, L.H. (1981). Experiments on stressed and unstressed bolt groups subject to torsion and shear, *Conference Proceedings, Joints in Structural Steelwork.* Teeside Polytechnic.

Bahia, C.S., Graham, J. and Martin, L.H. (1981). Experiments on rigid beam-to-column connections subject to shear and bending forces, *Conference Proceedings, Joints in Structural Steelwork.* Teeside Polytechnic.

BS 7668 (1994), BSEN 10029 (1991), BSEN 10113-1 to 3 (1993) and BSEN 10210-1 (1994). *Specification for weldable structural steels.* BSI.

BS 5400 (2000). *Steel concrete and composite bridges, Pt3 Code of practice for the design of steel bridges.* BSI.

BS 3692 (2001). *ISO Metric precision hexagon bolts, screws and nuts.* BSI.

BS 4190 (2001). *ISO Metric black hexagon bolts, screws and nuts.* BSI.

BSEN 499 (1995). *Covered electrodes for the manual metal arc welding of carbon and carbon manganese steels.* BSI.

BSEN ISO 4320 (1998). *Metal washers for general engineering purposes.* BSI.

BSEN 1011-1 (1998) and 2 (2001). *Specification for the process of arc welding of carbon and carbon manganese steels.* BSI.

BSEN 14399-1 to 5 (2005). *High strength friction grip bolts and associated nuts and washers for structural engineering: Pt1 General grade, Pt2 Higher grade bolts and nuts and general grade washers.* BSI.

Biggs, M.S.A.B., Crofts, M.R., Higgs, J.D., Martin, L.H. and Tzogius, A. (1981). Failure of fillet welded connections subject to static loading, *Conference Proceedings, Joints in Steelwork*, Teeside Polytechnic.

Chesson, Jr., E. Faustino, N.L. and Munse, W.H. (1965). *High strength bolts subject to torsion and shear*, A.S.C.E. (Structural Division), V91, ST5.

Clarke, A. (1970). *The strength of fillet welded connections*. MSc Thesis, Imperial College, University of London.

Clarke, P.J. (1971). Basis for the design of fillet welded joints under static loading, *Conference Proceedings*. Welding Institution, Improving Welding Design Paper 10, V1.

Croxton, P.C.L. and Martin, L.H. (1987 and 1989). *Solving Problems in Structures Vols. 1 and 2*. Longman Scientific and Technical.

Davies, G. (1981). Estimating the strength of some welded lap joints formed from rectangular hollow section members, *Conference Proceedings, Joints in Structural Steelwork*, Teeside Polytechnic.

Elzen, L.W.A. (1966). *Welding beams in beam-to-column connections without the use of stiffening plates*. Report 6-66-2. I.I.W. document XV-213-66.

EN 1993-1-1 (2005). *General rules and rules for buildings*. BSI.

EN 1993-1-8 (2005). *Design of joints*. BSI.

European Convention for Structural Steelwork (1981). *European recommendations for steel construction*. Construction Press.

Farrar, J.C.M. and Dolby, R.E. (1972). *Lamellar tearing in welded steel fabrication*. Welding Institute, Cambridge, England.

Fisher, J.W. and Struik, J.H.A. (1974). *Guide to Design Criteria for Bolted and Riveted Joints*. John Wiley and Sons.

Gourd, L.M. (1980) *Principles of Welding Technology*. Edward Arnold.

Holmes, M. and Martin, L.H. (1983). *Analysis and Design of Structural Connections*. Ellis Horwood Ltd.

Jones, S.W., Kirby, P.A. and Nethercot, D.A. (1981). Modelling of semi-rigid connection behaviour and its influence on steel column behaviour, *Conference Proceedings, Joints in Steelwork*, Teeside Polytechnic.

Kato, B. and Morita, K. (1974). *Strength of transverse fillet welded joints*. Welding Research.

Ligtenberg, F.K. (1968). *International test series, final report*, Stevin Laboratory, Technological University of Delft, Doc XV-242-68.

Mann, A.P. and Morris, L.J. (1979). Limit state design of extended plate connections, ASCE *Journal of Structural Engineering*, **105(ST3)**: 511–526.

Martin, L.H. (1979). Methods for limit state design of triangular steel gusset plates, *Building and Environment*.

Martin, L.H. and Robinson, S. (1981). Experiments to investigate parameters associated with the failure of gusset plates, *Conference Proceedings, Joints in Steelwork*, Teeside Polytechnic.

Maxwell, S.M., Jenkins, W.M. and Howlett, J.H. (1981). A theoretical approach to the analysis of connection behaviour. *Conference Proceedings, Joints in Steelwork*. Teeside Polytechnic.

Morris, L.J. and Newsome, C.P. (1981). Bolted corner connection subjected to an out of balance moment – the behaviour of the web panel, *Conference Proceedings, Joints in Structural Steelwork*, Teeside Polytechnic.

Owens, G.W., Driver, P.J. and Kriege, G.J. (1981). Punched holes in structural steelwork, Constructional Steel Research, **1(3)**.

Pillinger, A.H. (1988). Structural steel work: a flexible approach to the design of simple construction, *Structural Eng* **66(19/4)**, October.

Purkiss, J.A. and Croxton, P.C.L. (1981). Design of eccentric welded connections in rolled hollow sections, *Conference Proceedings, Joints in Structural Steelwork*. Teeside Polytechnic.

Rolloos, A. (1969). *The effective weld length of beam-to-column connections with stiffening plates,* Final report, I.I.W. Document XV-276-69.

Stamenkovic, A. and Sparrow, K.D. (1981). A review of existing methods for the determination of the static axial strength of welded T, Y, N, K, and X joints in circular hollow steel sections, *Conference Proceedings, Joints in Structural Steelwork*. Teeside Polytechnic.

Stark, J.W.B. and Bijlaard, F.S.K. (1988). *Design rules for beam-to-column connections in Europe, Steel Beam-to Column Building Connections*. pub. Elsevier Applied Science.

Chapter 8 / Frames and Framing

The previous chapters have dealt with design of beams, columns and connections. This chapter, and the next, deal with the way the individual components are assembled together and also with design problems associated with the whole structure.

In the first place it is necessary to give some consideration to how the choice is made of the structural form employed to carry the primary loading. For convenience this survey is divided into single and multi-storey structures.

8.1 SINGLE STOREY STRUCTURES

Typical examples of such structures include sports complexes, exhibition halls, factory units or assembly buildings. Unless architectural considerations prevail, the most economic solution will be obtained using one-way spanning structural systems rather than space frame structures. Systems which at first sight appear two-way spanning often comprise a number of overlaid one-way systems. Roof systems can be conveniently divided into flat and pitched roof systems.

8.1.1 Flat Roof Systems

For spans up to around 15 m rolled sections form the most economic solution. However, it should be noted that the potential extra cost of beams over 12 m long needs to be taken into consideration, and the use of 12–15 m long beams should only be contemplated if a large number are required. At around 14 m and up to 20 m the use of castellated or beams with circular openings in the web become economic. Although this type of beam incurs high fabrication costs and requires a higher construction depth than ordinary rolled sections, the holes in the web allow services to be contained within the beam depth. Above 20 m it is usual to use parallel chord lattice trusses fabricated either from rolled hollow sections or from lightweight cold formed sections (top and bottom chords) and bar members for the web. The roof decking is then generally lightweight steel sheeting with suitable finishes and insulation, although timber decking with asphaltic waterproofing, woodwool slabs or pre-cast concrete decking units may be used. One problem with flat roofs is drainage and thus sufficient cross-fall must be provided to give adequate run-off and avoid local ponding. The cross-fall is generally provided by variable depth purlins and joists or by adjusting screed depths. To some extent these problems are alleviated by employing a sloping top chord to the roof system.

8.1.2 Pitched Roof Systems

These fall into two main divisions: trusses with a sloping top chord and pitched roof portal frames.

Modern trusses have a relatively low slope to the top chord. A pitch of between 4° and 10° is adequate to allow run off and to allow the joints in the roof sheeting to remain watertight. This type of truss is generally fabricated from square or rectangular rolled hollow sections with fully welded nodal connections, with circular hollow sections sometimes being used for the web members in Grade S355 steel, owing to availability, Pitches greater than 10° are only seen where an existing building with traditional large pitch trusses designed for tiles or slates as roofing materials is being extended, or where a large pitch is required for architectural reasons in, for example, shopping malls.

The most common method of single storey construction is the pitched roof portal frame whether for factory units, small sports complexes or warehouses. This is basically due to the high speed and simplicity of construction. The internal bays are designed as rigid jointed frames. The end frames, unless there is likely to be an extension to the structure, are much lighter and have the rafters designed as spanning across the gable end posts which are also used to support the sheeting rails for the cladding. The most economic frame spacing is generally 7.5 m or 9.0 m for much higher frame spans. For spans below 20 m a frame spacing of 6.0 m may be adopted (Horridge, 1985; Horridge and Morris, 1986).

Both truss systems and portal frames can be used for spans up to 60 m, although for spans over 30 m multi-bay structures become an option. Multi-bay construction requires internal columns which may reduce the flexibility of usage. However, this situation may be mitigated by the use of internal lattice girder support systems. Note that where the roof system comprises trusses the internal support system can be within the depth of the truss. This is not possible with portal frames, so unless reduced headroom is acceptable internal columns must be used. It should also be remembered that where multi-bay construction is used, there will be the need to supply valley guttering and associated drainage. Such valleys cause build-up in snow loading and can be potential areas of leakage and cause problems with access for maintenance.

8.2 MULTI-STOREY CONSTRUCTION

8.2.1 Multi-storey Steel Skeleton

In UK practice the steel skeleton is generally designed to carry vertical loading due to the permanent and variable actions only, with the horizontal loading from wind and the notional horizontal loading taken by a bracing system, or more commonly the lift shaft(s) and stair well(s). The economics of various types of multi-storey construction is discussed in Gibbons (1995). When using lift shafts or stairwells as bracing care must be taken in their layout as torsional effects from lateral loading on an asymmetric layout must be avoided.

8.2.2 Flooring Systems

The flooring system is generally required to act as a horizontal diaphragm to carry horizontal forces from their point of application to the bracing or lift core. Thus it is essential that adequate lateral stiffness exists in the plane of the floor and the flooring system is adequately tied to the frame.

Historically the floors of multi-storey steel frames have comprised cast *in-situ* normal weight concrete slabs. This system is now little used partly owing to excess weight of the concrete and partly owing to the slow construction time and the need for propping the deck often over two storeys. The problems with propping can be reduced by the use of proprietary falsework systems involving lightweight trusses which allow the actual formwork to be struck before the props are removed. The use of *in-situ* concrete floor systems except for small areas not otherwise able to be handled has been superseded by pre-cast pre-stressed concrete units or by composite steel–concrete decking.

Pre-cast pre-stressed concrete units may be placed on the top flange of the steel beam, supported on shelf angles welded to the web of the beam, or placed between the flanges of a column section used as a beam. In all cases the flange of the supporting beam system may be considered to be restrained against lateral torsional buckling, although for shelf angle floors consideration may need to be given to the torsional load on the beam during construction. An alternative for ease of construction is to use asymmetric sections with the bottom flange larger than the top (Mullett, 1992; Lawson *et al.*, 1997). These alternatives are illustrated in Fig. 8.1.

For pre-cast units placed on the top flange small discrete vertical plates are welded to the top flange to give shear anchorage. For floor systems where the pre-cast unit sits on a bottom flange or shelf angle, the top corners of the units will need chamfering to ensure the units can be placed in position. Such floors also have the advantage of increased fire performance as the top flange is shielded by the concrete section from the effects of a fire on the underside. Equally the bottom flange of the steel section will be either partially or totally adjacent to a heat sink formed by the concrete slab, thus also inducing lower temperatures than would otherwise exist.

For most types of flooring a screed is required to give a proper finish and it is recommended that a light structural mesh reinforcement is placed over the support beams to help control cracking in the screed. One potential disadvantage of using pre-cast units is the large amount of hook time required to place the units which could otherwise have been used to hoist materials required for finishes, etc.

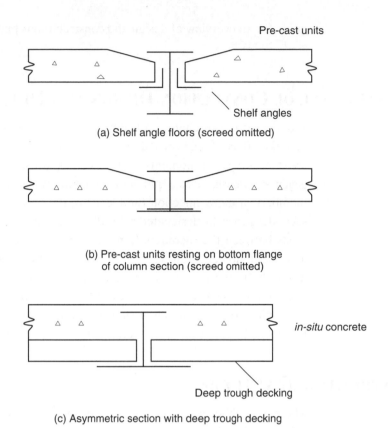

(a) Shelf angle floors (screed omitted)

(b) Pre-cast units resting on bottom flange
of column section (screed omitted)

(c) Asymmetric section with deep trough decking

FIGURE 8.1 Types of Flooring

Composite steel–concrete floors use thin gauge trough steel sheeting which initially acts as permanent shuttering to the concrete before acting compositely after the concrete has set to provide both tensile and shear 'reinforcement' to the finished concrete slab. The concrete used is lightweight concrete with a specific weight of around $18–20\,kN/m^3$, and designed to be placed using concrete pumps rather than skips. The only cranage required is that to lift the bundled steel sheets to the correct level, as the sheets are individually light enough to be manhandled. To provide shear continuity between the deck and the steel support beams *in-situ* through deck shear stud welding is employed. Reinforcement will be required in the form of mesh in the top face of the slab in areas of hogging moments if only to control the effects of cracking. This mesh will also contribute to the fire performance of the deck. Any propping can be avoided by limiting the spans of the decking whilst acting non-compositely by additional support beams. This option is more cost effective than propping. A recent development is to use deep decking over the whole depth of a UC used as a beam with the decking supported on flange plates welded to the lower flange of the UC or specially rolled asymmetric sections. This system is known as 'Slimflor' construction (Mullett and Lawson, 1993), in which the outer beams are downstand UB's or rolled hollow sections with plates welded to the soffit (Mullett, 1997). A very useful practical guide for an overview of design and construction is published by the SCI (Couchman *et al.*, 2000).

8.3 INFLUENCE OF CONNECTION DESIGN AND DETAILING

From the previous chapter it will be noticed that connections are either designed to take the effect of beam reactions in the form of shear (with nominal moments) or to resist the effects of moments, and the coincident shear and axial force. The first type is generally designated a 'pin' joint which allows large relative rotations between the members in the connection (and leads to the concept of simple construction). The second is generally designated a 'fixed' or 'rigid' joint in which rotational compatibility exists between the members framing into the connection. This type of connection allows full moment transfer and should ideally be welded, although in UK practice a heavily bolted stiff connection is taken as rigid. Obviously the type of connection in the structure will markedly influence the behaviour of the structure under whatever actions are applied to it. A fuller discussion of the implications of the above paragraph is given in Section 8.12, but it is first necessary to consider the actions applied to a structure.

8.4 STRUCTURAL ACTIONS

These may either be physical loading or imposed deformations due, for example, to differential settlement.

8.4.1 Physical Loading

For convenience this is divided into two categories: gravity and non-gravity loading.

8.4.1.1 Gravity Loading

This covers the self-weight of the structure, the finishes on the structure, the actions due to the usage of the structure (variable actions) and roof loading whether as a nominal load or as snow loading (uniform or drifting as appropriate). Values of loads are given in EN 1991-1-1. It should however be noted that snow drift loading given in EN 1991-1-3 constitutes an accidental load case and therefore takes lower partial safety factors, and thus may not be critical. It is advisable to design the structure under uniform roof loading and then check the structure, if appropriate, under non-uniform snow drift loads.

8.4.1.2 Non-gravity Loading

This can be considered under a series of sub-headings:

- *Wind loading*: This is covered by EN 1991-1-4.
- *Inertia and impact loads*: These have to be considered where dynamic loading is considered, for example cranes and supporting systems (EN 1991-3). Impact loading

needs considering in the case of car parks where columns can be damaged due to collisions or for bridge piers where vehicular impact may occur (EN 1991-1-7).

- *Seismic loads*: These are not of general importance in the UK except for nuclear installations and other similar structures, and are covered in EN 1998.
- *Accidental loads*: These can be due to snow drift loading (EN 1991-1-3), fire (EN 1991-1-2) or explosions (EN 1991-1-7). In the case of explosions either the implications of progressive collapse needs to be considered or the structure must be tied together and the resultant tying forces considered.

8.4.2 Deformations

Structural deformations can be either due to differential settlement (Section 8.11.1) or thermal movements (EN 1991-1-5) (Section 8.11.2).

8.4.3 Load Combinations

Load combinations for single variable loads have been covered in Chapter 3, and it therefore remains to cover load combinations for multiple variable loads. The classic example of this is where there are variable loads due to structure usage (e.g. office loading) and variable loads due to wind. The basis behind the combination rules in EN 1990 is that it is deemed statistically unlikely that all the variable loads will be acting at their maximum intensity at the same time. It should be noted that roof loading and floor loading taken for the structure as a whole are considered as separate variable actions. The rules take account of the non-simultaneity of maximum effects by introducing ψ factors on the non-principal variable actions.

EN 1990 allows a number of combination rules – this is reflected by the UK National Annexe to EN 1990.

At ultimate limit state, the following combination rules are available:

$$1{,}35G_k + 1{,}5Q_{k,1} + \sum_2^n 1{,}5\psi_{0,j}Q_{k,j} \tag{8.1}$$

where G_k is the permanent load, $Q_{k,1}$ is the principal (or leading) variable load and $Q_{k,2}$ to $Q_{k,n}$ are the accompanying variable actions. Note however that where the permanent load is unfavourable (i.e. is of opposite sign to the permanent load, the partial safety factor γ_G is set equal to 1,0, and where the variable loading is unfavourable its partial safety factor γ_Q is taken as *zero*. Equation (8.1) must be applied taking each variable load as the principal variable load in turn and the remainder as accompanying actions. Experience, however, may be used in reducing the calculations when it is clear which leading action is critical. Note it is possible for different types of actions to have different values of $\psi_{0,j}$.

$$1{,}35G_k + 1{,}5\psi_{0,1}Q_{k,1} + \sum_2^n 1{,}5\psi_{0,j}Q_{k,j} \tag{8.2}$$

It is necessary with this combination to examine the effects of different $\psi_{0,j}$ values, although it is likely to give lower loads than those determined from Eq. (8.1).

$$1{,}35\xi G_k + 1{,}5Q_{k,1} + \sum_{2}^{n} 1{,}5\psi_{0,j}Q_{k,j} \qquad (8.3)$$

The additional factor on the permanent load ξ is subject to the limit $0.85 \leq \xi \leq 1.0$. The UK National Annexe specifies a value of 0,925. Equation (8.3) will give the lowest total load provided the permanent load is unfavourable. It would appear permissible to use the ξ factor where there is only a single variable load.

For serviceability loading the combination rules are similar except the maximum values of γ_G and γ_Q are set equal to one as is ξ. The other change is that ψ_0 may be replaced by ψ_1 or ψ_2. The factor ψ_1 is used to determine effects under reversible limits states (known as the frequent combination) and ψ_2 is used to determine long-term effects or the appearance of the structure (known as quasi-permanent). The use of ψ_0 is only required for irreversible limit states.

Deflection is generally checked under the quasi-permanent combination although it would be advisable to check it under full variable load together with permanent load in order to assess any maximum instantaneous deflection.

It is now necessary to consider how structures resist the forces due to actions applied to the structure. The major concern is with the transfer of horizontally applied forces, although some consideration is given to the distribution to supporting members of vertically applied forces. Single storey structures will be considered in detail, before continuing by looking at multi-storey structures.

8.5 SINGLE STOREY STRUCTURES UNDER HORIZONTAL LOADING

Consider initially a basic structure comprising a single bay flat roof portal type frame with encastré feet under vertical loading only (Fig. 8.2). It does not at first sight matter whether the connections at B and C are rigid or pinned, as the structure can be analysed and the members designed under the resultant forces. However if the joints at B and C are pinned the rafter will be a heavier section as no 'fixing' moment will exist at the connections, and there will be an increased deflection at the centre of the rafter. The column section will be lighter and the connection detail simplified.

However, the bottom ends of the column are fixed with respect to the bases. This condition will cause problems for the foundation design (Section 8.11.1) as the ratio of moment to axial load will be too high to give an economic design. The general custom on such a frame is therefore to use non-moment resisting connections between the feet of the stanchion and foundations, thus if non-moment resisting connections are used between the rafter and the columns the frame is inherently unstable. This

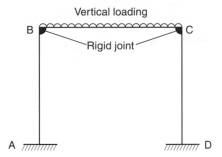

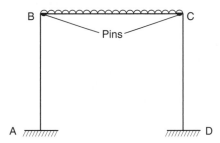

FIGURE 8.2 Simple single bay flat roof portal

statement is clearly true for horizontal loading, but is also true for vertical loading due to inherent imperfections in both the members and through construction. Two solutions, not considering moment resistant connections at the feet of the columns, are possible. The first is to restore the moment connections at B and C; the second is to find an alternative method to resist any horizontal forces.

The first method has the drawback that for single bay frames the total moments at the connections may reverse in sign due to the change in direction of the application of wind forces. The moment due to the wind force alone must change sign (Fig. 8.3). Allowing the connections to resist wind forces is possible in multi-bay structures where the wind force is adsorbed through a large number of connections, thus reducing the moment on each individual connection (see Section 8.12).

The second method is to adopt a bracing system within the frame. So for the wind force P (Fig. 8.4(a)), a diagonal member is placed such that the diagonal member is in tension (i.e. capable of taking full design load). This has the effect of triangulating the structure. When the wind blows from the reverse direction (force P' in Fig. 8.4(b)), the tie AC now becomes a strut with much reduced load capacity owing to buckling. Thus a further tie BD is now inserted giving rise to cross-bracing. The cross-bracing is designed by only considering the forces in tension members (i.e. the compression members are assumed to have zero load capacity, a conservative assumption). It should also be noted that since the structure is triangulated the frame analysis (and design) is much simplified.

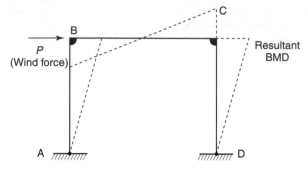

(a) Wind from left to right

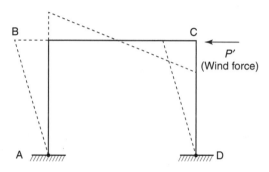

(b) Wind from right to left

FIGURE 8.3 Effect of wind reversal

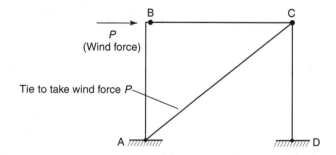

(a) Triangulation to take effect of wind force P

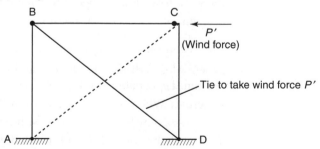

Note: AC now acts as a strut, and for calculating the force in BD is assumed to have zero strength

(b) Triangulation to take effect of wind force P′

FIGURE 8.4 Elementary wind bracing

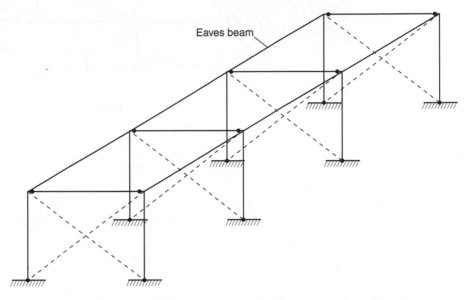

(a) Single storey multi-bay frame – each frame braced

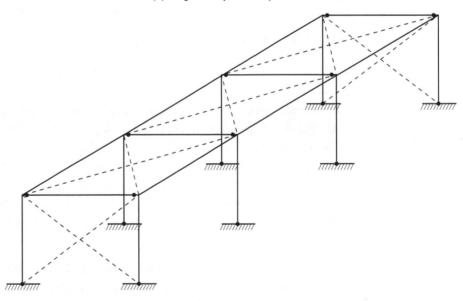

(b) Single storey multi-bay frame with wind girder

FIGURE 8.5 Wind girder bracing

No structure exists as a single frame. Where a set of such frames described above, complete with cross-bracing, to be assembled as multi-bay single span structure, it would be unusable (Fig. 8.5(a)). This state of affairs can be made acceptable by retaining the cross-bracing in each of the end frames only and by supplying full diagonal cross-bracing between each of the frames at rafter level (Fig. 8.5(b)). Such a bracing system is known as a wind girder, and transmits any horizontal forces through the girder to

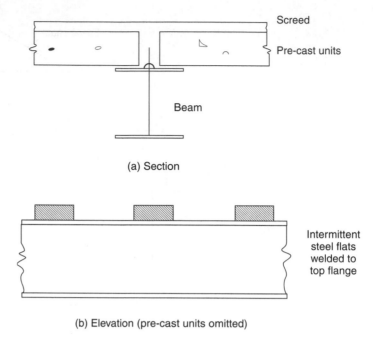

(a) Section

(b) Elevation (pre-cast units omitted)

FIGURE 8.6   Shear key detail

the bracing in the end frames. A wind girder is always needed where lightweight roofing systems are used in conjunction with light gauge purlin systems. Where stressed skin construction is used the wind girder can be omitted as the sheeting is designed to resist the shear due to the horizontal loadings (Davies and Bryan, 1982). If pre-cast concrete units are used then these may replace the wind girder provided adequate shear connection between the rafters and the units is provided (Fig. 8.6). Composite decking with through deck shear studs will not generally need additional bracing (but see Section 8.9.3).

For multi-bay multi-span structures a wind girder is needed around the periphery of the structure and full diagonal bracing in each of the corners of the structure (Fig. 8.7). For portal frame systems, the wind forces are taken by moment resisting connections in the plane of the frame, but for wind applied along the structure, i.e., normal to the plane of the portals, wind bracing must be provided. It may also be necessary to brace the end or gable frame for wind in the plane of the frame if it is designed as rafters spanning over intermediate gable end columns. Although this section has referred to discrete bracing it is possible to replace the bracing by alternatives. The most common replacement for vertical bracing is masonry shear wall construction. For single storey construction this need be no more than conventional masonry cladding provided such cladding is fully ties to the steel frame using, for example, half wall ties spot welded to the stanchions or approved proprietary tying systems. The design of such cladding is to EN 1996-1-1. It should however be remembered that in the case of masonry cladding

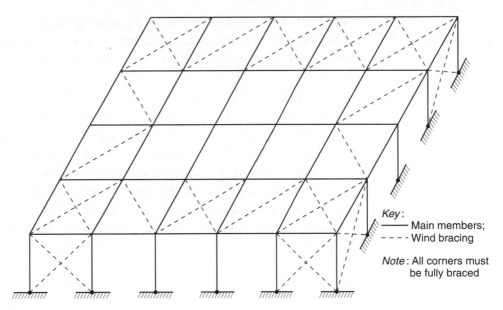

FIGURE 8.7 Schematic layout of wind bracing for a multi-bay structure

the frame is erected first and that the frame will need temporary bracing during this stage. It is the designer's responsibility to ensure this is present (Construction 'Design and Management' Regulations) (HMSO, 1994).

8.6 MULTI-STOREY CONSTRUCTION

The situation with respect to vertical and horizontal planes can be discussed separately.

8.6.1 Vertical Plane

The effect of horizontal forces in the vertical plane can be handled in a number of ways:

* Bracing
  The structure can be braced either with the bracing left as an external feature and hidden behind lightweight cladding. Such bracing can be at each corner of the structure, or provided the floors act as stiff diaphragms, in the centre of each face of the structure. The latter layout is only really suitable for structures symmetric about their centre lines as otherwise torsional effects are introduced which cannot easily be resisted.
* Shear walls
  This solution is generally adopted for the smaller end panels of structures which have a high aspect ratio with few columns in the smaller direction. The shear walls in order to act properly must be relatively unpierced, i.e., free from dominant or significant openings. This can produce architectural constraints. Irwin (1984) provides

information on design of shear walls. In the longitudinal direction the wind force is lower (as it acts on the smaller face) and there are a large number of bays to take the wind forces on the columns and the column–beam connections.

- Cores

  In the UK, the most common solution is to allow the wind forces to be adsorbed by the core(s) provided for the lift shaft(s) and/or stairwell(s). Such cores are generally of reinforced concrete construction, although the walls surrounding stairwells may be masonry. The walls surrounding lifts or stairwells do not have large openings as they generally provide access to fire escape routes and protected access areas for fire fighting, and thus stiff enough to provide the lateral restraint to the structure necessary. It should be noted that tolerance problems can arise when marrying up steelwork and reinforced concrete construction, and that shrinkage and creep effects in the concrete should not be ignored (Irwin, 1984).

- Rigid/semi-rigid construction

  The case where the frame is taken as rigid for both the vertical loading and horizontal loading is rare in the UK. However, a hybrid method of design is becoming popular in which the frame is consider pin jointed (simple construction) for the vertical loading but as rigid jointed for the horizontal loading. There are restrictions on this method mostly related to the size and shape of the structure (Salter *et al.*, 1999).

8.6.2 Horizontal Plane

Composite steel–concrete flooring systems will, when the concrete has hardened, prove a very stiff diaphragm to transmit horizontal forces provided adequate shear connection between the deck and the steel skeleton is available. The shear stud requirement to provide flexural composite action will generally be adequate. When pre-cast concrete units are used a shear key detail such as that illustrated in Fig. 8.6 should be used, and structural mesh provided in the screed in order to give full diaphragm action and prevent cracking over the beam.

8.7 BEHAVIOUR UNDER ACCIDENTAL EFFECTS

Accidental actions should be considered in a number of possible circumstances:

- Explosions whether due to gas or terrorism.
- Impact due to vehicles or aircraft.

  Should the risk of such an incident be high and the effects be catastrophic, or in certain circumstances the need to check be mandatory (e.g. vehicular impact on bridge piers), then the designer must ensure that the structure is designed and detailed to ensure that should an accidental situation occur, the structure does not suffer complete or partial collapse from either the accidental situation itself or subsequent events, such as spread of fire.

Four now classic cases where collapse occurred due to accidental or terrorist action are:

(1) Ronan Point block of flats where a gas explosion blew out a wall panel causing progressive collapse of a corner of the structure (Wearne, 1999).
(2) The Alfred P. Murrah Building in Oklahoma City where collapse was caused by a terrorist bomb blast (Wearne, 1999).
(3) World Trade Centre Towers in New York on 11 September 2001 due to impact from deliberate low flying aircraft and subsequent fire spread. The towers were steel framed structures and it has since been demonstrated that the prime cause of collapse was due to loss of fire protection on the floor members, as it was considered that the fire load from the impacting aircraft was not unduly high (Dowling, 2005). Information on considerations of the design of high rise construction is given in ISE (2002).
(4) Pentagon Building on 11 September 2001 following the deliberate crash of a commercial airline. The concrete structure suffered partial but not extensive collapse due to the impact and subsequent fire (Mlakar *et al.*, 2003).

8.7.1 Progressive Collapse

EN 1990 identifies the need to consider that in the event of an accident such as explosion or fire the structure should not exhibit disproportionate damage. This is reinforced by the requirements of the relevant Building Regulations within the UK, for example the recently revised Approved Document A of the England and Wales Building Regulations.

Such damage can be mitigated by:

• Attempting to reduce or limit the hazard
  In the case of fire this could be done by full consideration being given to the use of non-flammable materials within the structure and by the provisional of relevant active fire protection measures such as sprinklers. In the case of industrial processes where explosions are a risk then such potential risks need to be taken into account by, say, enclosing the process in blast proof enclosures.
• Maintenance
  The recent Pipers Row Car Park collapse of a lift slab concrete structure (Wood, 2003) indicated the need for adequate design especially where punching shear in a concrete slab may be critical as this is a quasi-brittle failure. The need for adequate inspection and maintenance where it is known that environmental effects will be severe was also highlighted. Although the failure was due to poor maintenance, it was exacerbated by uneven reaction distribution at the slab–column interface and no tying through the lower face of the slab. Admittedly the construction technique used would have made the latter extremely difficult although it contributed to the former.

- Consideration of the structural form

  A case where this is relevant is when the structure could be subject to externally provoked explosions. One reason why the damage was extensive following the Oklahoma City explosion was that the blast was amplified by an overhanging portion of the structure above ground floor level. A solution for structures known to be at potential risk is to provide an external curtain which is not structural and is easily blown out whilst ensuring the structure is stabilized by a core which is in the centre of the building. Additionally the risk can be mitigated by ensuring vehicles cannot come within certain limiting distances of the structure. More information on this, including assessment of blast forces is given in a SCI publication (Yandzio and Gough, 1999).

- Ensuring the structure is adequately tied to resist collapse

  This needs considering on two levels. The first is to determine the magnitudes of the likely levels of tying force required. The second is to provide properly detailed connections between elements of the structure. This is needed for both connections in the horizontal plane: beam-to beam or beam to column and in the vertical plane: column splices.

- Removal of key elements to establish stability

  In this approach a risk analysis is carried out to identify the key elements in a structure which are vital for its stability. A structural analysis is carried out with any one of these elements removed to establish the stability of the remaining structure and its ability to carry the loading so induced. As this is an accidental limit state lower partial safety factors are used on both the loading and the material strengths determining the member strengths. It is not necessary to check any serviceability limit states.

Approved Document A divides structures into four categories: 1, 2A, 2B and 3, with Category 1 being the least onerous. Category 1 constitutes low-rise housing, agricultural buildings and buildings with restricted access. These require no specific checks. For Category 2A which are medium consequence risk structures, the consideration of tying forces is adequate in most cases. For Category 2B defined as high consequence risk structures, then the requirement must be satisfied by using removal of key elements. For Category 3 structures which are very high consequence risk structures such as large capacity grandstands, hospitals, structures over 15 storeys or structures in which hazardous materials are involved require special consideration (Way, 2005). Alexander (2004) gives examples on risk assessment with the hazards grouped into two categories: those that must be considered for any Class 3 structure and those that only need considering for specific location dependent structure such as flooding.

8.7.2 Structure Stability

Any structure with high lateral loading or where vertical loading can be applied outside the frame envelope must be checked for overturning. Also any continuous member with a cantilever must be checked for the possibility of uplift on any support.

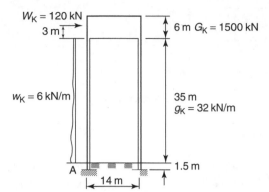

FIGURE 8.8 Design data for Example 8.1

In all these cases the loading which causes overturning is deemed to be unfavourable, and the loading tending to restore the situation is known as favourable. This is defined in EN 1990 as verification of static equilibrium EQU. For this situation the partial safety factors corresponding to EQU in EN 1990 (Table A1.2 A) must be used.

EXAMPLE 8.1   High wind load.

Consider the possibility of overturning for the water tower shown in Fig. 8.8.

Taking moments about point A:

Unfavourable effects due to wind:

$$1,5 \times 120(3 + 35 + 1,5) + 1,5 \times 6 \times 35 \left(\frac{35}{2} + 1,5 \right) = 13095 \, \text{kNm}$$

Favourable effects due to self-weight:

$$0,9 \times 1500 \frac{14}{2} + 0,9 \times 32 \times 35 \frac{14}{2} = 16506 \, \text{kNm}$$

The moment due to the favourable effects exceeds that due to the unfavourable effect, thus the structure will not overturn.

EXAMPLE 8.2   External loading.

Consider the cantilevering frame shown in Fig. 8.9.

Here the check is more complex, as the permanent load of 220 kN per storey is both favourable and unfavourable, in that the load acting on the cantilever is unfavourable as it contributes to the overturning effect. The permanent load per storey may be expressed as 27,5 kN/m run.

The favourable part of the permanent loading on the internal span takes a factor of 0,9, and that on the cantilever of 1,1.

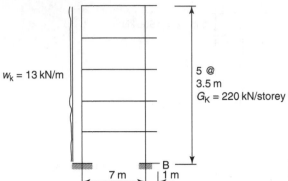

$w_k = 13$ kN/m

5 @
3.5 m
$G_K = 220$ kN/storey

B
7 m   1 m

FIGURE 8.9 Design data for Example 8.2

Taking moments about B:

Effect due to the wind loading (unfavourable):

$$1,5 \times 13\frac{17,5^2}{2} = 2986 \text{ kNm}$$

Effect due to permanent action (per storey):

$$0,9 \times 27,5\frac{7^2}{2} - 1,1 \times 27,5\frac{1^2}{2} = 591 \text{ kNm}$$

Total restoring (or favourable) moment:

$$5 \times 591 = 2955 \text{ kNm}$$

The structure will just overturn as the disturbing moment is marginally greater than the restoring moment. Thus either the structure needs tying down using, say, tension piles or ground anchors, or the permanent loading could be marginally increased. The latter is the cheaper option owing to the small margin involved.

8.8 TRANSMISSION OF LOADING

8.8.1 Transmission of Loading from Flooring Systems

When the loading is considered as a UDL, then the distribution of such loading to any supporting beam system will depend whether the decking system may be considered as one or two way spanning. Deck systems comprising lightweight proprietary systems (roofing only), pre-cast concrete slabs, composite steel–concrete decks and timber are all one way spanning. Thus half the load is taken to either end of the pre-cast unit, purlin system, joists or composite deck (in the direction of the troughs) (Fig. 8.10). Where a decking system is continuous over the supporting beam system then the reactions should be determined using a suitable analysis, although for approximate design continuity may be ignored. Where *in-situ* concrete is used then loading is distributed to

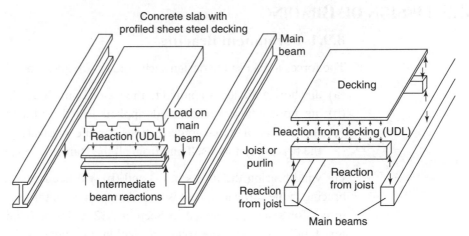

Concrete slab with
profiled sheet steel decking

Note: The loading on the main beam will be appiled
through web cleats (not shown)

(a) Transfer of load from one way spanning
   concrete deck

(b) Load transfer from decking through a
   joist or purlin system

FIGURE 8.10 Load transfer for one way spanning system

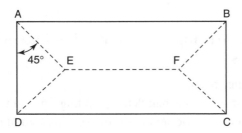

Note: (a) Load on area AED is supported by beam AD
      (b) Load on area AEFB is supported by beam AB

FIGURE 8.11 Approximate load
transfer from two way spanning
slabs

all the supporting beams. The loads on individual beams should be determined using
the same assumptions as in the slab design, such as Johansen's yield line method or
the Hillerborg strip approach. However, for approximate calculations a 45° dispersion from each corner of the slab may be used to partition the loading to the beams
(Fig. 8.11).

8.8.2 Lintels

In the absence of other data a 60° dispersion of the load above the lintel may be used,
that is, the lintel only carries the loading within the complete 60° equilateral triangle.
Where the triangle is incomplete due to openings then the whole load above should
be taken.

8.9 DESIGN OF BRACING

8.9.1 Permanent Bracing

The forces to be used to design such bracing are given in cl 5.3:

(a) any horizontal loads applied to the frame being braced,

(b) any horizontal or vertical loads applied directly to the bracing system

(c) the effects of initial imperfections, or the equivalent horizontal forces, from the bracing system itself and from all the frames it braces.

For wind bracing the forces in (a) and (b) will need considering, whereas for sway bracing in the vertical plane it is a combination of all three. The equivalent horizontal forces for sway are covered in Section 8.12.2. There is an additional requirement for stability forces at the splice in a column or beam which is covered in the next section.

The design of diagonal bracing is generally simplified by ignoring diagonal members in compression and assuming all joints are pinned. In the case of wind bracing the member should not be slender enough to induce an acceptable degree of sag under its own self-weight.

8.9.2 Restraint Bracing to Compression Flanges and Column Splices

- Compression flange bracing

  The problem can either be handled by adding an additional bow in the members to be restrained and designing for the resultant additional moments, or using an equivalent stabilizing force. In the first case the initial bow imperfection e_0 is given by

$$e_0 = \alpha_m \frac{L}{500} \tag{8.4}$$

where L is the span of the bracing system and α_m is given by

$$\alpha_m = \sqrt{0.5\left(1 + \frac{1}{m}\right)} \tag{8.5}$$

where m is the number of members to be restrained.

For convenience the effective bow imperfection in the members to be restrained by a bracing system may be replaced by the equivalent stabilizing force q_d is given by

$$q_d = \sum N_{Ed} 8 \frac{e_0 + \delta_q}{L^2} \tag{8.6}$$

where δ_q is the in-plane deflection due to the load q and any external loads calculated from a first order analysis. If second order analysis is used then $\delta_q = 0$.

The notional force N_{Ed} in the flange is defined as M_{Ed}/h where M_{Ed} is the maximum moment in the flange and h is the overall depth of the beam.

- Restraint to splices

   An additional requirement at splices is that the bracing should be able to resist a local force of $\alpha_m N_{Ed}/100$ together with any other applied forces. Note that the application of this local force is not co-existent with the force q defined above.

8.9.3 Temporary or Erection Bracing

Under the Temporary Workplaces Directive enforced in the UK by the Construction (Design and Management) Regulations, the client must appoint a planning supervisor at the *Design* stage on all health and safety matters throughout the execution of the project. It thus falls to the supervisor to ensure the structure is capable of being erected safely and that the requisite safety measures are in force. This means that the specification and design of temporary bracing should not be left to the steelwork erector without approval by the planning supervisor. The necessity for such bracing is often due to the intended construction sequence where construction starts at the centre of a structure in order to reduce cumulative tolerance errors whereas the permanent bracing designed to give stability to the completed structure is in the end bays. Further reasons are that support conditions or the effect of temporary loadings may be such that tension members are in compression or the signs of bending moments are reduced. There is also the requirement to ensure that any forces applied to the structure from safety equipment used during erection are considered. Where *in-situ* concrete is being used the compression flange will not receive full restraint until the concrete is hardened, and thus during construction the compression flange of the support beam system is unrestrained and therefore prone to lateral torsional buckling. Where profiled sheet steel decking is being used provided through deck welded shear studs are used, the beams running normal to the deck profiles can be considered as fully restrained, but those running parallel with the profiles will need checking for buckling though the deck will provide some degree of restraint (Lawson and Nethercot, 1985). An additional point to be noted is that large members may not be stable when being lifted into position partly due to the change in support conditions and partly due to the changes in restraints. Such members may need to be braced together at the correct spacing and lifted in pairs. Such bracing should if possible form part of the permanent bracing.

8.10 FIRE PERFORMANCE

It is recommended that the reader is referred to Ham *et al.* (1999) for an overview of the basic procedures and of methods of protection. Whilst it is not intended to cover detail fire design of steel structures, a brief overview is necessary. For full details reference should be made, for example, to Purkiss (2007).

8.10.1 Single Storey Structures

For most single storey structures the situation with respect to fire performance is relatively straightforward as there is generally no specific requirement within the England and Wales Building Regulations (Approved Document B) (DTLR, 2000). The only problem tends to be caused by outward collapse of walls and columns of single storey frames. This tends to be worse with pitched roof portals (see Simms and Newman, 2002). In addition there is the problem of ensuring that fire fighters have complete safe access to fight the fire during its complete duration.

8.10.2 Multi-storey Structures

Traditionally design has been by considering single elements with no interaction between such elements. However, fire tests on the steel frame structure at Cardington have demonstrated that with structural interaction between beams and columns it is possible to leave all beams unprotected (except at connections) and achieve a very good fire performance. Interim design data allowing some beams within floor systems being left unprotected has been published by the SCI (Newman *et al.*, 2006). It is also possible to encase columns in brickwork or blockwork or to fill rolled hollow section columns with concrete to achieve fire performance without additional measures (Bailey *et al.*, 1999).

Although a complete discussion of the Cardington tests is given elsewhere (Purkiss, 2007), an overview of the tests will be presented. The tests at Cardington demonstrated that temperatures in the unprotected steel beams which were composite with the floor slab reached temperatures on the bottom flange in excess of 1000°C. This is around 400°C higher than the limiting temperatures allowed after correcting for load level in design codes. It should be noted that the loading applied on the floors was around one-third of the ambient design value of 2,5 kPa giving a load ratio lower than that of most office type structures. Although the structure retained its load carrying capacity the deflections were extremely high with values of up to 640 mm being recorded. Although the tests showed far better performance than would be expected from the standard furnace test on individual elements, a number of points need to be raised:

(1) The high deflections reached would probably mean part at least of the structure would need replacing.
(2) Early tests in the series were performed with unprotected columns which suffered severe buckling just below the beam columns connections.
(3) Deflections of some beams were more than sufficient to cause internal lightweight compartment walls to fail leading to loss of compartmentation and thus increased fire spread (Bailey, 2004).
(4) Buckling in the lower flanges of the beams occurred at the ends owing to high-induced compressive forces as the moments were redistributed away from mid-span. This has led to the recommendation that design allowing for increased

moment capacity at the connection (Lawson, 1990a, b) should not be used where the beam is restrained against lateral movement during a fire (Bailey *et al.*, 1999).

(5) During the cooling phase the beams do not recover plastic deformations and potentially thus place the connections in tension. This in fact resulted in the partial failure of some connections due to shear in the bolts, excessive flexure of end plates or failure in the welds connecting the end plates to the beams (Bailey, 2004).

Design methods have, however, been introduced which allow for some secondary beams in composite construction to be left unprotected (Newman *et al.*, 2006) and which allow a better analysis of composite floor slab action under the compressive membrane force system induced in the fire limit state (Bailey (2001, 2003)).

Even with the traditional approach shelf angle beams will give 30 min standard fire performance and may give 60 min (Newman, 1993). Slim floor construction should give 60 min with no additional fire protection. Under no circumstances should columns be left unprotected. For fire performance periods of 30 min infilling the web space of a UC with non-structural blockwork can be sufficient (Newman, 1992).

8.11 ADDITIONAL DESIGN CONSTRAINTS

8.11.1 Ground Conditions and Foundations

This section is intended to be neither a comprehensive survey of foundation design or construction nor of problems associated with ground conditions. It is offered as an aide-memoir and covers the implications of any problematic areas. For a full consideration, the reader is referred to Henry (1986). Also the design of concrete foundations is covered in Martin and Purkiss (2006).

The selection of foundation type generally depends on the potential bearing capacity of the ground and its susceptibility to absolute or differential settlement. The most economic form of foundation is the simple pad foundation which in order to be of a reasonable size requires a reasonably high bearing capacity. If bearing capacities are low, and the loads applied to the foundations are high, the individual pad foundations will start to overlap and will become either combined foundations or in the extreme case raft foundations. Where the bearing capacity is low then either piled foundations, ground stabilization, or the use of replacement imported fill should be considered. The decision will be primarily based on cost, although environmental considerations such as noise, dust and extra traffic must be taken into account.

Ideally pad or raft foundations should be designed for uniform bearing pressure at ultimate limit state under imposed loading other than wind or accidental actions. This will generally be possible for raft foundations, but may not be for pad foundations. It is recommended that for pad foundations the ratio of the maximum to minimum bearing pressures should not exceed around 1,5. In order to achieve this it is possible to offset any stanchion away from the centre line of the foundation except where severe moment reversals may occur. It also leads to the adoption of nominally pinned feet where the axial load in single storey structures is low as the required ratio cannot be satisfied with fixed feet. Uplift should never be allowed below a foundation under normal conditions. It should be noted that it may be necessary, in practice, to supply a degree of fixity to bases of columns in single bay portal frames when considering such frames at site boundaries in the case of fire and thereby to accept a high ratio of maximum to minimum bearing pressures, even uplift, as this is an accidental situation.

Any horizontal reaction at the base of a column will also apply a moment to the underside of the foundation and must be included. If the structure is tied together by ground beams then only the net horizontal force need be considered, otherwise each reaction must be considered independently. For multi-storey structures it is general practice to employ tower cranes to facilitate erection. These need substantial foundations which are often incorporated into the structure as the foundations to the lift shafts.

Differential settlement may either be caused by non-uniform loading imposed by the structure on its foundations or by variable ground conditions beneath the foundations. The effect of non-uniform loading can be mitigated by placing the whole structure on a single raft foundation, and thus producing sensibly uniform bearing pressures. Also to avoid differential movement between any masonry cladding and the frame, the masonry can be supported on ground beams tied to the pad foundations below the columns. Alternatively the components of the structure can be allowed to settle independently by the provision of movement joints, in which case the cladding may not be considered to act as bracing and full wind bracing will need to be designed.

Where the structure is such that differential settlement cannot be avoided then either the structure must be designed to accept these movements (note, differential settlement does not affect moments when determined using plastic analysis; they do when an elastic analysis is used), or the structure can be articulated such that the movements do not affect the internal forces by the provision of hinges, often making the structure iso-static. Typically such hinges should be at the points of contraflexure, and must be checked for rotational capacity expected through such movements (e.g. Fig. 8.12). This type of articulation is frequently used on bridge decks where there may be substantial differential settlement due often to mining subsidence. Often in this case where such subsidence could be large, provision is built in to jack up the structure, in which case a realistic estimate will be needed of such movements.

8.11.2 Expansion and Contraction Joints

It can be a matter of some debate on whether these should be provided and if so at what spacing. It is suggested that masonry should have movement joints at around 7 m centres (or the frame spacing), and concrete floor slabs at around 20–30 m (Deacon, 1986; Bussell and Cather, 1995; Concrete Society, 2003). This suggests that continuous multi-bay frames should have movement joints at around the same spacing. Thus either

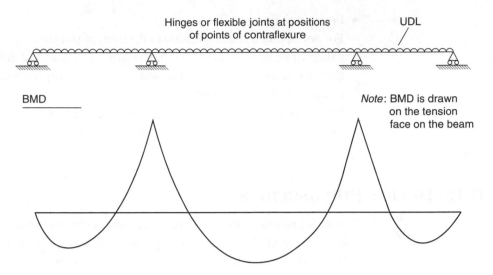

FIGURE 8.12   Cantilever-suspended span structure

movement joints should be provided through the whole height of the structure, in which case full wind bracing needs providing in each portion of the structure, or the structure will need designing to take the actions caused by such movements. For information on movement joints see Alexander and Lawson (1981).

8.11.3 Stability

Stability can be considered at two levels: member stability and frame stability.

- Member stability

  This effectively ensures that there will be sufficient ductility in the member to ensure that plastic rotational capacity is available when required by the design/analysis synthesis. The ductility check is made by considering limiting values of flange and web slenderness. It is essential that where plastic analysis is used Class 1 sections are mandatory, although where only a single hinge is needed for collapse, as in a simply supported beam, Class 2 sections may be used. It should be noted that virtually all UKBs are Class 1 whether Grade S255 or S335.

  The assumption is made in plastic analysis that the member can achieve its full plastic moment capacity before the onset of elasto-plastic buckling. For members in simple construction or isolated beam elements, this condition is not necessary as the load carrying capacity of the member can be checked using reduced strengths which allow for buckling. For rigid jointed frames premature elasto-plastic buckling is not permissible and must be counteracted either by bracing which reduces effective, or system, lengths below critical values or by increasing member sizes.

• Frame stability

For portal frame systems this takes the form of two checks. The first is the determination of vertical deflections at the ridges and horizontal deflections at the top of the stanchions at the eaves. The second check is only for multi-span portals and is designed to avoid 'snap through' of the rafters.

For multi-storey structures both the relative lateral deflections on each storey and the overall lateral deflection of the whole structure is subject to limits.

8.12 DESIGN PHILOSOPHIES

The basic analysis methods allowed are elastic or plastic (the use of elasto-plastic methods may be needed where sway is important). Traditionally the UK has adopted simple framing where the frame carries the vertical loading and, the bracing system, the horizontal loading. Continuous framing (except for portal frames) and semi-continuous framing are rarer.

8.12.1 Frame Classification

A frame may be classified as sway or non-sway. A non-sway frame is one where horizontal deformations are limited by the provision of substantial bracing members often in the form of lift shafts or stairwells often as part of a central core. A frame with bracing may still be classified as a sway frame.

The criteria for neglecting global second order effects are for

• Elastic analysis

$$\alpha_{cr} = \frac{F_{cr}}{F_{Ed}} \geq 10,0 \tag{8.7}$$

• Plastic analysis

$$\alpha_{cr} = \frac{F_{cr}}{F_{Ed}} \geq 15,0 \tag{8.8}$$

where α_{cr} is the ratio by which the design loading would have to be increased to cause global elastic instability, F_{Ed} is the design load on the structure and F_{cr} is the elastic critical global buckling load. The higher limit for plastic analysis is due to the issue that imperfections become more important in plastic analysis due to $P-\delta$ effects.

For shallow pitch portal frames (roof slope less than 26°) or beam and column type structures α_{cr} is given by

$$\alpha_{cr} = \frac{H_{Ed}}{V_{Ed}} \frac{h}{\delta_{h,Ed}} \tag{8.9}$$

where H_{Ed} is the total horizontal reaction at the bottom of a storey (including notional loads), V_{Ed} is total vertical loading at the bottom of a storey, $\delta_{h,Ed}$ is the horizontal displacement due to horizontal loading including notional horizontal loads and h is the storey height.

The value of α_{cr} may also be determined by applying nominal geometric rotations, or equivalent horizontal static forces, to the nodes in a frame and determining the resultant horizontal deformations. The disturbing forces are small so that any sway is low and will not cause large deflections. The method is based on the theory proposed by Horne (1975) for the determination of the elastic critical load factor. A braced frame cannot *a priori* be considered as a non-sway frame. Determination of sway classification is either carried out by imposing a rotation of φ' at the foot of each stanchion or by imposing horizontal forces of $\varphi'W$ at each level of the frame where W is the total factored vertical loading at that level. For each level the sawy index Δ/h is determined. From Horne and Morris (1981) the value of α_{cr} is given by the minimum value of

$$\alpha_{cr} = \frac{0,009}{\left(\frac{\Delta}{h}\right)_{max}} \tag{8.10}$$

For multi-storey frames with diagonal bracing Zalka (1999) provides a very quick way of estimating α_{cr}. The critical frame buckling load N_{cr} for a frame loaded with uniformly distributed loading is given by

$$N_{cr} = \lambda K \tag{8.11}$$

where the parameter λ is tabulated in terms of a parameter β_s defined by

$$\beta_s = \frac{K}{N_g} \tag{8.12}$$

where N_g is the global bending critical load given by

$$N_g = \frac{n}{n+1,588} \frac{7,837 E_c I_g}{H^2} \tag{8.13}$$

Where n is the number of storeys, H is the height of the building and $E_c I_g$ is the global flexural rigidity of the columns determined using the parallel axis theorem with the second moment of the columns themselves being neglected.

The shear stiffness K is dependent upon the type of bracing.

For single direction bracing,

$$K = \left(\frac{d^3}{A_d E_d h l^2} - \frac{1}{A_h E_h h}\right)^{-1} \tag{8.14}$$

and for complete diagonal cross-bracing,

$$K = A_d E_d \frac{h l^2}{d^3} \tag{8.15}$$

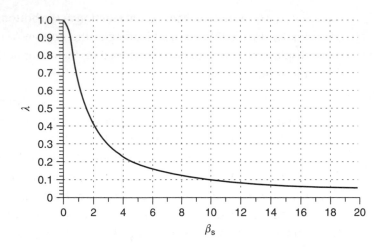

FIGURE 8.13 Critical load parameter λ (Zalka, 1999) by permission

where A_dE_d is the axial rigidity of the bracing member, A_hE_h is the axial rigidity of the horizontal member, d is the length of the diagonal member, l is the width of the braced bay and h is the storey height.

The formulations of K for other bracing layouts are given in Zalka (1999). Figure 8.13 from Zalka (1999) gives the relationship between β_s and λ (the data are also given in tabular form in Zalka).

8.12.2 Frame Imperfections

For frames sensitive to sway buckling, an allowance in the frame analysis should be made by introducing and initial sway imperfection and an initial bow.

8.12.2.1 Initial Sway Imperfection

The imperfection factor ϕ is determined from

$$\phi = \alpha_h\alpha_m\phi_0 \tag{8.16}$$

where the reduction factor for the building height α_h is given by

$$\alpha_h = \frac{2}{\sqrt{h}} \tag{8.17}$$

where h is the height of the building (in m) and $2/3 \leq \alpha_h \leq 1{,}0$.

The factor α_m depends on the number of columns m in a row,

$$\alpha_m = \sqrt{0{,}5\left(1 + \frac{1}{m}\right)} \tag{8.18}$$

The basic imperfection factor ϕ_0 is equal to 1/200.

Note, the effects of this may be ignored if $H_{Ed} \geq 0,15V_{Ed}$, provided the normalized slenderness ratio $\overline{\lambda}$ complies with

$$\overline{\lambda} \geq 0,3\sqrt{\frac{Af_y}{N_{Ed}}} \tag{8.19}$$

8.12.2.2 Initial Bow

This is expressed as the initial bow e_0 divided by the length l. It is a function of the strut buckling curve and whether elastic or plastic methods of analysis are used. The values are given in Table 5.1 of EN 1993-1-1. These effects should however be ignored if the members are checked for buckling.

8.12.3 Simple Framing

In simple framing the joints between the members may be assumed to offer negligible moment transfer, and the beams should be designed as simply supported. This means any horizontal loading must be taken by bracing or any equivalent substitute. Simple framing can be used to design conventional multi-storey beam and column structures or triangulated truss systems where there will be no significant secondary moments due to joint fixity.

8.12.4 Continuous Framing

In this case the connections are designed to give full rigidity between members at a joint. Note on sway and non-sway frames. Elastic, elasto-plastic or full rigid plastic analyses may be used. It is suggested that reference should be made to Ghali and Neville (1989), Coates *et al.* (1988), Horne and Morris (1981), Neale (1977) or Moy (1996).

8.12.5 Semi-continuous Framing

There are two basic approaches to semi-continuous framing: the first is to employ moment capacity/rotation characteristics and the second is a hybrid between simple and continuous.

The basic approach is to use an analysis which incorporates the connection characteristics in the analysis (Roberts, 1981; Taylor, 1981). In an elastic analysis the actual moment–rotation characteristic is input as an equivalent spring at the joint, in a plastic analysis the moment capacity is used as a local plastic moment capacity, provided the connection has sufficient ductility.

In the hybrid method, the frame is designed as simple for the purpose of determining the member sizes required to carry the imposed vertical loading, but as continuous when determining the effects of lateral, or horizontal, loading on the structure (Salter *et al.*, 1999). There are limitations on the method based on number of storeys and layout, but it can be clearly advantageous as it will avoid the need to provide any wind bracing, but at the expense of providing connections capable of transmitting the moments due to wind.

An alternative approach is to use an elastic analysis and then redistribute the moments to simulate plastic action. EN 1993-1-1, cl 5.4 4(B) places restrictions by specifying that the maximum redistribution is limited to 15%, the section classifications of members where moments are reduced must be no worse than Class 2, and that lateral torsional buckling must be prevented.

8.13 DESIGN ISSUES FOR MULTI-STOREY STRUCTURES

Brown *et al.* (2004) provide a useful overview of constructional details for multi-storey frames. A distinction needs to be drawn between frames which are rigid jointed under all applied loading and frames in which the connections are either semi-rigid or pinned. With pinned connections it should be reiterated that the frame requires bracing to resist the effects of any horizontal actions on the frame, and that even if a frame is braced it cannot be assumed that the frame is a no-sway frame (Section 8.12).

8.13.1 Simple Construction Frames

In simple construction frames, the beams are designed as simply supported and the columns designed to take the beam reactions together with nominal fixing moments. The nominal moments should be calculated assuming the beam reactions are applied at a minimum of 100 mm from the face of the column for major axis loading. For minor axis loading, which should be much smaller, the load is applied at either 100 mm from the web or the edges of the flanges depending upon the connection detail. If the reaction at the connection acts at a fixed point then this distance should be taken if it exceeds 100 mm. The nominal fixing moments may be distributed to the lengths of the columns immediately above and below the level being designed. There is no carry over to other storeys outside those above and below. The distribution should be in proportion to the stiffness of the column segments.

8.13.2 Semi-rigid Frames

Two methods are here available, either the frame is analysed using actual or design connection moment–rotation characteristics and the frame designed in the same manner as a rigid frame, or a hybrid method is used combining simple and continuous construction. The latter method has already been briefly outlined (Section 8.11.3), but further comment is necessary. The method has the effect of reducing the beam cross-section and the beam deflection. This will ensure that the combined effects of wind, permanent and variable loads will not be critical on the beam itself. The columns need to be designed to resist the worst effects of either the permanent and variable loads or the combination of wind, permanent and variable. In both cases the nominal fixing moments from the beams together with the moments induced by the nominal frame imperfection loads must be considered, and may well produce a heavier column cross-section than for simple construction.

8.13.3 Continuous Construction

For continuous construction either elastic or plastic methods may be used to carry out the analysis. There is a restriction that the compression flange at the hinge must be Class 1, and where a transverse force exceeding 10% of the shear resistance is applied to the web at the position of the plastic hinge then web stiffeners need to be supplied at a distance of $h/2$ from the hinge where h is the beam depth (cl 5.6 2b). For varying cross-section members, the web thickness should not be reduced for a distance $2d$ either side of the hinge (where d is the depth between fillets), the flanges must be Class 1 for the greater distance of $2d$ or the point where the moment drops to 80% of the plastic moment at the hinge.

Restraints must be provided at rotated plastic hinges (i.e. all but the last one to form) (cl. 6.3.5.2). This may be provided by diagonal bracing from the lower flange to beams or purlins fixed to the top flange. Where members carry moment only, or moment with axial tension, and are connected to a concrete slab by shear connectors, then this is deemed sufficient for rolled 'I' or 'H' sections.

At each hinge the restraint and connections should be capable of resisting a force equal to $0,0025N_{f,Ed}$ where $N_{f,Ed}$ is the axial force in the compression flange. In addition to the imperfection force given by Eq. (8.6), the bracing system should be able to carry a force Q_m is given by

$$Q_m = 1,5\alpha_m \frac{N_{f,Ed}}{100} \tag{8.20}$$

The lateral torsional buckling check of segments between restraints need not be carried out if the length between restraints is less than L_{stable} and the moment gradient is linear, where L_{stable} is given as

for $0,625 \leq \psi \leq 1,0$

$$L_{stable} = 35i_z\sqrt{\frac{235}{f_y}} \tag{8.21}$$

and for $-1,0 \leq \psi \leq 0,625$

$$L_{stable} = (60 - 40\psi)i_z\sqrt{\frac{235}{f_y}} \tag{8.22}$$

where the moment ratio ψ is given by

$$\psi = \frac{M_{Ed,mim}}{M_{pl,Rd}} \tag{8.23}$$

Note that where a rotated plastic hinge location occurs immediately adjacent to the end of a haunch, the tapered section need not be checked for stability if the restraint is placed at $h/2$ along the tapered section (not the uniform section) and the haunch remains elastic along its length.

8.13.4 Effective Length Factors for Continuous Construction

For either elastic or plastic analyses, the design of the columns takes account of the relative joint stiffnesses to determine the effective length factor for the column being designed. The two distribution factors k_1 and k_2 at the top and bottom of the column are defined as

$$k_1 = \frac{K_c + K_1}{K_c + K_1 + K_{11} + K_{12}} \tag{8.24}$$

$$k_2 = \frac{K_c + K_2}{K_c + K_2 + K_{21} + K_{22}} \tag{8.25}$$

where K_c is the stiffness of the column, K_1 and K_2 the stiffnesses of the columns at ends 1 and 2, K_{11} and K_{12} the stiffnesses of the beams framing in at end 1 and K_{21} and K_{22} the stiffnesses of the beams framing in at end 2 (Fig. 8.14).

- Non-sway frames

  The available information to calculate the effective length l is given in BS 5950 Part 1: 2000.

  The ratio of the effective length l to the actual length L is given by either

$$\frac{l}{L} = 0.5 + 0.14(k_1 + k_2) + 0.055(k_1 + k_2)^2 \tag{8.26}$$

Note that the moments due to the nominal frame imperfections must be included in the design moments for the frame in addition to those caused by any imposed actions.
- Sway frames

  The sway mode effective lengths are calculated from

$$\frac{l}{L} = \left[\frac{1 - 0.2(k_1 + k_2) - 0.12k_1k_2}{1 - 0.8(k_1 + k_2) + 0.6k_1k_2} \right]^{1/2} \tag{8.27}$$

Williams and Sharp (1990) provide alternative formulations:

$$\frac{l}{L} = \frac{\pi}{\sqrt{12 - 36\frac{k_1 + k_2 - k_1k_2}{4 - k_1k_2}}} \tag{8.28}$$

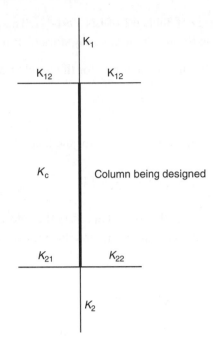

Note: The K values are the stiffnesses of the upper and lower beams, and the upper and lower columns in *consistent* units.

FIGURE 8.14   Data for determination of column effective lengths

or

$$\frac{l}{L} = \frac{\pi(1{,}05 - 0{,}05k_1k_2)}{\sqrt{12 - 36\frac{k_1+k_2-k_1k_2}{4-k_1k_2}}} \tag{8.29}$$

Note that for a sway frame l/L may exceed 1.

Sway moments should be increased using an amplification factor given by $1/(1 - 1/\alpha_{cr})$, provided $\alpha_{cr} \geq 3{,}0$.

8.13.5 Column Loads in Multi-storey Structures

In a multi-storey structure, the frame analysis will provide the reactions in the columns. However, such analyses assume that all floors are fully loaded at all times. Statistically this is conservative, thus the axial loads due to variable loading may be reduced in the lower columns of a multi-storey structure. Note, there is clearly no reduction in loads due to permanent (or quasi-permanent) loading, and each individual floor must be designed under full variable and permanent actions. The variable loading may also be adjusted for loaded area, as the loading will be more concentrated over smaller floor areas (and less concentrated over large areas). This means that column reactions may be reduced for this reason also. However, reductions may be made either for the number of floors *or* loaded area. These reductions are only for loading categories

A to D of Table 6.1 of EN 1991-1-1. This prohibits any reduction on situations where the loads are predominantly storage or industrial.

The reduction factor α_n for the number of stories n ($n > 2$) is given by

$$\alpha_n = \frac{2 + (n - 2)\,\psi_0}{n} \tag{8.30}$$

The reduction factor α_A for the loaded area is given by

$$\alpha_A = \frac{5}{7}\psi_0 + \frac{A_0}{A} \leq 1{,}0 \tag{8.31}$$

where A is the loaded area, A_0 the reference value of $10\,\mathrm{m}^2$. The reduction factor is also subject to the restriction that for load categories C and D it may not be less than 0,6.

8.14 PORTAL FRAME DESIGN

8.14.1 Single Bay Portal

To illustrate the design procedure an internal frame with full rigid connections is considered (EXAMPLE 8.3).

- A heavier section will be used for the stanchion than the rafter. Although this requires haunches at both the eaves and the ridge to accommodate the connections, the solution is still more economic than the use of a uniform section throughout.
- To avoid stability problems in the eaves haunch the stresses in the eaves are kept elastic by extending the haunch for a distance of span/10 from the face of the stanchion. The haunch is taken at its maximum depth of twice that of the rafter in order to ease fabrication.
- Plastic hinges are assumed to occur in the stanchion at the base of the haunch, that is, 1,5 times the rafter depth below the intersection of the rafter–stanchion centre lines, and at the second purlin point below the ridge. It is then usual to check the moment at the first purlin point.
- The frame will be designed under variable and permanent actions. The frame will not be checked for wind, as this case is rarely critical on a portal frame.
- Stability will be checked by applying notional forces of φN at the top of the stanchion. The approach given by Davies (1990) will also be checked.
- The connections at the eaves, ridge and base will not be designed (see Chapter 7).
- The background to the stanchion and rafter stability checks is given in Horne and Ajmani (1971a, b), Horne *et al.* (1979).

EXAMPLE 8.3   Portal frame design.

Prepare a design in grade S355 steel for the frame whose basic geometry is given in Fig. 8.15.

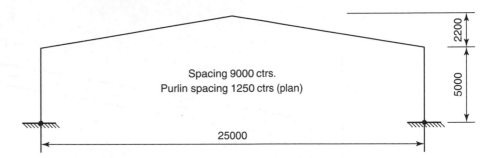

FIGURE 8.15 Basic frame geometry for Example 8.3

Permanent loading (kPa):	Sheeting	0,18
	Purlins	0,06
	Frame	0,18
	Total	0,42

UDL due to permanent loading: $0,42 \times 9,0 = 3,78 \,\text{kN/m}$

Nominal variable load (EN 1991-1-1): 0,60 kPa (or 5,40 kN/m)

Total factored design load: $q = 1,35 \times 3,78 + 1,5 \times 5,40 = 13,20 \,\text{kN/m}$

Since the purlins are at 1,25 m centres, it will be accurate enough to apply the roof loading as a UDL.

Preliminary design (geometry in Fig. 8.16):

Assume $h_h = 0,65$ m;

$h_2 = h_c - h_h = 5,00 - 0,65 = 4,35$ m

Distance from ridge to second purlin point, $b' = 2,5$ m

$l = L/2 - b' = 12,5 - 2,5 = 10,0$ m

Gradient of rafter, $s_r = 2,2/12,5 = 0,176$

Height to second purlin point, r:

$$r = h_c + l s_r = 5,0 + 10 \times 0,176 = 6,76 \,\text{m}$$

Moment at base of haunch, M_{B1}:

$$M_{B1} = H h_2 = 4,35H \tag{8.32}$$

Moment at second purlin point, M_2:

The vertical reaction at the base of the stanchion V is given by

$$V = \frac{qL}{2} = 13,2\frac{25}{2} = 165 \,\text{kN}$$

$$M_2 = Vl - Hr - \frac{qL^2}{2} = 10 \times 165 - 6,76H - \frac{13,2 \times 10^2}{2} = 990 - 6,76H \tag{8.33}$$

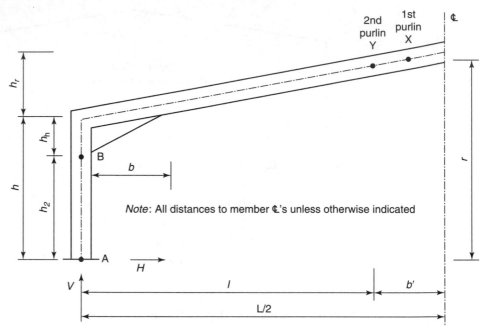

FIGURE 8.16 Detail frame geometry

By setting M_2 equal to M_{b2} (i.e. equal strength rafter and column) Eqs (8.32) and (8.33) may be solved to give

$$H = H_{bal} = 89{,}1 \, \text{kN}$$

This will give equal plastic moment capacities in the column and rafter. This is not an optimal solution as it is general practice to use a larger section column than rafter. An optimal solution can be obtained by increasing H_{bal} by around 15–20%. Increase H_{bal} to 104 kN (i.e. an increase of 16,7%).

Thus from Eq. (8.32), M_{B1} ($=M_{pl,stanchion}$) is given as

$$M_{B1} = Hh_2 = 4{,}35H = 4{,}35 \times 104 = 452{,}4 \, \text{kNm}$$

$$W_{pl} = M_{pl}\frac{\gamma_{M0}}{f_y} = 452{,}4 \times 10^3 \frac{1{,}0}{355} = 1274 \, \text{cm}^3$$

Select a 457 × 191 × 67 UKB (section classification Class 1: $M_{pl} = 522$ kNm)

and from Eq. (8.33) M_2 ($=M_{pl,rafter}$):

$$M_2 = 990 - 6{,}76H = 990 - 6{,}76 \times 104 = 287 \, \text{kNm}$$

$$W_{pl} = M_{pl}\frac{\gamma_{M0}}{f_y} = 287 \times 10^3 \frac{1{,}0}{355} = 808 \, \text{cm}^3$$

Select a 356 × 171 × 57 UKB (section classification is Class 1: $M_{pl} = 359$ kNm)

Note both the rafter and stanchion have been overdesigned to allow for any reduction in carrying capacity due to sway classification and any slight changes in frame geometry over those assumed in the preliminary assessment of dimensions.

Check haunch depth:

$h = 358$ mm, so depth to base of haunch, $h_h = 1,5 \times 358 = 537$ mm

Hinge at B_1 occurs at $5,00 - 0,537 = 4,463$ m above base.

Thus $H = M_{pl,stanchion}/h_2 = 475/4,463 = 106,4$ kN

The slight change in H can be ignored as it is small (2,3%).

Check load required to give collapse (q_1) at the second purlin point, as the first hinge to occur is at the eaves (from an elastic distribution of bending moments).

$$M_{pl,rafter} = 10 \times 12,5q_1 - 106,4 \times 6,76 - \frac{q_1 10^2}{2} = 75q_1 - 719 \tag{8.34}$$

Equate the value of $M_{pl,rafter}$ for Eq. (8.34) with the actual value to give

$$75q_1 - 719 = 326$$

or $q_1 = 13,9$ kN/m

This is greater than the design load of 13,2 kN/m, thus collapse does not occur at the second purlin point.

Check moment at first purlin point from the apex:

$$l = 11,25 \text{ m}; \ r = 5,00 + 0,176 \times 11,25 = 6,98 \text{ m}$$

$$M_1 = Vl - Hr - \frac{ql^2}{2} = 165 \times 11,25 - 106,4 \times 6,98 - \frac{13,2 \times 11,25^2}{2} = 278 \text{ kNm}$$

This is less than $M_{pl,rafter}$, and is therefore satisfactory.

Check moment at end of haunch:

$$b = 2,5 + 0,5h_{stanchion} = 2,5 + 0,5 \times 0,4534 = 2,727 \text{ m}$$

$$M_H = Vb - (5,00 + 0,176b)H - \frac{qb^2}{2}$$

$$= 165 \times 2,727 - 106,4(5,00 + 0,176 \times 2,727) - \frac{13,2 \times 2,727^2}{2} = -182 \text{ kNm}$$

Morris and Randall (1979) suggested this moment should be limited numerically to $0,87M_{pl,rafter}/\gamma$, where γ is the load factor between the total ultimate load and the characteristic loads. The 0,87 factor accounts for the approximate ratio between the moment to first yield and the plastic moment capacity (i.e. the shape factor).

Total load factor, γ:

Load at ULS $= 13,2$ kN/m and total characteristic load $= 3,78 + 5,40 = 9,18$ kN/m, so $\gamma = 13,2/9,18 = 1,44$.

$$\frac{0,87 M_{pl,rafter}}{\gamma} = \frac{0,87 \times 359}{1,44} = 217 \text{ kNm}$$

An alternative is to calculate the moment capacity to first yield M_{yield} divided by the total load factor:

$$\frac{M_{yield}}{\gamma} = \frac{1}{\gamma} W_{el} \frac{f_y}{\gamma_{M0}} = \frac{1}{1,44} 896 \frac{355}{1,0} \times 10^{-3} = 221 \text{ kNm}$$

Both methods give similar results, are less than the applied moment, and are therefore satisfactory.

The maximum load the frame can carry q_{max} assuming hinges occur simultaneously at the ridge and the base of the haunch is given by (Fig. 8.16):

$$q_{max} = \frac{8}{L^2} \left[M_{pl,R} + \left(\frac{1 + \frac{h_r}{h}}{1 - \frac{h_h}{h}} \right) M_{pl,c} \right] \tag{8.35}$$

or

$$q_{max} = \frac{8}{L^2} \left[M_{pl,R} + \left(\frac{1 + \frac{h_r}{h}}{1 - \frac{h_h}{h}} \right) M_{pl,c} \right] = \frac{8}{25^2} \left[359 + \left(\frac{1 + \frac{2,2}{5}}{1 - \frac{0,537}{5}} \right) 522 \right]$$

$$= 15,37 \text{ kN/m}$$

Thus the actual load factor γ_{max} on the applied loading is

$$\gamma_{max} = \frac{15,37}{9,18} = 1,67$$

The required load factor γ from above is 1,44.

Horizontal sway stability

The deflection δ_h under a force φN applied at the top of the (unhaunched) frame is given by

$$\delta_h = \frac{\varphi N h^3}{3 E I_c} \left[\left(\frac{\frac{I_r h}{I_c L_r} + \cos \theta}{\frac{I_r h}{I_c L_r}} \right) \left(1 - \frac{2 \frac{I_r h}{I_c L_r} + 6 + 3 \frac{h_1}{h}}{2 \left(\frac{I_r h}{I_c L_r} + 3 + 3 \frac{h_1}{h} + \left(\frac{h_1}{h} \right)^2 \right)} \right. \right.$$
$$\left. \left. + 2 \left(\frac{2 \frac{I_r h}{I_c L_r} + 6 + 3 \frac{h_1}{h}}{4 \left(\frac{I_r h}{I_c L_r} + 3 + 3 \frac{h_1}{h} \left(\frac{h_1}{h} \right)^2 \right)} \right) \right) - \frac{0,25 - \left(\frac{h_1}{h} \right)^2}{\frac{I_r h}{I_c L_r}} \cos \theta \right] \tag{8.36}$$

The notation is given in Fig. 8.17

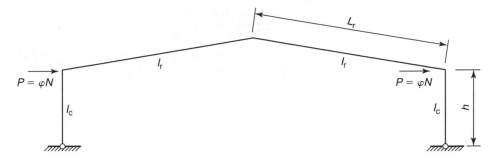

FIGURE 8.17 Idealized frame for stanchion deflection

Calculation of φ from Eq. (8.16):

Determine α_m from Eq. (8.18):

$$\alpha_m = \sqrt{0,5\left(1 + \frac{1}{m}\right)} = \sqrt{0,5\left(1 + \frac{1}{2}\right)} = 0,866$$

Determine α_h from Eq. (8.17) taking h as the height to eaves (as this is more conservative)

$$\alpha_h = \frac{2}{\sqrt{h}} = \frac{2}{\sqrt{5}} = 0,894$$

The value α_h is within the limits of 2/3 and 1,0, thus

$$\phi = \alpha_m \alpha_h \phi_0 = 0,866 \times 0,894 \frac{1}{200} = \frac{1}{258}$$

If N is taken as the axial load in the column, then it may be taken equal to V ($=165\,\text{kN}$).

The length of the rafter L_r is given by

$$L_r = \frac{L}{2\cos\theta} = \frac{25}{2 \times \cos 10} = 12,69\,\text{m}$$

Equation (8.36) is best evaluated in sections:

$$\frac{I_r h}{I_c L_r} = \frac{16040 \times 5}{29380 \times 12,69} = 0,215$$

$$2\frac{I_r h}{I_c L_r} + 6 + 3\frac{h_1}{h} = 2 \times 0,215 + 6 + 3\frac{2,2}{5} = 7,75$$

$$\frac{I_r h}{I_c L_r} + 3 + 3\frac{h_1}{h} + \left(\frac{h_1}{h}\right)^2 = 0,215 + 3 + 3\frac{2,2}{5} + \left(\frac{2,2}{5}\right)^2 = 4,729$$

$$\delta_h = \frac{\frac{1}{258} 165 \times 5^3}{3 \times 210 \times 10^6 \times 29380 \times 10^{-8}}$$

$$\times \left(\left(\frac{0{,}215 + \cos 10}{0{,}215} \right) \left(1 - \frac{7{,}75}{2 \times 4{,}729} + 2 \left(\frac{7{,}75}{4 \times 4{,}729} \right)^2 \right) \right.$$

$$\left. - \frac{0{,}25 - \left(\frac{2{,}2}{5} \right)^2}{0{,}215} \cos 10 \right)$$

$$= 1{,}133 \times 10^{-3} \, \text{m}$$

Horne and Morris (1981) suggested that under a horizontal load of 1% of the total applied load the deflection should be limited to $0{,}0009h_c$. This is equivalent to $0{,}00045h_c$ under 0.5% (1/200) of the load applied to a single column. If the initial imperfection φ_0 is modified by $\alpha_m \alpha_h$ then the limiting deflection δ_{lim} is reduced to $0{,}0045\alpha_m \alpha_h h_c$.

$$\delta_{\text{lim}} = 0{,}00045\alpha_m \alpha_h h_c = 0{,}00045 \times 0{,}866 \times 0{,}894 \times 5 = 1{,}74 \times 10^{-3} \, \text{m}$$

The actual deflection is lower and has been determined on an unhaunched frame.

An elastic analysis on the haunched frame gives a horizontal deflection of 0,823 mm, or $h/6075$ (=$164{,}6 \times 10^{-3}$ h). The deflection determined using the unhaunched frame is approximately 40% higher. This is acceptable.

Horne (1975) originally suggested that for multi-storey frames the elastic critical load factor $\alpha_{\text{cr,H}}$ was given by

$$\alpha_{\text{cr,H}} = \frac{H_{\text{nom}}}{\delta_{\text{nom}}} \frac{h_c}{V_{\text{ULS}}} \tag{8.37}$$

where δ_{nom} is the horizontal deflection due to a nominal horizontal force H_{nom}, and V_{ULS} is the axial force in the column.

Thus using the results determined above for H_{nom} (as φN) and δ_{nom} as δ_h, α_{cr} is given as

$$\alpha_{\text{cr,H}} = \frac{\frac{1}{258} 165}{1{,}133 \times 10^{-3}} \frac{5}{165} = 17{,}1$$

The actual value of the notional horizontal load is not critical as only the load-deflection ratio is required, but clearly the applied load should not cause any type of failure.

However, Lim et al. (2005) have suggested that Horne's approach to calculating α_{cr} is unconservative for portal frames, that is, it overestimates α_{cr}.

Lim et al. (2005) suggest a better estimate of $\alpha_{\text{cr,est}}$ is given by

$$\alpha_{\text{cr,est}} = 0.8 \left[1 - \left(\frac{N_{\text{R,ULS}}}{N_{\text{R,cr}}} \right)_{\text{max}} \right] \alpha_{\text{cr,H}} \tag{8.38}$$

where $N_{R,ULS}$ is the axial load in the rafter and $N_{R,cr}$ is the Euler buckling load of the rafter based on the span of the frame, and that the ratio should be taken at its maximum.

The maximum axial load in the rafter is at the haunch, and equals 133,5 kN.

Owing to the purlin restraints at 1,25 m centres means the rafter as a whole can only buckle about the major axis. The nominal buckling load N_{cr} assuming pinned ends and neglecting the effect of the haunches is given by

$$N_{cr} = \frac{\pi^2 EI_r}{L^2} = \frac{\pi^2 210 \times 10^6 \times 29380 \times 10^{-8}}{25^2} = 974,3\,\text{kN}$$

Determine $\alpha_{cr,est}$ from Eq. (8.38)

$$\alpha_{cr,est} = 0,8 \left[1 - \left(\frac{N_{R,ULS}}{N_{R,cr}}\right)_{max}\right]\alpha_{cr,H} = 0,8\left[1 - \frac{133,5}{974,3}\right]17,1 = 11,8$$

From Lim *et al.* (2005), the second order plastic collapse load factor for a Category A type frames, the second order plastic collapse load factor α_{p2} may be calculated using the Merchant–Rankine equation and the first order plastic collapse load factor α_{p1} as

$$\alpha_{p2} = \alpha_{pl}\frac{\alpha_{cr}-1}{\alpha_{cr}} \tag{8.39}$$

From earlier the load factor α_{p1} (determined as γ_{max}) is 1,67, thus

$$\alpha_{p2} = \alpha_{pl}\frac{\alpha_{cr}-1}{\alpha_{cr}} = 1,67\frac{11,8-1}{11,8} = 1,53$$

The required load factor on the frame is 1.44, thus the frame is satisfactory.

Check the method proposed by Davies (1990) for determining λ_{cr}:

$$\lambda_{cr} = \frac{3EI_r}{L_r(h_c N_{c,char} + 0.3L_r N_{r,char})} \tag{8.40}$$

where $N_{r,char}$ and $N_{c,char}$ are the forces in the rafter and column under characteristic frame loading, L_r is the rafter length, h_c is the height to eaves, and I_c and I_r are the major axis second moments of area of the column and rafter, respectively.

$L = 25,0\,\text{mm}$; $L_r = 12,69\,\text{m}$; $h_c = 5,0\,\text{m}$; $I_c = 29380\,\text{cm}^4$; $I_r = 16040\,\text{cm}^4$ (as only relative values are required).

Characteristic load/per unit run $= 3,78 + 5,40 = 9,18\,\text{kN/m}$,

$$N_{c,char} = \frac{qL}{2} = \frac{25 \times 9,18}{2} = 114,8\,\text{kN}$$

To estimate $N_{r,char}$ the following equation may be used

$$N_{r,char} = \frac{qL^2(3+5m)\cos\theta}{16Nh_c} + 0,25qL\sin\theta \tag{8.41}$$

where θ is the roof slope and the parameters N and m are given by

$$N = 2\left(1 + m + m^2 + \frac{I_r h_c}{L_r I_c}\right) \tag{8.42}$$

and

$$m = 1 + \frac{h_1}{h_c} \tag{8.43}$$

where h_1 is the rise from eaves to ridge.

$$m = 1 + \frac{h_1}{h_c} = 1 + \frac{2{,}2}{5} = 1{,}44$$

$$N = 2\left(1 + m + m^2 + \frac{I_r h_c}{L_r I_c}\right) = 2\left(1 + 1{,}44 + 1{,}44^2 + \frac{5 \times 16040}{12{,}69 \times 29380}\right) = 9{,}457$$

$$N_{r,char} = \frac{qL^2(3 + 5m)\cos\theta}{16Nh_c} + 0{,}25qL\sin\theta$$

$$= \frac{9{,}18 \times 25^2(3 + 5 \times 1.44)\cos 10}{16 \times 9{,}475 \times 5} + 0{,}25 \times 9{,}18 \times 25 \sin 10 = 86{,}0 \text{ kN}$$

$$\lambda_{cr} = \frac{3EI_r}{L_r(h_c N_{c,char} + 0{,}3L_r N_{r,char})} = \frac{3 \times 210 \times 10^6 \times 16040 \times 10^{-8}}{12{,}69(5 \times 114{,}8 + 0{,}3 \times 12{,}69 \times 86{,}0)} = 8{,}83$$

This is a lesser value than that derived from the imposition of a 0,5% sway load.

Adopt the criterion in Horne and Merchant (1965) for the plastic load factor derived from a modified Rankine type interaction equation with the lesser value of λ_{cr}.

As $4{,}6 < \lambda_{cr} < 10{,}0$:

$$\lambda_p = 0{,}9\frac{\lambda_{cr}}{\lambda_{cr} - 1} = 0{,}9\frac{8{,}83}{8{,}83 - 1} = 1{,}015$$

Design UDL at ULS $= 13{,}2$ kN/m: actual carrying capacity $= 15{,}37$ kN/m (from above); surplus factor $= 15{,}37/13{,}2 = 1{,}16$. This is greater than that required by the Horne and Merchant check, so frame is satisfactory.

An alternative approach to checking stability is given in Horne and Morris (1981), where a span/depth ratio check is carried out. The notation has been slightly changed to give

$$\frac{L - b}{h} \leq \frac{50W_0}{\lambda W}\frac{L}{h_c}\frac{\frac{I_c L_r}{I_r h}}{2 + \frac{I_c L_r}{I_r h}}\frac{240}{f_y}\cos\theta \tag{8.44}$$

W_0 is the load required to cause collapse of the rafter assuming it is a straight member of exactly the same section as the original rafter. Any stiffening caused by the ridge haunch can be ignored as it will have little effect. The eaves haunches cannot be ignored, however. There are two possible collapse mechanisms to give W_0 whereby a hinge always forms at the centre of the rafter beam, and the remaining two hinges can occur either at the ends of the haunch in the parent rafter section, or at the end of the haunch in the enlarged section.

For the general case W_0 is given by

$$W_0 = \frac{8}{(L - 2b)^2}(M_{\text{pl,end}} + M_{\text{pl,centre}}) \qquad (8.45)$$

Where L is the span, b is the lengthy of the haunch (taken as zero for hinges forming adjacent to the stanchions), $M_{\text{pl,end}}$ is the numerical value of the plastic moment capacity at the end of rafter and $M_{\text{pl,centre}}$ that of the numerical value of that at the eaves.

Consider first the hinges at the haunch–rafter intersection:

In this case, $b = 2{,}5\,\text{m}$ and $M_{\text{pl,end}} = M_{\text{pl,centre}} = 359\,\text{kNm}$, thus W_0 is given by

$$W_0 = \frac{8}{(L - 2b)^2}(M_{\text{pl,end}} + M_{\text{pl,centre}}) = \frac{8}{(25 - 2 \times 2{,}5)^2}(359 + 359) = 14{,}36\,\text{kN/m}$$

Now, consider hinges at the end of the haunch and the centre:

In this case, the hinge at the end of the haunch must occur in the weaker member, that is, the column with $b = 0$;

$$W_0 = \frac{8}{(L - 2b)^2}(M_{\text{pl,end}} + M_{\text{pl,centre}}) = \frac{8}{25^2}(359 + 522) = 11{,}28\,\text{kN/m}$$

Take the lesser value of W_0, that is, $11{,}28\,\text{kN/m}$.

$$\frac{50W_0}{\lambda W}\frac{L}{h_c}\frac{\frac{I_cL_r}{I_rh}}{2+\frac{I_rL_r}{I_ch}}\frac{240}{f_y}\cos\theta$$

$$= \frac{50 \times 11{,}28}{13{,}2}\left(\frac{25}{5}\right)\left(\frac{\frac{29380 \times 12{,}69}{5 \times 16040}}{2 + \frac{12{,}69 \times 29380}{16040 \times 5}}\right)\left(\frac{240}{355}\right)\cos 10 = 99{,}5$$

$$\frac{L - b}{h} = \frac{25 - 2{,}5}{0{,}358} = 62{,}9$$

The criterion is therefore satisfied, and the frame is stable.

Deflection check:

For an unhaunched frame with pinned feet under a UDL the vertical deflection δ_v at the ridge is given by

$$\delta_v = \frac{qL^4}{768EI_r}\left[10 - \frac{\left(8 + 5\frac{h_1}{h_c}\right)\left(3 + 2\left(\frac{h_1}{h_c}\right)\right)}{\left(3 + \left(\frac{h_1}{h_c}\right)^2 + 3\frac{h_1}{h_c} + \frac{I_rh_c}{I_cL_r}\right)^2}\right.$$

$$\left. \times \left(\frac{I_rh}{I_cL_r}\frac{\cos\theta - 1}{\cos\theta} + 3 + \left(\frac{h_1}{h_c}\right)^2 + 3\frac{h_1}{h_c} + \frac{I_rh_c}{I_cL_r}\right)\right] \qquad (8.46)$$

In Eq. (8.46) the term $\frac{I_{\mathrm{r}}h}{I_{\mathrm{c}}L_{\mathrm{r}}}\frac{\cos\theta-1}{\cos\theta}$ is negligible compared with the remainder of the terms in the last set of round parentheses, and thus Eq. (8.46) reduces to

$$\delta_{\mathrm{v}} = \frac{qL^4}{768EI_{\mathrm{r}}}\left[10 - \frac{\left(8+5\frac{h_1}{h_{\mathrm{c}}}\right)\left(3+2\left(\frac{h_1}{h_{\mathrm{c}}}\right)\right)}{\left(3+\left(\frac{h_1}{h_{\mathrm{c}}}\right)^2+3\frac{h_1}{h_{\mathrm{c}}}+\frac{I_{\mathrm{r}}h_{\mathrm{c}}}{I_{\mathrm{c}}L_{\mathrm{r}}}\right)}\right]$$

The horizontal deflection at the eaves δ_{h} is given by

$$\delta_{\mathrm{h}} = \delta_{\mathrm{v}} \tan\theta \qquad\qquad (8.47)$$

Determine δ_{v} and δ_{h} in terms of q:

$$\delta_{\mathrm{v}} = \frac{qL^4}{768EI_{\mathrm{r}}}\left[10 - \frac{\left(8+5\frac{h_1}{h_{\mathrm{c}}}\right)\left(3+2\left(\frac{h_1}{h_{\mathrm{c}}}\right)\right)}{3+\left(\frac{h_1}{h_{\mathrm{c}}}\right)^2+3\frac{h_1}{h_{\mathrm{c}}}+\frac{I_{\mathrm{r}}h_{\mathrm{c}}}{I_{\mathrm{c}}L_{\mathrm{r}}}}\right]$$

$$= \frac{25^4 q}{768\times210\times10^6\times16040\times10^{-8}}\left[10 - \frac{\left(8+5\frac{2.2}{5}\right)\left(3+2\frac{2.2}{5}\right)}{3+\left(\frac{2.2}{5}\right)^2+3\frac{2.2}{5}+\frac{16040\times5}{29380\times12.69}}\right]$$

$$= 0,0246q$$

$$\delta_{\mathrm{h}} = \delta_{\mathrm{v}} \tan\theta = 0,0246q \tan 10 = 4,34\times10^{-3}q$$

Vertical deflections:

Variable load of 5,4 kN/m:

$$\delta_{\mathrm{v}} = 0,0246q = 0,0246\times5,4 = 0,133\,\mathrm{m}$$

This is equivalent to span/188.

Permanent load of 3,78 kN/m

$$\delta_{\mathrm{v}} = 0,0246q = 0,0246\times3,78 = 0.093\,\mathrm{m}$$

This is equivalent to span/270.

Deflection under variable and permanent loading:

$$\delta_{\mathrm{v}} = 0,093 + 0,133 = 0,226\,\mathrm{m}$$

This is equivalent to span/110.

Horizontal deflections:

Variable load of 5,4 kN/m:

$$\delta_{\mathrm{h}} = 4,34\times10^{-3}q = 4,34\times10^{-3}\times5,4 = 0.023\,\mathrm{m}$$

This is equivalent to height/217.

Permanent load of 3,78 kN/m

$$\delta_h = 4{,}34 \times 10^{-3}q = 4{,}34 \times 10^{-3} \times 3{,}78 = 0{,}016 \,\text{m}$$

This is equivalent to height/313

Deflection under variable and permanent loading:

$$\delta_h = 0{,}016 + 0{,}023 = 0{,}039 \,\text{m}$$

This is equivalent to height/128.

Woolcock and Kitipornchai (1986) recommended limits under service permanent loading of span/360, variable loading span/240 for vertical deflection and height/150 for horizontal deflection. It appears unclear which loads are to be used for horizontal deflections.

The horizontal deflection under variable and permanent loading marginally exceeds their recommendation but the vertical deflection well exceeds them. However, it should be noted that the frame is stiffer than assumed in the calculations owing to the presence of haunches and the stiffness of the roof sheeting has been ignored. Thus the vertical deflection calculated above will be accepted as satisfactory.

A computer analysis allowing for the haunches gives a vertical deflection at the eaves under

Variable load: 79,2 mm (or span/316)

Permanent plus variable load: 134,7 mm (or span/186)

The analysis without haunches overpredicts the vertical deflections by almost 70%. Thus the estimates with no allowance for haunches are extremely conservative. The deflections determined with haunches will be reduced even further if the effects of cladding are considered.

Column stability:

There are two possible approaches, the first is to supply a torsional restraint at a distance L_m below the hinge position and then to check the remainder of the column below this restraint for the combined effects of strut buckling and lateral torsional buckling. The second is to check the whole length of the column below the hinge making use of restraint to the tension flange by the sheeting rails.

Method 1

The maximum distance to point of restraint in a column L_m is given by

$$L_m = \frac{38i_z}{\sqrt{\dfrac{1}{57.4}\dfrac{N_{Ed}}{A} + \dfrac{1}{756C_1^2}\dfrac{W_{ply}^2}{AI_t}\left(\dfrac{f_y}{235}\right)^2}} \tag{8.48}$$

Note, Eq. (8.48) must be solved iteratively as the moment gradient factor C_1 is dependent upon the value of the bending moment at a distance L_m below the hinge. A conservative solution may be obtained by setting $C_1 = 1.0$ (uniform moment).

As an alternative, the length to the restraint may be taken equal to L_s where L_s is given by

$$L_s = L_k \sqrt{C_m} \frac{M_{\text{pl},y,\text{Rk}}}{M_{\text{N}y,\text{Rk}} + aN_{\text{Ed}}} \tag{8.49}$$

where C_m is the modification factor for a linear moment gradient, $M_{\text{N}y,\text{Rd}}$ is the moment capacity of the column reduced due to axial load and L_k is given

$$L_k = \frac{\frac{h}{t_f}\left(5,4 + \frac{600 f_y}{E}\right) i_z}{\sqrt{5,4 \frac{f_y}{E}\left(\frac{h}{t_f}\right)^2 - 1}} \tag{8.50}$$

Use Eq. (8.49):

Force in the column $(N_{\text{Ed}}) = 165$ kN;

$$\frac{N_{\text{Ed}}}{A} = \frac{165 \times 10^3}{8550} = 19,3 \text{ MPa}$$

Suitable sheeting rails for this frame are 202 mm deep, thus

$$a = 0,5(453,4 + 202) = 327,7 \text{ mm}$$

Determine $M_{\text{N}y,\text{Rk}}$:

Determine the ratio n between the applied force N_{Ed} and the plastic axial resistance $N_{\text{pl},\text{Rd}}$

$$n = \frac{N_{\text{Ed}}}{N_{\text{pl},\text{Rd}}} = \frac{165 \times 10^3}{8550 \frac{355}{1,0}} = 0,054$$

Determine the ratio a between the total web area (excluding just the flanges) and the cross-section area:

$$a = \frac{A - 2bt_f}{A} = \frac{8550 - 2 \times 189,9 \times 12,7}{8550} = 0,436$$

$$M_{\text{N}y,\text{Rd}} = M_{\text{pl},y,\text{Rd}} \frac{1 - n}{1 - 0,5a} = M_{\text{pl},y,\text{Rd}} \frac{1 - 0,054}{1 - 0,5 \times 0,436} = 1,21 M_{\text{pl},y,\text{Rd}}$$

Thus $M_{\text{N}y,\text{Rd}} = M_{\text{pl},y,\text{Rd}} = 522$ kNm.

$$L_m = \frac{38 i_z}{\sqrt{\frac{1}{57,4}\frac{N_{\text{Ed}}}{A} + \frac{1}{756 C_1^2}\frac{W_{\text{pl},y}^2}{A I_t}\left(\frac{f_y}{235}\right)^2}}$$

$$= \frac{38 \times 41,2}{\sqrt{\frac{1}{57,4}19,3 + \frac{1}{C_1^2}\frac{(1471 \times 10^3)^2}{756 \times 8550 \times 37,1 \times 10^4}\left(\frac{355}{235}\right)^2}} = \frac{1566}{\sqrt{0,336 + \frac{2,059}{C_1^2}}}$$

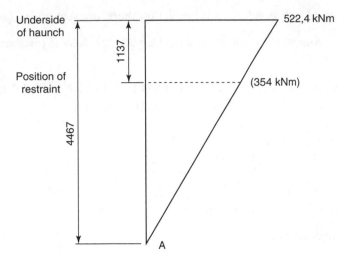

FIGURE 8.18 BMD in stanchion AB

Use Eq. (5.40) to determine C_1. Under a constant moment gradient, $C_1 = 1,0$, thus $L_m = 1012$ mm. After three iterations, $\psi = 0,678$, $C_1 = 1,148$ and $L_m = 1137$ mm.

Check column buckling below this level under axial load and moment (Fig. 8.18):

$M_{z,Ed} = 0$: $N_{Sd} = 165$ kN:

$M_{y,Ed} = 106,4 \times (4,463 - 1,137) = 354$ kNm.

Section classification:

Flanges: Class 1.

Web:

Actual slenderness:

$$\frac{c}{t_w} = \frac{d}{t_w} = \frac{407,6}{8,5} = 48,0$$

Length of web x_w required to carry axial force:

$$x_{web} = \frac{N_{Ed}}{t_w \frac{f_y}{\gamma_{M0}}} = \frac{165 \times 10^3}{8,5 \frac{355}{1,0}} = 54,7 \text{ mm}$$

Depth of web in compression αc:

$$\alpha c = \frac{d}{2} + \frac{x_{web}}{2} = \frac{407,6 + 54,7}{2} = 231,2 \text{ mm}$$

or

$$\alpha = \frac{(\alpha c)}{c} = \frac{231,2}{407,6} = 0,567$$

Limit for Class 1, with $\alpha > 0,5$:

$$\frac{d}{t_w} = \frac{396\sqrt{\frac{235}{f_y}}}{13\alpha - 1} = \frac{396\sqrt{\frac{235}{355}}}{13 \times 0,567 - 1} = 50,6$$

The actual web slenderness is below this, therefore section is Class 1.

As the compression flange is unrestrained, the lateral torsional buckling must be checked.

As the section is Class 1, Eqs (6.61) and (6.62) of EN 1993-1-1 may be simplified as $\Delta M_{y,Ed}$ and $\Delta M_{y,Ed}$ are both zero, as is $M_{z,Ed}$ to give

$$\frac{N_{Ed}}{\chi_y A \frac{f_y}{\gamma_{M1}}} + k_{yy} \frac{M_{y,Ed}}{\chi_{LT} W_{ply} \frac{f_y}{\gamma_{M1}}} \leq 1,0$$

$$\frac{N_{Ed}}{\chi_z A \frac{f_y}{\gamma_{M1}}} + k_{zy} \frac{M_{y,Ed}}{\chi_{LT} W_{ply} \frac{f_y}{\gamma_{M1}}} \leq 1,0$$

Buckling curves:

$$\frac{h}{b} = \frac{453,4}{189,9} = 2.39 > 1.2$$

y–y axis curve 'a' ($\alpha = 0.21$) and z–z axis curve 'b' ($\alpha = 0,34$).

Calculation of χ_y:

For yy axis buckling, the system length is taken as base of the haunch and the bottom of the column as restraints are present at both these points, that is, a length of 4,463 m

$$N_{cr,y} = \frac{\pi^2 EI_y}{L_y^2} = \frac{\pi^2 \times 210 \times 10^6 \times 29380 \times 10^{-8}}{4,463^2} = 30572 \text{ kN}$$

$$\bar{\lambda}_y = \sqrt{\frac{Af_y}{N_{cr,y}}} = \sqrt{\frac{8550 \times 355}{30572 \times 10^3}} = 0,315$$

$$\Phi_y = 0,5[1 + \alpha(\bar{\lambda}_y - 0,2) + (\bar{\lambda}_y)^2] = 0,5[1 + 0.21(0,315 - 0,2) + 0,315^2] = 0,562$$

$$\chi_y = \frac{1}{\Phi_y + \sqrt{\Phi_y^2 - (\bar{\lambda}_y)^2}} = \frac{1}{0,562 + \sqrt{0,562^2 - 0,315^2}} = 0,973$$

Calculation of χ_z:

For yy axis buckling, the system length is taken between the restraint 1,137 m below the haunch and the bottom of the column as restraints are present at both these points, that is, a length of 3,362 m

$$N_{cr,z} = \frac{\pi^2 EI_z}{L_z^2} = \frac{\pi^2 \times 210 \times 10^6 \times 1452 \times 10^{-8}}{3,362^2} = 2663 \text{ kN}$$

$$\bar{\lambda}_z = \sqrt{\frac{Af_y}{N_{cr,y}}} = \sqrt{\frac{8550 \times 355}{2663 \times 10^3}} = 1,064$$

$$\Phi_z = 0,5[1 + \alpha(\bar{\lambda}_z - 0,2) + (\bar{\lambda}_z)^2] = 0,5[1 + 0,34(1,064 - 0,2) + 1,064^2] = 1,213$$

$$\chi_z = \frac{1}{\Phi_z + \sqrt{\Phi_z^2 - (\bar{\lambda}_z)^2}} = \frac{1}{1,213 + \sqrt{1,213^2 - 1,064^2}} = 0,567$$

Calculation of χ_{LT}:

The system length for lateral torsional buckling is taken as the distance between the restraint 1,137 m below the haunch and the bottom of the column, that is, a length of 3,362 m

Determine M_{cr} from Eq. (5.5)

$$\frac{I_w}{I_z} = \frac{0{,}705 \times 10^{-6}}{1452 \times 10^{-8}} = 0{,}0486\,\text{m}^2$$

$$\frac{\pi^2 EI_z}{L^2} = \frac{\pi^2 \times 210 \times 10^6 \times 1452 \times 10^{-8}}{3{,}362^2} = 2663\,\text{kN}$$

$$\frac{L^2 GI_t}{\pi^2 EI_z} = \frac{81 \times 10^6 \times 37{,}1 \times 10^{-8}}{2663} = 0{,}0113\,\text{m}^2$$

$$M_{cr} = \frac{\pi^2 EI_z}{L^2}\left[\frac{I_w}{I_z} + \frac{L^2 GI_t}{\pi^2 EI_z}\right]^{1/2} = 2663[0{,}0486 + 0{,}0113]^{1/2} = 652\,\text{kNm}$$

Determine C_1 from Eq. (5.40):

$\psi = 0$,

$$\frac{1}{C_1} = 0{,}6 + 0{,}4\psi = 0{,}6 + 0{,}4 \times 0 = 0{,}6$$

or, $C_1 = 1{,}667$.

From Eq. (5.12) with M_{cr} modified by C_1,

$$\bar{\lambda}_{LT} = \sqrt{\frac{W_{pl} f_y}{C_1 M_{cr}}} = \sqrt{\frac{1471 \times 10^3 \times 355 \times 10^{-6}}{1{,}667 \times 652}} = 0{,}693$$

Use the general case for lateral torsion buckling, as $h/b > 2$, $\alpha_{LT} = 0{,}34$ (from Table 5.1)

$$\Phi_{LT} = 0{,}5[1 + \alpha_{LT}(\bar{\lambda}_{LT} - 0{,}2) + (\bar{\lambda}_{LT})^2] = 0{,}5[1 + 0{,}34(0{,}693 - 0{,}2) + 0{,}693^2]$$

$$= 0{,}824$$

$$\chi_{LT} = \frac{1}{\Phi_{LT} + \sqrt{\Phi_{LT}^2 - (\bar{\lambda}_{LT})^2}} = \frac{1}{0{,}824 + \sqrt{0{,}824^2 - 0{,}693^2}} = 0{,}788$$

Use Annex B of EN 1993-1-1 to determine k_{yy} and k_{zy}:

As the moment gradient is linear with the least value of moment equal to zero, $\psi = 0$, so from Table B.3, all the C_m values are 0,6.

$$k_{yy} = C_{my}\left(1 + (\bar{\lambda}_y - 0{,}2)\frac{N_{Ed}}{\chi_y A \frac{f_y}{\gamma_{M1}}}\right) \leq C_{my}\left(1 + 0{,}8\frac{N_{Ed}}{\chi_y A \frac{f_y}{\gamma_{M1}}}\right)$$

$$= 0{,}6\left(1 + (0{,}315 - 0{,}2)\frac{165 \times 10^3}{0{,}973 \times 8550\frac{355}{1{,}0}}\right) \leq 0{,}6\left(1 + 0{,}8\frac{165 \times 10^3}{0{,}973 \times 8550\frac{355}{1{,}0}}\right)$$

$$= 0{,}604 \leq 0{,}627$$

Thus, $k_{yy} = 0,604$.

As torsional deformations can occur (although they will be in practice slightly limited by the sheeting rails)

$$k_{zy} = 1 - \frac{0,1\bar{\lambda}_z}{C_{mLT} - 0,25} \frac{N_{Ed}}{\chi_z A \frac{f_y}{\gamma_{M1}}} \geq 1 - \frac{0,1}{C_{mLT} - 0,25} \frac{N_{Ed}}{\chi_z A \frac{f_y}{\gamma_{M1}}}$$

$$= 1 - \frac{0,1 \times 1,064}{0,6 - 0,25} \frac{165 \times 10^3}{0,567 \times 8550 \frac{355}{1,0}} \geq 1 - \frac{0,1}{0,6 - 0,25} \frac{165 \times 10^3}{0,567 \times 8550 \frac{355}{1,0}}$$

$$= 0,971 \geq 0,973$$

$k_{xy} = 0,973$

$$\frac{N_{Ed}}{\chi_y A \frac{f_y}{\gamma_{M1}}} + k_{yy} \frac{M_{y,Ed}}{\chi_{LT} W_{pl,y} \frac{f_y}{\gamma_{M1}}} = \frac{165 \times 10^3}{0,973 \times 8550 \frac{355}{1,0}} + 0,604 \frac{354 \times 10^6}{0,788 \times 1471 \times 10^3 \frac{355}{1,0}}$$

$$= 0,576 \leq 1,0$$

$$\frac{N_{Ed}}{\chi_z A \frac{f_y}{\gamma_{M1}}} + k_{zy} \frac{M_{y,Ed}}{\chi_{LT} W_{pl,y} \frac{f_y}{\gamma_{M1}}} = \frac{165 \times 10^3}{0,567 \times 8550 \frac{355}{1,0}} + 0,973 \frac{354 \times 10^6}{0,788 \times 1471 \times 10^3 \frac{355}{1,0}}$$

$$= 0,942 \leq 1,0$$

Thus the column is satisfactory below the torsional restraint.

Method 2

Determine the elastic critical torsional buckling load. This is given by Eq. (5.100) together with an additional term $\pi^2 EI_z a^2/L_t^2$ to allow for the restraint from the sheeting rails.

$a = 327,7$ mm (from Method 1)

$$N_{crT} = \frac{1}{i_s^2} \left(\frac{\pi^2 EI_z a^2}{L_t^2} + \frac{\pi^2 EI_w}{L_t^2} + GI_t \right)$$

where L_t is the distance between restraints to both flanges. Try L_t as the height from the underside of the haunch to the base, that is, 4,463 m, and i_s from (Eq. (5.101)) modified by an additional term a^2 to give

$$i_s^2 = i_z^2 + i_y^2 + a^2$$

or

$$i_s^2 = i_z^2 + i_y^2 + a^2 = (41,2^2 + 185^2 + 327,7^2) \times 10^{-6} = 0,1433 \, \text{m}^2$$

or, $i_s = 379$ mm.

$$N_{crT} = \frac{1}{i_s^2}\left(\frac{\pi^2 E I_z a^2}{L_t^2} + \frac{\pi^2 E I_w}{L_t^2} + G I_t\right)$$

$$= \frac{1}{0,1433}\left(\frac{\pi^2 \times 210 \times 10^6 \times 1452 \times 10^{-8}}{4,463^2}\right.$$

$$\left. + \frac{\pi^2 \times 210 \times 10^6 \times 0,705 \times 10^{-6}}{4,463^2} + 81 \times 10^6 \times 37,1 \times 10^{-8}\right)$$

$$= 11265 \text{ kN}$$

$$N_{crE} = \frac{\pi^2 E I_z}{L_t^2} = \frac{\pi^2 \times 210 \times 10^6 \times 1452 \times 10^{-8}}{4,463^2} = 1511 \text{ kN}$$

$$\eta = \frac{N_{crE}}{N_{crT}} = \frac{1511}{11265} = 0,134$$

$$B_0 = \frac{1 + 10\eta}{1 + 20\eta} = \frac{1 + 10 \times 0,134}{1 + 20 \times 0,134} = 0,636$$

$$B_1 = \frac{5\sqrt{\eta}}{\pi + 10\sqrt{\eta}} = \frac{5\sqrt{0,134}}{\pi + 10\sqrt{0,134}} = 0,269$$

$$B_2 = \frac{0,5}{1 + \pi\sqrt{\eta}} - \frac{0,5}{1 + 20\eta} = \frac{0,5}{1 + \pi\sqrt{0,134}} - \frac{0,5}{1 + 20 \times 0,134} = 0,097$$

Determine the ratio β_t defined as the ratio of the algebraically smaller end moment to the larger end moment. As the smaller end moment (at the base) is zero, $\beta_t = 0$.

$$C_m = \frac{1}{B_0 + B_1\beta_t + B_2\beta_t^2} \qquad (8.51)$$

As $\beta_t = 0$, Eq. (8.51) reduces to

$$C_m = \frac{1}{B_0} = \frac{1}{0,636} = 1,572$$

L_k is given by Eq. (8.50):

$$L_k = i_z \frac{\frac{h}{t_f}\left(5,4 + \frac{600 f_y}{E}\right)}{\sqrt{5,4 \frac{f_y}{E}\left(\frac{h}{t_f}\right)^2 - 1}} = 41,2\frac{\frac{453,4}{12,7}\left(5,4 + \frac{600 \times 355}{210 \times 10^3}\right)}{\sqrt{5,4\frac{355}{210 \times 10^3}\left(\frac{453,4}{12,7}\right)^2 - 1}} = 2893 \text{ mm}$$

The reduced moment capacity $M_{Ny,Rk}$ due to the axial load equals the plastic moment capacity $M_{pl,Rk}$.

$$L_s = \sqrt{C_m}L_k\left(\frac{M_{pl,y,Rk}}{M_{n,y,Rk} + a N_{Ed}}\right) = 2893\sqrt{1,572}\left(\frac{522}{522 + 165 \times 0,3277}\right) = 3286 \text{ mm}$$

This is less than the height to the underside of the haunch, and slightly less than the height to the restraint of 3362 mm.

Recheck using $L_t = 3362$ mm.

$$N_{crT} = \frac{1}{i_s^2}\left(\frac{\pi^2 EI_z a^2}{L_t^2} + \frac{\pi^2 EI_w}{L_t^2} + GI_t\right)$$

$$= \frac{1}{0,1433}\left(\frac{\pi^2 \times 210 \times 10^6 \times 1452 \times 10^{-8}}{3,362^2}\right.$$

$$\left. + \frac{\pi^2 \times 210 \times 10^6 \times 0,705 \times 10^{-6}}{3,362^2} + 81 \times 10^6 \times 37,1 \times 10^{-8}\right)$$

$$= 19692\,\text{kN}$$

$$N_{crE} = \frac{\pi^2 EI_z}{L_t^2} = \frac{\pi^2 \times 210 \times 10^6 \times 1452 \times 10^{-8}}{3,362^2} = 2663\,\text{kN}$$

$$\eta = \frac{N_{crE}}{N_{crT}} = \frac{2663}{19692} = 0,135$$

$$B_0 = \frac{1 + 10\eta}{1 + 20\eta} = \frac{1 + 10 \times 0,135}{1 + 20 \times 0,135} = 0,635$$

Hence, $L_k = 3290$ mm. This is probably adequate.

Rafter

(a) Haunch between connection and bracing at first purlin point.

Relevant dimensions are given in Fig. 8.19.

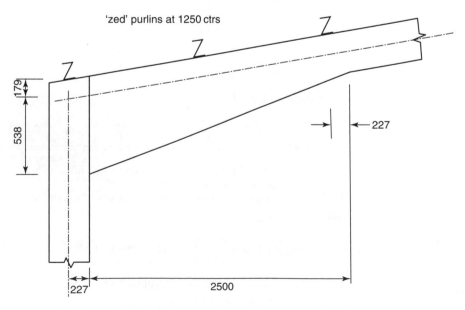

FIGURE 8.19 Haunch geometry

At any point along the rafter the value of the applied moment M_{Sd} is given by

$$M_{Ed,x} = 165x - 106,4(5 + 0,176x) - \frac{13,2x^2}{2}$$

At the connection, $x = 0,227$ m (ignoring end plate) and

$$M_{Sd} = -499 \text{ kNm}.$$

At the restraint $x = 1,25$ m, $M_{Sd} = -359,5$ kNm.

Section classification (with axial force)

Flanges are Class 1.

Web:

Determine the axial force in the member:

$$N_{Sd} = H \cos 10° + V \sin 10° = 106,4 \cos 10 + 165 \sin 10 = 133,4 \text{ kN}.$$

Web classification check is not needed as the web is restrained from buckling due to the end plate welded to the web and flanges forming the end plate for the connection.

Determine section properties ignoring both the central flange of the haunch and the fillets.

$$A = 2bt_f + dt_w = 2 \times 0,1722 \times 0,013 + 0,0081(0,716 - 2 \times 0,013) = 0,010066 \text{ m}^2$$

$$I_y = \frac{1}{12}\left[bh^3 - (b - t_w)(h - 2t_f)^3\right]$$

$$= \frac{1}{12}[0,1722 \times 0,716^3 - (0,1722 - 0,0081)(0,716 - 2 \times 0,013)^3]$$

$$= 0,775 \times 10^{-3} \text{ m}^4$$

$$I_z = \frac{1}{12}\left[2t_f b^3 + (h - 2t_f)t_w^3\right]$$

$$= \frac{1}{12}\left[2 \times 0,013 \times 0,1722^3 + (0,716 - 2 \times 0,013)0,0081^3\right] = 11,09 \times 10^{-6} \text{ m}^4$$

Plastic section modulus, $W_{pl,y}$:

$$W_{pl,y} = \frac{1}{4}\left[bh^2 - (b - t_w)(h - 2t_f)^2\right]$$

$$= \frac{1}{4}\left[0,1722 \times 0,716^2 - (0,1722 - 0,0081)(0,716 - 2 \times 0,013)^2\right]$$

$$= 2,538 \times 10^{-3} \text{ m}^3$$

Warping constant, I_w:

$$I_w = \frac{I_z h_s^2}{4} = \frac{11,09 \times 10^{-6}(0,716 - 0,013)^2}{4} = 1,370 \times 10^{-6} \text{ m}^6$$

where h_s is the depth between the centroids of the flanges.

To determine the torsional constant, I_t, it will be sufficiently accurate to treat the flanges and webs as thin, thus,

$$I_t = \frac{1}{3} \sum bt^3 = \frac{1}{3}\left[2 \times 0,1722 \times 0,013^3 + (0,716 - 2 \times 0,013)0,0081^3\right]$$

$$= 0,374 \times 10^{-6}\,\text{m}^4$$

The distance between restraints, L, is the slope distance between the end of the rafter and the first purlin, and is given by

$$L = \frac{1,250 - 0,277}{\cos 10°} = 0,988\,\text{m}$$

Determine M_{cr} from Eq. (5.5)

$$\frac{I_w}{I_z} = \frac{1,370 \times 10^{-6}}{11,09 \times 10^{-6}} = 0,1235\,\text{m}^2$$

$$\frac{\pi^2 E I_z}{L^2} = \frac{\pi^2 \times 210 \times 10^6 \times 11,09 \times 10^{-6}}{0,988^2} = 23547\,\text{kN}$$

$$\frac{L^2 G I_t}{\pi^2 E I_z} = \frac{81 \times 10^6 \times 0,374 \times 10^{-6}}{23547} = 1,287 \times 10^{-3}\,\text{m}^2$$

$$M_{cr} = \frac{\pi^2 E I_z}{L^2}\left[\frac{I_w}{I_z} + \frac{L^2 G I_t}{\pi^2 E I_z}\right]^{1/2} = 23547[0,1235 + 1,287 \times 10^{-3}]^{1/2}$$

$$= 8318\,\text{kNm}$$

Determine C_1 from Eq. (5.40),

$$\psi = \frac{359,5}{492} = 0,731$$

$$\frac{1}{C_1} = 0,6 + 0,4\psi = 0,6 + 0,4 \times 0,731 = 0,892$$

or, $C_1 = 1,121$.

$$\bar{\lambda}_{LT} = \sqrt{\frac{W_{pl}f_y}{C_1 M_{cr}}} = \sqrt{\frac{2,538 \times 10^{-3} \times 355}{1,121 \times 8318 \times 10^{-3}}} = 0,311$$

As $\bar{\lambda}_{LT} < 0,4$, lateral torsional buckling cannot occur.

Haunch instability

An alternative approach is to calculate the limiting length between lateral torsion restraints (i.e. between the haunch and the purlin immediately after the end of the haunch) (cl BB.3).

$$L_s = \frac{L_k\sqrt{C_n}}{c} \tag{8.52}$$

where L_k is the basic critical length, and c is a factor allowing for the shape of the haunch given by

$$c = 1 + \frac{3}{\frac{h}{t_f} - 9} \left(\frac{h_h}{h_s}\right)^{\frac{2}{3}} \sqrt{\frac{L_h}{L_y}} \tag{8.53}$$

where h/t_f is the ratio of the depth to flange thickness t_f for the basic section, $L_h L_y$ is the proportion of the length between the restraints taken up by the taper, and h_h/h_s is the ratio of additional depth of the beam at the haunch to that of the basic section. The basic length L_k is given by Eq. (8.50).

Determine L_h/L_y:

Length of taper 2500 mm, and horizontal distance from inside of stanchion to next purlin beyond end of taper is $3 \times 1250 - 227 = 3523$ mm, so

$$\frac{L_h}{L_y} = \frac{2500}{3523} = 0,710$$

As the haunch is fabricated using the same section size as the parent section minus one flange, $h_h = 358 - 13 = 345$ mm, so

$$\frac{h_h}{h_s} = \frac{345}{358} = 0,964$$

$$c = 1 + \frac{3}{\frac{h}{t_f} - 9} \left(\frac{h_h}{h_s}\right)^{\frac{2}{3}} \sqrt{\frac{L_h}{L_y}} = 1 + \frac{3}{\frac{358}{13} - 9} 0,964^{\frac{2}{3}} \sqrt{0,710} = 1,133$$

Determine L_k from Eq. (8.50):

$$L_k = i_z \frac{h}{t_f} \frac{5,4 + 600\frac{f_y}{E}}{\left[5,4\frac{f_y}{E}\left(\frac{h}{t_f}\right)^2 - 1\right]^{1/2}} = 39,1 \frac{358}{13} \frac{5,4 + 600\frac{355}{210 \times 10^3}}{\left[5,4\frac{355}{210 \times 10^3}\left(\frac{358}{13}\right)^2 - 1\right]^{1/2}} = 2838 \text{ mm}$$

Determination of C_n:

To determine C_n values of the parameter R are required which is defined by

$$R = \frac{M_{y,\text{Ed}} + aN_{\text{Ed}}}{W_{\text{pl}y} f_y}$$

From safe load tables a suitable depth purlin is 232 mm, thus the depth a between the centroid of the purlins and the centroid of the member is given by

$$a = 0,5h + 0,5 \times 232$$

Take the value of N_{Ed} as at the end of the member, that is, 134,4 kN.

The values of R are determined in Table 8.1

From Table 8.1 the maximum value of R, $R_S = 0,633$: R_E is max $(R_1, R_5) = 0,627$.

$$C_n = \frac{12}{R_1 + 3R_2 + 4R_3 + 3R_4 + R_5 + 2(R_S - R_E)}$$

$$R_S - R_E = 0{,}633 - 0{,}627 = 0{,}006 > 0$$

So,

$$C_n = \frac{12}{R_1 + 3R_2 + 4R_3 + 3R_4 + R_5 + 2(R_S - R_E)}$$

$$= \frac{12}{0{,}625 + 3 \times 0{,}629 + 4 \times 0{,}632 + 3 \times 0{,}633 + 0{,}627 + 2 \times 0{,}006} = 1{,}584$$

$$L_s = \frac{L_k\sqrt{C_n}}{c} = \frac{2838\sqrt{1{,}584}}{1{,}13} = 3158 \, \text{mm}$$

This exceeds the length of the haunch so there need only be restraints at haunch connection and the end of the haunch.

Check between haunch and first purlin point beyond point of contraflexure.

Point of contraflexure occurs when $M = 0$, or

$$165x - 106{,}4(5 + 0{,}176x) - 6{,}6x^2 = 0, \text{ or } x = 4{,}584 \, \text{m}$$

The distance from the end of the haunch to the point of contraflexure is $4{,}584 - 2{,}727 = 1{,}857 \, \text{m}$.

The distance to the next purlin point is $5{,}00 - 2{,}727 = 2{,}273 \, \text{m}$.

This is less than the value of L_k and therefore no additional restraint is required.

(c) Ridge

As the moment gradient is non-linear, the stable length is given by $L_k = 2833 \, \text{mm}$ (calculated above). This is greater than the slope length between purlins.

8.14.2 Notes on the Design of a Gable End Frame

• The rafter is usually designed as a continuous beam spanning over the gable end columns, with simple, non-moment transferring connections to the frame. This will

TABLE 8.1 Calculation of the values of R for the haunch.

	1	2	3	4	5
Distance x (mm)	227	852	1477	2102	2727
Depth h (mm)	716	626,5	537	447,5	358,0
A (mm^2)	10156	9431	8706	7981	7256
I_y ($\times 10^8$ mm^4)	7,854	5,556	4,031	2,658	1,603
W_{pl} ($\times 10^6$ mm^3)	2,538	2,104	1,702	1,332	0,996
M_x (kNm)	499,1	412,2	330,4	253,7	182,2
aN_{Ed} (kNm)	63,7	57,7	51,7	45,7	39,6
R	0,625	0,629	0,632	0,633	0,627

mean that if the spans are equal, the first bay will be critical since it will only be continuous at one end.

- The centre lines of the gable end columns should be coincident with the spacing of the roof purlins in order to resist the reaction from the wind loading on the gable end columns without subjecting the rafters to bi-axial bending. It is also preferable if the spacing of the gable end columns can be similar to the frame spacing in order to keep the same section for the sheeting rails, although this is not essential.
- Often the loading on the gable end will require the use of only relatively small sections, but detailing requirements and compatibility with the rest of the frame may make it necessary to use larger sections.

8.14.3 Multi-bay portal frames

It is not intended to present a design example of such a frame since the principles and methods are similar to those for single bay frames. It is intended, rather, to point out additional factors that need to be taken into account.

- The critical bay, assuming as in normal practice, that all the spans are equal, is the end bay. However, it must be noted that it is no longer possible to assume uniform snow loading. The frame should be first checked under the nominal roof loading of 0,6 kPa (modified for slope if necessary) as a variable load together with the permanent loading. It should then be checked for the accidental snow drift loading together with the permanent loading considered as an *accidental* load combination with reduced partial safety factors.

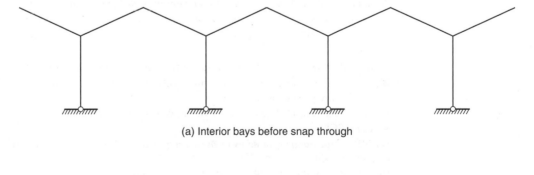

(a) Interior bays before snap through

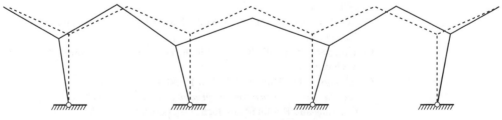

(b) Exterior bays after snap through

FIGURE 8.20 Snap through failure of multi-bay portal frames

- Although with balanced loading the interior stanchions carry axial load only, and the internal rafters have a higher reserve of strength, it is normal practice to detail *all* bays identically. This is partly to achieve fabrication economies, and to avoid problems on site with varying rafter sizes, and partly to avoid deflection and 'snap through' failure.
- Sway stability is either handled by the imposition of a nominal sway force as for single bay frames, or if the empirical method of Horne and Morris (1981) is used as in Eq. (8.44), then the ratio of the column stiffness (I_c/h_c) to the rafter stiffness (I_r/L) should be halved.
- In frames with more than two bays it is possible for snap through to occur when instability happens as the top of the stanchions spread followed by inversion of the rafter (Fig. 8.20)

This can be checked using the following empirical equation (Horne and Morris, 1981)

$$\frac{L-b}{h} = \frac{25\left(4+\frac{L}{h_c}\right)\left(1+\frac{I_c}{I_r}\right)}{\frac{\lambda W}{W_0}\left(\frac{\lambda W}{W_0}-1\right)}\frac{240}{f_y}\tan 2\theta \tag{8.54}$$

where symbols are defined in Eq. (8.44). If the arching ratio is less than one, snap through cannot occur, so there is no limit on the effective length to depth ratio.

REFERENCES

Alexander, S.W. (2004). New approach to disproportionate collapse, *Structural Engineer*, **82(23/24)**, 14–18.

Alexander, S.J. and Lawson, R.M. (1981). *Movement design in buildings*. Technical Note 107. CIRIA.

Bailey, C.G. (2001). *Steel structures supporting composite floor slabs: design for fire*. Digest 462. BRE.

Bailey, C.G. (2003). *New fire design method for steel frames with composite floor slabs*. FBE Report 5. BRE.

Bailey, C.G. (2004). Structural fire design: core or specialist subject? *Structural Engineer*, **82(9)**, 32–38.

Bailey, C.G., Newman, G.M. and Simms, W.I. (1999). Design of steel framed buildings without applied fire protection. P186. SCI.

Brown, D.G., King, C.M., Rackham, J.W. and Way, A.G.J. (2004). *Design of multi-storey braced frames*. P334. SCI.

BS 5950-1. *Structural use of steelwork in building – Part 1: Code of practice for design – rolled and welded sections*. BSI.

Bussell, M.N. and Cather, R. (1995). *Design and construction of joints in concrete structures*. Report 146. CIRIA.

Coates, R.C., Coutie, M.G. and Kong, F.K. (1988). *Structural Analysis* (3rd edition). Van Nostrand Reinhold.

Concrete Society (2003). Concrete industrial ground floors (3rd edition). Report TR 43. Concrete Society.

Couchman, G.H., Mullett, D.L. and Rackham, J.W. (2000). *Composite slabs and beams using steel decking: Best practice for design and construction*. Technical Paper No.13 (SCI P300). Metal Cladding and Roofing Manufacturers Association/SCU.

Davies, J.M. (1990). In plane stability in portal frames, *Structural Engineer*, **68(8)**, 141–147.

Davies, J.M. and Bryan, E.R. (1982). *Manual of stressed skin diaphragm construction*. Granada Irwin.

Deacon, R.C. (1986). *Concrete ground floors: Their design, construction and finish.* Cement and Concrete Association.

Dowling, J. (2005). United States can learn from worldwide fire engineering expertise, *New Steel Construction*, **13(9)**, 26–26.

DTLR (2000). *The Building Regulations: Approved Document B: Fire safety.* DTLR.

EN 1990. *Eurocode – basis of structural design.*

EN 1991-1-1. *Eurocode 1: Actions on structures: Part 1–1: General actions – densities, self-weight, imposed loads for buildings.* CEN/BSI.

EN 1991-1-2. *Eurocode 1: Actions on structures: Part 1–2: Actions on structures exposed to fire.* CEN/BSI.

EN 1991-1-3. *Eurocode 1: Actions on structures: Part 1–3: Snow loads.* CEN/BSI.

EN 1991-1-4. *Eurocode 1: Actions on structures: Part 1–4: Wind loads.* CEN/BSI.

EN 1991-1-5. *Eurocode 1: Actions on structures: Part 1–5: General actions – thermal actions.* CEN/BSI.

EN 1991-1-7. *Eurocode 1: Actions on structures: Part 1–7: Accidental actions due to impact and explosions.* CEN/BSI.

EN 1991-3. *Eurocode 1: Actions on structures: Part 3: Actions induced by cranes and machinery.* CEN/BSI.

EN 1996-1-1. *Eurocode 6: Design of masonry structures: Part 1–1: Rules for reinforced and unreinforced masonry.* CEN/BSI.

EN 1998. *Eurocode 8: Design of structures for earthquake resistance.* CEN/BSI.

Ghali, A. and Neville, A.M. (1989). *Structural analysis – a unified classical and matrix approach* (3rd edition). Chapman and Hall.

Gibbons, C. (1995). Economic steelwork design, *Structural Engineer*, **73(15)**, 250–253.

Ham, S.J., Newman, G.M., Smith, C.I. and Newman, L.C. (1999). *Structural fire safety: A handbook for architects and engineers.* P197. SCI.

Henry, F.D.C. (ed) (1986). *Design and construction of engineering foundations.* Chapman and Hall.

HMSO (1994). *Construction (Design and Management) Regulations.* HMSO.

Horne, M.R. (1975). An approximate method for calculating the elastic critical loads of multi-storey plane frames, *The Structural Engineer*, **53**, 242–8.

Horne, M.R. and Ajmani, J.L. (1971a). Design of columns restrained by side-rails, *Structural Engineer*, **49(8)**, 339–345.

Horne, M.R. and Ajmani, J.L. (1971b). The post-buckling behaviour of laterally restrained columns, *Structural Engineer*, **49(8)**, 346–352.

Horne, M.R. and Merchant, W. (1965). *The stability of frames.* Pergammon Press.

Horne, M.R. and Morris, L.J. (1981). Plastic design of low rise frames. Constrado Monographs (republished Collins, 1985). Granada publishing.

Horne, M.R., Shakir-Khalil, H. and Akhtar, S. (1979). The stability of tapered and haunched beams, *Proceedings of the Institution of Civil Engineers*, **67**, 677–694.

Horridge, J.F. (1985). The design of industrial buildings, *Civil Engineering Steel Supplement*, 13–16.

Horridge, J.F. and Morris, L.J. (1986). Comparative costs of a single-storey steel framed structures, *Structural Engineer*, **64A(7)**, 177–181.

Irwin, A.W. (1984). *Design of shear wall buildings.* Report 112. CIRIA.

ISE (2002). *Safety in tall buildings and other buildings with large occupancy.* Institution of Structural Engineers.

Lawson, R.M. (1990a). Behaviour of steel beam to column connections in fire, *Structural Engineer*, **68**, 263–271.

Lawson, R.M. (1990b). *Enhancement of fire resistance of beams by beam to column connections.* TR-086. SCI.

Lawson, R.M. and Nethercot, D.J. (1985). Lateral stability of I-beams restrained by profiled sheeting, *Structural Engineer*, **63(B)**, 1–13.

Lawson, R.M., Mullett, D.L. and Rackham, J.W. (1997). *Design of Asymmetric Slimflor® Beams using deep composite decking.* Publication 175. SCI.

Lim, J.B.P., King, C.M., Rathbone, A.J., Davies, J.M. and Edmonson, V. (2005). Eurocode 3 and the in-plane stability of portal frames, *Structural Engineer*, **83(21)**, 43–49.

Martin, L.H. and Purkiss, J.A. (2006). *Concrete design to EN 1992* (2nd edition). Butterworth-Heinemann.

Mlakar, P.F., Dusenberry, D.O., Harris, J.R., Haynes, G., Phan, L.T. and Sozen, M.A. (2003). Pentagon building performance report, *Civil Engineering*, **73(2)**, 43–55.

Morris, L.J. and Randall, A.L. (1979). *Plastic Design.* Constrado.

Moy, S.S.J. (1996). *Plastic methods for steel and concrete structures* (2nd edition). MacMillan.

Mullett, D.L. (1992). *Slim floor design and construction.* Publication 110. SCI.

Mullett, D.L. (1997). *Design of RHS Slimflor® Edge Beams.* Publication 169. SCI.

Mullett, D.L. and Lawson, R.M. (1993). *Slim floor construction using deep decking.* Technical Report 120. SCI.

Neale, B.G. (1977). *The plastic methods of structural analysis* (3rd edition). Chapman and Hall.

Newman, G.M. (1992). *The fire resistance of web-infilled steel columns.* Publication 124. SCI.

Newman, G.M. (1993). *The fire resistance of shelf angle floor beams to BS 5950: Part 8.* Publication 126. SCI.

Newman, G.M., Robinson, J.T. and Bailey, C.G. (2006). *Fire safe design: A new approach to multi-storey steel-framed buildings*, P. 288. Steel Construction Institute.

Purkiss, J.A. (2007). *Fire safety engineering design of structures* (2nd edition). Butterworth-Heinemann.

Roberts, E.H. (1981). Semi-rigid design using the variable stiffness method of column design, In *Joints in Structural Steelwork* (eds J.H. Howlett, W.M. Jenkins, and R. Stainsby); *Proceedings of the International Conference.* Teesside Polytechnic (now Teesside University), 6–9 April, 1981, Pentech Press, 5.36–5.49.

Salter, P.R., Couchman, G.H. and Anderson, D. (1999). *Wind-moment design of low-rise frames.* P-263. SCI.

Simms, W.I. and Newman, G.M. (2002). *Single storey steel framed buildings in fire boundary conditions.* P313. SCI.

Taylor, J.C. (1981). Semi-rigid beam connections: Effects on column design: B20 Code Method, In *Joints in structural steelwork* (eds J.H. Howlett, W.M. Jenkins, and R. Stainsby); *Proceedings of the International Conference.* Teesside Polytechnic (now Teesside University), 6–9 April, 1981, Pentech Press, 5.50–5.57.

Way, A.G.J. (2005) *Guidance on meeting the robustness requirements in Approved Document A* (2004 Edition), Publication P341. SCI.

Wearne, P. (1999). *Collapse – Why buildings fall down.* Channel 4 Books.

Williams, F.W. and Sharp, G. (1990). Simple elastic critical load and effective length calculations for multi-storey rigid sway frames, *Proceedings of the Institution of Civil Engineers*, **90**, 279–287.

Wood, J.G.M. (2003). Pipers row car park collapse: Identifying risk. *Concrete*, **37(9)**, 29–31.

Woolcock, S.T. and Kitipornchai, S. (1986). Deflection limits for portal frames, *Steel Construction*, **20(3)**, 2–10.

Yandzio, E. and Gough, M. (1999) *Protection of buildings against explosions.* Publication 244. SCI.

Zalka, K.A. (1999). Full-height buckling of frameworks with cross-bracing, *Structures and Buildings, Proceedings of the Institution of Civil Engineers*, **134**, 181–191.

Chapter 9 / Trusses

This chapter is concerned with the design of triangulated and non-triangulated trusses. With the advent of rolled hollow sections the design of both types of truss has been revolutionized.

The use of rolled hollow sections for trusses provides a far more efficient use of material as the buckling strengths are higher as radii of gyration are larger and lateral torsional buckling is either non-existent in the case of square or circular sections, or the effects are much reduced for rectangular sections. Rolled section trusses are much easier to fabricate as the welding is generally fillet welds or full strength butt welds. However, it needs remembering that room must be provided to allow the welds to be properly formed. This can be achieved by limiting the number of members at a node to the main chord member and at most two subsidiary web members and limiting the minimum angle between members to 30°. A problem however with rolled sections is forming joints to enable large trusses to be transported. If circular sections are used as main chord members fabrication problems and resultant high costs may ensue. Maintenance, and painting, is much easier with rolled sections as spray techniques can be employed. Note, rolled hollow sections are commonly available in Grade S355, but it may be difficult to obtain such sections in S275.

9.1 TRIANGULATED TRUSSES

It is generally acceptable to analyze triangulated trusses on the basis of pinned joints at the nodes together with loading from the roof system applied at nodes. However, where purlins exist between the nodes then the relevant members need to designed under the effects of flexure form the loads concerned. It is conservative to consider the main chord members as simply supported between nodes, although use of continuity can be made. Where it is necessary to consider the existence of services within the roof space, such loading can be considered as nodal loads on the bottom chord of the truss. The effect of service loads must be neglected when considering the effects of wind uplift.

It is not generally necessary to consider the effects of secondary moments due to joint fixity as these are usually low. Note, however if the original pin-jointed analysis is performed using a plane frame computer analysis package then it is relatively easy to check the effect of secondary moments by using the sections determined as suitable to carry the pinned forces and simply altering the end conditions of the members. Also, provided the secondary moments are negligible, the deflections can be determined from a pin-joint analysis. It should be noted that the actual deflections in a truss will be less than the calculated deflections as any stiffening effect of the roof cladding and purlins has been ignored. Note that even if the applied actions due to wind uplift and permanent loads are lower than those due to wind downthrust, roof loading and permanent loads, consideration still needs to be given to bracing on the lower boom of the truss as this is now in compression. This will almost certainly mean the provision of longitudinal bracing to ensure system lengths are kept reasonably small to ensure buckling does not occur out of plane.

EXAMPLE 9.1   Triangulated truss design.

Prepare a design using square and rectangular rolled hollow sections in Grade S355 steel for the truss whose geometry is given in Fig. 9.1.

The load due to initial imperfections has been omitted partly because the value of the load is small and partly because the truss is triangulated and thus will exhibit virtually no sway effects.

Actions (kPa):

Permanent: Sheeting   0,22
             Purlins   0,08
             Total   0,30

UDL for each truss $= 0,30 \times 9,0 = 2,7\,kN/m$.

Assume self-weight is 1,5 kN/m, thus total permanent action is 4,2 kN/m.

Since the purlins are at nodal points, nodal permanent action $= 3,0 \times 4,2 = 12,60\,kN$ (on an internal node).

Variable action:

From EN 1991-1-1 variable action (with no access, snow drift ignored) is 0,6 kPa, so nodal load $= 0,6 \times 9,0 \times 3,0 = 16,2\,kN$

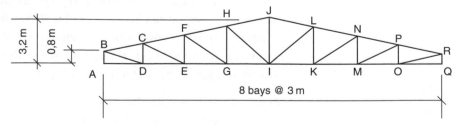

FIGURE 9.1   Truss geometry for EXAMPLE 9.1

Although wind loading will need considering as it may cause uplift due to suction, in this example only permanent and variable actions due to roof loading will be considered. Thus the only combination of actions needed is $1{,}35g_k + 1{,}5q_k$.

So the total nodal action is $1{,}35 \times 12{,}6 + 1{,}5 \times 16{,}2$, that is, 41,31 kN per internal node.

The truss will be designed using a pin-joint analysis, and then checked using a fixed-joint analysis. The effect of notional horizontal loading will be ignored for this design example as its effect will be negligible.

Table 9.1 shows the results from a pin-joint analysis under factored variable and permanent actions for half the truss.

Member design:

The top and bottom chord together with the end posts will be fabricated from the same section.

Maximum force (compressive) N_{Sd} is in member CF and is 379 kN.

(The maximum tensile force is 320 kN and will not be critical.)

From cl BB.1.3 the buckling length may be taken as $0{,}9L$ in both planes where L is the system length.

Effective length:

$L_{FC} = 3{,}105$ m (slope length).

TABLE 9.1   Member forces for EXAMPLE 9.1.

	Pin-joint analysis	Fixed-joint analysis			
Member	Axial force kN	Axial force kN	Shear force kN	Moment End 1 kNm	End 2 kNm
AB	−165,24	−163,38	−7,19	2,85	−2,93
CD	−82,62	−82,12	−1,57	−1,07	1,16
EF	−28,92	−29,33	0,04	0,04	−0,04
GH	9,54	9,17	0,12	−0,12	0,17
IJ	82,62	81,63	0	0	0
AD	0	7,19	1,86	−2,85	2,77
DE	309,83	307,14	−0,50	1,90	0,87
EG	371,79	371,29	0,04	0,50	0,62
GI	357,50	356,79	−0,37	0,83	0,29
BC	−315,98	−311,97	1,82	0,29	−0,29
CF	−379,14	−378,60	−0,37	−1,61	−0,45
FH	−364,56	−364,02	0,12	−0,45	−0,78
HJ	−315,98	−316,14	−0,54	0,66	−0,99
BD	320,64	308,46	0,17	0,29	−0,29
CE	68,37	73,28	−0,08	0,04	0,25
FG	−17,18	−17,27	0	0,04	0,08
HI	−63,08	−61,76	0	0,04	0

Out-of-plane buckling:

Buckling effective length is the length of the member, that is, 3,105 m.

In-plane buckling

Buckling length is $0,9L_{FC} = 2,795$ m, as there is a large degree of restraint owing to the welded joints (and continuity of the chord member).

Try a $100 \times 100 \times 8$ HFRHS.

$A = 2910$ mm^2, $i = 37,4$ mm, $I = 4,08 \times 10^{-6}$ m^4.

Section classification:

$$\frac{c}{t} = \frac{b - 2r - 2t}{t} = \frac{100 - 2 \times 8 - 2 \times 8}{8} = 8,5$$

Limit for Class 1:

$$33\varepsilon = 33\sqrt{\frac{235}{f_y}} = 33\sqrt{\frac{235}{355}} = 26,8$$

Section is Class 1.

Use the greatest system length of 3,105 to calculate N_{cr}:

$$N_{cr} = \frac{\pi^2 EI}{L^2} = \frac{\pi^2 \times 210 \times 10^6 \times 4,08 \times 10^{-6}}{3,105^2} = 877 \text{ kN}$$

$$\bar{\lambda} = \sqrt{\frac{Af_y}{N_{cr}}} = \sqrt{\frac{2910 \times 355 \times 10^{-3}}{877}} = 1,085$$

For a rolled hollow section, buckling curve 'a' is used, $\alpha = 0,21$:

$$\Phi = 0,5\left[1 + \alpha(\bar{\lambda} - 0,2) + (\bar{\lambda})^2\right] = 0,5\left[1 + 0,21(1,085 - 0,2) + 1,085^2\right] = 1,182$$

$$\chi = \frac{1}{\Phi + \sqrt{\Phi^2 - (\bar{\lambda})^2}} = \frac{1}{1,182 + \sqrt{1,182^2 - 1,085^2}} = 0,606$$

$$N_{b,Rd} = \chi A \frac{f_y}{\gamma_{M1}} = 0,606 \times 2910 \frac{355}{1,0} \times 10^{-3} = 626 \text{ kN}$$

All the remaining members will be the same section size in order to aid fabrication.

Maximum force N_{Sd} is in member BC and is 320 kN (tension).

Try a $60 \times 60 \times 6,3$ HFRHS.

$A = 1330$ mm^2, $i = 21,7$ mm, $I = 63,4 \times 10^8$ mm^4.

Section classification:

$$\frac{c}{t} = \frac{b - 2r - 2t}{t} = \frac{60 - 2 \times 6,3 - 2 \times 6,3}{6,3} = 5,52$$

Limiting value for Class 1:

$$\frac{c}{t} = 33\sqrt{\frac{235}{f_y}} = 33\sqrt{\frac{235}{355}} = 26{,}8.$$

Section is therefore Class 1.

$$N_{t,Rd} = A\frac{f_y}{\gamma_{M0}} = 1330\frac{355}{1{,}0} \times 10^{-3} = 472\,\text{kN}$$

However the compression members also need checking.

Member DC carries the largest compression force of $-82{,}6\,\text{kN}$.

Length of DC is $1{,}4\,\text{m}$.

Critical effective length is that for out-of-plane buckling, that is, $1{,}4\,\text{m}$.

$$N_{cr} = \frac{\pi^2 EI}{L^2} = \frac{\pi^2 \times 210 \times 10^6 \times 63{,}4 \times 10^{-8}}{1{,}4^2} = 670\,\text{kN}$$

$$\bar{\lambda} = \sqrt{\frac{Af_y}{N_{cr}}} = \sqrt{\frac{1330 \times 355}{670 \times 10^3}} = 0{,}839$$

For a hot rolled RHS, $\alpha = 0{,}21$, so

$$\Phi = 0{,}5\left[1 + \alpha(\bar{\lambda} - 0{,}2) + (\bar{\lambda})^2\right] = 0{,}5\left[1 + 0{,}21(0{,}839 - 0{,}2) + 0{,}839^2\right] = 0{,}987$$

$$\chi = \frac{1}{\Phi + \sqrt{\Phi^2 - (\bar{\lambda})^2}} = \frac{1}{0{,}987 + \sqrt{0{,}987^2 - 0{,}839^2}} = 0{,}664$$

$$N_{b,Rd} = \chi A\frac{f_y}{\gamma_{M1}} = 0{,}664 \times 1330\frac{355}{1{,}0} \times 10^{-3} = 314\,\text{kN} > N_{Sd}$$

The deflections were also assessed on a pin-jointed analysis and gave $0{,}018\,\text{m}$ under variable load (span/1333) and $0{,}033\,\text{m}$ (span/727) under variable together with permanent loading. Both these deflection ratios are acceptable.

Check on self-weight of the truss:

From the member sizes determined on the basis of the force distribution for the pin-joint analysis, the total mass of the truss was $1{,}6\,\text{ton}$ compared with the estimate of $3{,}6\,\text{ton}$. This means that both the total forces and the total deflection have been overestimated by around 10%. The check on self-weight is carried out in Table 9.2.

Web capacity in the bottom chord at A (Section 4.8).

The reaction is $165{,}2\,\text{kN}$.

Note as a rolled hollow section is being used, $f_{yw} = f_{yf} = 355\,\text{MPa}$. Since the length of stiff bearing s_s is not known, set $s_s = 0$ and determine the value required should the check fail.

For an end support with $c = s_s = 0$, $k_F = 2$ (type c)

TABLE 9.2 Check on self-weight estimate for EXAMPLE 9.1.

Member	Total length (m)	Sub total (m)	Mass/unit length (kg/m)	Mass (kg)
AQ	24			
AB and QR	1,6			
BJ and JR	24,48	50,08	22,9	1147
BC and OR	6,21			
DE and PH	6,62			
FG and NK	7,21			
HI and LI	7,68			
DC and PO	2,8			
EF and NH	4,0			
HG and LK	5,2			
JI	3,2	42,92	10,5	451
Total mass				1598

Determine m_1:

$$m_1 = \frac{f_{yf}b_f}{f_{yw}t_w} = \frac{b_f}{t_w} = \frac{50}{8} = 6,25$$

As m_2 is dependant upon $\bar{\lambda}_F$ initially assume $m_2 = 0$.

As s_s and c have been assumed to be zero, then $l_c = 0$, then the least value of l_y is given by

$$l_y = t_f\sqrt{\frac{m_1}{2}} = 8,0\sqrt{\frac{6,25}{2}} = 14,1\,\text{mm}$$

The depth of the web h_w has been taken as the clear depth of the web,

$$h_w = h - 2r - 2t = h - 4t = 100 - 4 \times 8 = 68\,\text{mm}$$

$$F_{CR} = 0,9k_F E\frac{t_w^3}{h_w} = 0,9 \times 2 \times 210\frac{8,0^3}{68} = 2846\,\text{kN}$$

$$\bar{\lambda}_F = \sqrt{\frac{l_y t_w f_{yw}}{F_{CR}}} = \sqrt{\frac{14,1 \times 8,0 \times 355}{2846 \times 10^3}} = 0,119$$

As $\bar{\lambda}_F < 0,5$, $m_2 = 0$

$$\chi_F = \frac{0,5}{\bar{\lambda}_F} = \frac{0,5}{0,119} = 4,2$$

The maximum value of χ_F is 1,0, thus

$$L_{eff} = \chi_F l_y = 1,0 \times 14,1 = 14,1\,\text{mm}$$

$$F_{Rd} = L_{eff}t_w\frac{f_{yw}}{\gamma_{M1}} = 14,1 \times 8,0\frac{355}{1,0} \times 10^{-3} = 40\,\text{kN}$$

The total resistance therefore is 80 kN (from both webs).

A stiff bearing is therefore required to supply a further $165,2 - 80 = 85,2$ kN.

Try a 25 mm stiff bearing.

$$k_F = 2 + 6\left(\frac{s_s + c}{h_w}\right) = 2 + 6\left(\frac{25 + 0}{68}\right) = 4,21$$

$$l_c = \frac{k_F E t_w^2}{2 f_{yw} h_w} \le s_s + c = \frac{3,33 \times 210 \times 10^3 \times 8^2}{2 \times 355 \times 68} \le 25 + 0 = 927 \le 25$$

The limiting value of l_c is greater than s_s, so $l_c = 25$ mm.

$$l_y = l_c + t_f\sqrt{\frac{m_1}{2} + \left(\frac{l_c}{t_f}\right)^2} = 25 + 8,0\sqrt{\frac{6,25}{2} + \left(\frac{25}{8}\right)^2} = 53,7 \text{ mm} \tag{a}$$

$$l_y = l_c + t_f\sqrt{m_1} = 25 + 8,0\sqrt{6,25} = 45 \text{ mm} \tag{b}$$

The value of l_y using the equation for cases (a) and (b) is clearly larger, thus the value of l_c is the lesser of those above, that is, 45 mm

$$F_{CR} = 0,9 k_F E \frac{t_w^3}{h_w} = 0,9 \times 4,21 \times 210\frac{8,0^3}{68} = 5991 \text{ kN}$$

$$\overline{\lambda}_F = \sqrt{\frac{l_y t_w f_{yw}}{F_{CR}}} = \sqrt{\frac{45,0 \times 8,0 \times 355}{5991 \times 10^3}} = 0,146$$

$$\chi_F = \frac{0,5}{\overline{\lambda}_F} = \frac{0,5}{0,146} = 3,42$$

The maximum value of χ_F is 1,0, thus

$$L_{eff} = \chi_F l_y = 1,0 \times 45 = 45 \text{ mm}$$

$$F_{Rd} = L_{eff} t_w \frac{f_{yw}}{\gamma_{M1}} = 45 \times 8,0\frac{355}{1,0} \times 10^{-3} = 128 \text{ kN}$$

The total resistance therefore is 256 kN (from both webs). This exceeds the reaction of 165,2 kN. Thus the length of stiff bearing could be reduced to around 20 mm. Note also that the end of the bottom member will be sealed by a thin plate welded over its end to prevent corrosive matter reaching the inside of the tubular section. This will have a slight stiffening effect.

The results from a fixed-joint analysis are also given in Table 9.1 where it will be observed that with the exception of the members close to the supports, the axial force resultants show negligible difference. The exception is member AD which now carries an axial force of 7 kN and a moment of 2,85 kNm ($M_{pl,Rd} = 35,5$ kNm). Thus it may be concluded the secondary forces induced by joint fixity are negligible, and may be ignored. In the fixed-joint analysis the deflections are reduced by around 1–2%.

9.2 NON-TRIANGULATED TRUSSES

Another common form of truss is the Vierendeel girder. This is non-triangulated, even though the top and bottom cords may not be parallel. As the loading is carried by sway, there are substantial moments at the nodes, although the shears and axial forces of reasonable size. As a result, the Vierendeel girder is automatically classified as a sway frame and is generally analyzed elastically with the system or effective lengths of the members determined for the sway case.

The notional horizontal load is applied at the end of the truss in the direction of the top chord.

EXAMPLE 9.2   Design of a parallel chord Vierendeel girder.

Prepare a design in Grade S355 steel using rectangular hollow section for the girder detailed in Fig. 9.2.

Loading:

Permanent (kPa):   Sheeting   0,22

                        Purlins   0,10

                        Total   0,32

UDL per truss $= 0,32 \times 8,5 \quad = 2,72 \,\text{kN/m}$
Self-weight (estimated) $\qquad = 2,00 \,\text{kN/m}$
Total permanent $\qquad\qquad = 4,72 \,\text{kN/m}$
or a nodal load of $3,0 \times 4,72 \ = 14,16 \,\text{kN}$ (internal node)
Variable action $= 0,6 \,\text{kPa}$, or $8,5 \times 3,0 \times 0,6 = 15,3 \,\text{kN}$ per node
Total factored ultimate load per internal node $= 1,35 \times 14,16 + 1,5 \times 15,3 = 42,07 \,\text{kN}$

For this example any load combination involving wind will not be considered. Note, however, where the load cases including wind to be considered then the moments due to wind would need multiplying by the sway amplification factor.

Determination of notional horizontal loading:

Determine α_h from Eq. (8.17):

$$\alpha_h = \frac{2}{\sqrt{h}} = \frac{2}{\sqrt{2}} = 1{,}414$$

The maximum value allowed for α_h is 1,0,

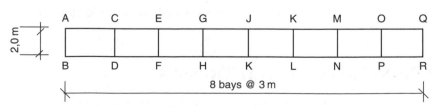

FIGURE 9.2   Truss geometry for EXAMPLE 9.2

thus $\alpha_h = 1,0$.

Determine α_m from Eq. (8.18):

$$\alpha_m = \sqrt{0,5\left(1 + \frac{1}{m}\right)} = \sqrt{0,5\left(1 + \frac{1}{9}\right)} = 0,745$$

Determine φ from Eq. (8.16):

$$\phi = \phi_0\alpha_h\alpha_m = 1,0 \times 0,745\frac{1}{200} = 3,725 \times 10^{-3}$$

Total factored vertical loading is $8 \times 42,07 = 336,6\,\text{kN}$.

Thus the notional horizontal load is $336,6 \times 3,725 \times 10^{-3} = 1,25\,\text{kN}$.

Determination of α_{cr}:

A notional load of 1% of the total factored load is applied at the top chord level. The load is $0,01 \times 336,6 = 3,37\,\text{kN}$.

The actual analysis for both these cases was carried out using a computer package with a 1 kN load, and the results were obtained pro-rata.

The results for the notional horizontal load of 1,25 kN is given in Table 9.3, and the deflections (absolute and relative) for the notional load of 3,37 kN are given in Table 9.4.

As only a single storey is being considered, then from Table 9.4 the maximum relative deflection is 0,058 mm. Thus

$$\frac{\Delta}{h} = \frac{0,058}{2000} = 29 \times 10^{-6}$$

TABLE 9.3 Internal stress resultants due to the notional horizontal load.

Member	Axial force kN	Shear force kN	Moment End 1 kNm	End 2 kNm
AB	−0,52	0,093	−0,0845	0,0845
CD	0	0,15	−0,1475	0,1475
EF	0	0,15	−0,154	0,154
GH	0	0,16	−0,155	0,155
IJ	0	0,16	−0,155	0,155
AC	−1,16	−0,05	−0,085	0,072
CE	−1,00	−0,05	−0,077	0,079
EG	−0,86	−0,05	−0,075	0,081
GI	−0,70	−0,05	−0,074	0,083
BD	1,16	0,05	0,085	−0,072
DF	1,0	−0,05	0,077	−0,079
FH	0,85	−0,05	0,075	−0,081
HJ	0,70	−0,05	0,074	0,083

TABLE 9.4 Determination of α_{cr}.

| Member | Deflection | | Net deflection |
	Top mm	Bottom mm	mm
AB	0,058	0	0,058
CD	0,053	−0,005	0,058
EF	0,049	−0,009	0,058
GH	0,045	−0,012	0,057
IJ	0,042	−0,016	0,058
KL	0,040	−0,018	0,058
MN	0,038	−0,020	0,058
OP	0,037	−0,021	0,058
QR	0,037	−0,021	0,058

From Eq. (8.10), the value of α_{cr} is given by

$$\alpha_{cr} = \frac{0,009}{\frac{\Delta}{h}} = \frac{0,009}{29 \times 10^{-6}} = 310$$

The minimum value for elastic analysis of α_{cr} is 10, thus second order effects may be ignored.

The magnification factor to be applied to moments causing sway (i.e. for this particular example only the moments due to the notional horizontal loads) is (cl 5.2.2 (5)B, EN 1993-1-1)

$$\frac{1}{1 - \frac{1}{\alpha_{cr}}} = \frac{1}{1 - \frac{1}{310}} = 1,003$$

The stress resultants due to the vertical loads are given in Table 9.5. The bending moment, shear force and axial force diagrams for the total loading are plotted in Fig. 9.3(a) to (c), respectively for half the frame

Try a 250 × 150 × 12,5 S355J2H

Member checks:

Section classification:

Web:

The web is subject to compression:

Maximum axial compressive force N_{Ed} is 479 kN.

Length of web χ_w to resist the compressive force is given by

$$\chi_w = \frac{N_{Ed}}{2t_w \frac{f_y}{\gamma_{M0}}} = \frac{479 \times 10^3}{2 \times 12,5 \frac{355}{1,0}} = 54 \text{ mm}$$

TABLE 9.5 Internal stress resultants due to the factored vertical load.

Member	Axial force kN	Shear force kN	Moment End 1 kNm	End 2 kNm
AB	−94,53	−116,11	−166,11	116,24
CD	−21,20	−175,56	−175,56	175,56
EF	−20,99	−62,52	−124,11	123,90
GH	−21,04	−62,52	−62,52	62,52
IJ	−21,04	0	0	0
AC	−116,11	73,50	115,95	104,59
CE	−291,67	52,63	−70,93	86,96
EG	−415,78	31,55	−35,04	57,51
GI	−478,29	10,52	−5,01	26,55
BD	116,11	73,75	116,24	−104,93
DF	291,67	52,55	70,72	−86,92
FH	415,78	31,55	37,15	−57,51
HJ	478,29	−10,52	5,01	−26,55

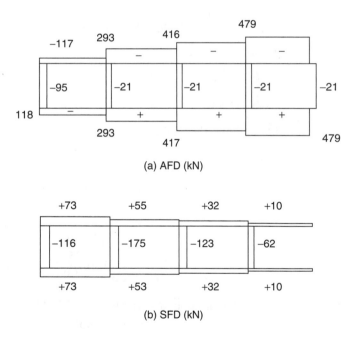

(a) AFD (kN)

(b) SFD (kN)

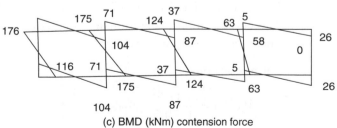

(c) BMD (kNm) contension force

FIGURE 9.3 Final AF, SF and BM diagrams for EXAMPLE 9.2

$$c = h - 2t_f - 2r = 250 - 2 \times 12,5 - 2 \times 12,5 = 200 \, \text{mm}$$

$$\alpha c = 0,5c - 0,5a_x = 100 + 27 = 127 \, \text{mm}$$

$$\alpha = \frac{127}{200} = 0,635$$

$$\frac{c}{t_w} = \frac{200}{12,5} = 16$$

Class 1 limit for $\alpha > 0,5$:

$$\frac{c}{t_w} = 396 \frac{\sqrt{\frac{235}{f_y}}}{13\alpha - 1} = 396 \frac{\sqrt{\frac{235}{355}}}{13 \times 0,635 - 1} = 44,4$$

Web is Class 1.

Flange:

$$c = b - 2t_f - 2r = 150 - 2 \times 12,5 - 2 \times 12,5 = 100 \, \text{mm}$$

$$\frac{c}{t_f} = \frac{100}{12,5} = 8$$

Limit for Class 1 (Table 5.2 Sheet 1):

$$\frac{c}{t_f} = 72\sqrt{\frac{235}{f_y}} = 72\sqrt{\frac{235}{355}} = 58,6$$

Flange is Class 1.

The section is therefore Class 1.

Shear:

The shear area A_v for a load parallel to the depth is given in cl 6.2.6 (EN 1993-1-1) as

$$A_v = A \frac{h}{h+b} = 9300 \frac{250}{250+150} = 5813 \, \text{mm}^2$$

$$V_{pl,Rd} = \frac{1}{\sqrt{3}} A_v \frac{f_y}{\gamma_{M0}} = \frac{1}{\sqrt{3}} 5813 \frac{355}{1,0} \times 10^{-3} = 1191 \, \text{kN}$$

The maximum applied shear force is 175,4 kN. Thus the section is satisfactory and there is no moment capacity reduction due to shear ($V_{Ed}/V_{pl,Rd} = 0,15$).

Members BD, DF, FH and HJ need checking under combined tension and bending. This check reduces to determining whether the reduced flexural capacity due to axial load exceeds the applied moment.

For box sections:

$$M_{N,y,Rd} = \frac{1-n}{1-0,5a_w} M_{pl,y,Rd}$$

$$a_w = \frac{A - 2bt}{A} = \frac{9300 - 2 \times 150 \times 12,5}{9300} = 0,597$$

Limiting maximum value of $a_w = 0,5$, therefore $a_w = 0,5$.

$$1 - 0,5a_w = 1 - 0,5 \times 0,5 = 0,75$$

$$M_{pl,y,Rd} = W_{pl,y}\frac{f_y}{\gamma_{M0}} = 751000\frac{355}{1,0} \times 10^{-6} = 267\,\text{kNm}$$

$$N_{pl,Rd} = A\frac{f_y}{\gamma_{M0}} = 9300\frac{355}{1,0} \times 10^{-3} = 3302\,\text{kN}$$

$$n = \frac{N_{Ed}}{N_{pl,Rd}}$$

The checks are carried out in Table 9.6, where it should be noted that in no case is the plastic moment capacity reduced, and that the values quoted for $M_{y,Ed}$ are the larger *absolute* values within the member.

Members AB, CD, DF, GH, AC, CE, EG, GI need checking under combined bending and compressive axial force.

The check for reduction in moment capacity for members AC to GI is exactly the same as for BD to HJ. Members AB and CD carry lower axial forces and will not therefore be critical for moment capacity reductions.

Lateral torsional buckling:

The system length may be taken as the worst case of member length, that is, 3 m. Lateral torsional buckling is checked using Section 5.1.8.1.

Determine ϕ_b from Eq. (5.63):

$$\phi_b = \sqrt{\frac{W_{pl,y}^2\left[1 - \frac{I_z}{I_y}\right]\left[1 - \frac{GI_t}{EI_y}\right]}{AI_t}}$$

$$= \sqrt{\frac{(751 \times 10^3)^2\left[1 - \frac{3310}{7518}\right]\left[1 - \frac{81 \times 7317}{210 \times 7518}\right]}{9300 \times 7317 \times 10^4}} = 0,538$$

The moment gradient factor C_1 should be determined for the greatest ratio of end moments which occurs in member GI.

$$\psi = \frac{-5,08}{26,63} = -0,191$$

TABLE 9.6 Member capacity checks for members in tension.

Member	N_{Ed} kN	n	$M_{N,y,Rd}$ kNm	$M_{pl,y,Rd}$ kNm	$M_{y,Ed}$ kNm
BD	117,3	0,036	343	267	116,3
DF	292,7	0,089	324	267	87,0
FH	416,6	0,126	311	267	57,6
HJ	479,0	0,145	304	267	26,6

Determine C_1 from Eq. (5.40):

$$\frac{1}{C_1} = 0{,}6 + 0{,}4\psi = 0{,}6 + 0{,}4(-0{,}191) = 0{,}524 > 0{,}4$$

Thus $C_1 = 1{,}908$

The slenderness λ is given by

$$\lambda = \frac{L}{i_z} = \frac{3000}{59{,}7} = 50{,}3$$

Determine the lateral torsional buckling slenderness ratio λ_{LT} from Eq. (5.62):

$$\lambda_{LT} = \frac{1}{C_1^{1/2}}\left[\pi\sqrt{\frac{E}{G}}\right]^{1/2}(\phi_b\lambda)^{1/2} = \frac{1}{\sqrt{1{,}908}}\left[\pi\sqrt{\frac{210}{81}}\right]^{1/2}(0{,}538 \times 50{,}3)^{1/2} = 8{,}47$$

Determine the normalized slenderness ratio $\overline{\lambda}_{LT}$ from Eq. (5.60):

$$\overline{\lambda}_{LT} = \frac{\lambda_{LT}}{\pi\sqrt{\frac{E}{f_y}}} = \frac{8{,}47}{\pi\sqrt{\frac{210\times10^3}{355}}} = 0{,}111$$

As $\overline{\lambda}_{LT} < 0{,}4$, lateral torsional buckling cannot occur.

Check the critical length for lateral torsional buckling l_{crit} from Eq. (5.64a):

$$l_{crit} = \frac{113400\,(h-t)}{f_y}\frac{\left(\frac{b-t}{h-t}\right)^2}{1+3\frac{b-t}{h-t}}\sqrt{\frac{3+\frac{b-t}{h-t}}{1+\frac{b-t}{h-t}}}$$

$$= \frac{113400\,(250-12{,}5)}{355}\frac{\left(\frac{150-12{,}5}{250-12{,}5}\right)^2}{1+3\frac{150-12{,}5}{250-12{,}5}}\sqrt{\frac{3+\frac{150-12{,}5}{250-12{,}5}}{1+\frac{150-12{,}5}{250-12{,}5}}}$$

$$= 14000\,\text{mm}$$

This is well in excess of the system length of 3 m. This also confirms lateral torsional buckling will not occur.

Check Eq. (5.64b) from Kaim (2006),

$$\overline{\lambda}_{z,\lim} = \frac{25}{\frac{h}{b}}\sqrt{\frac{235}{f_y}}$$

$$\lambda_{z,\lim} = \overline{\lambda}_{z,\lim}\lambda_1 = \frac{25}{\frac{h}{b}}\sqrt{\frac{235}{f_y}}93{,}9\sqrt{\frac{235}{f_y}} = \frac{55167}{\frac{h}{b}f_y} = \frac{55167}{\frac{250}{150}355} = 93{,}2$$

$$L = \lambda_{z,\lim}i_z = 93{,}2 \times 59{,}7 = 5560\,\text{mm}$$

The actual length of 3 m is below the critical value, therefore lateral torsional buckling will not occur.

Thus as lateral torsional buckling is not critical Table B.1 can be used to determine the values of k_{yy} and k_{zy}.

TABLE 9.7 Member capacity checks for compression members.

Calculation	Member AC	CE	EG	GI
K_C	0,3333	0,3333	0,3333	0,3333
K_1	0	0,3333	0,3333	0,3333
K_{11}	0	0	0	0
K_{12}	0,5	0,5	0,5	0,5
k_1	0,4	0,571	0,571	0,571
K_2	0,3333	0,3333	0,3333	0,3333
K_{21}	0	0	0	0
K_{22}	0,5	0,5	0,5	0,5
k_2	0,571	0,571	0,571	0,571
$l_{cr,y}/L$ (Eq. (8.22))	1,470	1,612	1,612	1,612
$l_{cr,y}/L$ (Eq. (8.23))	1,418	1,571	1,571	1,571
$l_{cr,y}/L$ (Eq. (8.24))	1,473	1,624	1,624	1,624
$M_{Ed,max}$	116,03	87,04	57,59	26,63 kNm
$M_{Ed,min}$	−104,66	−71,01	−35,12	−5,08 kNm
N_{Ed}	117,27	292,68	416,63	478,99 kN
$N_{cr,y}$	7979	6565	6565	6565 kN
$\bar{\lambda}_y$	0,643	0,709	0,709	0,709
χ_y	0,873	0,843	0,843	0,843
$N_{cr,z}$	7623	7623	7623	7623 kN
$\bar{\lambda}_z$	0,658	0,658	0,658	0,658
χ_z	0,867	0,867	0,867	0,867
k_{yy}	0,407	0,421	0,430	0,570
k_{zy}	0,244	0,253	0,258	0,342
$\dfrac{N_{Ed}}{\chi_y A f_y}$	0,041	0,105	0,150	0,172
$\dfrac{k_{yy} M_{Ed,max}}{W_{pl,y} f_y}$	0,177	0,138	0,093	0,057
Total	0,218	0,243	0,243	0,229
$\dfrac{N_{Ed}}{\chi_z A f_y}$	0,041	0,102	0,146	0,168
$\dfrac{k_{zy} M_{Ed,max}}{W_{pl,y} f_y}$	0,106	0,083	0,056	0,034
Total	0,147	0,185	0,201	0,202

The calculations for member capacity are carried out in Table 9.7 where it will be seen that the member capacities are overgenerous. The 2 m verticals (AB, CD, EF and GH) have not been checked as they are shorter than the top chord compression members (therefore have a higher buckling load) and carry lower axial forces.

The strut buckling lengths have been determined assuming the frame can sway. This is conservative. The buckling length co-efficients have been determined using Eqs (8.22)–(8.24) for comparison. As will be noted in Table 9.7 There is little difference between the results. In this example those from Eq. (8.24) have been used for buckling in-plane. For out-of-plane buckling, the actual length has been used, although this is conservative. To determine the effective lengths, the EI value of the section has been taken as unity as all the members are fabricated from the same section. Thus EI/L reduces to $1/L$. The axial forces although compression are given as positive, as is the larger end moment. The smaller one is given the appropriate sign.

Deflection:

Permanent load deflection: 22,4 mm (span/1070)

Variable load deflection: 24,2 mm (span/992)

Total deflection: 44,6 mm (span/538)

These are satisfactory.

Web check (Section 4.8):

The web needs checking at B with the capacity calculated for a single web and then doubled.

Reaction $= 168,3$ kN, moment $= 116$ kNm.

For an end support with $c = s_s = 0$, $k_F = 2$ (type c)

Determine m_1:

$$m_1 = \frac{f_{yf}b_f}{f_{yw}t_w} = \frac{b_f}{t_w} = \frac{75}{12,5} = 6,0$$

As m_2 is dependant upon $\overline{\lambda}_F$ initially assume $m_2 = 0$.

As s_s and c have been assumed to be zero, then $l_c = 0$, then the least value of l_y is given by

$$l_y = t_f\sqrt{\frac{m_1}{2}} = 12,5\sqrt{\frac{6,0}{2}} = 21,7 \text{ mm}$$

The depth of the web h_w has been taken as

$$h_w = d - 2t - 2r = 250 - 2 \times 12,5 - 2 \times 12,5 = 200 \text{ mm}$$

$$F_{CR} = 0,9k_F E \frac{t_w^3}{h_w} = 0,9 \times 2 \times 210 \frac{12,5^3}{200} = 3691 \text{ kN}$$

$$\overline{\lambda}_F = \sqrt{\frac{l_y t_w f_{yw}}{F_{CR}}} = \sqrt{\frac{21,7 \times 12,5 \times 355}{3691 \times 10^3}} = 0,162$$

As $\overline{\lambda}_F < 0,5$, $m_2 = 0$.

$$\chi_F = \frac{0,5}{\overline{\lambda}_F} = \frac{0,5}{0,162} = 3,1$$

The maximum value of χ_F is 1,0, thus

$$L_{eff} = \chi_F l_y = 1,0 \times 21,7 = 21,7 \, \text{mm}$$

$$F_{Rd} = L_{eff} t_w \frac{f_{yw}}{\gamma_{M1}} = 21,7 \times 12,5 \frac{355}{1,0} \times 10^{-3} = 96,3 \, \text{kN}$$

The total for both webs is 192,6 kN.

This exceeds the reaction of 168,3 kN.

$$\eta_2 = \frac{F_{Ed}}{L_{eff} t_w \frac{f_{yw}}{\gamma_{M1}}} = \frac{F_{Ed}}{F_{Rd}} = \frac{168,3}{192,6} = 0,874$$

$\eta_2 = {<}1,0$, therefore the web resistance at A is satisfactory without a stiff bearing.

However, an interaction equation needs checking owing to the co-existence of shear and bending moment:

$$\eta_2 + 0,8\eta_1 \leq 1,4$$

As there is no axial force and no shift in the neutral axis as the section is Class 1, the equation for η_1 reduces to

$$\eta_1 = \frac{M_{Ed}}{W_{pl} \frac{f_y}{\gamma_{M1}}} = \frac{116 \times 10^6}{751 \times 10^3 \frac{355}{1,0}} = 0,435$$

$$\eta_2 + 0,8\eta_1 = 0,874 + 0,8 \times 0,435 = 1,22 \leq 1,4$$

The web at B is therefore satisfactory.

Check the self-weight:

Member lengths (m):	Top chord:	24
	Bottom chord:	24
	Verticals:	18

Total length = 66 m, mass/unit run = 73,0 kg/m, total mass = 4,82 ton, or very slightly over the estimated value of 2,0 kN/m. The very slight underestimate will not be critical as all the members are well overdesigned.

REFERENCES

Kaim, P. (2006). Buckling of members with rectangular hollow sections, In *Tubular structures XI* (eds J.A. Packer and S. Willibald.) Taylor and Francis, 443–449.

EN 1991-1-1. *Eurocode 1: Actions on structures – Part 1–1: General loads – Densities, self weight, imposed loads for buildings*. British Standards Institution/Comité Européen de Normalization.

EN 1993-1-1. *Eurocode 3: Design of steel structures – Part 1–1: General rules and rules for buildings*. British Standards Institution/Comité Européen de Normalization.

Chapter 10 / Composite Construction

As mentioned in the introduction to Chapter 8, composite construction for slabs and beams has now become widespread, although some use is still made of pre-cast slabs which whether supported on the top flange of the beam grid or on shelf angles do not act compositely with the supporting beams (although they will provide lateral torsional restraint to the beam). To ensure composite action between the beam and the concrete slab, the two must have adequate shear coupling. In general this is achieved by through-deck stud welding.

A further use of composite construction is concrete filled rolled hollow section columns. This has the effect of both increasing the load carrying capacity and also the fire resistance as the concrete core provides a heat sink, and enables load to be transferred from the steel outer to the core. Construction is fast as the steel section itself acts as formwork for the concrete.

Due to developments in alternative, lighter and less time consuming methods of fire protection, concrete for encasement of steel sections, whether beams or columns, is now little used in the UK, even though one of the drawbacks, namely extended construction time and the need for formwork, can be countered by pre-casting the concrete encasement off-site. There will still exist the problems due to the large additional self-weight. Also earlier design methods did not traditionally make full use of the concrete in determining the load carrying capacity (this has changed in EN 1994-1-1).

10.1 COMPOSITE SLABS

A composite slab comprises profile sheet steel decking which acts both as permanent shuttering to the slab and as tension reinforcement for the sagging moments in the slab. There are essentially two basic patterns for profile sheet steel decking; on open trapezoidal section (Fig. 10.1(a)), and a re-entrant trapezoidal section (Fig. 10.1(b)). There will be variations from these basic shapes dependant upon the specific manufacturer.

The re-entrant profile has a slight advantage in that proprietary wedge fixings placed in the re-entrant slot can be used to support suspended ceilings and lightweight services.

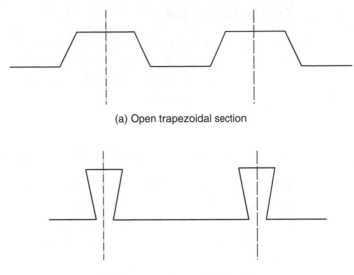

(a) Open trapezoidal section

(b) Re-entrant trapezoidal section

FIGURE 10.1 Types of decking.

The design of such decking when used compositely is considered under two headings, namely those of the decking acting as formwork and the composite slab (Bunn and Heywood, 2004).

10.2 DESIGN OF DECKING

10.2.1 Actions

The actions to be applied to the decking are:

- the weight of the concrete and the self-weight of the deck,
- construction and storage loads,
- any additional load due to ponding of the deck.

The construction loads are to be taken as 1,5 kPa over an area of 3 m by 3 m (or the span if less) placed such as to cause the maximum value of the load effect being considered (flexure, shear or deflection) and 0,75 kPa over the remainder of the deck (EN 1991-1-6, cl 4.11.2). These loads are additional to the weight of the concrete. The unit weight of wet concrete should be taken as 1 kN/m^3 greater than the dry unit weight (Table A.1 of EN 1991-1-1).

Where the deck is continuous over the support beams, the internal forces should be determined using an elastic analysis with constant values of section properties. Although the properties of the steel sheeting may be calculated from first principles, the calculations are tedious and the information required is generally tabulated by the manufacturer.

10.2.2 Deflection

If the deflection under self-weight and the wet concrete is less than the least of span/180 then any additional effects due to ponding can be ignored (cl. 6.6(2)). If the central deflection δ is less than one-tenth of the slab thickness, ponding may be ignored. If this limit is exceeded then ponding can be allowed for by increasing the nominal thickness of the concrete by $0,7\delta$ over the whole span.

10.2.3 Design of Composite Slab (EN 1994-1-1)

The analysis may be carried out for ultimate limit state by any recognized method, including simple plastic analysis (cl 9.4.2). If elastic analysis based on uncracked section properties is used, then up to 30% redistribution is allowed. However, it must be noted that plastic analysis or elastic analysis with a large amount of redistribution will reduce the reinforcement in the top face which may then give problems with achieving an adequate fire performance for other than low fire endurance periods. Any top steel is increasing mobilized during the course of a fire as the decking and the bottom steel, if provided, loses strength. This situation will be most critical in end spans where the sagging moment can only redistribute to one support during the fire. The designer must effectively choose either to allow redistribution under ambient ultimate load design or in the fire design. Information from tests given are likely to have been based on an elastic distribution of design moments.

Three basic checks are required:

(1) Flexural failure
It is generally sufficient to design the slab as simply supported between the support beam system, although advantage can be taken of continuity. Where the slab is designed as simply supported then reinforcement having a cross-sectional area 0,2% of the gross concrete section must be provided if the slab is constructed as unpropped, and 0,4% if propped (cl 9.8.1 (2)).
(2) Shear bond failure
Most profile decking is manufactured with indents or small shear keys to generate bond to give strain compatibility between the concrete and the steel sheet. The parameters required to check this are determined from tests, and are made available by manufacturers for a specific deck. It should be noted that these parameters are not defined in the same manner as those for BS 5950 Part 4, and thus it is important that current manufacturer's information is available (Evans and Wright, 1988; Johnson, 2004; Johnson and Anderson, 2004).
The position of the regression line in the two standards is not compatible owing to values of confidence limits used. Equally the parameters derived from British Standard tests are concrete strength independent, whereas the intercept defined in the test described in EN 1994-1-1 is not. It would appear that the slopes (m values) are similar, but the intercept (k values) need multiplying by approximately $1/(0,8f_{cu})^{1/2}$ ($= 1,12\sqrt{f_{ck}}$) to convert from BS values to EN values with allowance

for conversion of cube strengths to cylinder strengths (Johnson and Anderson, 2004).

(3) Vertical shear

This is rarely critical except in short spans carrying high local loading. Also where local high loads exist punching shear will need checking.

10.2.4 Flexural Design (cl 9.7.2)

10.2.4.1 Sagging

The flexural capacity is determined using a rectangular stress block with a compressive stress in the concrete of $0,85f_{ck}/\gamma_c$. The moment capacity of the slab is then given by

$$M_{Rd} = N_p \left(d_p - \frac{x}{2} \right) \tag{10.1}$$

where N_p is the force in the profile decking given by $A_p f_{yp}/\gamma_{ap}$ where A_p is the area of the profile decking, f_{yp} is the characteristic yield strength of the decking, γ_{ap} is the partial safety factor ($= 1,0$), d_p is the depth to the centroid of the profile decking measured from the extreme compression fibre, and x the depth to the neutral axis is given by

$$x = \frac{N_p}{0,85b\frac{f_{ck}}{\gamma_c}} \tag{10.2}$$

where b is the width of the section.

10.2.4.2 Hogging

The concrete deck should be treated as normal reinforced concrete, although it is possible to take the decking can be taken into account if it is continuous over the support.

10.2.5 Shear Bond (Longitudinal Shear) (cl 9.7.3)

The shear bond strength $V_{l,Rd}$ should be taken as

$$V_{l,Rd} = \frac{bd_p}{\gamma_{vs}} \left(\frac{mA_p}{bL_s} + k \right) \tag{10.3}$$

where m and k are empirical design factors obtained from tests, L_s is the shear span and in this case γ_{vs} is taken as 1,25.

The shear span L_s is taken as $L/4$ for a UDL, for two equal symmetric loads, the distance from the support to the nearest load. For other cases L_s can be determined approximately by dividing the maximum moment by the greater vertical shear force adjacent to the support for the span being considered. For internal spans of a continuous slab, L_s may be taken as $0,8L$, and fore external spans $0,9L$ where L is the span.

10.2.5.1 Longitudinal Shear for Slabs with End Anchorage (cl 9.7.4 (3))

The steel sheet must be anchored at the end span by shear studs designed to resist the tensile force in the steel decking at ultimate limit state. The strength of the stud should be the lesser of its design shear capacity (cl 6.6.4.2, Section 10.3.5) or the bearing resistance $P_{pb,Rd}$ is given by

$$P_{pb,Rd} = \frac{k_\phi d_{d0} t f_{yp}}{\gamma_{ap}} \tag{10.4}$$

where t is the thickness of the sheeting, d_{d0} is the diameter of the weld collar (which may be taken as 1,1 times the stud diameter), k_φ is a factor allowing for the influence of the deck geometry on the shear strength given by

$$k_\phi = 1 + \frac{a}{d_{d0}} \le 4{,}0 \tag{10.5}$$

where a is the distance from the centre of the stud to the end of the deck which is to be not less than $1{,}5 d_{d0}$.

10.2.6 Flexural Shear Capacity (cl 9.7.5)

The flexural shear capacity $V_{v,Rd}$ in accordance with EN 1992-1-1 is given by

$$V_{v,Rd} = C_{Rd,c} k \left(100 \rho_1 f_{ck}\right)^{\frac{1}{3}} b_0 d_p \tag{10.6}$$

where b_0 is the mean width of the ribs, k is a shear factor $(= 1 + \sqrt{(200/d_p)} \le 2)$, ρ_1 is the steel ratio $(= A_p / b_0 d_p < 0{,}02)$, and $C_{Rd,c}$ is given by $0{,}18/\gamma_c$ for normal weight concrete and $0{,}15/\gamma_c$ for lightweight concrete $(\gamma_c = 1{,}25)$.

10.2.7 Punching Shear

The punching shear resistance $V_{p,Rd}$ is given by

$$V_{p,Rd} = C_{Rd,c} k \left(100 \rho_1 f_{ck}\right)^{\frac{1}{3}} c_p d_p \tag{10.7}$$

where c_p is the punching shear perimeter defined in Fig. 10.2 and h_c is the depth of the concrete above the ribs.

10.2.8 Deflection (cl 9.8.2)

Deflection checks may be omitted if the span depth ratio does not exceed the appropriate limit for lightly loaded members (EN 1992-1-1, Table 7.4), and for external spans provided the load required to produce an end slip between the decking and the concrete of 0,5 mm exceeds 1,2 times the required service load on the slab.

EXAMPLE 10.1 Composite slab design

A floor system for an office block is to be designed using composite construction. The beam layout is given in Fig. 10.3.

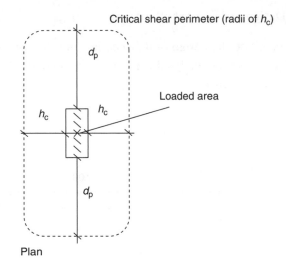

Critical shear perimeter (radii of h_c)

d_p

h_c h_c

Loaded area

d_p

Plan

d_p h_c

Section

FIGURE 10.2 Critical shear perimeter. (from EN 1994-1-1)

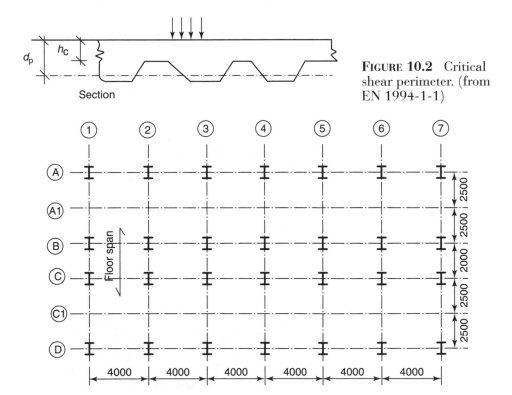

FIGURE 10.3 Grid layout for composite deck design.

The profile decking chosen is Richard Lees Holorib, 0,9 mm gauge. The slab is 105 mm thick with lightweight concrete Grade LC25/30. As the usual available length of sheeting is 12,5 m and the total span is only 12 m, then the sheeting will be continuous and the design can consider the beneficial effects of this with two spans of 2,5 m either side of a central span of 2 m.

a) Design of deck as shuttering.

To facilitate the design of the sheeting and the composite deck, a series of unit load UDL cases were analysed with each span loaded. The resultant BMDs and SFDs are given in Fig. 10.4.

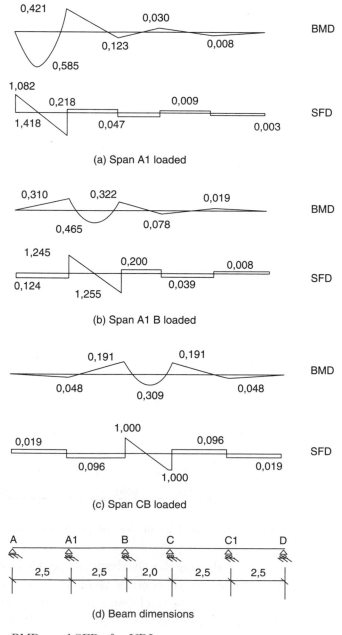

FIGURE 10.4 BMDs and SFDs for UDL

In order to asses the maximum moments on the decking, a series of load cases need considering:

(1) A constant UDL over the whole system.
(2) A UDL of 2,5 m length over the end span (although strictly the construction UDL should be 3 m long, any extension on to the next interior span will be ignored),
(3) A UDL over 2,5 m length symmetric about the first internal support (this is not quite the worst case, which could be established by using influence lines for continuous beams). The BMD and SFD for this case are given in Fig. 10.5.

Assume the dry specific weight of concrete is $19 \, kN/m^3$. The wet unit weight is then $20 \, kN/m^3$.

Take the depth of concrete 105 mm with no allowance for voids.

In determining the moment co-efficients, it is assumed the sagging moment is at mid-span where there is no loading on that span.

UDL due to self-weight of concrete $= 0,105 \times 20 = 2,1 \, kPa$.

From Fig. 10.4, the sagging moment co-efficients are given in Table 10.1.

Moments due to self-weight of concrete,

Maximum sagging moment $= 0,440 \times 2,1 = 0,924 \, kNm/m$,

Maximum hogging moment $= 0,694 \times 2,1 = 1,458 \, kNm/m$.

The construction load can be taken as 0,75 kPa over the whole system and 0,75 kPa applied locally either over span AA1 or over the support at A1.

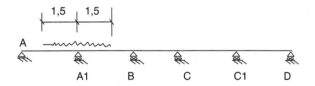

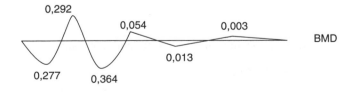

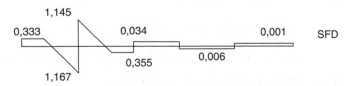

FIGURE 10.5 UDL over support A1

TABLE 10.1 Bending moment co-efficients.

Span loaded	Sagging AA1	Hogging Support A1
AA1	0,585	0,421
A1B	0,155	0,310
BC	−0,024	−0,048
CC1	−0,030	0,019
C1D	0,004	−0,008
Total	0,440	0,694

Uniform UDL:

Sagging: $M = 0{,}440 \times 0{,}75 = 0{,}330$ kNm/m

Hogging: $M = 0{,}694 \times 0{,}75 = 0{,}521$ kNm/m

Local UDL:

Sagging: $M = 0{,}585 \times 0{,}75 = 0{,}439$ kNm/m

Hogging (Fig. 10.5): $M = 0{,}292 \times 0{,}75 = 0{,}219$ kNm/m

Totals due to construction loading:

Sagging: $M = 0{,}330 + 0{,}439 = 0{,}769$ kNm/m

Hogging: $M = 0{,}521 + 0{,}219 = 0{,}740$ kNm/m

Due to permanent and construction loading:

$M_{Sd+} = 0{,}924 + 0{,}769 = 1{,}693$ kNm/m (unfactored)

$M_{Sd-} = 1{,}458 + 0{,}740 = 2{,}198$ kNm/m (unfactored)

From manufacturer's catalogue:

$M_{Rd+} = 6{,}05$ kNm/m (ultimate)

$M_{Rd-} = 6{,}23$ kNm/m (ultimate)

With a partial safety factor of 1,35 applied to the permanent and construction loading, the capacity of the sheeting to support such loading is more than adequate.

Maximum shear:

This is with the construction load on the end 2,5 m span:

Due to the extra 0,75 kPa:

$$V = 1{,}418 \times 0{,}75 = 1{,}064 \text{ kN/m}$$

Due to self-weight and the global 0,75 kPa load:

$$V = (1{,}418 + 0{,}124 - 0{,}008 + 0{,}003)(2{,}1 + 0{,}75) = 4{,}38 \text{ kN/m}$$

$$V_{Sd} = 1{,}35 \times (4{,}38 + 1{,}064) = 7{,}35 \text{ kN/m}$$

Allowable shear $V_{Rd} = 95,8$ kN/m, satisfactory

Deflection:

Initially consider the end span as simply supported under concrete weight (self-weight of sheeting is negligible).

$q = 2,1$ kPa; $L = 2,5$ m; $I = 644 \times 10^3$ mm^4; $E = 210$ GPa.

The deflection δ on a unit width of slab is given by

$$\delta = \frac{5}{384} \frac{qL^4}{EI} = \frac{5}{384} \frac{2,1 \times 2,5^4}{210 \times 10^6 \times 644 \times 10^{-9}} = 0,008 \text{ m}$$

$$\frac{\text{Span}}{180} = \frac{2,5}{180} = 0,014 \text{ mm}$$

Slab thickness/10 $= 0,105/10 = 0,011$ m, therefore ponding need not be considered.

The deflection is satisfactorily based on simply supported conditions and need not be checked more accurately.

b) Slab design:

Since the structure of which this slab is a part requires a 90 min fire resistance, redistribution will be used in the fire limit state rather than at the ambient ultimate limit state, thus the distribution of moments at ultimate will be taken from an elastic analysis with no redistribution. The concrete self-weight is based on the dry density of 19 kN/m^3.

Actions (kPa): Self-weight 2,0

> Finishes 2,5

> Variable 4,0

Factored permanent load: $1,35(2,0 + 2,5) = 6,08$ kPa

Factored variable load: $1,5 \times 4,0 = 6,0$ kPa.

Moments due to permanent load on all spans:

Sagging: $0,445 \times 6,08 = 2,706$ kNm/m

Hogging: $0,694 \times 6,08 = 4,220$ kNm/m

Maximum hogging (Spans AA1, A1B and CC1 loaded):

Moment co-efficient is $0,421 + 0,310 + 0,019 = 0,75$

Hogging moment $= 0,75 \times 6,0 = 4,5$ kNm/m.

Maximum sagging (spans AA1, A1B and C1D loaded):

Moment co-efficient is $0,585 + 0,155 + 0,004 = 0,744$

Sagging moment $= 0,744 \times 6,0 = 4,464$ kNm/m.

ULS moments:

Hogging: $4{,}22 + 4{,}5 = 8{,}72\,\text{kNm/m}$

Sagging: $2{,}706 + 4{,}464 = 7{,}17\,\text{kNm/m}$.

Flexural design:

Sagging:

$$N_p = \frac{A_p f_{yp}}{\gamma_p} = 1597 \times \frac{280}{1{,}0} = 447200\,\text{N/m}.$$

From Eq. (10.2)

$$x = \frac{N_p}{0{,}85b\frac{f_{ck}}{\gamma_c}} = \frac{447200}{0{,}85 \times 1000 \times \frac{25}{1{,}5}} = 31{,}6\,\text{mm}$$

Effective depth, $d_p =$ overall depth − depth of NA of sheet $= 105 - 16{,}7 = 88{,}3\,\text{mm}$

From Eq. (10.1)

$$M_{Rd} = N_p \left(d_p - \frac{x}{2} \right) = 447200 \left(88{,}3 - \frac{31{,}6}{2} \right) = 32{,}4\,\text{kNm/m} > 7{,}17\,\text{kNm/m}$$

Hogging:

Design as reinforced concrete ignoring sheeting:

The depth allowing for the trapezoidal indents is given by

$$105 - 0{,}5 \times 51(38 + 12) = 96\,\text{mm}.$$

Assume 25 mm cover and 10 mm bars, $d = 96 - 25 - 5 = 66\,\text{mm}$.

Assuming $\alpha_{cc} = 0{,}85$, then using Eq. (6.22) in Martin and Purkiss (2006),

$$\frac{M_{Sd}}{bd^2 f_{ck}} = \frac{8{,}72 \times 10^6}{1000 \times 66^2 \times 25} = 0{,}080$$

$$\frac{A_s f_{yk}}{bd f_{ck}} = 0{,}652 - \sqrt{0{,}425 - 1{,}5\frac{M_{Sd}}{bd^2 f_{ck}}} = 0{,}652 - \sqrt{0{,}452 - 1{,}5 \times 0{,}080} = 0{,}0997$$

$$A_s = \frac{0{,}0997 \times 1000 \times 66 \times 25}{500} = 329\,\text{mm}^2/\text{m}.$$

Fix B385 mesh [385 mm²/m]

Flexural shear:

Maximum shear is at A1 on span AA1:

Co-efficient for complete UDL is

$$1{,}418 + 0{,}124 - 0{,}019 + 0{,}019 - 0{,}008 + 0{,}002 = 1{,}537$$

Shear due to permanent loading $= 1{,}537 \times 6{,}08 = 9{,}345\,\text{kN/m}$

Co-efficient for UDL on AB and BC is $1{,}418 + 0{,}124 = 1{,}542$

Shear due to variable loading $= 1{,}542 \times 6{,}0 = 9{,}252$ kN/m.

Total shear $= 9{,}345 + 9{,}252 = 18{,}6$ kN/m.

b_0 is the distance between rib centres, that is, 150 mm.

Applied shear per rib $= 18{,}6(150/1000) = 2{,}79$ kN/m

Using Eq. (10.6) to calculate $V_{v,Rd}$,

$$\rho_1 = \frac{385}{1000 \times 66} = 5{,}83 \times 10^{-3}$$

$$k = 1 + \sqrt{\left(\frac{200}{d}\right)} = 1 + \sqrt{\left(\frac{200}{66}\right)} = 2{,}74 > 2{,}0$$

$$V_{v,Rd} = C_{Rd,c}k \left(100\rho_1 f_{ck}\right)^{\frac{1}{3}} b_0 d = \frac{0{,}15}{1{,}25}(2 \times 150 \times 66)\sqrt[3]{100 \times 5{,}83 \times 10^{-3} \times 25}$$

$$= 5{,}8 \text{ kN/rib}$$

This is greater than the applied shear.

Shear bond:

For an end span $L_s = 0{,}9\,L = 0{,}9 \times 2{,}5 = 2{,}25$ m, as the loading is a UDL $L_s/4$ is used, that is, $2250/4 = 562.5$ mm.

The values of m ($=166{,}6$) and k ($=0{,}005$) were from British standard tests, thus whilst the value of m remains unchanged the value of k has to be amended.

$$k = 0{,}005 \times 1{,}12 \times f_{ck}^{1/2} = 0{,}005 \times 1{,}12 \times 25^{1/2} = 0{,}028$$

Use Eq. (10.3) to calculate $V_{l,Rd}$:

$$V_{l,Rd} = \frac{bd_p}{\gamma_{vs}}\left(\frac{mA_p}{bL_s} + k\right) = \frac{1000 \times 88{,}3}{1{,}25}\left(\frac{166{,}6 \times 1507}{1000 \times 625} + 0{,}028\right)$$

$$= 32{,}0 \text{ kN/m} > V_{Sd}$$

Deflection:

Initially assume there will be no problems with end slip.

From Table 7.4 EN 1992-1-1, basic span depth ratio for the end span of a continuous beam (or slab) is 26 (lightly stressed).

$$\frac{\text{Actual span}}{\text{effective depth ratio}} = \frac{2500}{105 - 16{,}7} = 28{,}3$$

This is higher than the allowable of 26, but since the mid-span capacity is much higher than the applied moment, the section will be relatively uncracked thus reducing deflections.

End anchorage:

Use 20 mm shear studs.

Tensile force required $= N_p = 434000$ N/m.

Bearing resistance check Eq. (10.4):

For a single shear stud, take a at its minimum distance, that is, $2d_0$, thus from Eq. (10.5),

$$k_\phi = 1 + \frac{a}{d_{d0}} = 1 + \frac{2 \times d_{d0}}{d_{d0}} = 3 < 4$$

From Eq. (10.4), the resistance per stud is

$$P_{pb,Rd} = \frac{k_\phi d_{a0} t f_{yp}}{\gamma_{ap}} = \frac{3 \times (1,1 \times 20) \times 0,9 \times 280}{1,0} = 16632 \text{ N}$$

No. of studs $= N_{cf}/P_{pb,Rd} = 447200/16632 = 26,9$ studs.

No. of troughs per metre width $= 1000/150 = 6,67$.

No. of studs per trough $= 26,9/6,67 = 4,03$. Use 4 per trough.

10.3 COMPOSITE BEAMS

Only the common case of rolled sections with profile steel sheet decking running either parallel or transversely to the steel beam will be considered. Through-deck welded shear studs only will be treated. Continuous beam systems will not themselves be considered although both sagging and hogging flexural resistance for composite beams will be described. For other configurations of floor systems, reference should be made to the appropriate monograph produced by the Steel Construction Institute (e.g. Mullett, 1992; Mullett and Lawson, 1992; Lawson, et al., 1997; Mullett, 1997).

Although lateral torsional buckling in the sagging zone cannot occur owing to the restraint offered by the continuous composite concrete deck, it is possible for lateral buckling to occur in the hogging zone at the support. Reference should be made to Johnson (2004) or Johnson and Anderson (2004) for the background to cl 4.6.2 (4). Hogging buckling will not be considered further as it is generally only critical in composite bridge beams, other than to comment that the likelihood of this is reduced by cross-bracing between adjacent beams at supports.

Although traditionally beams have been considered to be fully connected at the concrete deck to beam interface, that is, there is no slip, in practice the degree of shear connection is not full and that some relative slip will occur. This small amount of slip will not significantly affect either the beam deflection or its ultimate load behaviour (Yam and Chapman, 1968). The acceptance of slip has led to the concept that shear studs operate at 80% of their design capacity at full interaction. It is possible to reduce this value further to a level of only partial interaction, in which case both the deflection and ultimate load capacity are affected (Johnson and May, 1975). Full interaction may require excessive transverse reinforcement for shear and also large numbers of connectors giving construction problems.

10.3.1 Section Classification

The section classification limits are determined on the same basis to those for plain steelwork. It is suggested, however, that the sections used are limited to Class 1, or, if necessary, Class 2. If a compression flange is restrained from buckling by attachment to a concrete flange by shear connectors, it may be assumed to be Class 1 (cl 5.5.2 (1)).

The web slenderness ratio must be checked with allowance being made for the fact that the neutral axis is not at the mid-height of the web. For Class 1 sections the d/t_w ratio should not exceed $396\varepsilon/(13\alpha - 1)$ when $\alpha > 0,5$ or $36\varepsilon/\alpha$ when $\alpha < 0,5$. For Class 2 sections the constants are 456 and 41,5. The ratio α is the normalized depth of the compression zone of the web. Where partial interaction is used, the web classification must be carried out using full interaction.

10.3.2 Design Criteria

In addition to the expected checks for flexure, vertical shear (carried by the web only), web capacity under in plane forces and deflection, the transverse shear and the shear connector capacity also need checking.

10.3.3 Flexural Design

For stresses during construction the difference between propped and unpropped methods need to be noted.

- Propped construction
  The load due to the concrete is not transferred to the composite section until the concrete has hardened. However, additional forces due to the release of the props may also need considering.
- Unpropped construction
  The effect of the wet concrete is taken on the steel beam alone before the onset of composite action.

In both cases the effects of finishes and variable loads are taken on the composite section.

It should be noted that in unpropped construction the steel beam may still need checking for lateral torsional buckling during the construction stage. Where the profile decking runs parallel to the span of the beam, the sheeting does not provide full restraint. In the case of decking perpendicular to the span, full restraint is likely to be available (Lawson and Nethercot, 1985).

At ultimate limit state, it is not relevant whether the construction was propped or unpropped.

Owing to the relatively high flange widths of composite beams, the whole flange cannot be taken in calculating the moment capacity of the beam owing to shear lag. The effective width either side of the web centre line should be considered separately. The partial effective width $b_{e,i}$ should be taken as $L_e/8$, where L_e is the span for simply supported beams. The total effective width b_{eff} is then given as $2b_{e,i} + b_0$, where b_0 is the transverse distance between the shear connectors. In practice the contribution of b_0 may be neglected. The total effective width b_{eff} for continuous beams, see cl 5.4.1 (EN 1994-1-1). The partial effective width should not exceed either the distance to the free edge or half the spacing to the adjacent beam centre line.

The following assumptions are made when determining the flexural capacity at ULS:

- The concrete in between the ribs is ignored, and the remainder is stressed with a uniform strength of $0{,}85f_{ck}/\gamma_c$.
- The profile decking may be included if it is in tension and is then stressed to f_y/γ_{ap}, otherwise it is ignored.
- Reinforcement is stressed to f_{sk}/γ_s where f_{sk} is the characteristic strength of the reinforcement (compression reinforcement may be ignored).
- The steel beam is stressed to a uniform stress of f_y/γ_a.

10.3.3.1 Sagging

All symbols are defined in Fig. 10.6(a).

The force in the concrete N_c is given by

$$N_c = \frac{0{,}85f_{ck}}{\gamma_c}b_{eff}\left(h_f - h_p\right) \tag{10.8}$$

The force in the steel section N_a is given by

$$N_a = A_a\frac{f_y}{\gamma_a} \tag{10.9}$$

where A_a is the cross-sectional area of the steel beam.

There are three possibilities for the position of the plastic neutral axis, in the concrete slab, the top flange of the steel beam or the web of the steel beam.

If the neutral axis lies in the concrete slab, then $N_c > N_a$, otherwise the neutral axis is in the steel beam.

(a) Neutral axis in concrete slab (Fig. 10.6(b)).
   Depth of neutral axis in the slab x is given by

$$x = \frac{N_a}{\frac{0{,}85f_{ck}}{\gamma_c}b_{eff}} \tag{10.10}$$

   The moment capacity $M_{pl,Rd}$ is then given by

$$M_{pl,Rd} = N_a\left[\frac{h}{2} + h_f - \frac{x}{2}\right] \tag{10.11}$$

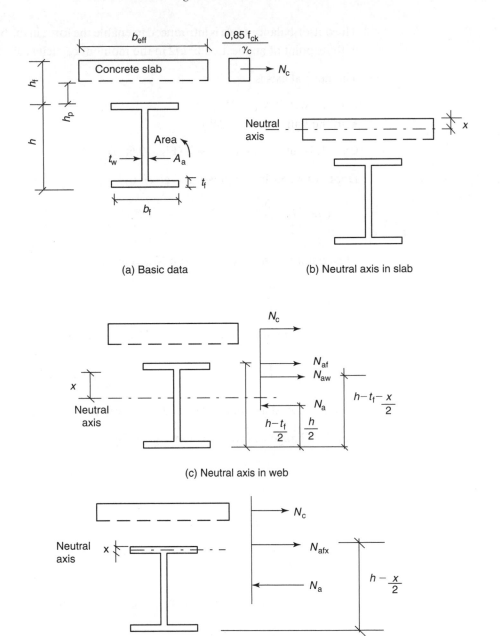

FIGURE 10.6 Determination of plastic moment capacity (sagg mg).

(b) Neutral axis in the steel beam ($N_a > N_c$)

Define an out-of-balance steel compression force N_{ac} as

$$N_{ac} = N_a - N_c \tag{10.12}$$

and the capacity of the flange N_{af} (ignoring fillets) as

$$N_{af} = b_f t_f \frac{f_y}{\gamma_a} \tag{10.13}$$

The out-of-balance force is introduced to enable the force in the beam N_a to be retained with its point of application of $h/2$ in the moment capacity calculations.

The neutral axis is

- in the web if $N_{ac} > 2N_{af}$ or
- in the flange if $N_{ac} < 2N_{af}$.

Case 1: Neutral axis in the web (Fig. 10.6(c))

Depth of web x in compression is given by

$$x = \frac{N_{ac} - N_{af}}{2t_w \frac{f_y}{\gamma_a}} \tag{10.14}$$

The compression force in the web N_{aw} is given by

$$N_{aw} = xt_w \frac{f_y}{\gamma_a} \tag{10.15}$$

and the design moment of resistance $M_{pl,Rd}$ is given by

$$M_{pl,Rd} = N_c \left(h + \frac{h_p + h_s}{2}\right) + 2N_{af}\left(h - \frac{t_f}{2}\right) + 2N_{aw}\left(h - t_f - \frac{x}{2}\right) - N_a\frac{h}{2} \tag{10.16}$$

Case 2: Neutral axis in the flange (Fig. 10.6(d))

Depth of flange x in compression is given by

$$x = \frac{N_{ac}}{2b \frac{f_y}{\gamma_a}} \tag{10.17}$$

The force in the flange N_{afx} is given by

$$N_{afx} = xb \frac{f_y}{\gamma_a} \tag{10.18}$$

and the design moment of resistance $M_{pl,Rd}$ is given by

$$M_{pl,Rd} = N_c \left(h + \frac{h_p + h_f}{2}\right) + 2N_{afx}\left(h - \frac{x}{2}\right) - N_a\frac{h}{2} \tag{10.19}$$

10.3.3.2 Hogging (negative moment)

Conservatively the contribution of the steel profile decking will be ignored.

All symbols are defined in Fig. 10.7(a).

The force in the reinforcement N_r is given by

$$N_r = A_s \frac{f_{sk}}{\gamma_s} \tag{10.20}$$

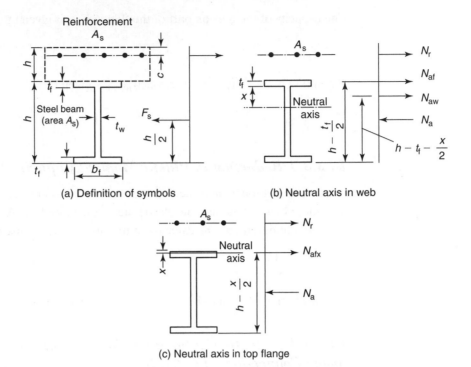

FIGURE 10.7 Calculation of the plastic moment capacity (hogging).

The force in the steel beam N_a is given by Eq. (10.9), and the force in a flange N_{af} is given by Eq. (10.13). The neutral axis is in

- the web, if $N_a > N_r + 2N_{af}$,
- the flange, if $N_a < N_r + 2N_{af}$.

Case 1: Neutral axis in web (Fig. 10.7(b))

The depth of the web x in tension is given by

$$x = \frac{N_a - N_r - 2N_{af}}{2t_w \frac{f_y}{\gamma_a}} \tag{10.21}$$

The capacity of the tension part of the web N_{aw} is given by

$$N_{aw} = xt_w \frac{f_y}{\gamma_a} \tag{10.22}$$

The moment capacity of the section $M_{pl,Rd}$ is given by

$$M_{pl,Rd} = N_r \left(h + h_s - c \right) + N_{af} \left(h - \frac{t_f}{2} \right) + N_{aw} \left(h - t_f - \frac{x}{2} \right) - N_a \frac{h}{2} \tag{10.23}$$

Case 2: Neutral axis in web (Fig. 10.7(c))

The depth of the flange x in tension is given by

$$x = \frac{N_a - N_r}{2b \frac{f_y}{\gamma_a}} \tag{10.24}$$

The capacity of the tension part of the flange N_{afx} is given by

$$N_{\text{afx}} = x b_{\text{f}} \frac{f_y}{\gamma_a} \tag{10.25}$$

The moment capacity of the section $M_{\text{pl,Rd}}$ is given by

$$M_{\text{pl,Rd}} = N_{\text{r}} \left(h + h_{\text{f}} - c \right) + N_{\text{afx}} \left(h - \frac{x}{2} \right) - N_{\text{a}} \frac{h}{2} \tag{10.26}$$

10.3.3.3 Reduction in Plastic Moment Capacity Due to Shear

In a similar fashion to plain steel beams, the moment capacity is reduced when the vertical shear V_{Ed} exceeds half the plastic shear resistance V_{Rd}. A reduced design strength $(1 - \rho) f_{\text{yd}}$ to determine the capacity of the steel section is used where ρ is defined as

$$\rho = \left(\frac{2 V_{\text{Ed}}}{V_{\text{Rd}}} - 1 \right)^2 \tag{10.27}$$

For calculation of plastic shear capacity see Section 10.3.4.

10.3.3.4 Reduction in Sagging Moment Capacity with Partial Shear Connection (cl 6.2.1.3)

The section capacity should be calculated using a force in the concrete section of $N_{\text{c,f}}$ where $N_{\text{c,f}}$ is given by

$$N_{\text{c,f}} = \eta N_{\text{c}} \tag{10.28}$$

where η is the degree of shear connection.

The moment capacity $M_{\text{pl,Rd}}$ is then conservatively given by

$$M_{\text{Rd}} = M_{\text{pl,a,Rd}} + \frac{N_{\text{c}}}{N_{\text{c,f}}} \left(M_{\text{pl,Rd}} - M_{\text{pl,a,Rd}} \right) \tag{10.29}$$

where $M_{\text{pl,Rd}}$ is the moment capacity of the composite section with full shear connection and $M_{\text{pl,a,Rd}}$ is the plastic moment capacity of the steel section alone.

10.3.3.5 Elastic Capacity

For elastic resistance conventional elastic theory is used with the slab taking its effective width b_{eff} and where appropriate account is taken of creep (cl 6.2.1.5). The limiting stresses for $M_{\text{el,Rd}}$ are $0{,}85\, f_{\text{ck}}/\gamma_c$ for concrete, f_y/γ_a for structural steel (Class 1,2 or 3 cross-sections) and f_{sk}/γ_s for reinforcing steel.

Account should be taken of creep and shrinkage in determining the elastic capacity. However, for beams with only one flange composite this may be achieved by using an appropriate modular ratio n_{L} (cl 5.4.2.2.2). The modular ratio n_L is defined as

$$n_{\text{L}} = n_0 \left(1 + \psi_{\text{L}} \phi_{\text{t}} \right) \tag{10.30}$$

where n_0 is the short-term modular ratio defined as E_a/E_{cm}, φ_t is the creep co-efficient defined as $\varphi(t,t_0)$ in EN 1992-1-1, and ψ_L is creep multiplier depending on the type of loading (1,1 for permanent loads, 0,55 for shrinkage and 1,5 for prestressing by imposed deformations). For cases where any amplification of internal forces is less than 10% due to deformations, the structure is not mainly intended for storage nor prestressed by imposed deformations, the effect of creep can be taken into account for both short- and long-term loading by using a nominal modular ratio n determined using an effective concrete elastic modulus $E_{c,eff}$ given by $0,5E_{cm}$.

Note, there appears to be no explicit requirement to check serviceability elastic stresses if the plastic moment capacity is used to determine the strength of the beam.

10.3.4 Flexural Shear

The flexural shear capacity $V_{pl,Rd}$ is calculated exactly as for a normal steel beam. Additionally for an unstiffened and uncased web d/t_w should not exceed 69ε, for a cased web the constant is 124.

10.3.5 Design of Shear Connectors

The strength of shear connectors is dependant upon both the strength and elasticity of the concrete and the ultimate tensile strength of the connector itself.

Following a large series of tests Olgaard *et al*. (1971) proposed the following equation for the strength of shear stud connectors

$$P_{Rd} = \frac{kd^2\sqrt{f_{cu}E_c}}{1,25} \tag{10.31}$$

where d is the diameter of the stud, E_c Young's Modulus of the concrete, f_{cu} concrete cube strength and k is an empirical constant allowing for the height to diameter ratio of the stud. Using the 80% utilization factor proposed by Yam and Chapman (1968), and converting the cube strength to cylinder strength by a factor of 0,8 gives the following strength formula,

$$P_{Rd} = \frac{0,29\alpha d^2\sqrt{f_{ck}E_{cm}}}{\gamma_v} \tag{10.32}$$

where γ_v takes a value of 1,25 and

for $3 < h_{sc}/d < 4$

$$\alpha = 0,2\left(\frac{h_{sc}}{d} + 1\right) \tag{10.33}$$

for $h/d > 4$

$$\alpha = 1,0 \tag{10.34}$$

Oehlers and Johnson (1987) carried out further examination of shear stud capacity and proposed an upper bound which was dependant upon the ultimate strength of the shear stud. Their results have been simplified in the code and are given as

$$P_{Rd} = \frac{0,8 f_u}{\gamma_v} \frac{\pi d^2}{4} \qquad (10.35)$$

where f_u is the ultimate tensile strength of the stud material (taken as not greater than 500 MPa). For decks with ribs parallel to the beam the strength of the shear studs needs to be reduced by the parameter k_1 as the containment by the concrete is incomplete (Mottram and Johnson, 1990),

$$k_1 = 0,6 \frac{b_0}{h_p} \left(\frac{h_{sc}}{h_p} - 1 \right) \le 1,0 \qquad (10.36)$$

where h_p is the height of the profile and h_{sc} is the overall height of the stud ($< h_p + 75$).

For ribs transverse to the beam the value of f_u should be taken as not greater than 450 MPa and modified by the parameter k_t,

$$k_t = \frac{0,7}{\sqrt{n_r}} \frac{b_0}{h_p} \left(\frac{h_{sc}}{h_p} - 1 \right) \qquad (10.37)$$

where n_r is the number of studs in one rib (not greater than 2). For studs with diameters not exceeding 20 mm welded through the deck, the maximum values of k_t, $k_{t,max}$ depend on the sheeting thickness and n_r. For $n_r = 1$, $k_{t,max} = 0,85$ for sheeting thicknesses less than 1,0 mm, and 1,0 for sheeting thicker than 1,0 mm. For $n_r = 2$, $k_{t,max} = 0,70$ for sheeting thicknesses less than 1,0 mm, and 0,8 for sheeting thicker than 1,0 mm.

10.3.6 Biaxial Loading of Shear Connectors

Where the shear connectors are required to produce composite action for both the beam and the slab, then the following relationship should be satisfied

$$\left(\frac{F_l}{P_{l,Rd}} \right)^2 + \left(\frac{F_t}{P_{t,Rd}} \right)^2 \le 1,0 \qquad (10.38)$$

where F_l is the design longitudinal force caused by composite action in the beam, F_t is the design transverse force caused by composite action in the slab, $P_{l,Rd}$ and $P_{t,Rd}$ are the corresponding design shear resistances of the stud.

10.3.7 Partial Shear Connection

The code places limits on the amount of partial shear connection that may be used (cl 6.6.1.2).

For headed studs having a length exceeding $4d$ with $16 \leq d \leq 25$ mm, and the steel beam has equal top and bottom flanges, the minimum value of the ratio η is given by

for $L_e \leq 25$ m

$$\eta \geq 1 - \frac{355}{f_y}\,(0{,}75 - 0{,}03 L_e) \geq 0{,}4 \qquad (10.39)$$

for $L_e \geq 25$ m

$$\eta \geq 1{,}0 \qquad (10.40)$$

where η is defined by

$$\eta = \frac{n}{n_f} \qquad (10.41)$$

where n is the actual number of shear connectors, n_f is the number required with full shear connection and L_e is the length of the beam subject to sagging bending.

Where the studs have an overall length, after welding of 76 mm, a diameter of 19 or 20 mm, are placed centrally in the rib of continuous profiled steel sheeting (with $b_0/h_p \geq 2$ and $h_p \leq 60$ mm) running perpendicular to a rolled equal flange I or H beam and N_c is calculated from Eq. (10.8), then the following values for η may be used

for $L_e \leq 25$ m

$$\eta \geq 1 - \frac{355}{f_y}\,(1{,}0 - 0{,}04 L_e) \geq 0{,}4 \qquad (10.42)$$

for $L_e \geq 25$ m

$$\eta \geq 1{,}0 \qquad (10.43)$$

10.3.8 Shear Connector Spacing

The spacing should be such that longitudinal shear is adequately transferred and separation between the slab and beam is prevented.

Ductile shear connectors may be spaced uniformly between critical cross-sections if

- all critical sections are Class 1 or 2,
- η satisfies the limits for partial interaction,
- the plastic moment resistance of the section does not exceed 2,5 times the plastic moment resistance of the steel section alone.

10.3.9 Longitudinal Shear Force

This is based on the capacity of the shear studs within the length being considered.

The longitudinal shear force/per unit run V_{Sd} is given by

$$V_{Sd} = \frac{n_r P_{Rd}}{s_L} \qquad (10.44)$$

where P_{Rd} is shear capacity of the stud, n_r is the number of studs in a cross-section and s_L is the longitudinal spacing of the studs along the beam.

10.3.10 Transverse Shear

This arises due to the transmission of the longitudinal shear in the beam to the slab. The Eurocode for steel–concrete composite design refers to EN 1992-1-1 cl 6.2.4 and uses a truss analogy to determine the reinforcement required.

The critical planes that need checking are illustrated in Fig. 10.8.

For the surface 2–2 the length of the shear surface is taken as $2h$ plus the head diameter for a single row of studs or $2h + s_t$ where s_t is the transverse spacing of the studs (h is the height of the stud). The shear stress V_{Sd} that needs to be resisted is given by the shear force per unit length (V_{Sd}) divided by the length of the shear plane (s).

The amount of reinforcement per unit run A_{sf}/s_f is determined from

$$\frac{A_{sf}f_{yd}}{s_f} > \frac{v_{Ed}h_f}{\cot \theta_f} \qquad (10.45)$$

where θ_f is the angle of inclination of the concrete truss member such that $1{,}0 \le \cot \theta_f \le 2{,}0$.

For normal weight concrete the shear resistance v_{Ed} is given by

$$v_{Ed} = 0{,}5 v f_{cd} \qquad (10.46)$$

where v is given by $0{,}6\,(1 - f_{ck}/250)$.

For lightweight concrete the shear resistance v_{Ed} is given by

$$v_{Ed} = 0{,}5 \eta_1 v_1 f_{cd} \qquad (10.47)$$

where v_1 is given by $0{,}5 \eta_1 (1 - f_{ck}/250)$, and η_1 is given by $0{,}4 + 0{,}6 \rho / 2200$ where ρ is the density of the concrete.

A limit is placed on v_{Ed} such that

$$v_{Ed} < v f_{cd} \sin \theta_f \cos \theta_f \qquad (10.48)$$

It would appear acceptable to take the traditional approach and use an angle of $45°$ in the truss analogy giving $\cot \theta_f = 1{,}0$ and $\sin \theta_f = \cos \theta_f = 1/\sqrt{2}$ (Johnson, 2004).

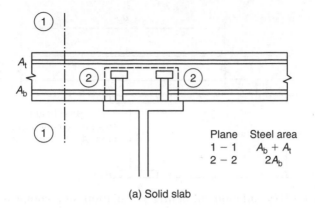

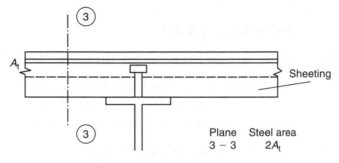

(a) Solid slab

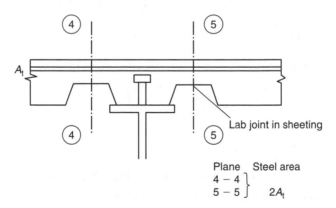

(b) Composite slab (sheeting perpendicular to span)

(c) Composite slab (sheeting parallel to span)

FIGURE 10.8 Critical transverse shear planes.

Where the profile sheet decking is normal to the span and is continuous over the beam then Eq. (10.45) applied to vertical shear planes may be enhanced by the effect of the decking to give

$$\frac{A_{sf}f_{yd}}{s_f} + A_{pe}f_{yp,d} > \frac{v_{Ed}h_f}{\cot\theta_f} \tag{10.49}$$

where A_{pe} is the effective cross-sectional area of the decking and $f_{yp,d}$ is its design strength.

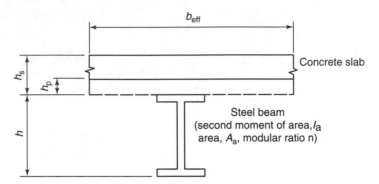

FIGURE 10.9   Second moment of area calculation for a composite beam.

10.3.11 Deflection (cl 5.2.2)

For the deflection of the steel member alone the principles of EN 1993-1-1 are applied.

No account need be taken of partial interaction if $\eta \geq 0,5$ or the force in the shear connector does not exceed P_{Rd} at the serviceability limit state, and for ribbed slabs transverse to the beam the rib height does not exceed 80 mm.

The composite second moment of area may be determined as follows (Fig. 10.9)

$$I_{\text{c}} = nI_{\text{a}} + \frac{b_{\text{eff}} \left(h_{\text{s}} - h_{\text{p}}\right)^3}{12} + nA_{\text{a}} \left(\frac{h}{2} - x\right)^2 + b_{\text{eff}} \left(h_{\text{s}} - h_{\text{p}}\right) \left(h + \frac{h_{\text{s}} + h_{\text{p}}}{2} - x\right)^2 \tag{10.50}$$

where x is given by

$$x = \frac{nA_{\text{a}} + \left(h_{\text{s}} - h_{\text{p}}\right) b_{\text{eff}} \left(h + \frac{h_{\text{s}} + h_{\text{p}}}{2}\right)}{nA_{\text{a}} + \left(h_{\text{s}} - h_{\text{p}}\right) b_{\text{eff}}} \tag{10.51}$$

10.3.12 Vibration

This is only likely to be critical on lightly loaded, long span beams. It is suggested that the following formula is used (Wyatt, 1989)

$$f = \frac{18}{\sqrt{\delta_{\text{sw}}}} \tag{10.52}$$

where f is the frequency (Hz) and δ_{sw} is the instantaneous deflection (mm) due to self-weight and permanent loading. The suggested limits for f are 4 Hz for most buildings, 3 for car parks and 5 for sports halls (Lawson and Chung, 1994).

10.3.13 Detailing

10.3.13.1 Cover

This should be the greater of 20 mm or the values specified in EN 1992-1-1 (Table 4.4) less 5 mm (cl 6.6.5.2).

10.3.13.2 Spacing (cl 6.6.5.5)

This should be not less than $22t_f\varepsilon$ for a solid slab or $15t_f\varepsilon$ for a ribbed slab where the flange becomes a Class 1 or Class 2 by virtue of restraint due to shear connection with the edge distance not exceeding $9t_f\varepsilon$. The maximum spacing should not exceed the lesser of 800 mm or six times the slab thickness. The edge distance should also not exceed 20 mm.

The overall height of a stud connector should be not less than $3d$ (where d is the shank diameter), the spacing in the direction of the shear force should be not less than $5d$, and transverse to the shear force $2,5d$ for solid slabs and $4d$ for other cases. Unless the stud is directly over the web the diameter of the stud should not exceed $2,5t_f$ (cl 6.6.5.7).

For through-deck welding, the profile sheet steel decking should not exceed 1,25 mm thick if galvanized and 1,5 if not. The studs should extend $2d$ above the deck after welding.

Johnson (2005) expresses concern over the detailing rules (and stud capacities) when applied to open trapezoidal decks.

EXAMPLE 10.2   Composite beam (decking transverse to span)

Design a composite beam (ref C1/23 (Fig. 10.3)) in Grade S355 steel to carry the composite slab designed in Example 10.1. The concrete is lightweight LC25/30 and a dry specific weight of 19 kN/m³. The span is 4 m and carries the two 2,5 m deck spans.

Loading due to wet concrete: $0,105 \times 19 = 2,0\,kPa$

Finishes: 2,5 kPa

Variable loading: 4,0 kPa

The loading co-efficients for the reaction at A1 (from SFDs in Fig. 10.4) are given in Table 10.2 for all spans loaded.

For the variable loading, spans AA1, A1B, CC1 and C1D are loaded. This gives an overall co-efficient of 3,062

TABLE 10.2   Loading co-efficients for reactions A1.

Span	Individual components	Total
AA1	$1,418 + 0,218$	1,636
A1B	$0,124 + 1,254$	1,369
BC	$-0,019 - 0,096$	$-0,115$
CC1	$0,039 + 0,008$	0,047
C1D	$0,009 + 0,003$	0,012
Overall total		2,949

Load per unit run on the beam:

Permanent: $2,949(2,0 + 2,5) = 13,27\,\text{kN/m}$

Variable: $3,062 \times 4 = 12,25\,\text{kN/m}$

Ultimate limit state design:

$M_{\text{Sd}} = (1,35 \times 13,27 + 1,5 \times 12,25) \times 4^2/8 = 72,6\,\text{kNm}$

Effective width of slab, b_{eff} (ignoring spacing between connectors as it is likely only a single row will be needed)

$b_{\text{e,i}} = L_{\text{e}}/8 = 4000/8 = 500\,\text{mm}$

$b_{\text{eff}} = 2b_{\text{e,i}} = 1000\,\text{mm}.$

Try a $203 \times 133 \times 25$ UKB (Grade S355)

Since the flange is restrained by through deck studs, the beam flanges are automatically Class 1. The web slenderness satisfies the shear buckling check.

Actual width between beam centre lines $= 2500\,\text{mm} > b_{\text{eff}}$

Dimensions to calculate $M_{\text{pl,Rd}}$ assuming full shear connection are given in Fig. 10.10.

From Eq. (10.8), the force in the concrete flange N_{c} is given by

$$N_{\text{c}} = \frac{0,85 f_{\text{ck}}}{\gamma_{\text{c}}} b_{\text{eff}} \left(h_{\text{f}} - h_{\text{p}} \right) = \frac{0,85 \times 25}{1,5} 1000 \times (105 - 51) \times 10^{-3} = 765\,\text{kN}$$

From Eq. (10.9) the force in the steel beam N_{a} is given by

$$N_{\text{a}} = A_{\text{a}} \frac{f_{\text{y}}}{\gamma_{\text{a}}} = 3200 \frac{355}{1,0} \times 10^3 = 1136\,\text{kN}$$

From Eq. (10.12) the out-of-balance force N_{ac} is given by

$$N_{\text{ac}} = N_{\text{a}} - N_{\text{c}} = 1136 - 765 = 371\,\text{kN}$$

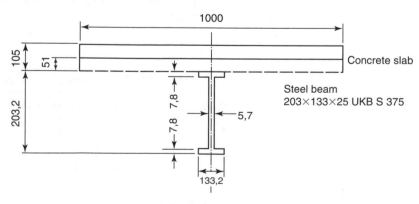

FIGURE 10.10 Design dimensions for EXAMPLE 10.2

From Eq. (10.13) the capacity of the flange N_{af} is given by

$$N_{af} = bt_f \frac{f_y}{\gamma_a} = 133,2 \times 7,8 \frac{355}{1,0} \times 10^{-3} = 369\,\text{kN}$$

Since $N_{ac} < 2N_{af}$, the neutral axis is in the flange.

From Eq. (10.17) the depth of flange x is given by

$$x = \frac{N_{ac}}{2b\frac{f_y}{\gamma_a}} = \frac{371 \times 10^3}{2 \times 133,2 \times \frac{355}{1,0}} = 3,92\,\text{mm}$$

From Eq. (10.18) the compression force in the flange N_{afx} is given by

$$N_{afx} = xb\frac{f_y}{\gamma_a} = 3,92 \times 133,2 \frac{355}{1,0} \times 10^{-3} = 185\,\text{kN}$$

From Eq. (10.19) $M_{pl,Rd}$ is given by

$$M_{pl,Rd} = N_c \left(h + \frac{h_p + h_f}{2}\right) + 2N_{afx}\left(h - \frac{x}{2}\right) - N_a \frac{h}{2}$$

$$= 0,765\left(203,2 + \frac{105 + 51}{2}\right) + 2 \times 0,185\left(203,2 - \frac{3,78}{2}\right) - 1,136\frac{203,2}{2}$$

$$= 174,2\,\text{kNm}$$

Note, for convenience all the forces have been expressed in MN. Then if the dimensions are left in mm, the moment is in kNm.

Since the neutral axis for full shear connection is in the flange, the only web check is for shear buckling (see above).

As $M_{pl,Rd}$ exceeds M_{Ed} by a substantial margin, partial interaction may be used.

For the beam alone the section is Class 2 (which is satisfactory).

Check shear capacity:

$$V_{pl,Rd} = \frac{1}{\sqrt{3}} A_v \frac{f_y}{\gamma_a} = \frac{1}{\sqrt{3}} 1280 \frac{355}{1,0} \times 10^{-3} = 262\,\text{kN}$$

$$V_{Ed} = \frac{(1,35 \times 13,27 + 1,5 \times 12,25) \times 4}{2} = 72,6\,\text{kN}$$

Since $V_{Sd} < 0,5V_{pl,Rd}$, there is no reduction in moment capacity.

Shear connectors:

Use 100 mm long by 19 mm diameter stud connectors.

The load capacity of a single stud, Eq. (10.32):

$h/d = 100/19 = 5,3$, so from Eq. (10.34), $\alpha = 1,0$

$f_{ck} = 25\,\text{MPa}$

For normal weight concrete,

$$E_{cm} = 22 \left(\frac{f_{ck} + 8}{10} \right)^{1/3} = 22 \left(\frac{25 + 8}{10} \right)^{1/3} = 31,5 \, \text{GPa}$$

Modification factor for lightweight concrete η_E (from cl 11.3.2 of EN 1992-1-1)

$$\eta_E = \left(\frac{\rho}{2200} \right)^2 = \left(\frac{1900}{2200} \right)^2 = 0,746$$

So, $E_{cm} = 0,746 \times 31,5 = 23,5 \, \text{GPa}$

From Eq. (10.32)

$$P_{Rd} = \frac{0,29 \alpha d^2 \sqrt{f_{ck} E_{cm}}}{\gamma_v} = \frac{0,29 \times 1 \times 19^2 \times \sqrt{25 \times 23,5 \times 10^3}}{1,25} \times 10^{-3} = 64,2 \, \text{kN}$$

Limiting capacity:

For ribs transverse to beam the ultimate connector strength ($f_u = 450 \, \text{MPa}$) needs to be multiplied by the factor k_t.

Determine k_t from Eq. (10.37)

In view of the large overdesign on moment capacity, number of connectors per rib will be 1, that is, $n_r = 1$.

For a dovetail deck b_0 is the distance between the top of the dovetails, that is $b_0 = 150 - 38 = 112 \, \text{mm}$

Profile height $h_p = 55 \, \text{mm}$, height of connector, $h = 100 \, \text{mm}$,

$$k_t = \frac{0,7}{\sqrt{n_r}} \frac{b_0}{h_p} \left(\frac{h_{sc}}{h_p} - 1 \right) = \frac{0,7}{\sqrt{1}} \frac{112}{55} \left(\frac{100}{55} - 1 \right) = 1,17$$

However the maximum value allowed for k_t is 1,0.

Limiting capacity of studs from Eq. (10.35)

$$P_{Rd} = \frac{0,8 f_u}{\gamma_v} \frac{\pi d^2}{4} = \frac{0,8 \times 450}{1,25} \frac{\pi \times 19^2}{4} \times 10^{-3} = 81,7 \, \text{kN}$$

Shear stud capacity is the lower of the two values, that is 64,2 kN.

No. of connectors for full shear connection:

$N_c = 765 \, \text{kN}$, $P_{Rd} = 64,2 \, \text{kN}$, thus n_f for the half beam span is given by

$n_f = N_c / P_{Rd} = 765/64,2 = 11,9$

Use 1 connector per rib, thus n for the whole beam is given by span over distance between centre lines of ribs $= 4000/300 = 13,3$.

The ratio between η between n and n_r is given by

$$\eta = \frac{n}{n_r} = \frac{13,3}{2 \times 11,9} = 0,56$$

The limiting value of η is given by Eq. (10.39) with $L_e = 4,0$ m, thus

$$\eta \geq 1 - \frac{355}{f_y} \left(0,75 - 0,03L_e\right) \geq 0,4 = 1 - \frac{355}{355} \left(0,75 - 0,03 \times 4\right) = 0,37$$

The limiting value is 0,4, as the actual value is 0,56, it is therefore satisfactory.

The moment capacity M_{Rd} is given by Eq. (10.29).

The moment capacity of the steel section $M_{pl,a,Rd}$ alone is given by

$$M_{pl,a,Rd} = 258 \times \left(\frac{355}{1,0}\right) \times 10^{-3} = 91,6 \text{ kNm}$$

$$M_{Rd} = M_{pl,a,Rd} + \frac{n_c}{n_{c,f}} \left(M_{pl,Rd} - M_{pl,a,Rd}\right) = 91,6 + 0,56 \, (174,2 - 91,6)$$

$$= 137,9 \text{ kNm}$$

This is greater than M_{Ed}.

Longitudinal shear:

The only plane requiring checking is 4–4 (or 5–5) of Fig. 10.8

Determine V_{Sd} from Eq. (10.44),

$$V_{Sd} = \frac{n_r P_{Rd}}{s_L} = \frac{1,0 \times 64,2}{0,3} = 214 \text{ kN/m}$$

The shear stress V_{Sd} is determined using the thickness of the concrete above the ribs, that is, $105 - 51 = 54$ mm. Also the shear may be equally divided between the two shear planes, thus V_{Ed} is given by

$$v_{Ed} = 0,5 \times \frac{214}{54} = 1,98 \text{ MPa}$$

As the sheeting is continuous across the beam with the ribs running normal to the beam, the contribution of the sheeting may be mobilized.

Using a 45° angle for the truss analogy, $\cos \theta_f = \sin \theta_f = 1/\sqrt{2}$ and $\cot \theta_f = 1,0$.

For lightweight concrete v_{Ed} is given by Eq. (10.47).

Determine v_1:

The reduction factor for lightweight concrete η_1 is given by

$$\eta_1 = 0,4 + 0,6 \frac{\rho}{2200} = 0,4 + 0,6 \frac{1900}{2200} = 0,918$$

$$v_1 = 0,5 \eta_1 \left(1 - \frac{f_{ck}}{250}\right) = 0,5 \times 0,918 \left(1 - \frac{25}{250}\right) = 0,413$$

$$V_{Ed} = 0,5v_1\frac{f_{ck}}{\gamma_c} = 0,5 \times 0,413 \times \frac{25}{1,5} = 3,44\,\text{MPa}$$

From Eq. (10.48) the maximum value of V_{Ed} is given by

$$V_{Ed} = vf_{cd}\sin\theta_r\cos\theta_r = 0,6\left(1 - \frac{25}{250}\right)\frac{25}{1,5}\frac{1}{\sqrt{2}}\frac{1}{\sqrt{2}} = 4,5\,\text{MPa}$$

Use Eq. (10.49) to determine the requirement for $A_{sf}s_f$ as the sheeting runs normal to the span.

Evaluate the right hand side of Eq. (10.49),

$$\frac{v_{Ed}h_f}{\cot\theta_f} = \frac{3,44 \times 55}{1,0} = 1892,2\,\text{N/mm}$$

Evaluate the contribution of the sheeting:

$$A_{pe}f_{yp,d} = \frac{1597}{1000}\frac{280}{1,0} = 447,2\,\text{N/mm}$$

The sheeting overprovides the required resistance, therefore only minimum reinforcement is necessary.

Deflections and service stresses:

a) Wet concrete:

This is taken on the steel beam alone. The uniformly distributed load due to the wet concrete q is given by

$$q = 20 \times 0,105 \times 2,5 = 5,25\,\text{kN/m}$$

Deflection, δ_{conc} is given by:

$$\delta_{conc} = \frac{5}{384}\frac{qL^4}{EI} = \frac{5}{384}\frac{5,25 \times 4^4}{210 \times 10^6 \times 2340 \times 10^{-4}} = 3,6 \times 10^{-3}\,\text{m}$$

$$M_{Ed} = \frac{qL^2}{8} = \frac{5,25 \times 4^2}{8} = 10,5\,\text{kNm}$$

The numerical value of the stress σ_{conc} is given by

$$\sigma_{conc} = \frac{M_{Ed}}{W_{el}} = \frac{10,5 \times 10^3}{230} = 46\,\text{MPa}$$

b) Variable and permanent loads:

As the ratio n/n_f of the number of shear studs provided to that required for full connection is 0,57 (and is therefore greater than the critical ratio of 0,5), no account need be taken of partial shear connection in determining the deflection.

Assume $E_{c,eff} = E_{cm}/2 = 23,5/2 = 11,75\,\text{GPa}$

The value of n from cl 5.4.2.2 (11) is $E_s/E_{c,eff}(=210/11,75 = 17,9)$

Use Eq. (10.51) to determine the elastic centroidal axis x,

$$x = \frac{nA_a + (h_s - h_p)\ b_{eff} \left(h + \frac{h_s + h_p}{2}\right)}{nA_a + (h_s - h_p)\ b_{eff}}$$

$$= \frac{17,9 \times 3200\frac{203,2}{2} + (105 - 51)\ 2500\left(203,2 + \frac{105+51}{2}\right)}{17,9 \times 3200 + (105 - 51)\ 2500} = 227,7\,\text{mm}$$

From Eq. (10.50)

$$I_c = nI_a + \frac{b_{eff}\ (h_s - h_p)^3}{12} + nA_a\left(\frac{h}{2} - x\right)^2 + b_{eff}\ (h_s - h_p)\left(h + \frac{h_s + h_p}{2} - x\right)^2$$

$$= 17,9 \times 23,4 \times 10^6 + \frac{2500\ (105 - 51)^3}{12} + 17,9 \times 3200\left(\frac{203,2}{2} - 227,7\right)^2$$

$$+ 2500\ (105 - 51)\left(203,2 + \frac{105 + 51}{2} - 227,7\right)^2 = 1,75 \times 10^9\,\text{mm}^4$$

$$I_c E_{c,eff} = 1,75 \times 10^9 \times 11,75 = 20,45 \times 10^9\,\text{kNmm}^2 = 20,56 \times 10^3\,\text{kNm}^2$$

Variable load is 13,27 kN/m, and the moment M_{Ed} is 26,54 kNm.

Thus the deflection δ_v is given by

$$\delta_v = \frac{5}{384}\frac{qL^4}{EI} = \frac{5}{384}\frac{13,27 \times 4^4}{20,56 \times 10^3} = 2,15 \times 10^{-3}\,\text{m}$$

This is equivalent to span/1860 which is acceptable.

Although not strictly necessary, determine the stresses due to the variable load:

Top of the concrete slab $\sigma_{v,top,c}$:

$$\sigma_{v,top,c} = -\frac{M_{Ed}\ (h + h_s - x)}{Ic} = -\frac{26,54 \times 10^6\ (203,2 + 105 - 227,7)}{1,75 \times 10^9} = -1,22\,\text{MPa}$$

Stress at the soffit of the steel beam, $\sigma_{v,soffit}$:

$$\sigma_{v,soffit} = \frac{nM_{Ed}x}{Ic} = \frac{17,9 \times 26,54 \times 10^6 \times 227,7}{1,75 \times 10^9} = 61,8\,\text{MPa}$$

Stress at the top of the steel beam, $\sigma_{v,top,A}$:

$$\sigma_{v,top,c} = -\frac{nM_{Ed}(h - x)}{Ic} = -\frac{17,9 \times 26,54 \times 10^6\ (203,2 - 227,7)}{1,75 \times 10^9} = 6,65\,\text{MPa}$$

Permanent load is 12,07 kN/m, and the moment $M_{Ed} = 24,14$ kNm.

$$\delta_v = \frac{5}{384}\frac{qL^4}{EI} = \frac{5}{384}\frac{12,07 \times 4^4}{20,45 \times 10^3} = 2,0 \times 10^{-3}\,\text{m}$$

Total deflection:

$$\delta_{total} = \delta_p + \delta_v + \delta_{conc} = 2,2 + 2,0 + 3,6 = 7,8\,\text{mm}$$

This is equivalent to span/513 which is acceptable.

Although not strictly necessary determine the stresses due to the variable load:

Top of the concrete slab $\sigma_{p,top,c}$:

$$\sigma_{p,top,c} = -\frac{M_{Ed}\left(h + h_s - x\right)}{I_c} = -\frac{24{,}14 \times 10^6 \left(203{,}2 + 105 - 227{,}7\right)}{1{,}75 \times 10^9}$$
$$= -1{,}11\,\text{MPa}$$

Stress at the soffit of the steel beam, $\sigma_{p,soffit}$:

$$\sigma_{p,soffit} = \frac{nM_{Ed}x}{I_c} = \frac{17{,}9 \times 24{,}14 \times 10^6 \times 227{,}7}{1{,}75 \times 10^9} = 56{,}2\,\text{MPa}$$

Stress at the top of the steel beam, $\sigma_{p,top,A}$:

$$\sigma_{p,top,c} = -\frac{nM_{Ed}\left(h - x\right)}{I_c} = -\frac{17{,}9 \times 24{,}14 \times 10^6 \left(203{,}2 - 227{,}7\right)}{1{,}75 \times 10^9} = 6{,}05\,\text{MPa}$$

Final total stresses:

Top of concrete slab: $-1{,}11 - 1{,}22 = -2{,}33\,\text{MPa}$

Top of steel beam: $-46 + 6{,}65 + 6{,}05 = -33{,}3\,\text{MPa}$

Soffit of steel beam: $46 + 61{,}8 + 56{,}2 = 164\,\text{MPa}$

All these stresses are acceptable.

Vibration:

Use $\psi_2 = 0{,}3$, so quasi-permanent load is $12{,}07 = 0{,}3 \times 13{,}27 = 16{,}05\,\text{kN/m}$

Deflection under this load δ_{sw} is given by

$$\delta_{sw} = \frac{5}{384}\frac{qL^4}{EI} = \frac{5}{384}\frac{16{,}05 \times 4^4}{20{,}45 \times 10^3} = 0{,}0026\,\text{m} = 2{,}6\,\text{mm}$$

Determine the frequency f from Eq. (10.52)

$$f = \frac{18}{\sqrt{\delta_{sw}}} = \frac{18}{\sqrt{2{,}6}} = 11{,}2\,\text{Hz}$$

This is well above the recommended limit of 3 Hz.

If the full variable load is taken, then $f = 8{,}7\,\text{Hz}$ (and is still acceptable).

EXAMPLE 10.3   Composite beam design (decking parallel to the span).

Prepare a design in Grade S355 steel for the beam Mark 3AD of Fig. 10.3 where there are no intermediate columns. The actions on the beam are given in Fig. 10.11. The composite deck is that of EXAMPLE 10.1.

Due to the long span, the beam is designed as propped to eliminate the high deflections under the permanent loading due to the wet concrete. The resultant shear force and bending moment diagrams for the applied loading are given in Fig. 10.12.

The effective width, either side of the beam centre line, $b_{e,i} = L_e/8 = 12000/8 = 1500$ mm. Thus total effective width $b_{eff} = 2b_{e,i} = 3000$ mm. (The actual beam spacing is 4 m.)

To determine the capacity of the beam, the effect of the voids in the deck will be ignored as these run in the direction of the beam and are small.

Try a $838 \times 292 \times 194$ Grade S355 UKB.

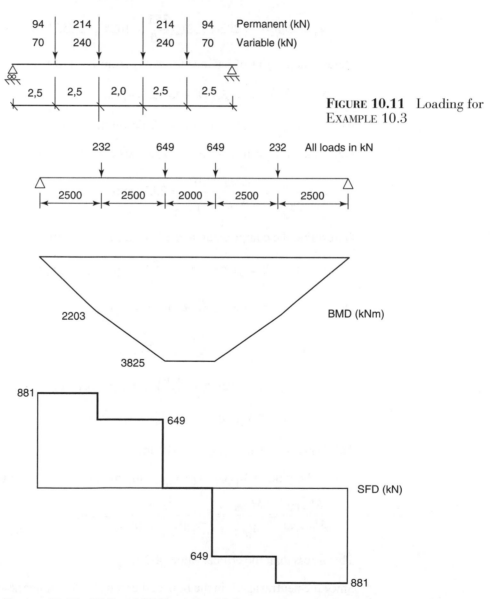

FIGURE **10.11** Loading for EXAMPLE 10.3

FIGURE **10.12** BM and SF EXAMPLE 10.3

Moment capacity with full shear connection:

Determine the force in the concrete N_c from Eq. (10.8),

$$N_c = \frac{0{,}85 f_{ck}}{\gamma_c} b_{eff} (h_f - h_p) = \frac{0{,}85 \times 25}{1{,}5} 3000 \times 105 \times 10^{-6} = 4{,}463 \text{ MN}$$

Determine the force in the steel section N_a from Eq. (10.9):

$$N_a = A_a \frac{f_y}{\gamma_a} = 24700 \frac{355}{1{,}0} \times 10^{-6} = 8{,}769 \text{ MN}$$

Determine the maximum force in a flange N_{af} from Eq. (10.13):

$$N_{af} = b_f t_f \frac{f_y}{\gamma_a} = 292{,}4 \times 21{,}7 \frac{355}{1{,}0} \times 10^{-6} = 2{,}253 \text{ MN}$$

Determine the out-of-balance force N_{ac} from Eq. (10.12):

$$N_{ac} = N_a - N_c = 8{,}769 - 4{,}463 = 4{,}306 \text{ MN}$$

As $N_{ac} < 2N_{af}$, therefore neutral axis lies in the flange.

Determine the position of the neutral axis x from Eq. (10.17):

$$x = \frac{N_{ac}}{2b \frac{f_y}{\gamma_a}} = \frac{4{,}306 \times 10^6}{2 \times 292{,}4 \frac{355}{1{,}0}} = 20{,}74 \text{ mm}$$

Determine the compression force N_{afx} in the flange from Eq. (10.19):

$$N_{afx} = xb \frac{f_y}{\gamma_a} = 20{,}74 \times 292{,}4 \frac{355}{1{,}0} \times 10^{-6} = 2{,}153 \text{ MN}$$

Determine the plastic moment capacity $M_{pl,Rd}$ from Eq. (10.19):

$$M_{pl,Rd} = N_c \left(h + \frac{h_p + h_f}{2} \right) + 2N_{afx} \left(h - \frac{x}{2} \right) - N_a \frac{h}{2}$$

$$= 4{,}463 \left(840{,}7 + \frac{105}{2} \right) + 2 \times 2{,}153 \left(840{,}7 - \frac{20{,}74}{2} \right) - 8{,}769 \frac{840{,}7}{2}$$

$$= 3876 \text{ kNm}$$

This is greater than M_{Ed} (=3825 kNm).

Check the ratio of $M_{pl,Rd}$ to $M_{pl,a,Rd}$ (the moment capacity of bare steel section):

$$\frac{M_{pl,Rd}}{M_{pl,a,Rd}} = \frac{M_{pl,Rd}}{W_{pl} \frac{f_y}{\gamma_a}} = \frac{3876}{7640 \frac{355}{1{,}0} \times 10^{-3}} = 1{,}43$$

This is less than the critical value of 2,5.

Since the neutral axis is in the flange, the web check is unnecessary and thus the section is Class 1.

Flexural shear:

$$V_{pl,Rd} = \frac{1}{\sqrt{3}} A_v \frac{f_y}{\gamma_a} = \frac{1}{\sqrt{3}} 13100 \frac{355}{1,0} \times 10^{-3} = 2685 \, kN$$

$V_{Ed} = 881 \, kN \, (<0,5 V_{pl,Rd})$, thus there is no reduction of moment capacity.

Shear connector design:

As the margin between $M_{pl,Rd}$ and M_{Ed} is small $(M_{pl,Rd}/M_{Ed} = 3876/3817 = 1,015)$, there will be full shear connection.

Use 100 mm long by 19 mm diameter stud connectors.

Load capacity of a single stud:

$h/d = 100/19 = 5,3$, so from Eq. (10.34) $\alpha = 1,0$

With $f_{ck} = 25 \, MPa$ (as in EXAMPLE 10.2), $E_{cm} = 23,5 \, GPa$.

From Eq. (10.32)

$$P_{Rd} = \frac{0,29\alpha d^2 \sqrt{f_{ck} E_{cm}}}{\gamma_v} = \frac{0,29 \times 1 \times 19^2 \times \sqrt{25 \times 23,5 \times 10^3}}{1,25} \times 10^{-3} = 64,2 \, kN$$

Determine the reduction factor, k_1 from Eq. (10.36):

With $b_0 = 150 - 38 = 112 \, mm$; $h_p = 55 \, mm$; $h = 100 \, mm$,

$$k_1 = 0,6 \frac{b_0}{h_p} \left(\frac{h_{sc}}{h_p} - 1 \right) = 0,6 \frac{112}{55} \left(\frac{100}{55} - 1 \right) = 1,0$$

Limiting capacity:

Limiting capacity of studs from Eq. (10.35):

For ribs parallel to beam $f_u = 500 \, MPa$,

$$P_{Rd} = \frac{0,8 f_u}{\gamma_v} \frac{\pi d^2}{4} = \frac{0,8 \times 500}{1,25} \frac{\pi \times 19^2}{4} \times 10^{-3} = 90,8 \, kN$$

Shear stud capacity is the lower of the two values, that is, 64,2 kN.

Number of connectors for full shear connection:

$n = N_c/P_{Rd} = 4,463 \times 10^3/64,2 = 70$ per half beam.

Use studs in pairs transversely, so $s_L = 6000/(70/2) = 171 \, mm$.

Use a spacing of 150 mm.

Transverse Shear:

The only plane requiring checking is 4–4 (or 5–5) of Fig. 10.8

Determine V_{Sd} from Eq. (10.44),

$$V_{Sd} = \frac{n_r P_{Rd}}{s_L} = \frac{2 \times 64.2}{0.150} = 856\,\text{kN/m}$$

Assume the sheeting rib is along the centre line of the beam, v_{Ed} is determined using the total thickness of the concrete. Also the shear may be equally divided between the two shear planes, thus v_{Ed} is given by $v_{Ed} = 0.5 \times 856/105 = 4.08\,\text{MPa}$.

This must be resisted entirely by reinforcement as the ribs of the sheeting are running parallel with the beam.

Using a 45° angle for the truss analogy, $\cos\theta_f = \sin\theta_f = 1/\sqrt{2}$ and $\cot\theta_f = 1.0$.

As the concrete is the same as EXAMPLE 10.2, $v_{Ed} = 3.44\,\text{MPa}$, and the maximum value is again 4,5 MPa.

Use Eq. (10.49) to determine the requirement for $A_{st}s_f$:

$$\frac{A_{sf}}{s_f} > \frac{1}{f_{yd}}\frac{v_{Ed}h_f}{\cot\theta_f} = \frac{1.15}{500}\frac{4.08 \times 105}{1.0} = 0.985\,\text{mm}$$

This can be divided between top and bottom, so A_{st}/S_r on each face is 0,493 mm²/mm.

Use B503 Mesh [503 mm²/m]

Deflection:

The loading for determining variable and total deflections is given in Fig. 10.11.

Use the formula $\delta = (WL_3/48EI)(3(a/L) - 4(a/L)^3)$

To simplify calculations determine $L^3/48EI$ as a constant for the composite and steel sections, and then determine $W(3(a/l) - 4(a/L)^3)$ for each load.

Determination of I_c:

Neglect effect of profiles and take $h_p = 0$, $b_{eff} = 3000\,\text{mm}$ and $\alpha_e = 17.9$ (as EXAMPLE 10.2).

Summarizing results: $x = 396.3\,\text{mm}$, $I_c = 0.162 \times 10^{12}\,\text{mm}^4$.

$$E_{c,eff}I_c = 11.75 \times 10^6 \times 0.162 = 1.904 \times 10^6\,\text{kNm}^2$$
$$L^3/48E_{c,eff}I_c = 12^3/48 \times 1.904 \times 10^6 = 18.9 \times 10^{-6}\,\text{m/kN}$$

Loads at A1:

$$a/L = 2.5/12 = 0.208; 3(a/L) - 4(a/L)^3 = 0.589$$

Total load:

$$W = 94 + 70 = 164\,\text{kN},$$
$$\delta = 164 \times 0.589 \times 18.9 \times 10^{-6} = 0.0018\,\text{m}$$

Loads at B:

$$a/L = 5/12 = 0,417; 3(a/L) - 4(a/L)^3 = 0,961$$

Total load:

$$W = 214 + 240 = 454 \text{ kN},$$

$$\delta = 454 \times 0,961 \times 18,9 \times 10^{-6} = 0,0082 \text{ m}$$

Final deflection under total loads:

$$\delta = 2(0,0018 + 0,0082) = 0,0202 \text{ m}$$

Span/deflection ratio is $12/0,0202 = 594$. This is acceptable.

Vibration:

Total variable deflection $= 10,2 \times 10^{-3}$ m

Total permanent deflection $= 9,87 \times 10^{-3}$ m

Variable deflection due to a value of $\psi_2 = 0,3$:

$$0,3 \times 10,2 \times 10^{-3} = 0,0031 \text{ m}$$

$$\delta_{\text{sw}} = 0,0031 + 0,0102 = 0,0133 \text{ m} = 13,3 \text{ mm}$$

From Eq. (10.52), $f = 18/\sqrt{13,3} = 4,94$ Hz

This is higher than the minimum recommended value of 3 Hz.

If the total variable load is taken then $f = 4,0$ Hz, which is still acceptable.

10.4 COMPOSITE COLUMNS

These can take a variety of forms but fall essentially into two categories; partially or totally encased Universal Columns (or H sections) and filled rolled hollow sections, with or without additional reinforcement. Typical configurations are given in Fig. 10.13. This text only considers composite columns which are symmetric about both axes.

The methods given in EC 1994-1-1 only hold if

(a) the steel contribution ratio δ defined as

$$\delta = \frac{A_a f_{yd}}{N_{pl,Rd}} \tag{10.53}$$

satisfies the limits $0,2 \leq \delta \leq 0,9$. For $\delta < 0,2$ the column should be designed as reinforced concrete, and for $\delta > 0,9$ designed as non-composite steel.

(b) the normalized slenderness ratio $\bar{\lambda}$ is less than 2,0.

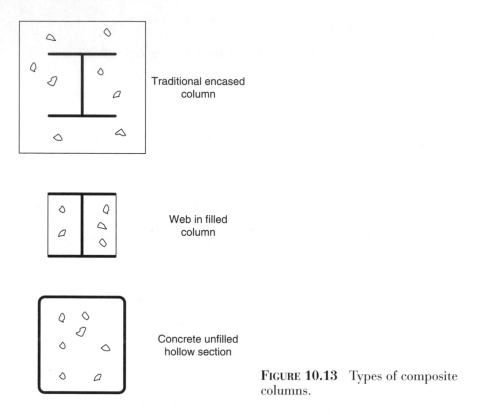

Traditional encased column

Web in filled column

Concrete unfilled hollow section

FIGURE 10.13 Types of composite columns.

10.4.1 Axial Compression

The design condition for axial compression is that the design resistance $\chi N_{pl,Rd}$ should exceed the applied load N_{Ed}. The buckling co-efficient χ is determined using a non-dimensionalized slenderness ratio $\bar{\lambda}$ (defined in Eq. (10.73)) in combination with buckling curve 'a' for concrete filled hollow sections with $\rho_s \leq 3\%$ or 'b' $3\% < \rho_s < 6\%$ (where ρ_s is the percentage of reinforcement), 'b' for partially or fully encased I sections bending about the major axis, and 'c' for partially or fully encased sections bending about the minor axis (the buckling curve designations are those used in EN 1993-1-1).

10.4.2 Uniaxial Bending and Axial Compression

Initially, the resistance of the cross-section is determined using an interaction diagram between the axial load resistance and bending moment resistance in a similar fashion to reinforced concrete columns. The diagram is shown schematically in Fig. 10.14, where $N_{pl,Rd}$ is the axial squash capacity (Point A), and $M_{pl,Rd}$ is the plastic moment capacity (Point B). Although the interaction diagram which is of a similar shape to that for reinforced concrete columns is strictly curved, it may be approximated to a series of straight lines. Point C is established by the application of the moment $M_{pl,Rd}$ and a resultant axial capacity of the concrete alone $N_{pm,Rd}$, Point D by the moment capacity $M_{max,Rd}$ under $0{,}5N_{pm,Rd}$.

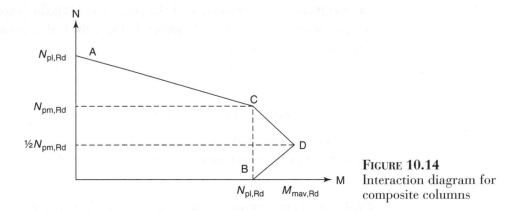

FIGURE 10.14
Interaction diagram for
composite columns

The moment capacity $\mu_{dy}M_{pl,y,Rd}$ (or $M_{pl,N,Rd}$) is determined from the interaction diagram under an axial load of N_{Ed}. The member is deemed to have sufficient capacity when

$$\frac{M_{Ed}}{M_{pl,N,Rd}} = \frac{M_{Ed}}{\mu_d M_{pl,Rd}} = \alpha_M \tag{10.54}$$

The co-efficient α_M takes a value of 0,9 for steel grades of S235 and S355 and 0,8 for grades S420 and S460. The factor is partially needed to compensate for the assumption over the depth of the rectangular stress block between EN 1992-1-1 where it is taken as extending over $0,8x$ but over x in EN 1994-1-1, since the same stress level of $0,85 f_{ck}/\gamma_c$ is used in both, and partially to allow for the adverse effect of the higher yield strain on the crushing on the concrete.

Should the bending moment be entirely due to the eccentricity of the axial load, then μ_d can be greater than unity.

10.4.3 Biaxial Bending

The procedure is similar to uniaxial bending except that two parameters μ_{dz} and μ_{dy} now need to be determined. However, the effect of imperfections needs only to be considered for the likely failure axis (usually the minor or zz axis). The column is then satisfactory if the following conditions are satisfied,

$$\frac{M_{y,Ed}}{\mu_{dy}M_{pl,y,Rd}} \leq \alpha_{M,y} \tag{10.55}$$

$$\frac{M_{z,Ed}}{\mu_{dz}M_{pl,z,Rd}} \leq \alpha_{M,z} \tag{10.56}$$

and

$$\frac{M_{y,Ed}}{\mu_{dy}M_{pl,y,Rd}} + \frac{M_{z,Ed}}{\mu_{dz}M_{pl,z,Rd}} \leq 1,0 \tag{10.57}$$

The co-efficients $\alpha_{M,y}$ and $\alpha_{M,z}$ are to be taken as α_M. The increase of the interaction co-efficient to 1 in the combined equation is due to the higher crushing strength of the concrete under biaxial bending.

10.4.4 Determination of Member Capacities

The formulae for the flexural capacity of concrete filled rectangular tubes (with reinforcing steel ignored) are taken from Johnson and Anderson (2004).

10.4.4.1 Axial Squash Capacity, $N_{pl,Rd}$ [Point A]

This is given by the sum of the individual components due to the steel section, the concrete and the reinforcement (Fig. 10.15(a)). So in general $N_{pl,Rd}$ is given by

$$N_{pl,Rd} = A_a \frac{f_y}{\gamma_a} + A_c \frac{0{,}85 f_{ck}}{\gamma_c} + A_s \frac{f_{sk}}{\gamma_s} \tag{10.58}$$

where A_a is the area of the steel section and f_y is the yield strength, A_c is the concrete area and f_{ck} is the characteristic cylinder strength, A_s is the area of the reinforcement and f_{sk} is the characteristic strength.

For concrete filled hollow sections the full cylinder strength f_{ck} may be used owing to the containment of the concrete.

For concrete filled circular tubes $N_{pl,Rd}$ is modified to take account of the triaxial stresses in the concrete due to its containment, and of the reduction in the allowable strength of the steel cross-section owing to the induced hoop tension from the concrete triaxial stresses. The modified value of $N_{pl,Rd}$ is subject to two conditions:

(1) $\bar{\lambda} < 0{,}5$ and
(2) $e = M_{Ed}/N_{Ed} \leq d/10$ (where d is the diameter)

The equation for $N_{pl,Rd}$ becomes

$$N_{pl,Rd} = A_a \eta_a \frac{f_y}{\gamma_a} + A_c \left[1 + \eta_c \frac{t}{d} \frac{f_y}{f_{ck}}\right] \frac{f_{ck}}{\gamma_c} + A_s \frac{f_{sk}}{\gamma_s} \tag{10.59}$$

where t is the thickness of the tube and η_a and η_c are co-efficients determined as follows,

$$\eta_c = \eta_{c0}\left(1 - 10\frac{e}{d}\right) \tag{10.60}$$

where η_{c0} is given by

$$\eta_{c0} = 4{,}9 - 18{,}5\bar{\lambda} + 17(\bar{\lambda})^2 \geq 0 \tag{10.61}$$

and

$$\eta_a = \eta_{a0} + (1 - \eta_{a0})10\frac{e}{d} \tag{10.62}$$

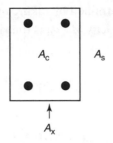

(a) Point A

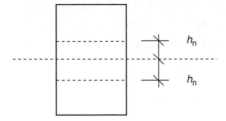

(b) Point B and C

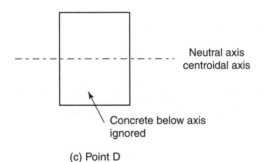

(c) Point D

FIGURE 10.15 Calculation of section capacities

where η_{a0} is given by

$$\eta_{a0} = 0{,}25(3 + 2\bar{\lambda}) \le 1{,}0 \tag{10.63}$$

For $e > d/10$, $\eta_c = 0$ and $\eta_a = 1{,}0$.

10.4.4.2 Calculation of $M_{pl,Rd}$ [Point B]

For the typical section given in Fig. 10.15(b), the forces in the flanges and the reinforcement cancel out, and thus the tension force in the web must balance the compression block in the concrete. It is therefore straightforward to determine the height of the web h_n above the centroidal axis. Equating compressive and tensile forces gives

$$h_n = \frac{N_{pm,Rd}}{2b\frac{f_{ck}}{\gamma_c} + 4t_w\left(2\frac{f_y}{\gamma_a} - \frac{f_{ck}}{\gamma_c}\right)} \tag{10.64}$$

where $N_{pm,Rd}$ is given by Eq. (10.67).

The plastic moment capacity is determined by taking moments about the centroidal axis of the section. The moment capacity $M_{pl,Rd}$ is thus given by

$$M_{pl,Rd} = M_{max,Rd} - M_{n,Rd} \tag{10.65}$$

The values of $M_{max,Rd}$ and $M_{n,Rd}$ are given by Eqs (10.71) and (10.68), respectively.

10.4.4.3 Determination of $N_{pm,Rd}$ [Point C]

This can be done by noting that the neutral axis shifts from h_n above the centroidal axis to the same distance below it (Fig 10.15(b)). $N_{pm,Rd}$ is determined from horizontal force equilibrium recognizing that the forces in the steel flanges and reinforcement cancel out, that is, the axial load is carried by the concrete alone.

The area of the concrete A_c is given by

$$A_c = (b - 2t)(h - 2t) - (4 - \pi)r^2 \tag{10.66}$$

where r is the corner radius taken equal to t, and $N_{pm,Rd}$ is given by

$$N_{pm,Rd} = A_c \frac{f_{ck}}{\gamma_c} \tag{10.67}$$

The moment capacity $M_{n,Rd}$ is given by

$$M_{n,Rd} = W_{p,a,n}\frac{f_y}{\gamma_a} + 0,5W_{p,c,n}\frac{f_{ck}}{\gamma_c} \tag{10.68}$$

where $W_{p,a,n}$ and $W_{p,c,n}$ are the plastic section moduli for the portions of the steel tube and concrete contained within $\pm h_n$, and are given by

$$W_{p,c,n} = (b - 2t)h_n^2 \tag{10.69}$$

and

$$W_{p,a,n} = bh_n^2 - W_{p,c,n} \tag{10.70}$$

10.4.4.4 Determination of $M_{max,Rd}$ [Point D]

For this case when an axial force of $0,5N_{pm,Rd}$ acts the neutral axis coincides with the centroidal axis (Fig. 10.15(c)), and thus $M_{max,Rd}$ is simply given by the sum of the plastic moment capacities of the reinforcement, the steel section and the concrete above the centroidal axis. $M_{max,Rd}$ is given by

$$M_{max,Rd} = W_{pa}\frac{f_y}{\gamma_a} + 0,5W_{pc}\frac{f_{ck}}{\gamma_c} \tag{10.71}$$

where W_{pa} is the plastic section modulus for the steel section (taken from tables) and W_{pc} is calculated from

$$W_{pc} = \frac{(b - 2t)(h - 2t)^2}{4} - \frac{2}{3}r^3 - (4 - \pi)(0,5h - t - r)r^2 \tag{10.72}$$

Note, the equations have been derived for bending about the major (yy) axis. For minor axis (zz) bending, h and b are simply interchanged.

10.4.5 Buckling

The non-dimensionalized slenderness ratio $\bar{\lambda}$ is defined by

$$\bar{\lambda} = \sqrt{\frac{N_{\text{pl,Rk}}}{N_{\text{cr}}}} \tag{10.73}$$

where $N_{\text{pl,Rk}}$ is the characteristic plastic axial load capacity from Eq. (10.58) or Eq. (10.59) with all the materials' partial safety factors set equal to unity and the effective buckling load N_{cr} is given by

$$N_{\text{cr}} = \frac{\pi^2 (EI)_{\text{eff}}}{l^2} \tag{10.74}$$

where l is the buckling length with the effective flexural stiffness is given by

$$(EI)_{\text{eff}} = E_a I_a + K_e E_{\text{cm}} I_c + E_s I_s \tag{10.75}$$

where $E_a I_a$ is the flexural rigidity of the steel section alone, $E_{\text{cm}} I_c$ is the flexural rigidity of the concrete, and $E_s I_s$ is the flexural rigidity of the reinforcement and K_e is a correction factor taken as 0,6. E_{cm} is given in Table 3.1 of EN 1992-1-1.

Where appropriate, account should be taken of the influence of long-term loading by using an effective concrete modulus $E_{\text{c,eff}}$ determined from

$$E_{\text{c,eff}} = \frac{E_{\text{cm}}}{1 + \phi_t \dfrac{N_{\text{G,Ed}}}{N_{\text{Ed}}}} \tag{10.76}$$

where N_{Ed} is the total design normal force and $N_{\text{G,Ed}}$ is that portion which is permanent and φ_t is the creep co-efficient determined from EN 1992-1-1.

10.4.6 Design Moments

Second order effects within the column length may be allowed for by increasing the larger design bending moment determined from a first order analysis by a factor k given by

$$k = \frac{\beta}{1 - \dfrac{N_{\text{Ed}}}{N_{\text{cr,eff}}}} \geq 1,0 \tag{10.77}$$

where the moment ratio factor β for end moments is given by

$$\beta = 0,66 + 0,44r \geq 0,44 \tag{10.78}$$

and $N_{\text{cr,eff}}$ is an effective buckling load determined using the actual column length and an effective stiffness $(EI)_{\text{eff,II}}$ given by

$$(EI)_{\text{eff,II}} = K_0 (E_a I_a + K_{\text{e,II}} E_{\text{cm}} I_c + E_s I_s) \tag{10.79}$$

where K_0 is a calibration factor ($=0,9$) and $K_{e,II}$ is a correction factor representing the effects of cracking in the concrete ($=0,5$).

Johnson (2004) indicates that where $N_{cr,eff} \geq 10N_{Ed}$, then the second order effects need considering and there is an additional moment induced by the additional imperfection.

Thus the design moment M_{Ed} is given by

$$M_{Ed} = k_{end}M + k_{imp}N_{Ed}e_0 \tag{10.80}$$

where k_{end} is magnification factor due to the moment gradient, M is the larger end moment, e_0 is the eccentricity due to imperfections, k_{imp} is the magnification factor determined from Eq. (10.78) with $\beta = 1,0$.

10.4.7 Other Checks and Detailing

10.4.7.1 Local Buckling

For circular hollow sections: $d/t \leq 90\varepsilon^2$;

Rectangular hollow sections: $h/t \leq 52\varepsilon$;

Partially encased sections: $b/t_f \leq 44\varepsilon$.

10.4.7.2 Cover

For reinforcement, this is governed by the requirements in EN 1992-1-1. For encased steel sections, this should be a maximum of 40 mm or $b/6$, where b is the flange width.

10.4.7.3 Shear

The shear bond between the steel section and the concrete should be checked using an elastic distribution of forces on the uncracked section with a transmission length not exceeding twice the relevant transverse direction. The values of shear bond should not exceed 0,3 MPa for fully encased sections, 0,55 MPa for circular concrete filled sections, 0,40 MPa for circular concrete filled sections and 0,2 MPa for the flanges only in partially encased sections. Where necessary shear studs should be used on encased I sections to resist shear.

10.4.7.4 Fire

For concrete filled hollow sections it is essential that two vent holes of 20 mm diameter should be drilled through the steel section at the top and bottom of each storey subject to a maximum spacing of 5 m (Newman and Simms, 2000). These holes must not be within the depth of the floor construction. The purpose of these holes is to allow the build up of water vapour to escape whilst the moisture within the concrete is driven off in the early stages of heating in the fire.

It is clear that the method of designing composite columns under uniaxial or biaxial bending is complex, and leads itself readily to the use of spreadsheets or design charts. The analysis of given sections to determine their carrying capacity is much more straightforward. The first example illustrates the determination of axial carrying capacity. To avoid duplication of calculations the second and third both use the same section, one under uniaxial bending about the major axis, and the other with biaxial bending. In each case the loading is considered totally short term, that is, $E_{c,eff}$ is taken as E_{cm}. Also any reinforcement is considered negligible and is neglected.

EXAMPLE 10.4   Determination of axial load capacity of a composite column.

Determine the axial load carrying capacity of 4 m effective length $150 \times 150 \times 8$ Grade S355 rolled hollow section filled with Grade C25/30 concrete.

Check h/t:

Actual:

$$\frac{h}{t} = \frac{150}{8} = 18{,}75$$

Allowable:

$52\varepsilon = 52 \times (235/355)^{1/2} = 42{,}3$, therefore satisfactory

Determination of $N_{pl,Rd}$:

Use Eq. (10.58) with the concrete taken at full strength,

$$N_{pl,Rd} = A_a \frac{f_y}{\gamma_a} + A_c \frac{0{,}85 f_{ck}}{\gamma_c} + A_s \frac{f_{sk}}{\gamma_s} = 4510 \frac{355}{1{,}0} + \left[(150 - 16)^2 - (4 - \pi)8^2 \right] \frac{25}{1{,}5}$$

$$= 1{,}899 \text{ MN}$$

Determine the load contribution ratio δ from Eq. (10.53):

$$\delta = \frac{A_a \frac{f_y}{\gamma_a}}{N_{pl,Rd}} = \frac{4510 \frac{355}{1{,}0} \times 10^{-6}}{1{,}899} = 0{,}84$$

As δ lies between 0,2 and 0,9, the column may be designed as composite.

Use Eq. (10.75) to determine the effective stiffness $(E_s I_s = 0)$

$$E_{cm} = 22 \left(\frac{f_{ck} + 8}{10} \right)^{1/3} = 22 \left(\frac{25 + 8}{10} \right)^{1/3} = 31{,}5 \text{ GPa}$$

$$(EI)_{eff} = E_a I_a + 0{,}6 E_{cm} I_c = 210 \times 10^6 \times 1510 \times 10^{-8} + 0{,}6 \times 31{,}5 \times 10^6$$

$$\times \frac{(150 - 16)^4}{12} = 3678 \text{ kNm}^2$$

Determine the Euler critical load N_{cr} from Eq. (10.74)

$$N_{cr} = \frac{\pi^2 (EI)_{eff}}{l^2} = \frac{\pi^2 \times 3678}{4^2} = 2269 \text{ kN}$$

Determine $N_{\text{pl,Rk}}$ using Eq. (10.58) with the materials' partial safety factors set equal to 1,

$$N_{\text{pl,Rk}} = A_a \frac{f_y}{\gamma_a} + A_c \frac{0{,}85 f_{ck}}{\gamma_c} + A_s \frac{f_{sk}}{\gamma_s} = 4510 \frac{355}{1{,}0} + \left[(150-16)^2 - (4-\pi)8^2\right] \frac{25}{1{,}0}$$

$$= 2{,}048 \text{ MN}$$

Determine the normalized slenderness ratio $\bar{\lambda}$ from Eq. (10.73),

$$\bar{\lambda} = \sqrt{\frac{N_{\text{pl,Rk}}}{N_{cr}}} = \sqrt{\frac{2048}{2269}} = 0{,}95$$

This is less than the critical value of 2,0.

The strength reduction factor is determined using buckling curve 'a' with $\alpha = 0{,}21$,

$$\phi = 0{,}5(1 + \alpha(\bar{\lambda} - 0{,}2) + (\bar{\lambda})^2) = 0{,}5(1 + 0{,}21(0{,}95 - 0{,}2) + 0{,}95^2) = 1{,}03$$

$$\chi = \frac{1}{\phi + \sqrt{\phi^2 - (\bar{\lambda})^2}} = \frac{1}{1{,}03 + \sqrt{1{,}03^2 - 0{,}95^2}} = 0{,}70$$

Determine the axial capacity N_{Rd}:

$$N_{Rd} = \chi N_{\text{pl,Rd}} = 0{,}70 \times 1899 = 1329 \text{ kN}$$

EXAMPLE 10.5 Axial load and uniaxial bending about the major axis.

Determine whether a column having a system length of 4 m fabricated from $150 \times 100 \times 8$ Grade S355 RHS filled with Grade C25/30 normal weight concrete can carry an axial load at ULS of 400 kN and a moment at ULS about the major axis of 18 kNm.

$$A_a = 3710 \text{ mm}^2; \quad W_{\text{pl,y}} = 183 \times 10^3 \text{ mm}^3.$$

Determine W_{pc} from Eq. (10.72)

$$W_{pc} = \frac{(b-2t)(h-2t)^2}{4} - \frac{2}{3}r^3 - (4-\pi)(0{,}5h - t - r)r^2$$

$$= \frac{(100 - 2\times 8)(150 - 2\times 8)^2}{4} - \frac{2}{3}8^3 - (4-\pi)\left(\frac{150}{2} - 8 - 8\right)$$

$$= 373500 \text{ mm}^3$$

The area of the concrete A_c is given by Eq. (10.66)

$$A_c = (b-2t)(h-2t) - (4-\pi)r^2 = (150 - 2\times 8)(100 - 2\times 8) - (4-\pi)8^2$$

$$= 11200 \text{ mm}^2$$

Axial squash capacity $N_{\text{pl,Rd}}$ is given by Eq. (10.58):

$$N_{\text{pl,Rd}} = A_a \frac{f_y}{\gamma_a} + A_c \frac{0{,}85 f_{ck}}{\gamma_c} + A_s \frac{f_{sk}}{\gamma_s} = 3710 \frac{355}{1{,}0} + 11200 \frac{25}{1{,}5} = 1504 \text{ kN}$$

Determine the contribution ratio, δ from Eq. (10.53):

$$\delta = \frac{A_s f_{yd}}{N_{pl,Rd}} = \frac{3710\frac{355}{1,0} \times 10^{-3}}{1504} = 0,876$$

This is within the limits for design as a composite column.

Maximum moment capacity $M_{max,Rd}$ from Eq. (10.71):

$$M_{max,Rd} = W_{pa}\frac{f_y}{\gamma_a} + 0,5W_{pc}\frac{f_{ck}}{\gamma_c} = 183000\frac{355}{1,0} \times 10^{-6} + 0,5 \times 373500\frac{25}{1,5} \times 10^{-6}$$
$$= 68,1\,\text{kNm}$$

Determine $N_{pm,Rd}$ from Eq. (10.67):

$$N_{pm,Rd} = A_c\frac{f_{ck}}{\gamma_c} = 11200\frac{25}{1,5} \times 10^{-3} = 187\,\text{kN}$$

$$\frac{N_{pm,Rd}}{2} = 93,5\,\text{kN}$$

Determine h_n from Eq. (10.64) determine h_n:

$$h_n = \frac{N_{pm,Rd}}{2b\frac{f_{ck}}{\gamma_c} + 4t_w\left(2\frac{f_y}{\gamma_a} - \frac{f_{ck}}{\gamma_c}\right)} = \frac{18700}{2 \times 100 \times \frac{25}{1,5} + 4 \times 8\left(2\frac{355}{1,0} - \frac{25}{1,5}\right)} = 7,33\,\text{mm}$$

Determine $W_{p,c,n}$ from Eq. (10.69):

$$W_{p,c,n} = (b - 2t)h_n^2 = (100 - 2 \times 8)7,33^2 = 4513\,\text{mm}^3$$

Determine $W_{p,a,n}$ from Eq. (10.70):

$$W_{p,a,n} = bh_n^2 - W_{p,c,n} = 100 \times 7,33^2 - 4513 = 860\,\text{mm}^3$$

Determine $M_{n,Rd}$ from Eq. (10.68):

$$M_{n,Rd} = W_{p,a,n}\frac{f_y}{\gamma_a} + 0,5W_{p,c,n}\frac{f_{ck}}{\gamma_c} = 860\frac{355}{1,0} \times 10^{-6} + 0,5 \times 4513\frac{25}{1,5} \times 10^{-6}$$
$$= 0,3\,\text{kNm}$$

Determine $M_{pl,Rd}$ from Eq. (10.65):

$$M_{pl,Rd} = M_{max,Rd} - M_{n,Rd} = 68,1 - 0,3 = 67,8\,\text{kNm}$$

The values required to plot the interaction diagram are given in Table 10.3.

TABLE 10.3   Values required for major axis interaction diagram for EXAMPLE 10.5.

Point		Moment capacity (kNm)	Axial capacity (kN)
A	$(0, N_{pl,Rd})$	0	1504
B	$(M_{pl,Rd}, 0)$	67,8	0
C	$(M_{pl,Rd}, N_{pm,Rd})$	67,8	187
D	$(M_{max,Rd}, 0,5N_{pm,Rd})$	69,1	93,5

These values are plotted in Fig. 10.15.

Determine the resistance to axial buckling about the major axis:

$$E_a I_a = 210 \times 10^6 \times 1106 \times 10^{-8} = 2323 \, \text{kNm}^2$$

$$E_{cm} = 22 \left(\frac{(f_{ck} + 8)}{10} \right)^{1/3} = 31,5 \, \text{GPa}$$

$$I_c = \frac{(0,150 - 0,016)^3 (0,100 - 0,016)}{12} = 16,84 \times 10^{-6} \, \text{m}^4$$

$$E_{cd} I_c = 31,5 \times 10^6 \times 16,84 \times 10^{-6} = 530 \, \text{kNm}^2$$

Determine $(EI)_{eff,II}$ from Eq. (10.79):

$$(EI)_{eff,II} = 0,9(E_a I_a + 0,5 E_{cm} I_c) = 0,9(2323 + 0,5 \times 530) = 2329 \, \text{kNm}^2$$

Determine $N_{cr,eff}$ from Eq. (10.74) with $(EI)_{eff}$ replaced by $(EI)_{eff,II}$:

$$N_{cr,eff} = \frac{\pi^2 (EI)_{eff,II}}{l^2} = \frac{\pi^2 \times 2329}{4^2} = 1437 \, \text{kN}$$

$N_{Ed} = 400 \, \text{kN} > N_{cr,eff}/10$, therefore second order effects need to be considered.

$$N_{pl,Rk} = (3710 \times 355 + 11200 \times 25) \times 10^{-3} = 1597 \, \text{kN}.$$

Determine $\bar{\lambda}$ from Eq. (10.73):

$$\bar{\lambda} = \sqrt{\frac{N_{pl,Rk}}{N_{cr,eff}}} = \sqrt{\frac{1597}{1437}} = 1,054 > 2,0$$

Thus the column satisfies the limits for composite design.

Second order effects:

(a) Within the column length:
Assuming the column is in single curvature, $r = 1,0$, so $\beta = 1,0$.

$$k_{end} = \frac{\beta}{1 - \frac{N_{Ed}}{N_{cr,eff}}} = \frac{1,0}{1 - \frac{400}{1437}} = 1,386$$

(b) Due to initial bow:

From Table 6.3 of EN 1994-1-1, for a infilled hollow section, $e_0 = L/300 = 4/300$

As $\beta = 1,0$ for initial bow, $k_{imp} = 1,386$

The design moment M_{Ed} is given by Eq. (10.80):

$$M_{Ed} = k_{end} M + k_{imp} N_{Ed} e_0 = 1,386 \times 18 + 1,386 \times 400 \frac{4}{300} = 32,3 \, \text{kNm}$$

From the interaction diagram in Fig. 10.15, the moment M_{400} corresponding to an axial load of 400 kN is given by

$$M_{400} = M_{pl,Rd}\frac{N_{pl,Rd} - N_{Ed}}{N_{pl,Rd} - N_{pm,Rd}} = 67,8\frac{1504 - 400}{1504 - 187} = 56,5\,\text{kNm}$$

$$\mu_d = \frac{M_{400}}{M_{pl,Rd}} = \frac{56,5}{67,4} = 0,833$$

$$\frac{M_{Ed}}{\mu_d M_{pl,Rd}} = \frac{M_{Ed}}{M_{400}} = \frac{32,3}{56,5} = 0,572 < 0,9$$

The limiting value of α_M is 0,9 as Grade S355 steel is being used. Thus the column is therefore satisfactory.

EXAMPLE 10.6   Axial load and biaxial bending.

Determine whether a column having a system length of 4 m fabricated from $150 \times 100 \times 8$ Grade S355 RHS filled with Grade C25/30 normal weight concrete can carry an axial load at ULS of 350 kN and a moment at ULS about the major axis of 15 and 10 kNm about the minor axis.

Two interaction diagrams are required, one for each axis.

For the major axis the interaction diagram is as EXAMPLE 10.5. For the minor axis, the formulae in Eqs (10.64)–(10.73) are used but with b and h interchanged.

$$A_a = 3710\,\text{mm}^2;\ W_{pl,y} = 183 \times 10^3\,\text{mm}^3;\ W_{pl,z} = 133 \times 10^3\,\text{mm}^3.$$

Determine W_{pc} from Eq. (10.72):

$$W_{pc} = \frac{(h - 2t)(b - 2t)^2}{4} - \frac{2}{3}r^3 - (4 - \pi)(0,5b - t - r)r^2$$

$$= \frac{(150 - 2\times 8)(100 - 2\times 8)^2}{4} - \frac{2}{3}8^3 - (4 - \pi)\left(\frac{100}{2} - 8 - 8\right) = 234200\,\text{mm}^3$$

From EXAMPLE 10.5, $A_c = 11200\,\text{mm}^2$, $N_{pl,Rd} = 1504\,\text{kN}$ and $\delta = 0,876$.

Determine the maximum moment capacity $M_{max,Rd}$ from Eq. (10.71):

$$M_{max,Rd} = W_{pa}\frac{f_y}{\gamma_a} + 0,5W_{pc}\frac{f_{ck}}{\gamma_c} = 133000\frac{355}{1,0} \times 10^{-6} + 0,5 \times 234200\frac{25}{1,5} \times 10^{-6}$$

$$= 49,2\,\text{kNm}$$

From EXAMPLE 10.5, $N_{pm,Rd} = 187\,\text{kN}$, and $0,5N_{pm,Rd} = 93,5\,\text{kN}$

Determine h_n from Eq. (10.64):

$$h_n = \frac{N_{pm,Rd}}{2h\frac{f_{ck}}{\gamma_c} + 4t_w\left(2\frac{f_y}{\gamma_a} - \frac{f_{ck}}{\gamma_c}\right)} = \frac{187000}{2 \times 150\frac{25}{1,5} + 4 \times 8\left(2\frac{355}{1,0} - \frac{25}{1,5}\right)} = 6,41\,\text{mm}$$

Determine $W_{pc,n}$ from Eq. (10.69):

$$W_{pc,n} = (h - 2t)h_n^2 = (150 - 2 \times 8)6{,}41^2 = 5506\,\text{mm}^3$$

Determine $W_{pa,n}$ from Eq. (10.70):

$$W_{pa,n} = hh_n^2 - W_{pc,n} = 150 \times 6{,}41^2 - 5506 = 657\,\text{mm}^3$$

Determine $M_{n,Rd}$ from Eq. (10.68):

$$M_{n,Rd} = W_{pa,n}\frac{f_y}{\gamma_a} + 0{,}5W_{pc,n}\frac{f_{ck}}{\gamma_c} = 657\frac{355}{1{,}0} \times 10^{-6} + 0{,}5 \times 5506\frac{25}{1{,}5} \times 10^{-6}$$
$$= 0{,}3\,\text{kNm}$$

Determine $M_{pl,Rd}$ from Eq. (10.65):

$$M_{pl,Rd} = M_{max,Rd} - M_{n,Rd} = 49{,}2 - 0{,}3 = 48{,}9\,\text{kNm}$$

The values required to plot the interaction diagram are given in Table 10.4.

These values are plotted in Fig. 10.16.

TABLE 10.4 Values required for minor axis interaction diagram for EXAMPLE 10.6.

Point		Moment capacity (kNm)	Axial capacity (kN)
A	$(0, N_{pl,Rd})$	0	1504
B	$(M_{pl,Rd}, 0)$	48,9	0
C	$(M_{pl,Rd}, N_{pm,Rd})$	48,9	187
D	$(M_{max,Rd}, 0{,}5N_{pm,Rd})$	49,12	93,5

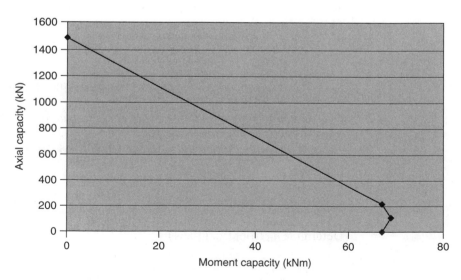

FIGURE 10.16 Major axis M–N interaction diagram (EXAMPLE 10.5)

Determine the resistance to axial buckling about the minor axis:

$$E_a I_a = 210 \times 10^6 \times 577 \times 10^{-8} = 1212 \, \text{kNm}^2$$

$E_{cm} = 31{,}5 \, \text{GPa (as EXAMPLE 10.5)}$

$$I_c = \frac{(0{,}100 - 0{,}016)^3 (0{,}150 - 0{,}016)}{12} = 7{,}11 \times 10^{-6} \, \text{m}^4$$

$$E_{cd} I_c = 31{,}5 \times 10^6 \times 7{,}11 \times 10^{-6} = 224 \, \text{kNm}^2$$

$$(EI)_{\text{eff,II}} = 0{,}9(E_a I_a + 0{,}5 E_{cm} I_c) = 0{,}9(1212 + 0{,}5 \times 224) = 1192 \, \text{kNm}^2$$

$$N_{\text{cr,eff}} = \frac{\pi^2 (EI)_{\text{eff,II}}}{l^2} = \frac{\pi^2 \times 1192}{4^2} = 735 \, \text{kN}$$

$N_{Ed} = 350 \, \text{kN} > N_{\text{cr,eff}}/10$, therefore second order effects need to be considered.

$$N_{\text{pl,Rk}} = A_a f_y + A_c f_{ck} = 3710 \times 355 \times 10^{-3} + 11200 \times 25 \times 10^{-3} = 1597 \, \text{kN}$$

$$\bar{\lambda} = \sqrt{\frac{N_{\text{pl,Rk}}}{N_{cr}}} = \sqrt{\frac{1597}{735}} = 1{,}474 < 2{,}0$$

Second order effects (major yy axis) using the appropriate critical buckling load:

(a) Within the column length:
Assuming the column is in single curvature, $r = 1{,}0$, so $\beta = 1{,}0$.

$$k_{\text{end,y}} = \frac{\beta}{1 - \dfrac{N_{Ed}}{N_{\text{cr,eff}}}} = \frac{1{,}0}{1 - \dfrac{350}{1437}} = 1{,}322$$

(b) Due to initial bow:

From Table 6.3 of EN 1994-1-1, for a infilled hollow section, $e_0 = L/300 = 4/300$

As $\beta = 1{,}0$ for initial bow, $k_{\text{imp,z}} = 1{,}322$.

Second order effects (minor zz axis):

(a) Within the column length:
Assuming the column is in single curvature, $r = 1{,}0$, so $\beta = 1{,}0$.

$$k_{\text{end,z}} = \frac{\beta}{1 - \dfrac{N_{Ed}}{N_{\text{cr,eff}}}} = \frac{1{,}0}{1 - \dfrac{350}{730}} = 1{,}921$$

(b) Due to initial bow:

From Table 6.3 of EN 1994-1-1, for a infilled hollow section, $e_0 = L/300 = 4/300$

As $\beta = 1{,}0$ for initial bow, $k_{\text{imp,z}} = 1{,}921$.

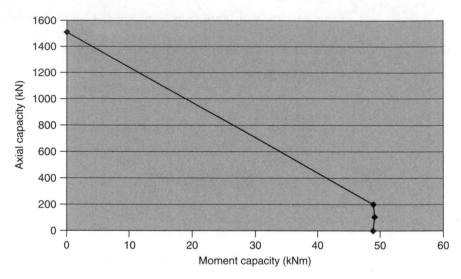

FIGURE 10.17 Minor axis M–N interaction diagram (EXAMPLE 10.6)

From the interaction diagram for the major axis in Fig. 10.16, the moment M_{350} corresponding to an axial load of 350 kN is given by

$$M_{350} = M_{pl,Rd} \frac{N_{pl,Rd} - N_{Ed}}{N_{pl,Rd} - N_{pm,Rd}} = 67,8 \frac{1504 - 350}{1504 - 187} = 59,4 \text{ kNm}$$

$$\mu_{dy} = \frac{M_{350}}{M_{pl,Rd}} = \frac{59,4}{67,4} = 0,881$$

From the interaction diagram in Fig. 10.17, for the minor axis the moment M_{350} corresponding to an axial load of 350 kN is given by

$$M_{350} = M_{pl,Rd} \frac{N_{pl,Rd} - N_{Ed}}{N_{pl,Rd} - N_{pm,Rd}} = 48,9 \frac{1504 - 350}{1504 - 187} = 42,8 \text{ kNm}$$

$$\mu_{dy} = \frac{M_{350}}{M_{pl,Rd}} = \frac{42,8}{48,9} = 0,875$$

The second order effect due the bow may only be applied on *one* axis. Use Eq. (10.80) to calculate the design moments.

Two cases, therefore, need considering:

(a) Bow on major axis.

$$M_{Ed,y} = k_{end,y} M_{Sd,y} + k_{imp,y} N_{Ed} e_0 = 1,322 \times 15 + 1,322 \times 350 \frac{4}{300} = 26,0 \text{ kNm}$$

$$M_{Ed,z} = k_{end,z} M_{Sd,z} = 1,921 \times 10 = 19,2 \text{ kNm}$$

$$\frac{M_{Ed,y}}{\mu_{dy}M_{pl,y,Rd}} = \frac{26,0}{59,4} = 0,438 < 0,9$$

$$\frac{M_{Ed,z}}{\mu_{dz}M_{pl,z,Rd}} = \frac{19,2}{42,8} = 0,45 < 0,9$$

$$\frac{M_{Ed,y}}{\mu_{dy}M_{pl,y,Rd}} + \frac{M_{Ed,z}}{\mu_{dz}M_{pl,z,Rd}} = \frac{26,0}{59,4} + \frac{19,2}{42,8} = 0,887 < 1,0$$

(b) Bow on minor axis.

$$M_{Ed,y} = k_{end,y}M_{Sd,y} = 1,322 \times 15 = 19,8\,\text{kNm}$$

$$M_{Ed,z} = k_{end,z}M_{Sd,z} + k_{imp,z}N_{Ed}e_0 = 1,921 \times 10 + 1,921 \times 350\frac{4}{300}$$

$$= 28,2\,\text{kNm}$$

$$\frac{M_{Ed,y}}{\mu_{dy}M_{pl,y,Rd}} = \frac{19,8}{59,4} = 0,333 < 0,9$$

$$\frac{M_{Ed,z}}{\mu_{sdz}M_{pl,z,Rd}} = \frac{28,2}{42,8} = 0,66 < 0,9$$

$$\frac{M_{Ed,y}}{\mu_{dy}M_{pl,y,Rd}} + \frac{M_{Ed,z}}{\mu_{dz}M_{pl,z,Rd}} = \frac{19,8}{59,4} + \frac{28,2}{42,8} = 0,992 < 1,0$$

It will be noted that the application of the initial bow to the minor axis is the critical case. This is due to both the lower moment capacity and the lower buckling load.

REFERENCES

Bunn, R. and Heywood, M. (2004) *Supporting services from structure*. Co-Construct (BSRIA).

EN 1991-1-1 *Eurocode 1: Actions on structures – Part 1–1: General actions – Densities, self-weight, imposed loads for buildings*. CEN/BSI.

EN 1991-1-6 *Eurocode 1: Actions on structures – Part 1–6: Actions during execution*. CEN/BSI.

EN 1992-1-1 *Eurocode 2: Design of concrete structures – Part 1–1: General rules and rules for buildings*. CEN/BSI.

EN 1993-1-1 *Eurocode 3: Design of steel structures – Part 1.1: General rules and rules for buildings*. CEN/BSI.

EN 1994-1-1 *Eurocode 4: Design of composite steel and concrete structures – Part 1.1: General rules and rules for buildings*. CEN/BSI.

Evans, H.R. and Wright, H.D. (1988). Steel–concrete composite flooring deck structures, In *Steel–concrete composite structures* (ed R. Narayanan). Elsevier.

Johnson, R.P. (2004). *Composite structures of steel and Concrete* (3rd edition). Blackwell Publishing.

Johnson, R.P. (2005). Shear connection in beams that support composite slabs – BS 5950 and EN 1994-1-1, *Structural Engineer*, **83(22)**, 21–24.

Johnson, R.P. and Anderson, D. (2004). *Designers' guide to EN 1994-1-1 Eurocode 4: Design of composite steel and concrete structures – Part 1.1: General rules and rules for buildings*. Thomas Telford.

Johnson, R.P. and May, I.M. (1975). Partial-interaction design of composite beams, *The Structural Engineer*, **53(8)**, 305–311.

Lawson, R.M. and Chung, K.F. (1994). *Composite beam design to eurocode 4*. Publication 121. SCI.

Lawson, R.M. and Nethercot, D.A. (1985). Lateral stability of I-beams restrained by profiled sheeting, *The Structural Engineer,* **63B(1)**, 1–7, 13.

Lawson, R.M., Mullett, D.L. and Rackham, J.W. (1997). *Design of asymmetric Slimflor® beams using deep composite decking.* Publication 175. SCI.

Martin, L.H. and Purkiss, J.A. (2006). *Concrete design to EN 1992.* Butterworth-Heinemann.

Mottram, J.T. and Johnson, R.P. (1990). Push tests on studs welded through profiled steel sheeting, *The Structural Engineer,* **68(10)**, 187–193.

Mullett, D.L. (1992). *Slim floor design and construction.* Publication 110. SCI.

Mullett, D.L. (1997). *Design of RHS Slimflor® Edge Beams.* Publication 169. SCI.

Mullett, D.L. and Lawson, R.M. (1992). *Slim floor construction using deep decking.* Technical Report 120. SCI.

Newman, G.M. and Simms, W.I. (2000). *The fire resistance of concrete filled tubes to Eurocode* 4. Technical Report 259. SCI.

Oehlers, D.J. and Johnson, R.P. (1987). The strength of shear stud connections in composite beams, *The Structural Engineer,* **65B(2)**, 44–48.

Olgaard, J.G., Slutter, R.G. and Fisher, J.W. (1971). Shear strength of stud connectors in lightweight and normal-weight concrete, *Engineering Journal, American Institute of Steel Construction,* **8(2)**, 55–64.

Wyatt, T.A. (1989). *Design guide on the vibration of floors.* Publication 076. SCI.

Yam, L.C.P. and Chapman, J.C. (1968). The inelastic behaviour of simply supported composite beams of steel and concrete, *Proceedings of the Institution of Civil Engineers,* **41**, 651–683.

Chapter 11 / Cold-formed Steel Sections

Thin-walled, cold-formed steel sections are widely used as purlins and rails, the intermediate members between the main structural frame and the corrugated roof or wall sheeting in buildings for farming and industrial use (see Fig. 11.1). Trapezoidal sheeting is usually fixed to these members in order to enclose the building. The most common sections are the zed, channel and sigma shapes, which may be plain or have lips. The lips are small additional elements provided to a section to improve its efficiency under compressive loads by enhancing the section ability against local buckling.

Cold-formed steel sections are fabricated by means of folding, press-braking of plates or cold-rolling of coils made from carbon steel. Sheet steel used in cold-formed sections is typically 0.9–8 mm thick. It is usually supplied pre-galvanized in accordance with European Standard EN 10142. Galvanizing gives adequate protection for internal members or those adjacent to the boundaries of the building envelope. Cold working of the steel increases its yield strength but also lowers its ductility (see Fig. 11.2). For example, a 20% reduction in thickness can increase yield strength by 50% but reduces elongation to as little as 7%, which probably represents the limit of formability for simple shapes.

The main benefits of using a cold-formed section are not only its high strength-to-weight ratio but also its lightness, which can save costs on transport, erection and the construction of foundation, and flexibility that the members can be produced in a wide variety of sectional profiles, which can result in more cost effective designs. Examples of the structural use of cold-formed sections include roof and wall members, steel framing, wall partitions, large panels for housing, lintels, floor joists, modular frames for commercial buildings, trusses, space frames, curtain walling, prefabricated buildings, frameless steel buildings, storage racking, lighting and transmission towers, motorway crash barriers, etc.

The prime difference between the behaviour of cold-formed sections and hot rolled structural sections is that cold-formed members involve thin plate elements which tend to buckle locally under compression. Cold-formed cross-sections are therefore

This chapter is contributed by Long-yuan Li (Aston University) and Xiao-ting Chu (Auckland University of Technology, New Zealand)

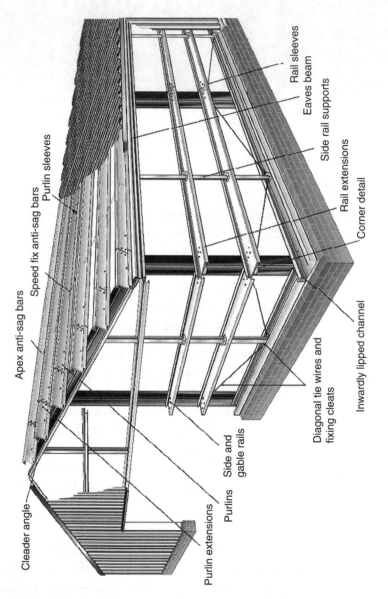

FIGURE 11.1 The building using zed and channel sections as purlins and rails
(Copy from Albion Sections Ltd design manual by permission)

usually classified as slender because they cannot generally reach their full strength based on the amount of material in the cross-section (Rhodes and Lawson, 1992). The secondary difference is that cold-formed members have low lateral stiffness and low torsional stiffness because of their open, thin, cross-sectional geometry, which gives great flexural rigidity about one axis at the expense of low torsional rigidity and low flexural rigidity about a perpendicular axis. This leads to cold-formed members being susceptible to distortional buckling and lateral–torsional buckling.

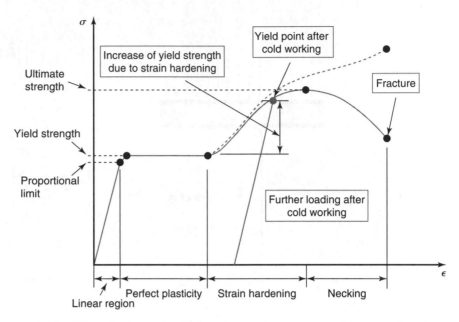

FIGURE 11.2 The influence of cold forming on the stress–strain curve of steel

11.1 ANALYTICAL MODEL

Most purlins and rails are laterally restrained by their supported cladding or sheeting either partially or completely. Hence, it is necessary to consider the influence of the lateral restraints when establishing an analytical model. Also, it is well known that, when a thin-walled beam has one or more cross-sections that are constrained against warping, a complex distribution of longitudinal warping stresses can be developed. These warping stresses together with the longitudinal stresses generated by bending moments may cause the beam to have local, distortional and lateral–torsional buckling.

Consider a zed section beam that is partially restrained by the sheeting on its upper flange. Without loss of generality, the restraints of the sheeting can be simplified by using a translational spring and a rotational spring, both of which are uniformly distributed along the longitudinal direction of the beam (see Fig. 11.3). Let the origin of the coordinate system (x, y, z) be the centroid of the cross-section, with x-axis being along the longitudinal direction of the beam, and y- and z-axes taken in the plane of the cross-section, as shown in Fig. 11.3. According to the bending theory of asymmetric beams (Vlasov, 1961; Oden, 1967) and noticing that for a zed section y- and z-axes used in Fig. 11.3 are not the principal axes, the constitutive relationships between moments and generalized strains can be expressed as

$$M_y = -EI_y \frac{\mathrm{d}^2 w}{\mathrm{d}x^2} - EI_{yz} \frac{\mathrm{d}^2 v}{\mathrm{d}x^2}$$

$$M_z = -EI_{yz} \frac{\mathrm{d}^2 w}{\mathrm{d}x^2} - EI_z \frac{\mathrm{d}^2 v}{\mathrm{d}x^2} \tag{11.1}$$

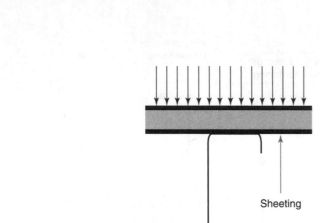

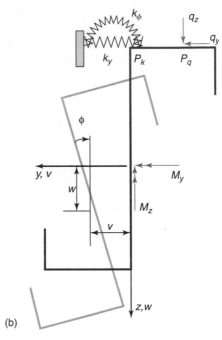

FIGURE 11.3 (a) Purlin-sheeting system and (b) a simplified analysis model

$$M_\omega = EI_w \frac{d^2\phi}{dx^2}$$

$$M_T = GI_T \frac{d\phi}{dx}$$

where M_y and M_z are the bending moments about y- and z-axes, M_ω is the warping moment, M_T is the twisting moment, E is the modulus of elasticity, G is the shear modulus, I_y and I_z are the second moments of the cross-sectional area about y- and z-axes, I_{yz} is the product moment of the cross-sectional area, I_w is the warping constant, I_T is the torsion constant, v and w are the y- and z-components of displacement of the centroid of the cross-section, ϕ is the angle of twisting.

The above four equations together with three equilibrium equations can be used to determine seven unknowns (four moments, M_y, M_z, M_ω and M_T and three displacements, v, w and ϕ). For the present problem it is convenient to derive the equilibrium equations by using the principle of minimum potential energy. For the partially restrained beam the total potential energy involves the strain energy of the beam, the strain energy of the two springs and the potential of the applied loads, that is

$$\Pi = U_b + U_s + W \tag{11.2}$$

in which,

$$U_b = \frac{1}{2}\int_0^l \left[M_y\left(-\frac{d^2w}{dx^2}\right) + M_z\left(-\frac{d^2v}{dx^2}\right) + M_\omega\frac{d^2\phi}{dx^2} + M_T\frac{d\phi}{dx} \right] dx$$

$$= \text{strain energy of the beam}$$

$$U_s = \frac{1}{2}\int_0^l \left[k_y(v+z_k\phi)^2 + k_\phi\phi^2\right]dx = \text{strain energy of the springs}$$

$$W = -\int_0^l \left[(v+z_q\phi)q_y + (w-y_q\phi)q_z\right]dx = \text{potential of the applied loads}$$

where l is the length of the beam, k_y and k_ϕ are the stiffness constants per unit length of the translational and rotational springs, q_y and q_z are the uniformly distributed loads in y- and z-directions, $(-y_k, -z_k)$ and $(-y_q, -z_q)$ are the coordinates of the spring and loading points P_k and P_q (see Fig. 11.3), respectively. Substituting Eq. (11.1) into Eq. (11.2) yields

$$\Pi = \frac{1}{2}\int_0^l \left[EI_y\left(\frac{d^2w}{dx^2}\right)^2 + 2EI_{yz}\frac{d^2v}{dx^2}\frac{d^2w}{dx^2} + EI_z\left(\frac{d^2v}{dx^2}\right)^2\right]dx$$

$$+ \frac{1}{2}\int_0^l \left[EI_w\left(\frac{d^2\phi}{dx^2}\right)^2 + GI_T\left(\frac{d\phi}{dx}\right)^2\right]dx + \frac{1}{2}\int_0^l \left[k_y(v+z_k\phi)^2 + k_\phi\phi^2\right]dx$$

$$- \int_0^l [(v+z_q\phi)q_y + (w-y_q\phi)q_z]dx \tag{11.3}$$

The following three equilibrium equations can be obtained by the variation of the total potential energy with respect to the displacement components, v and w, and the angle of twisting, ϕ

$$EI_z\frac{d^4v}{dx^4} + k_y(v+z_k\phi) + EI_{yz}\frac{d^4w}{dx^4} = q_y$$

$$EI_{yz}\frac{d^4v}{dx^4} + EI_y\frac{d^4w}{dx^4} = q_z$$

$$EI_w\frac{d^4\phi}{dx^4} - GI_T\frac{d^2\phi}{dx^2} + (z_k^2 k_y + k_\phi)\phi + z_k k_y v = q_y z_q - q_z y_q \tag{11.4}$$

For beams that have no restraints, that is, $k_y = k_\phi = 0$, Eq. (11.4) are simplified to

$$EI_z\frac{d^4v}{dx^4} + EI_{yz}\frac{d^4w}{dx^4} = q_y$$

$$EI_{yz}\frac{d^4v}{dx^4} + EI_y\frac{d^4w}{dx^4} = q_z$$

$$EI_w\frac{d^4\phi}{dx^4} - GI_T\frac{d^2\phi}{dx^2} = q_y z_q - q_z y_q \tag{11.5}$$

For beams that are fully restrained, that is, $v = \phi = 0$, Eq. (11.4) are simplified to

$$R_k + EI_{yz}\frac{\mathrm{d}^4 w}{\mathrm{d}x^4} = q_y$$

$$EI_y\frac{\mathrm{d}^4 w}{\mathrm{d}x^4} = q_z$$

$$R_k z_k + M_k = q_y z_k - q_z y_k \qquad (11.6)$$

where R_k and M_k are the reaction force and reaction moment at the restrained point P_k. Most cold-formed sections are supported by cleats bolted to the web of the section as shown in Fig. 11.4. The boundary conditions thus can be assumed as

$$v = 0 \quad M_z \approx 0$$

$$w = 0 \quad M_y \approx 0$$

$$\phi = 0 \quad M_\omega \approx 0 \qquad (11.7)$$

The cleats are designed so that the lower flange of the section does not bear directly on the rafter, and web crippling problems are avoided. However, the shear or bearing strength of the connecting bolts is critical to the design.

Governing Eqs (11.4), (11.5) or (11.6) together with boundary conditions (11.7) can be used to determine the displacements and angle of twisting of the beam under the action of external loads, q_y and q_z. The bending moments at any place can be calculated using Eq. (11.1). The bending and shear stresses thus can be calculated from the moments

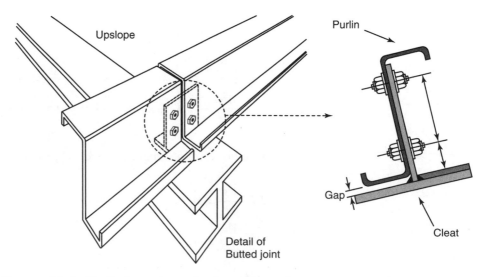

FIGURE 11.4 Purlin butted to rafter beam by a cleat

and shear force as follows:

$$\sigma_x = \frac{M_z I_y - M_y I_{yz}}{I_y I_z - I_{yz}^2} y + \frac{M_y I_z - M_z I_{yz}}{I_y I_z - I_{yz}^2} z + E(\overline{\omega} - \omega)\frac{d^2\phi}{dx^2}$$

$$\tau_{max} = \frac{3M_T}{L_s t^2} + \frac{V}{A_v} \tag{11.8}$$

where ω is the sectorial coordinate with respect to the shear centre, $\overline{\omega}$ is the average value of ω, L_s is the total length of the middle line section, t is the thickness, V is the shear force, A_v is the shear area. The sectorial coordinates are properties of the cross-section and are calculated as follows (Chu *et al.*, 2004a,b)

$$\omega = \int_0^s h_s ds \quad \text{and} \quad \overline{\omega} = \frac{1}{L_s}\int_0^{L_s} \omega ds \tag{11.9}$$

where h_s is the perpendicular distance from a tangent at the point under consideration to the shear centre, and s is the distance from any chosen origin to the same point measured along the middle line of the section.

Equation (11.8) indicates that when warping torsion is involved twisting produces not only the shear stress but also axial stress. More about warping torsion can be found in the books of Oden (1967) and Walker (1975). Ye *et al.* (2004) investigated the influences of restraints on the magnitude and distribution of the axial stress within the cross-section through varying the stiffness constants of two springs. They also used the stress pattern obtained from Eq. (11.8) as an input to the finite strip analysis program and investigated the influence of restraints on the behaviour of local, distortional and lateral–torsional buckling of channel and zed section beams (Ye *et al.*, 2002).

It is interesting to notice from Eq. (11.6) that, if it is fully restrained the zed section beam bends only in the plane of the web and the bending stress and deflection can be calculated simply based on the bending rigidity of the beam in the plane of the web although the section itself is point symmetric, that is,

$$\sigma_x = \frac{M_y z}{I_y}$$

$$\frac{d^4 w}{dx^4} = \frac{q_z}{EI_y} \tag{11.10}$$

11.2 LOCAL BUCKLING

Cold-formed members are usually very thin, and thus the thin plate elements tend to buckle locally under compression. The local buckling mode of a cold-formed member normally involves plate flexure along, with no transverse deformation of a line or lines of intersection of adjoining plates, and can be characterized by a relatively short

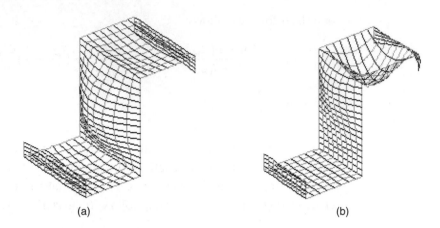

(a)   (b)

FIGURE 11.5 Local buckling modes of a zed section ($h = 120$ mm, $b = 75$ mm, $c = 20$ mm, $t = 2,5$ mm). (a) Under a pure compression (web buckle) and (b) under a pure bending (flange buckle)

half-wavelength of the order of magnitude of individual plate elements, as illustrated in Fig. 11.5.

11.2.1 Elastic Local Buckling Stress

It is known from the stability of plates that, a simply supported rectangular plate may buckle if it is subjected to compressive loads in the plane of its middle surface. The elastic critical compressive stress when the plate buckles is expressed as (Bulson, 1970)

$$\sigma_{\mathrm{cr,p}} = \frac{k_\sigma \pi^2 E}{12(1 - v^2)} \left(\frac{t}{b_\mathrm{p}} \right)^2 \tag{11.11}$$

where b_p is the width of the plate, v is Poisson's ratio and k_σ is the buckling coefficient determined from

$$k_\sigma = \left(\frac{l}{mb_\mathrm{p}} \right)^2 + 2 + \left(\frac{mb_\mathrm{p}}{l} \right)^2 \tag{11.12}$$

where l is the length of a plate and m is the number of half waves of the buckling mode in the longitudinal direction in which the plate is compressed. Note that, k_σ varies with m. When $m = l/b_\mathrm{p}$, k_σ has a minimum value of 4, which makes the compressive stress $\sigma_{\mathrm{cr,p}}$ critical. This indicates that the buckles approximate to square wave forms, as demonstrated in Fig. 11.5.

Equations (11.11) and (11.12) are only for a plate simply supported on the two long sides and subjected to uniform compressive stresses. For a compression element of width b_p with different support conditions and/or subjected to non-uniform compressive stresses the critical compressive stress can still be calculated using Eq. (11.11) but the buckling coefficient needs to take account the influence of both boundary conditions and stress pattern. When these factors have been taken into account k_σ is expressed as follows:

For doubly supported compression elements (Table 4.1 in EN 1993-1-5 (2006))

$$k_\sigma = \begin{cases} \dfrac{8,2}{1,05 + \psi} & 0 \leq \psi \leq 1 \\ 7,81 - 6,29\psi + 9,78\psi^2 & -1 < \psi < 0 \\ 5,98(1 - \psi)^2 & -3 < \psi \leq -1 \end{cases} \qquad (11.13a)$$

For outstand compression elements (support at σ_1) (Table 4.2 in EN 1993-1-5 (2006))

$$k_\sigma = \begin{cases} \dfrac{0,578}{0,34 + \psi} & 0 \leq \psi \leq 1 \\ 1,7 - 5\psi + 17,1\psi^2 & -1 \leq \psi < 0 \end{cases} \qquad (11.13b)$$

For outstand compression elements (support at σ_2) (Table 4.2 in EN 1993-1-5 (2006))

$$k_\sigma = 0,57 - 0,21\psi + 0,07\psi^2 \quad -3 < \psi \leq 1 \qquad (11.13c)$$

For single edge compression stiffener elements (lips) (cl 5.5.3.2.5 in EN 1993-1-3 (2006))

$$k_\sigma = \begin{cases} 0,5 & c_p/b_p \leq 0,35 \\ 0,5 + 0,83(c_p/b_p - 0,35)^{2/3} & 0,35 < c_p/b_p \leq 0,6 \end{cases} \qquad (11.13d)$$

where $\psi = \sigma_2/\sigma_1$ is the ratio of stresses at the two ends of the element (σ_1 is the larger compressive stress, σ_2 is the tensile stress or smaller compressive stress, and the compressive stress is assumed to be positive), c_p and b_p are the lengths of the middle lines of the lip and flange, respectively.

Equations (11.11) and (11.13) are used to calculate the critical stress of local buckling of a compression element. For channel and zed sections, the web and lipped flange may be treated as the doubly supported elements if the lip satisfies the requirement specified in Section 5.2 in EN 1993-1-3 (2006). Flanges that have no intermediate stiffeners and no edge lips are treated as the outstand elements. When a web or a flange has an intermediate stiffener, the actual width of the element should be taken as the width of the individual part separated by the stiffener. More details for dealing with elements with intermediate stiffeners can be found in EN 1993-1-3 (2006) and EN 1993-1-5 (2006).

11.2.2 Post-Buckling Behaviour and the Calculation of Effective Width

When an element buckles locally it does not necessarily mean that this element will collapse or loss its ability of carrying loads. In fact, a plate can be allowed to take a considerably increased load beyond initial buckling before any danger of collapse occurs. This is because the deflections due to buckling are accompanied by stretching of the middle surface of the plate.

It is not always possible for practical reasons to allow some elements of a structure to buckle, but if stable buckles can be tolerated, a considerable gain follows in structural efficiency. For a uniformly compressed rectangular plate, up to the buckling load, the stress distribution is uniform. With increase in load, the central unconstrained portion of the plate will start to deflect laterally and will therefore not support much additional load, whereas the portions close to the supported edges will be constrained to remain straight and will continue to carry increasing stresses. Figure 11.6 shows the typical variation of the stress distribution in a plate in pre- and post-buckling stages. The ultimate strength of the plate is when the maximum stress at the edges reaches the compressive yield strength of the material. Thus the ultimate load of the plate should be calculated based on the stress distribution at failure through the width of the plate. The problem, however, is that analysis of the post-buckled plate is a complicated process and no exact closed form results have been obtained for compressed plates. Therefore, instead of using the stress distribution in the post-buckling range, an alternative approach to assessing the ultimate load of the plate is to use an effective width concept.

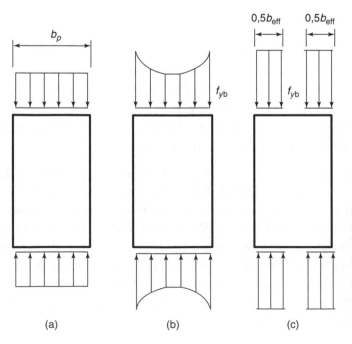

FIGURE 11.6 The concept of effective width. (a) Stress distribution up to buckling, (b) stress distribution at failure and (c) stress distribution in effective width

The concept of effective width was originally developed by Von Karman *et al.* (1932) and calibrated for cold-formed members by Winter (1968). The method assumes that when the ultimate stress is reached, the total load is carried by two fictitious strips adjacent to the edges of the plate (see Fig. 11.6c), which carry a uniform stress equal to the yield strength of the material, and the central region is unstressed. Obviously, the calculation of the effective width is dependent on the stress distribution at the time when the plate fails, which is influenced by a number of factors including the pattern of applied compressive stresses and the boundary conditions, relative slenderness and geometrical imperfections of the plate. Based on large numbers of tests, empirical functions have been developed. In EN 1993-1-5 (2006) the following equations have been recommended for calculating the effective width of a compression element.

For doubly supported compression elements (cl 4.4.2 in EN 1993-1-5 (2006))

$$\rho = \begin{cases} 1,0 & \bar{\lambda}_{\text{p,red}} \leq 0,673 \\ \dfrac{\bar{\lambda}_{\text{p,red}} - 0,055(3 + \psi)}{\bar{\lambda}^2_{\text{p,red}}} & 0,673 < \bar{\lambda}_{\text{p,red}} \end{cases} \tag{11.14a}$$

For outstand compression elements or single edge compression stiffener elements (cl 4.4.2 in EN 1993-1-5 (2006))

$$\rho = \begin{cases} 1,0 & \bar{\lambda}_{\text{p,red}} \leq 0,748 \\ \dfrac{\bar{\lambda}_{\text{p,red}} - 0,188}{\bar{\lambda}^2_{\text{p,red}}} & 0,748 < \bar{\lambda}_{\text{p,red}} \end{cases} \tag{11.14b}$$

in which,

$$\bar{\lambda}_{\text{p,red}} = \bar{\lambda}_{\text{p}} \sqrt{\frac{\sigma_{\text{com,Ed}}}{f_{yb}/\gamma_{\text{M0}}}} = \text{reduced slenderness}$$

$$\bar{\lambda}_{\text{p}} = \sqrt{\frac{f_{yb}}{\sigma_{\text{cr,p}}}} = \text{relative slenderness for local buckling}$$

where ρ is the reduction factor to determine the effective width of a compression element defined in Tables 11.1 and 11.2, f_{yb} is the basic yield strength of the material, $\sigma_{\text{com,Ed}}$ ($\sigma_{\text{com,Ed}} \leq f_{yb}/\gamma_{\text{M0}}$) is the largest compressive stress in the compression element, and γ_{M0} is the partial safety factor for resistance of the cross-section.

After the effective widths of individual compression elements have been determined, the effective area, second moments of the effective area and effective section modulus can be calculated, from which the design values of the resistance to bending moments can be determined.

TABLE 11.1   Doubly supported compression elements ($\psi = \sigma_2/\sigma_1$).

Stress distribution (compression positive)	Effective width, b_{eff}
	$\psi = 1$ $b_{\text{eff}} = \rho b$ $b_{e1} = b_{e2} = 0{,}5 b_{\text{eff}}$
	$1 > \psi \geq 0$ $b_{\text{eff}} = \rho b$ $b_{e1} = \dfrac{2 b_{\text{eff}}}{5 - \psi}$ $b_{e2} = b_{\text{eff}} - b_{e1}$
	$0 > \psi$ $b_{\text{eff}} = \rho b_c = \dfrac{\rho b}{1 - \psi}$ $b_{e1} = 0{,}4 b_{\text{eff}}$ $b_{e2} = 0{,}6 b_{\text{eff}}$

11.3 DISTORTIONAL BUCKLING

Distortional buckling involves both rotation and translation at the corners of the cross-section. Distortional buckling of flexural members such as channel and zed sections involves rotation of only the compression flange and lip about the flange–web junction as shown in Fig. 11.7. The web undergoes flexure at the same half-wavelength as the flange buckle, and the compression flange may translate in a direction normal to the web, also at the same half-wavelength as the flange and web buckling deformations. The elastic distortional buckling stress of cold-formed flexural members can be determined using either analytical methods, such as those suggested in AS/NZS 4600 (1996) and EN 1993-1-3 (2006) or numerical methods, such as the finite strip method (FSM) (Schafer, 1997) and the generalized beam theory (GBT) method (Davies *et al.*, 1993).

11.3.1 The Calculation Method in EN 1993-1-3 (2006)

In EN 1993-1-3 (2006), the design of compression elements with intermediate or edge stiffeners is based on the assumption that the stiffener behaves as a compression

TABLE 11.2 Outstand compression element ($\psi = \sigma_2/\sigma_1$).

Stress distribution (compression positive)	Effective width, b_{eff}
	$1 > \psi \geq 0$ $b_{\text{eff}} = \rho b$
	$0 > \psi$ $b_{\text{eff}} = \rho b_c = \dfrac{\rho b}{1 - \psi}$
	$1 > \psi \geq 0$ $b_{\text{eff}} = \rho b$
	$0 > \psi$ $b_{\text{eff}} = \rho b_c = \dfrac{\rho b}{1 - \psi}$

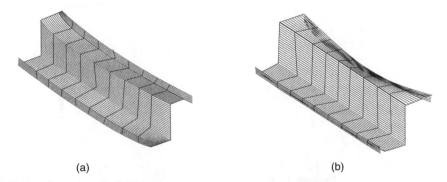

(a)   (b)

FIGURE 11.7 Distortional buckling modes of a zed section ($h = 120$ mm, $b = 75$ mm, $c = 20$ mm, $t = 2{,}5$ mm) (a) under a pure compression and (b) under a pure bending

member with continuous partial restraint, with a spring stiffness that depends on the boundary conditions and the flexural stiffness of the adjacent plane elements of the cross-section. The spring stiffness of the stiffener is determined by applying a unit load per unit length to the cross-section at the location of the stiffener, as illustrated in Fig. 11.8, and is determined from

$$K = \frac{1}{\delta} = \frac{Et^3}{4(1 - v^2)} \frac{1}{b_1^2(b_1 + h_p)} \tag{11.15}$$

where δ is the deflection of the centroid of the stiffener due to a unit load, b_1 is the horizontal distance from the web line to the centroid of the effective area of the edge stiffener, and h_p is the depth of the web. The elastic critical buckling stress for a long strut on an elastic foundation of a spring stiffness coefficient K is given by Timoshenko and Gere (1961) as follows:

$$\sigma_{cr,d} = \frac{\pi^2 E I_s}{A_s \lambda^2} + \frac{K \lambda^2}{A_s \pi^2} \tag{11.16}$$

where A_s and I_s are the area and second moment of the effective section of the stiffener, as illustrated in Fig. 11.9 for an edge stiffener, and $\lambda = l/m$ is the half-wavelength of the distortional buckling (l is the member length and m is the number of half waves). For a sufficiently long strut, the critical half-wavelength can be obtained by minimizing the critical stress as defined by Eq. (11.6) with respect to λ, to give,

$$\lambda_{cr} = \pi \left(\frac{E I_s}{K} \right)^{1/4} \tag{11.17}$$

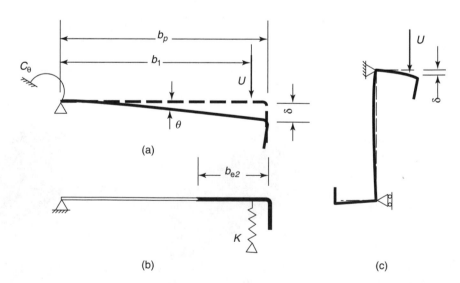

FIGURE 11.8 (a) Distortional buckling model used in EN 1993-1-3, (2006) (b) edge stiffener on an elastic foundation of a spring stiffness coefficient and (c) model used to determine spring stiffness coefficient (copy from EN 1993-1-3 (2006) by permission)

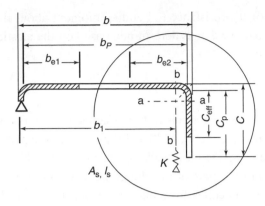

FIGURE 11.9 Effective cross-sectional area of an edge stiffener in EN 1993-1-3 (2006) (copy from EN 1993-1-3 (2006) by permission)

Substituting Eq. (11.17) into Eq. (11.16) yields,

$$\sigma_{\text{cr,d}} = \frac{2\sqrt{KEI_s}}{A_s} \tag{11.18}$$

Equation (11.18) is given in EN 1993-1-3 (2006) for calculating the critical stress of distortional buckling of the edge stiffener.

The design strength in EN 1993-1-3 (2006) (Section 5.5.3.1) for distortional buckling is considered by using a reduced thickness of the edge stiffener. The reduction factor is calculated in terms of the relative slenderness as follows,

$$\chi_{\text{d}} = \begin{cases} 1,0 & \overline{\lambda}_{\text{d}} \leq 0,65 \\ 1,47 - 0,723\overline{\lambda}_{\text{d}} & 0,65 < \overline{\lambda}_{\text{d}} < 1,38 \\ 0,66/\overline{\lambda}_{\text{d}} & 1,38 \leq \overline{\lambda}_{\text{d}} \end{cases} \tag{11.19}$$

in which,

$$\overline{\lambda}_{\text{d}} = \sqrt{\frac{f_{\text{yb}}}{\sigma_{\text{cr,d}}}} = \text{relative slenderness for distortional buckling}$$

The procedure for calculating χ_{d} can be summarized as follows (Section 5.5.3.2 in EN 1993-1-3 (2006)).

- **Step 1** Obtain an initial effective cross-section for the stiffener using effective widths determined by assuming that the stiffener gives full restraint and that $\sigma_{\text{com,Ed}} = f_{\text{yb}}/\gamma_{\text{M0}}$.
- **Step 2** Use the initial effective cross-section of the stiffener to determine the reduction factor for distortional buckling (flexural buckling of a stiffener), allowing for the effects of the continuous spring restraint.
- **Step 3** Optionally iterate to refine the value of the reduction factor for buckling of the stiffener; that is, re-calculate the effective widths of the lip and the part of the flange near the lip based on the compressive stress $\sigma_{\text{com,Ed}} = \chi_{\text{d}}f_{\text{yb}}/\gamma_{\text{M0}}$ and calculate the reduction factor again based on the newly calculated effective widths.

The design value of the resistance to bending moment about the y-axis due to both local and distortional buckling is determined based on the elastic section modulus of the effective section,

$$M_{c,Rd} = \frac{f_{yb} W_{eff,y}}{\gamma_{M0}}$$ (11.20)

where $W_{eff,y}$ is the section modulus of the effective section for bending about y-axis, in which, apart from the effective widths of the web and the part of the flange near to web are calculated using the local buckling formulae, the effective widths and thicknesses of the lip and the part of the flange near the lip are calculated using both local and distortional buckling formulae, based on the reduced compressive stress, $\sigma_{com,Ed} = \chi_d f_{yb}/\gamma_{M0}$.

11.3.2 The Calculation Methods in AS/NZS 4600 (1996)

In EN 1993-1-3 (2006) the critical stress of distortional buckling of a section is calculated based on the model of an edge stiffener on an elastic foundation and the effect of the distortional buckling on the section properties is taken into account by reducing the thickness of the stiffener. An alternative to determine the critical stress of distortional buckling of a cold-formed steel section is to use AS/NZS 4600 (1996) design code. The elastic distortional buckling formulae for channel and zed sections in AS/NZS 4600 (1996) are based on a simple flange buckling model where the flange is treated as a thin-walled compression member, as shown in Fig. 11.10, undergoing flexural–torsional buckling (Lau and Hancock, 1987; Hancock, 1997). The rotational spring stiffness k_θ represents the flexural restraint provided by the web which is in flexure, and the translational spring stiffness k_x represents the resistance to translational movement of the section in the buckling mode. The model includes a reduction in the flexural restraint provided by the web as a result of the compressive stress in the web.

Lau and Hancock (1987) showed that the translational spring stiffness has no significant influence on results and thus is assumed to be zero. The rotational spring stiffness and the critical stress at distortional buckling are given as

$$k_\theta = \frac{2Et^3}{5{,}46(h_p + 0{,}06\lambda)}\left[1 - \frac{1{,}11\sigma_{cr,d}}{Et^2}\left(\frac{h_p^4 \lambda^2}{12{,}56\lambda^4 + 2{,}192h_p^4 + 13{,}39\lambda^2 h_p^2}\right)\right]$$ (11.21)

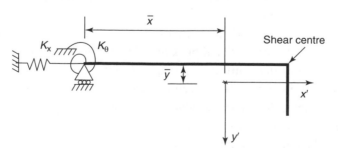

FIGURE 11.10 Flange elastically restrained along flange-web junction in AS/NZS 4600 (1996)

$$\sigma_{\mathrm{cr,d}} = \frac{E}{2A_{\mathrm{f}}}\left[(\alpha_1 + \alpha_2) - \sqrt{(\alpha_1 + \alpha_2)^2 - 4\alpha_3}\right] \tag{11.22}$$

in which,

$$\alpha_1 = \frac{\eta}{\beta_1}\left(I_{x\mathrm{f}}b_{\mathrm{p}}^2 + 0{,}039J_{\mathrm{f}}\lambda^2\right) + \frac{k_\theta}{\beta_1\eta E}$$

$$\alpha_2 = \eta\left(I_{y\mathrm{f}} + \frac{2}{\beta_1}\bar{y}b_{\mathrm{p}}I_{xy\mathrm{f}}\right)$$

$$\alpha_3 = \eta\left(\alpha_1 I_{y\mathrm{f}} - \frac{\eta}{\beta_1}I_{xy\mathrm{f}}^2 b_{\mathrm{p}}^2\right)$$

$$\beta_1 = \bar{x}^2 + \left(\frac{I_{x\mathrm{f}} + I_{y\mathrm{f}}}{A_{\mathrm{f}}}\right)$$

$$\lambda = 4{,}80\left(\frac{I_{x\mathrm{f}}b_{\mathrm{p}}^2 h_{\mathrm{p}}}{2t^3}\right)^{1/4}$$

$$\eta = \left(\frac{\pi}{\lambda}\right)^2$$

where A_{f} is the full cross-sectional area of the compression flange and lip, $I_{x\mathrm{f}}$ and $I_{y\mathrm{f}}$ are the second moments of the area A_{f} about x'- and y'-axes, respectively, where the x'- and y'-axes are located at the centroid of area A_{f} with x'-axis parallel with flange, $I_{xy\mathrm{f}}$ is the product moment of the area A_{f} about x'- and y'-axes, J_{f} is the St Venant torsion constant of the area A_{f}, $\bar{x}$ and $\bar{y}$ are the distances from the flange-web junction to the centroid of area A_{f} in the x'- and y'-directions, respectively. Due to the coupling of $\sigma_{\mathrm{cr,d}}$ and k_θ in Eqs (11.21) and (11.22), Hancock (1997) suggested that k_θ can be calculated based on an initial $\sigma_{\mathrm{cr,d}}$ obtained by assuming $k_\theta = 0$ and after then $\sigma_{\mathrm{cr,d}}$ can be calculated based on the obtained k_θ value. In the iteration, if $k_\theta < 0$, k_θ should be calculated using $\sigma_{\mathrm{cr,d}} = 0$.

The elastic critical moment for distortional buckling is calculated based on the critical buckling stress as follows,

$$M_{\mathrm{cr,d}} = \sigma_{\mathrm{cr,d}}W_y \tag{11.23}$$

where W_y is the elastic section modulus of the gross cross-section for the extreme compression fibre. The design value of the resistance to bending moment about y-axis due to distortional buckling which involves rotation of the compression flange and lip about the flange-web junction is calculated as follows:

$$M_{\mathrm{c,Rd}} = \begin{cases} M_{\mathrm{c}} & 0 \le k_\theta \\ M_{\mathrm{c}}\dfrac{W_{\mathrm{eff},y}}{W_y} & k_\theta < 0 \end{cases} \tag{11.24}$$

in which,

$$M_c = \begin{cases} M_{\text{yield}} \left(1 - \dfrac{M_{\text{yield}}}{4M_{\text{cr,d}}}\right) \dfrac{W_{\text{eff,}y}}{W_y} & 0,5M_{\text{yield}} < M_{\text{cr,d}} \\[3em] M_{\text{yield}} \left[0,055 \left(\sqrt{\dfrac{M_{\text{yield}}}{M_{\text{cr,d}}}} - 3,6\right)^2 + 0,237\right] \dfrac{W_{\text{eff,}y}}{W_y} & M_{\text{cr,d}} \leq 0,5M_{\text{yield}} \end{cases}$$

where $M_{\text{yield}} = f_{yb}W_y$ is the moment causing initial yield at the extreme compression fibre of the gross section.

A comparison of the critical stresses of distortional buckling by using EN 1993-1-3 (draft version of 2001) and AS/NZS 4600 (1996) with experimental data was made by Kesti and Davies (1999). It was found that Lau and Hancock's analytical expressions give a good prediction of the distortional buckling stress. The method given in EN 1993-1-3 does not correlate as well as Lau and Hancock's method. The error in the distortional buckling stress could lead to a consequential error in the effective cross-sectional area depending on the distortional buckling stress level.

11.4 LATERAL–TORSIONAL BUCKLING

In practice, purlins and rails are usually used together with their supported cladding or sheeting, and thus they are generally considered to be restrained against lateral deflections perpendicular to the line of action of the loading. If a beam is fully restrained on its translational and rotational degrees neither will the beam rotate nor deflect laterally. However, if the cladding or sheeting is not strong enough then it is possible for the beam to become unstable and for very large lateral deflections to occur at a critical value of the applied load. This type of behaviour is called lateral–torsional buckling. Chapter 5 has discussed the lateral–torsional buckling of unrestrained beams. In this section it is to deal with the lateral–torsional buckling of zed section beams with partial restraints from the sheeting. For the lateral–torsional buckling of partially restrained channel section beams readers can see the work of Chu *et al.* (2004).

11.4.1 Critical Moment of Lateral–Torsional Buckling

The model presented here for analysing the lateral–torsional buckling of partially restrained beams was originally developed by Li (2004) and lately expanded by Chu *et al.* (2004b), which is similar to that described in Section 11.1. Assume that the displacements and moments in a state of equilibrium are (v,w,ϕ) and (M_y,M_z,M_ω,M_T). Now let v_b and w_b be the y- and z-components of the buckling displacement of the centroid of the cross-section and ϕ_b be the buckling angle of twisting of the section. Assuming that the displacements in pre-buckling are very small, the increase of the strain energy of the system due to the lateral–torsional buckling thus can be expressed as

$$U_2 = \frac{1}{2} \int\limits_0^l \left[EI_y \left(\frac{d^2 w_b}{dx^2} \right)^2 + 2EI_{yz} \frac{d^2 v_b}{dx^2} \frac{d^2 w_b}{dx^2} + EI_z \left(\frac{d^2 v_b}{dx^2} \right)^2 \right] dx$$

$$+ \frac{1}{2} \int\limits_0^l \left[EI_w \left(\frac{d^2 \phi_b}{dx^2} \right)^2 + GI_T \left(\frac{d\phi_b}{dx} \right)^2 \right] dx$$

$$+ \frac{1}{2} \int\limits_0^l [k_y (v_b + z_k \phi_b)^2 + k_\phi \phi_b^2] dx \tag{11.25}$$

On the other hand, the lateral–torsional buckling leads to a decrease of the potential of the pre-buckling moments and loads, which can be expressed as (Chu, 2004 Li, 2004; Chu *et al.*, 2004b)

$$W_2 = \int\limits_0^l \left[(M_z \phi_b) \frac{d^2 w_b}{dx^2} - (M_y \phi_b) \frac{d^2 v_b}{dx^2} + \frac{1}{2} M_\omega \left(\frac{d\phi_b}{dx} \right)^2 \right] dx$$

$$+ \int\limits_0^l \frac{1}{2} [(q_y y_q + q_z z_q) \phi_b^2] dx \tag{11.26}$$

If the net change of the total potential is positive for any of possible buckling displacements, that is, $U_2 > W_2$, then the equilibrium of the pre-buckling state is said to be stable because the generation of the buckling displacements requires an energy input into the system. On the other hand, if the net change of the total potential is negative, that is, $U_2 < W_2$, then the equilibrium of the pre-buckling state is said to be unstable because the buckling displacements can be generated without any input of energy. A critical state between stable and unstable equilibria from which the critical load of the lateral–torsional buckling can be determined is

$$U_2 = W_2 \tag{11.27}$$

Equation (11.27) is an eigenvalue type equation. For given reference loads the smallest eigenvalue and corresponding eigenvector calculated from Eq. (11.27) represent the critical loading factor and corresponding buckling mode.

The above model can be applied directly to the purlins with intermediate lateral restraints such as provided by anti-sag bars if the pre-buckling moments are calculated based on the same model and the buckling displacements satisfy the displacement restraint conditions at the places where the anti-sag bars are placed. It is difficult to achieve closed form solutions of Eq. (11.27). In most cases only numerical solutions can be obtained. A general numerical computation procedure has been described by Li (2004) and Chu *et al.* (2004b) to obtain the critical buckling load, in which the following piecewise cubic spline functions are used to construct the displacement fields before and during buckling

$$
\begin{Bmatrix} v(x) \\ w(x) \\ \phi(x) \end{Bmatrix} = \sum \begin{bmatrix} N_i(x) & 0 & 0 \\ 0 & N_i(x) & 0 \\ 0 & 0 & N_i(x) \end{bmatrix} \begin{Bmatrix} \delta_{vi} \\ \delta_{wi} \\ \delta_{\phi i} \end{Bmatrix} \tag{11.28}
$$

$$
\begin{Bmatrix} v_b(x) \\ w_b(x) \\ \phi_b(x) \end{Bmatrix} = \sum \begin{bmatrix} N_i(x) & 0 & 0 \\ 0 & N_i(x) & 0 \\ 0 & 0 & N_i(x) \end{bmatrix} \begin{Bmatrix} \delta_{vi}^b \\ \delta_{wi}^b \\ \delta_{\phi i}^b \end{Bmatrix} \tag{11.29}
$$

where $N_i(x)$ is the spline interpolation function at node i and $\{\delta_{vi}, \delta_{wi}, \delta_{\phi i}\}$ and $\{\delta_{vi}^b, \delta_{wi}^b, \delta_{\phi i}^b\}$ are the nodal displacement vectors before and during buckling. By using the displacement expressions (11.28) and (11.29), the equilibrium Eq. (11.4) and buckling Eq. (11.27) can be simplified into the following algebraic matrix equations

$$
\begin{bmatrix} \mathbf{K}_{vv} & \mathbf{K}_{vw} & \mathbf{K}_{v\phi} \\ \mathbf{K}_{vw} & \mathbf{K}_{ww} & \mathbf{0} \\ \mathbf{K}_{v\phi} & \mathbf{0} & \mathbf{K}_{\phi\phi} \end{bmatrix} \begin{Bmatrix} \delta_v \\ \delta_w \\ \delta_\phi \end{Bmatrix} = \begin{Bmatrix} \mathbf{F}_v \\ \mathbf{F}_w \\ \mathbf{F}_\phi \end{Bmatrix} \tag{11.30}
$$

$$
\begin{bmatrix} \mathbf{K}_{vv} & \mathbf{K}_{vw} & \mathbf{K}_{v\phi} \\ \mathbf{K}_{vw} & \mathbf{K}_{ww} & \mathbf{0} \\ \mathbf{K}_{v\phi} & \mathbf{0} & \mathbf{K}_{\phi\phi} \end{bmatrix} \begin{Bmatrix} \delta_v^b \\ \delta_w^b \\ \delta_\phi^b \end{Bmatrix} = \frac{q_{cr}}{q_{ref}} \begin{bmatrix} \mathbf{0} & \mathbf{0} & \mathbf{K}_{v\phi}^0 \\ \mathbf{0} & \mathbf{0} & \mathbf{K}_{w\phi}^0 \\ \mathbf{K}_{v\phi}^0 & \mathbf{K}_{w\phi}^0 & \mathbf{K}_{\phi\phi}^0 \end{bmatrix} \begin{Bmatrix} \delta_v^b \\ \delta_w^b \\ \delta_\phi^b \end{Bmatrix} \tag{11.31}
$$

where $\mathbf{K}_{ij}$ $(i,j = v,w,\phi)$ is the stiffness matrix, $\mathbf{K}_{ij}^0$ $(i,j = v,w,\phi)$ is the geometric stiffness matrix, δ_j $(j = v,w,\phi)$ is the nodal displacement vector, $\mathbf{F}_j$ $(j = v,w,\phi)$ is the nodal force vector, δ_j^b $(j = v,w,\phi)$ is the nodal vector of buckling displacements, (q_{cr}/q_{ref}) represents the scale factor between the critical and referenced loads.

As an example, Fig. 11.11 shows the critical load factors of simply supported zed purlin beams (web depth $h = 202\,mm$, flange width $b = 75\,mm$, lip length $c = 20\,mm$ and thickness $t = 2,3\,mm$), with zero, one and two anti-sag bars, subjected to a uniformly distributed uplift load. It can be seen from the figure that the lateral restraint has remarkable influence on the lateral–torsional buckling of the beam with no anti-sag bars. The influence is found to decrease with the increase of the beam length. Interestingly, when the beam has one or two anti-sag bars, the influence of the lateral restraint on the lateral–torsional buckling becomes almost negligible. This implies that the anti-sag bar not only has the ability of increasing the critical load but also can reduce the influence of the lateral restraint provided by cladding. Practically, purlins are often used as continuous beams over two or more spans, in which case the boundary conditions of the beam can be regarded as simply supported at one end and fixed at the other end. Fig. 11.12 shows the critical load factors of the zed purlin beam with this kind of boundary conditions. The results show that there is a significant increase in critical load when the purlin has a fixed boundary condition.

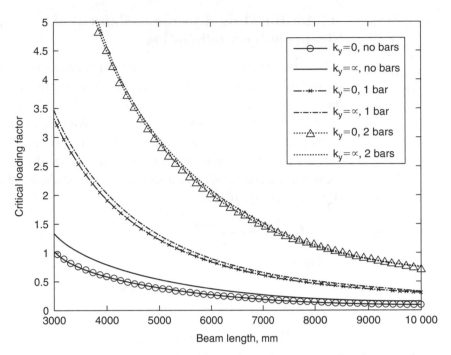

FIGURE 11.11 Lateral–torsional buckling of simply supported zed section beams ($h = 202$ mm, $b = 75$ mm, $c = 20$ mm, $t = 2,3$ mm) subjected to uniformly distributed uplift loading, with various different restraints

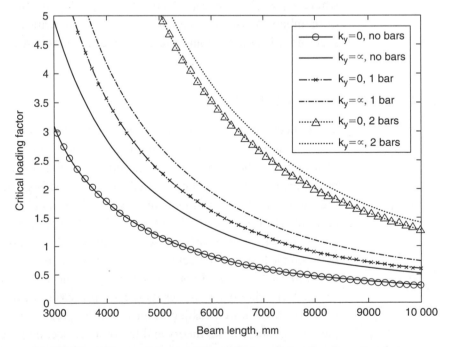

FIGURE 11.12 Lateral–torsional buckling of pinned-fixed zed section beams ($h = 202$ mm, $b = 75$ mm, $c = 20$ mm, $t = 2,3$ mm) subjected to uniformly distributed uplift loading, with various different restraints

11.4.2 Buckling Resistance Moment of Beams Subject to Bending

The critical load calculated from the lateral–torsional buckling is based on an idealized model in which the beam has no geometrical imperfections, the cross-section of the beam does not deform, and the beam does not buckle either locally or distortionally before the lateral–torsional buckling occurs. When designing a real member, however, these factors should be taken into account. The method for determining the design buckling resistance moment for the lateral–torsional buckling of cold-formed section beams is the same as that used for other steel section members (Section 6.3.2.2 in EN 1993-1-1 (2005)), that is,

$$M_{b,Rd} = \chi_{LT} \frac{W_{eff,y} f_{yb}}{\gamma_{M1}} \tag{11.32}$$

in which,

$$\chi_{LT} = \frac{1}{\phi_{LT} + \left(\phi_{LT}^2 - (\bar{\lambda}_{LT})^2\right)^{1/2}} = \text{reduction factor for lateral–torsional buckling } (\chi_{LT} \leq 1,0)$$

$$\phi_{LT} = 0,5[1 + \alpha_{LT}(\bar{\lambda}_{LT} - 0,2) + (\bar{\lambda}_{LT})^2] = \text{factor used to calculate the reduction factor}$$

$$\bar{\lambda}_{LT} = \left(\frac{M_{c,Rd}}{M_{cr,LT}}\right)^{1/2} = \text{relative slenderness for lateral–torsional buckling}$$

where $\alpha_{LT} = 0,34$ is the imperfection factor, $\gamma_{M1} = 1,0$ is the partial factor for resistance of members to instability and $M_{cr,LT}$ is the elastic critical moment of the gross cross-section for lateral–torsional buckling about the main axis.

11.5 Calculation of Deflections

The deflections of channel and zed section beams under uniformly distributed transverse loads can be evaluated using the analytical model presented in Section 11.1, provided that the load does not exceed the critical loads of local and distortional buckling. For the evaluation of deflections at loads greater than any critical load the influence of the local buckling and/or distortional buckling must be taken into account. The precise analysis of the post-buckling behaviour of a cold-formed section beam, however, is very difficult. A simple approach is to assume that the relationships between the load and deflection are linear for both pre- and post-buckling analyses as illustrated in Fig. 11.13. In pre-buckling region, the deflections can be calculated using simple beam theory and the gross section properties of the beam, since the beam is fully effective before buckling. In post-buckling region, the deflections can be evaluated using simple beam theory and the reduced section properties of the beam, since the beam is not fully effective after buckling. This assumption is only approximate since, in reality, the line in post-buckling region is not a straight line, but the errors introduced in the approximation are acceptable and conservative when fully reduced

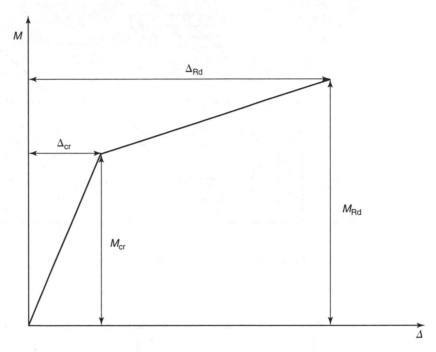

FIGURE 11.13 Simplified model used for the calculation of deflections

section properties are used. Note that, in Fig. 11.13 Δ in the pre-buckling region and $(\Delta - \Delta_{cr})$ in the post-buckling region are linearly proportional to M and $(M - M_{cr})$, respectively. Therefore, the deflection of the beam, Δ, at an applied moment, M, can be expressed as

$$
\Delta = \begin{cases} \Delta_{cr} \dfrac{M}{M_{cr}} & M \leq M_{cr} \\[3mm] \Delta_{cr} \left(1 + \dfrac{M - M_{cr}}{M_{cr}} \dfrac{I}{I_{eff}} \right) & M_{cr} < M \leq M_{Rd} \end{cases}
\tag{11.33}
$$

where M_{cr} and M_{Rd} are the critical moment and moment resistance of the beam, Δ_{cr} is the deflection of beam when it buckles, I is the second moment of the gross cross-sectional area, and I_{eff} is the second moment of the effective cross-sectional area.

11.6 FINITE STRIP METHODS

The FSM was originally developed by Cheung (1976) and it can be considered as a specialization of the finite element method (Zienkiewicz and Taylor, 2000). The method is mainly applied to structures whose geometries do not vary with at least one of the coordinate axes. The approach of the FSM is considered particularly favourable when dealing with the initial buckling or natural frequency characteristics of thin-walled prismatic structures. In the FSM, the prismatic structure is discretized into a number of longitudinal strips and the displacement fields associated with each strip

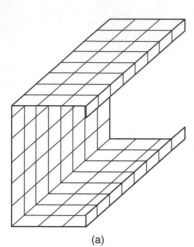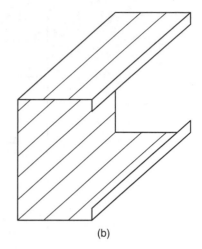

(a)   (b)

FIGURE 11.14 (a) The finite element analysis and (b) the finite strip analysis

vary sinusoidally along the strip length and algebraically across the strip width. Similar to the finite element method, shape functions are also used to define the variation of displacement fields along the strip width but they are only the functions of the cross-section coordinates of the strip (see Fig. 11.14).

The use of the FSM for understanding and predicting the behaviour of cold-formed steel members was pioneered by Lau and Hancock in Australia (1986, 1989). They modified the stiffness matrices derived in Cheung's book and created a commercial computer program for solution of the elastic buckling problem of open thin-walled members via finite strip called *Thin-Wall*. Similar programs were developed by Loughlan (1993) in the UK and by Schafer (1997) in *USA*. The program developed by Schafer is available in internet (http://www.ce.jhu.edu/bschafer/cufsm) and is particularly friendly to use. The codes were written in Matlab language and thus can easily be modified by users.

11.6.1 Element Stiffness Matrix of the Strip

Consider a strip shown in Fig. 11.15, in which the local coordinate system is defined as that, the x- and y-axes are the two axes within the plane of the strip and the z-axis is normal to the plane of the strip. The three components of buckling displacement of the strip at a point (x, y) can be expressed in terms of the nodal displacements as follows

$$\left\{ \begin{matrix} v(x,y) \\ u(x,y) \end{matrix} \right\} = \begin{bmatrix} \sin\dfrac{m\pi x}{a} & 0 \\ 0 & \cos\dfrac{m\pi x}{a} \end{bmatrix} \begin{bmatrix} 1-\dfrac{y}{b} & 0 & \dfrac{y}{b} & 0 \\ 0 & 1-\dfrac{y}{b} & 0 & \dfrac{y}{b} \end{bmatrix} \left\{ \begin{matrix} v_{1m} \\ u_{1m} \\ v_{2m} \\ u_{2m} \end{matrix} \right\} \quad (11.34)$$

$$w(x,y) = \sin \frac{m\pi x}{a} \left[1 - \frac{3y^2}{b^2} + \frac{2y^3}{b^3}, y - \frac{2y^2}{b} + \frac{y^3}{b^2}, \frac{3y^2}{b^2} - \frac{2y^3}{b^3}, \frac{y^3}{b^2} - \frac{y^2}{b} \right] \begin{Bmatrix} w_{1m} \\ \theta_{1m} \\ w_{2m} \\ \theta_{2m} \end{Bmatrix} \quad (11.35)$$

where $u(x,y)$ and $v(x,y)$ are the plane displacements, $w(x,y)$ is the deflection, $(u_{1m}, v_{1m}, w_{1m}, \theta_{1m})$ and $(u_{2m}, v_{2m}, w_{2m}, \theta_{2m})$ are the nodal displacements associated with wave number m, a and b are the length and width of the strip, respectively. The assumed displacement functions satisfy only the simply supported conditions at the two end sides of the strip. The strain energy of the strip is given as follows

$$U_2 = \frac{Et}{2(1 - v^2)} \int_0^a \int_0^b \left[\left(\varepsilon_x + \varepsilon_y \right)^2 - 2(1 - v)\left(\varepsilon_x \varepsilon_y - \frac{\gamma_{xy}^2}{4} \right) \right] dx \, dy$$

$$+ \frac{Et^3}{24(1 - v^2)} \int_0^a \int_0^b \left[\left(\kappa_x + \kappa_y \right)^2 - 2(1 - v)\left(\kappa_x \kappa_y - \kappa_{xy}^2 \right) \right] dx \, dy \quad (11.36)$$

The membrane and bending strains in Eq. (11.36) are defined as follows,

$$\varepsilon_x = \frac{\partial u}{\partial x} \qquad \varepsilon_y = \frac{\partial v}{\partial y} \qquad \gamma_{xy} = \frac{\partial u}{\partial y} + \frac{\partial v}{\partial x}$$

$$\kappa_x = -\frac{\partial^2 w}{\partial x^2} \qquad \kappa_y = -\frac{\partial^2 w}{\partial y^2} \qquad \kappa_{xy} = -\frac{\partial^2 w}{\partial x \, \partial y} \quad (11.37)$$

The element stiffness matrix can be obtained by substituting Eqs (11.34) and (11.35) into Eq. (11.37) and then into Eq. (11.36), that is,

$$U_2 = \frac{1}{2} \{\delta\}_m^T [\mathbf{K}]_m \{\delta\}_m \quad (11.38)$$

in which,

$$\{\delta\}_m = \lfloor u_{1m}, v_{1m}, u_{2m}, v_{2m}, w_{1m}, \theta_{1m}, w_{2m}, \theta_{2m} \rfloor^T = \text{element nodal vector}$$

$$[\mathbf{K}]_m = \begin{bmatrix} K_{11}^{uv} & K_{12}^{uv} & K_{13}^{uv} & K_{14}^{uv} & & & & \\ & K_{22}^{uv} & -K_{14}^{uv} & K_{24}^{uv} & & & [\mathbf{0}]_{4x4} & \\ & & K_{11}^{uv} & -K_{12}^{uv} & & & & \\ & & & K_{22}^{uv} & & & & \\ & & & & K_{11}^{w\theta} & K_{12}^{w\theta} & K_{13}^{w\theta} & K_{14}^{w\theta} \\ & & & & & K_{22}^{w\theta} & -K_{14}^{w\theta} & K_{24}^{w\theta} \\ & \text{symmetric} & & & & & K_{11}^{w\theta} & -K_{12}^{w\theta} \\ & & & & & & & K_{22}^{w\theta} \end{bmatrix}$$

$$= \text{element stiffness matrix}$$

$$K_{11}^{uv} = \frac{12D}{t^2} \left[\frac{1}{2} \frac{a}{b} + \frac{1 - v}{12} \frac{b}{a} (m\pi)^2 \right]$$

$$K_{12}^{uv} = \frac{12D}{t^2} \left[\frac{3v - 1}{8} (m\pi) \right]$$

$$K_{13}^{uv} = \frac{12D}{t^2}\left[-\frac{1}{2}\frac{a}{b} + \frac{1-\nu}{24}\frac{b}{a}(m\pi)^2\right]$$

$$K_{14}^{uv} = \frac{12D}{t^2}\left[\frac{1+\nu}{8}(m\pi)\right]$$

$$K_{22}^{uv} = \frac{12D}{t^2}\left[\frac{1-\nu}{4}\frac{a}{b} + \frac{1}{6}\frac{b}{a}(m\pi)^2\right]$$

$$K_{24}^{uv} = \frac{12D}{t^2}\left[-\frac{1-\nu}{4}\frac{a}{b} + \frac{1}{12}\frac{b}{a}(m\pi)^2\right]$$

$$K_{11}^{w\theta} = \frac{D}{a^2}\left[\frac{13}{70}\frac{b}{a}(m\pi)^4 + \frac{6}{5}\frac{a}{b}(m\pi)^2 + 6\left(\frac{a}{b}\right)^3\right]$$

$$K_{12}^{w\theta} = \frac{D}{a}\left[\frac{11}{420}\left(\frac{b}{a}\right)^2(m\pi)^4 + \frac{1+5\nu}{10}(m\pi)^2 + 3\left(\frac{a}{b}\right)^2\right]$$

$$K_{13}^{w\theta} = \frac{D}{a^2}\left[\frac{9}{140}\frac{b}{a}(m\pi)^4 - \frac{6}{5}\frac{a}{b}(m\pi)^2 - 6\left(\frac{a}{b}\right)^3\right]$$

$$K_{14}^{w\theta} = \frac{D}{a}\left[-\frac{13}{840}\left(\frac{b}{a}\right)^2(m\pi)^4 + \frac{1}{10}(m\pi)^2 + 3\left(\frac{a}{b}\right)^2\right]$$

$$K_{22}^{w\theta} = D\left[\frac{1}{210}\left(\frac{b}{a}\right)^3(m\pi)^4 + \frac{2}{15}\frac{b}{a}(m\pi)^2 + 2\frac{a}{b}\right]$$

$$K_{24}^{w\theta} = D\left[-\frac{3}{840}\left(\frac{b}{a}\right)^3(m\pi)^4 - \frac{1}{30}\frac{b}{a}(m\pi)^2 + \frac{a}{b}\right]$$

$$D = \frac{Et^3}{12(1-\nu^2)}$$

11.6.2 Element Geometric Stiffness Matrix of the Strip

The geometric stiffness matrix of a strip subjected to linearly varying edge traction can be derived by considering the change of the potential of the in-plane forces during buckling. Similar to the approach described in Section 11.4, the change of the potential of the in-plane forces can be expressed as,

$$W_2 = \frac{t}{2}\int_0^a\int_0^b\left[\sigma_1 - (\sigma_1-\sigma_2)\frac{y}{b}\right]\left[\left(\frac{\partial u}{\partial x}\right)^2 + \left(\frac{\partial v}{\partial x}\right)^2 + \left(\frac{\partial w}{\partial x}\right)^2\right]dx\,dy \quad (11.39)$$

where σ_1 and σ_2 are the compressive stresses at nodes 1 and 2 (see Fig. 11.15). Substituting Eqs (11.34) and (11.35) into Eq. (11.39) yields

$$W_2 = \frac{1}{2}\{\delta\}_m^T[\mathbf{K}_g]_m\{\delta\}_m \quad (11.40)$$

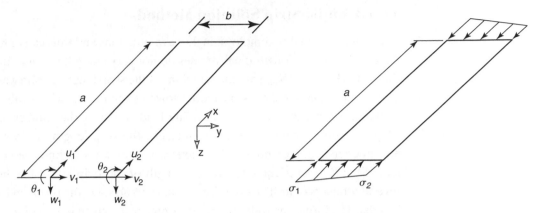

FIGURE 11.15 Local coordinates, degrees of freedom and stress distribution in a strip

in which,

$$[\mathbf{K}_g]_m = \frac{b(m\pi)^2}{1680a} \begin{bmatrix} [\mathbf{K}_g^{uv}] & [\mathbf{0}]_{4x4} \\ [\mathbf{0}]_{4x4} & [\mathbf{K}_g^{w\theta}] \end{bmatrix} = \text{element geometric stiffness matrix}$$

$$[\mathbf{K}_g^{uv}] = \begin{bmatrix} 70(3T_1 + T_2) & 0 & 70(T_1 + T_2) & 0 \\ & 70(3T_1 + T_2) & 0 & 70(T_1 + T_2) \\ & & 70(T_1 + 3T_2) & 0 \\ \text{symmetric} & & & 70(T_1 + 3T_2) \end{bmatrix}$$

$$[\mathbf{K}_g^{w\theta}] = \begin{bmatrix} 8(30T_1 + 9T_2) & 2b(15T_1 + 7T_2) & 54(T_1 + T_2) & -2b(7T_1 + 6T_2) \\ & b^2(5T_1 + 3T_2) & 2b(6T_1 + 7T_2) & -3b^2(T_1 + T_2) \\ & & 24(3T_1 + 10T_2) & -2b(7T_1 + 15T_2) \\ \text{symmetric} & & & b^2(3T_1 + 5T_2) \end{bmatrix}$$

For a member composed of multiple strips the global stiffness matrix and global geometric stiffness matrix can be obtained by the assembly of element stiffness matrices and element geometric stiffness matrices, that is

$$[\mathbf{K}_m] = \sum_k [\mathbf{K}]_m \qquad [\mathbf{K}_{gm}] = \sum_k [\mathbf{K}_g]_m \tag{11.41}$$

where the index k denotes the k-th element. The summation implies proper coordinate transformations and correct addition of the stiffness terms in the global coordinates and degrees of freedom. The elastic buckling problem is a standard eigenvalue problem of the following form:

$$[\mathbf{K}_m]\{\delta_m\} = \lambda_m[\mathbf{K}_{gm}]\{\delta_m\} \tag{11.42}$$

where eigenvalues, λ_m, and eigenvectors, $\{\delta_m\}$, are the buckling loads and corresponding buckling modes.

11.6.3 Finite Strip Solution Methods

The finite strip analysis employs single wave functions and thus can only be applied to the case where the longitudinal stresses do not vary along the longitudinal axis. Note that both $[\mathbf{K}_m]$ and $[\mathbf{K}_{gm}]$ are the functions of the strip length, a, and wave number, m. The buckling loads and buckling modes solved from Eq. (11.42) are also the functions of them. Since only the critical buckling load, which is the smallest eigenvalue for any of possible wave numbers, is of interest, the strip length and wave number are not independent each other. There are two quick ways to find the critical buckling load. One is to let a equal to the real length of the strip and solve the problem for several numbers of m. The other is to let m = 1 and solve the problem for a number of lengths. The former provides the results only for the given length of the strip, whereas the latter provides a complete picture of the critical buckling loads and modes at various different half-wavelengths, which has clear physical meanings. For this reason the latter method is often used.

Figure 11.16 shows the buckling curves of a laterally restrained (both displacement and rotation) zed section beam ($h = 202$ mm, $b = 75$ mm, $c = 20$ mm and $t = 2,3$ mm) subjected to pure bending, in which the three local minima represent the local, distortional and secondary distortional buckling. The secondary distortional buckling is sometimes called lateral–distortional buckling to distinguish it from flange distortional buckling. The secondary distortional buckling exists only when the section is restrained

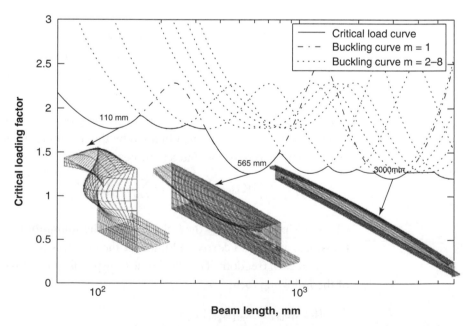

FIGURE 11.16 Buckling curves of a simply supported zed section beam restrained at the junction between web and tension flange subjected to pure bending ($h = 202$ mm, $b = 75$ mm, $c = 20$ mm, $t = 2,3$ mm) (copy from Chu *et al.* (2006) by permission)

laterally. Note that the buckling curves only provide information about how and when the beam buckles. The actual critical load of the beam for a given length should be taken as the lowest value from all of the buckling curves of the same length, and this critical load curve is the solid line plotted in Fig. 11.16.

The original FSM can only deal with the buckling problem of members subjected to either pure compression or pure bending. Recently, Chu *et al.* (2005, 2006) modified the geometric stiffness matrix in the FSM by allowing for the variation of the pre-buckling stress along the longitudinal axis, thus leading to a semi-analytical FSM which is able to deal with the buckling problem of cold-formed section beams subjected to uniformly distributed transverse loading. As an example, Fig. 11.17 shows the critical load curve of the restrained zed section beam ($h = 202$ mm, $b = 75$ mm, $c = 20$ mm and $t = 2{,}3$ mm) under uniformly distributed loading. In order to compare the difference in critical load between pure bending and uniformly distributed loading, the results of the beam under pure bending is also superimposed in the figure. It is evident that the critical load associated with uniformly distributed loading is significantly higher than that arising from pure bending although, for both local and distortional buckling, the differences between the two loading cases decrease with the beam length. For example, for a 5 m long beam, the critical load of a uniformly distributed loading beam is 12% higher than that of the pure bending beam compared to 20% higher for a 2 m long beam.

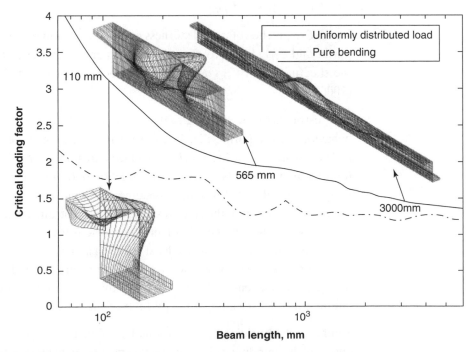

FIGURE 11.17 Critical load curves of a simply supported zed section beam restrained at the junction between web and tension flange subjected to a uniformly distributed uplift load and a pure bending ($h = 202$ mm, $b = 75$ mm, $c = 20$ mm, $t = 2{,}3$ mm) (copy from Chu *et al.* (2006) by permission)

11.7 DESIGN METHODS FOR BEAMS PARTIALLY RESTRAINED BY SHEETING

Purlins and rails are usually used together with their supported trapetzoidal sheeting. Thus it is generally assumed that the sheeting takes the load in the plane of the sheeting and the purlin takes the load normal to the plane of the sheeting, that is, in the plane of web. Also, the purlin may be regarded as being laterally restrained in the plane of the sheeting and partially restrained in twisting if the trapetzoidal sheeting is connected to a purlin and the connection meets the following condition (cl 10.1.1.6 in EN 1993-1-3 (2006))

$$S \geq \frac{70}{h^2}\left(\frac{\pi^2 EI_w}{l^2} + GI_T + \frac{\pi^2 h^2 EI_z}{4l^2}\right) \tag{11.43}$$

where S is the portion of the shear stiffness provided by the sheeting for the examined member connected to the sheeting at each rib (if the sheeting is connected to a purlin every second rib only, then S should be substituted by $0,2S$), h is the web depth and l is the span length. The partial torsional restraint may be represented by a rotational spring with a spring stiffness C_D, which can be calculated based on the stiffness of the sheeting and the connection between the sheeting and the purlin, as follows,

$$\frac{1}{C_D} = \frac{1}{C_{D,A}} + \frac{1}{C_{D,C}} \tag{11.44}$$

where $C_{D,A}$ is the rotational stiffness of the connection between the sheeting and the purlin and $C_{D,C}$ is the rotational stiffness corresponding to the flexural stiffness of the sheeting. Both $C_{D,A}$ and $C_{D,C}$ are specified in Section 10.1.5.2 in EN 1993-1-3 (2006).

The restraints of the sheeting to the purlin have important influence on the buckling behaviour of the purlin. Fig. 11.18 shows the buckling curves of a simply supported zed purlin beam ($h = 202$ mm, $b = 75$ mm, $c = 20$ mm and $t = 2,3$ mm) with various different lateral restraints applied at the junction between the web and the compression flange when subjected to a pure bending. The figure shows that, when the translational displacement of the compression flange is restrained the purlin does not buckle lateral torsionally. On the other hand, when the rotation of the compression flange is restrained the critical stresses of local buckling and distortional buckling are increased quite significantly. However, when the restraints are applied at the junction between the web and the tension flange, it is only the rotational restraint that influences the lateral–torsional buckling of the purlin (see Fig. 11.19).

Similar to the buckling behaviour, the lateral restraints also have considerable influence on the bending behaviour of the purlin. It has been found from both finite element analyses and experiments that, the bending behaviours are different when the restraints are applied at tension and compression flanges, and the free flange and the restrained flange are bent differently, particular when the free flange is under

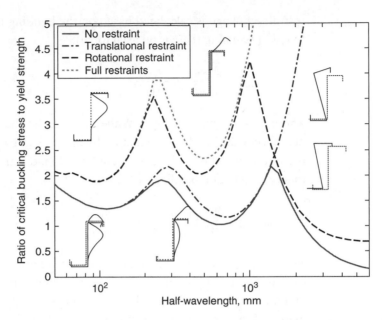

FIGURE 11.18 Buckling curves of a simply supported zed section beam with different restraint applied at the junction between web and compression flange subjected to pure bending ($h = 202$ mm, $b = 75$ mm, $c = 20$ mm, $t = 2,3$ mm)

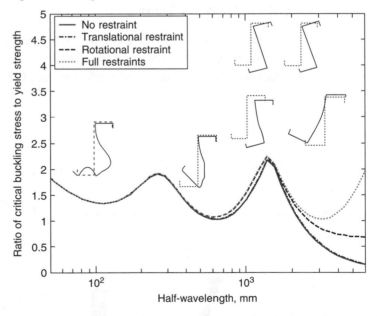

FIGURE 11.19 Buckling curves of a simply supported zed section beam with different restraint applied at the junction between web and tension flange subjected to pure bending ($h = 202$ mm, $b = 75$ mm, $c = 20$ mm, $t = 2,3$ mm)

compression (Lucas et al., 1997a, b; Vrany, 2002). This led to development of different treatments for sections when subjected to gravity loading and uplift loading, since, for instance, for a simply supported beam, the free flange is in tension for gravity loading but in compression for uplift loading. In EN 1993-1-3 (2006), the stress in the restrained flange is calculated based on the bending moment in the plane of web as follows (assuming there is no axial force),

$$\sigma_{max,Ed} = \frac{M_{y,Ed}}{W_{eff,y}} \leq \frac{f_y}{\gamma_{M0}}$$

(11.45)

where f_y is the yield strength. While the stress in the free flange is calculated based on not only the bending moment in the plane of web but also the bending moment in the free flange due to the equivalent lateral load acting on the free flange caused by torsion and lateral bending as follows,

$$\sigma_{max,Ed} = \frac{M_{y,Ed}}{W_{eff,y}} + \frac{M_{fz,Ed}}{W_{fz}} \leq \frac{f_y}{\gamma_{M0}}$$

(11.46)

where $M_{fz,Ed}$ is the bending moment in the free flange due to the equivalent lateral load as defined in Fig. 10.3 in EN 1993-1-3 (2006), and W_{fz} is the gross elastic section modulus of the free flange plus 1/5 of the web height for the point of web–flange intersection, for bending about the z–z axis (see Fig. 11.20). The determination of $M_{fz,Ed}$ is dependent on the section dimensions, loading position, span length, number of anti-sag bars, and spring stiffness C_D, the detail of which is specified in Section 10.1.4.1 in EN 1993-1-3 (2006).

It should be pointed out that Eq. (11.46) applies to only the case where the free flange is under compression. For the free flange under tension, where due to positive influence of flange curling and second order effect moment $M_{fz,Ed}$ may be taken equal to zero.

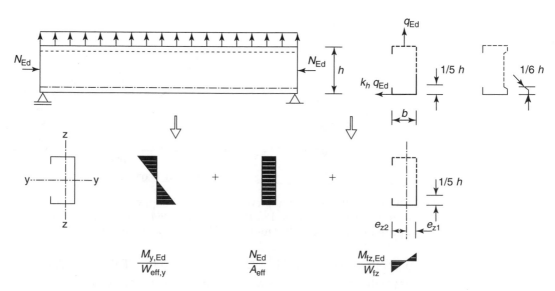

FIGURE 11.20 Superposition of stresses in free flange (copy from EN 1993-1-3 (2006) by permission)

Similar to any compression members, the flexural buckling of the free flange under compression also need to be considered. The buckling resistance of the free flange is verified by using the following formula

$$\frac{1}{\chi_{LT}} \frac{M_{y,Ed}}{W_{eff,y}} + \frac{M_{fz,Ed}}{W_{fz}} \leq \frac{f_y}{\gamma_{M1}} \tag{11.47}$$

in which, $\chi_{LT} = 1/[\phi_{LT} + (\phi_{LT}^2 - 0,75(\overline{\lambda}_{fz})^2)^{1/2}] = $ reduction factor for flexural buckling of the free flange ($\chi_{LT} \leq 1,0$ and $\chi_{LT} \leq 1/(\overline{\lambda}_{LT})^2$)

$\phi_{LT} = 0,5\lfloor 1 + \alpha_{LT}(\overline{\lambda}_{fz} - 0,4) + 0,75(\overline{\lambda}_{fz})^2\rfloor = $ factor used to calculate the reduction factor

where $\alpha_{LT} = 0,34$ is the imperfection factor and $\overline{\lambda}_{fz}$ is the relative slenderness for flexural buckling of the free flange and is determined from

$$\overline{\lambda}_{fz} = \frac{l_{fz}}{i_{fz}} \sqrt{\frac{f_{yb}}{\pi^2 E}} \tag{11.48}$$

where i_{fz} is the radius of gyration of the gross cross-section of the free flange plus the contributing part of the web for bending about the z–z axis and l_{fz} is the buckling length for the free flange which is specified in Section 10.1.4.2 in EN 1993-1-3 (2006).

11.8 WORKING EXAMPLES

In order to assist designers, manufactures of cold-formed steel sections often provide design manuals for their products, which, in general, include gross section properties, effective section properties, and load tables for a number of specified span lengths under various different sets and/or types of connection. The gross section properties are calculated based on the geometric dimensions of the section, whereas the effective section properties are calculated based on the effective widths by taking into account the effects of local and distortion buckling. The load tables can be obtained from either tests or calculations by considering the following modes of failure:

- Flexural failure involving local and distortional buckling in compression
- Lateral–torsional buckling due to insufficient lateral restraints
- Excessive deflection
- Shear failure
- Web crushing under direct loads or reactions
- Combined effects between bending and web crushing, and bending and shear.

The load tables design their sections for specific uses and give the performance of sections under various given circumstances.

The following is an example of calculation of gross and effective section properties for a cold-formed lipped zed section subjected to bending in the plane of web.

The calculation is based on the method recommended in EN 1993-1-3 (2006). The dimensions of the cross-section are (where the influence of rounding of the corners is neglected):

Depth of web	$h = 202\,\text{mm}$
Width of flange in compression	$b_c = 75\,\text{mm}$
Width of flange in tension	$b_t = 75\,\text{mm}$
Length of lip	$c = 20\,\text{mm}$
Thickness	$t = 2\,\text{mm}$

The material properties of the section are:

Modulus of elasticity	$E = 210000\,\text{N/mm}^2$
Poisson's ratio	$\nu = 0{,}3$
Basic yield strength	$f_{yb} = 390\,\text{N/mm}^2$
Partial factor	$\gamma_{M0} = 1{,}00$

Checking of geometrical proportions

The design method of EN 1993-1-3 (2006) can be applied if the following conditions are satisfied (Section 5.2):

$$
\begin{aligned}
b/t \leq 60 \quad & b_c/t = 75/2 = 37{,}5 < 60 \quad &&\rightarrow \quad \text{ok} \\
& b_t/t = 75/2 = 37{,}5 < 60 \quad &&\rightarrow \quad \text{ok} \\
c/t \leq 50 \quad & c/t = 20/2 = 10 < 50 \quad &&\rightarrow \quad \text{ok} \\
h/t \leq 500 \quad & h/t = 202/2 = 101 < 500 \quad &&\rightarrow \quad \text{ok}
\end{aligned}
$$

In order to provide sufficient stiffness and avoid primary buckling of the stiffener itself, the size of stiffener should be within the following range (Section 5.2 in EN 1993-1-3 (2006)):

$$
\begin{aligned}
0{,}2 \leq c/b \leq 0{,}6 \quad & c/b_c = 20/75 = 0{,}27 \quad &&\rightarrow \quad \text{ok} \\
& c/b_t = 20/75 = 0{,}27 \quad &&\rightarrow \quad \text{ok}
\end{aligned}
$$

For cold-formed steel sections the section properties are usually calculated based on the dimensions of the section middle line as follows (see Fig. 11.21),

Depth of web	$h_p = h - t = 202 - 2 = 200\,\text{mm}$
Width of flange in compression	$b_{p1} = b_c - t = 75 - 2 = 73\,\text{mm}$
Width of flange in tension	$b_{p2} = b_t - t = 75 - 2 = 73\,\text{mm}$
Length of lip	$c_p = c - t/2 = 20 - 2/2 = 19\,\text{mm}$

Calculation of gross section properties

Gross cross-section area:

$$A = t(2c_p + b_{p1} + b_{p2} + h_p) = 2 \times (2 \times 19 + 73 + 73 + 200) = 768\,\text{mm}^2$$

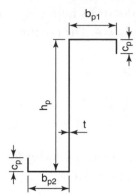

FIGURE 11.21 Symbols used for representing the dimensions of middle lines of a zed section

Position of the neutral axis with regard to the flange in compression:

$$z_c = \frac{h_p}{2} = \frac{200}{2} = 100 \, \text{mm}$$

Position of the neutral axis with regard to the flange in tension:

$$z_t = h_p - z_c = 200 - 100 = 100 \, \text{mm}$$

Second moment of the gross cross-sectional area:

$$I_y = \frac{(h_p^3 + 2c_p^3)t}{12} + \frac{(b_{p1} + b_{p2})t^3}{12} + z_c^2 b_{p1} t + z_t^2 b_{p2} t + c_p t \left(z_c - \frac{c_p}{2}\right)^2$$
$$+ c_p t \left(z_t - \frac{c_p}{2}\right)^2 = 4880000 \, \text{mm}^4$$

Gross section modulus with regard to the flange in compression:

$$W_{y,c} = \frac{I_y}{z_c} = \frac{4880000}{100} = 48800 \, \text{mm}^3$$

Gross section modulus with regard to the flange in tension:

$$W_{y,t} = \frac{I_y}{z_t} = \frac{4880000}{100} = 48800 \, \text{mm}^3$$

Calculation of effective section properties

The general (iterative) procedure is applied to calculate the effective properties of the compression flange and the lip (plane element with edge stiffener). The calculation should be carried out in three steps:

Step 1 Obtain an initial effective cross-section for the stiffener using effective widths of the flange and lip determined by assuming that the compression flange is doubly supported, the stiffener gives full restraint ($K = \infty$) and that the design strength is not reduced, that is, $\sigma_{com,Ed} = f_{yb}/\gamma_{M0}$.

• Effective width of the compressed flange

For the internal compression flange the stress ratio $\psi = 1$ (uniform compression), so the buckling coefficient is taken as $k_\sigma = 4$. The relative slenderness thus is:

$$\bar\lambda_{b,red} = \bar\lambda_b = \sqrt{\frac{f_{yb}}{\sigma_{cr,b}}} = \sqrt{\frac{390}{570}} = 0{,}827$$

where $\sigma_{cr,b} = \dfrac{\pi^2 E}{12(1-\nu^2)}\dfrac{k_\sigma}{(b_{p1}/t)^2} = \dfrac{3{,}14^2 \times 210000}{12(1-0{,}3^2)}\dfrac{4}{(73/2)^2} = 570\,\text{N/mm}^2$

Since $\bar\lambda_{b,red} = 0{,}827 > 0{,}673$, the width reduction factor for the doubly supported compression element is calculated by (Eq. (11.14a))

$$\rho = \frac{\bar\lambda_{b,red} - 0{,}055(3+\psi)}{\bar\lambda_{b,red}^2} = \frac{0{,}827 - 0{,}055(3+1)}{0{,}827^2} = 0{,}887$$

The effective width of the compressed flange thus is:

$$b_{eff} = \rho b_{p1} = 0{,}887 \times 73 = 64{,}8\,\text{mm}$$

$$b_{e1} = b_{e2} = 0{,}5 b_{eff} = 0{,}5 \times 64{,}8 = 32{,}4\,\text{mm}$$

- Effective length of the lip

  For the compression lip, the buckling coefficient should be taken as follows (Eq. (11.13d))

  $$k_\sigma = 0{,}5 \qquad\qquad\qquad\qquad \text{if } c_p/b_{p1} \le 0{,}35$$
  $$k_\sigma = 0{,}5 + 0{,}83(c_p/b_{p1} - 0{,}35)^{2/3} \quad \text{if } 0{,}35 \le c_p/b_{p1} \le 0{,}6$$

  For $c_p/b_{p1} = 19/73 = 0{,}260 < 0{,}35\ k_\sigma = 0{,}5$.

  The relative slenderness is:

  $$\bar\lambda_{c,red} = \bar\lambda_c = \sqrt{\frac{f_{yb}}{\sigma_{cr,c}}} = \sqrt{\frac{390}{1051}} = 0{,}609$$

  where $\sigma_{cr,c} = \dfrac{\pi^2 E}{12(1-\nu^2)}\dfrac{k_\sigma}{(c_p/t)^2} = \dfrac{3{,}14^2 \times 210000}{12(1-0{,}3^2)}\dfrac{0{,}5}{(19/2)^2} = 1050\,\text{N/mm}^2$

  Since $\bar\lambda_{c,red} = 0{,}609 < 0{,}673$, the width reduction factor for the outstand compression element thus is given by (Eq. (11.14b)):

  $$\rho = 1{,}0$$

  The effective length of the compression lip thus is:

  $$c_{eff} = \rho c_p = 1{,}0 \times 19 = 19\,\text{mm}$$

  The corresponding effective area of the edge stiffener is:

  $$A_s = t(b_{e2} + c_{eff}) = 2 \times (32{,}4 + 19) = 103\,\text{mm}^2$$

Step 2 Use the initial effective cross-section of the stiffener to determine the reduction factor, allowing for the effects of the distortional buckling. The elastic critical stress of the distortional buckling for the edge stiffener is (Eq. (11.18))

$$\sigma_{cr,d} = \frac{2\sqrt{KEI_s}}{A_s} = 343\,\text{N/mm}^2$$

where $K = \dfrac{Et^3}{4(1 - \nu^2)(b_1^2 h_p + b_1^3)} = 0{,}445\,\text{N/mm}^2$

$b_1 = b_{p1} - \dfrac{b_{e2}^2}{2(b_{e2} + c_{eff})} = 62{,}8\,\text{mm}$

$I_s = \dfrac{b_{e2}t^3}{12} + \dfrac{c_{eff}^3 t}{12} + b_{e2}t\left[\dfrac{c_{eff}^2}{2(b_{e2} + c_{eff})}\right]^2 + c_{eff}t\left[\dfrac{c_{eff}}{2} - \dfrac{c_{eff}^2}{2(b_{e2} + c_{eff})}\right]^2$

$\quad = 3330\,\text{mm}^4$

Thickness reduction factor for the edge stiffener is calculated based on the relative slenderness of the edge stiffener as follows (Eq. (11.19)):

$\bar{\lambda}_d = \sqrt{\dfrac{f_{yb}}{\sigma_{cr,d}}} = \sqrt{\dfrac{390}{343}} = 1{,}066$

$\chi_d = 1{,}0$ if $\bar{\lambda}_d \leq 0{,}65$
$\chi_d = 1{,}47 - 0{,}723\bar{\lambda}_d$ if $0{,}65 < \bar{\lambda}_d < 1{,}38$
$\chi_d = 0{,}66/\bar{\lambda}_d$ if $\bar{\lambda}_d \geq 1{,}38$

$0{,}65 < \bar{\lambda}_d < 1{,}38$ so $\chi_d = 1{,}47 - 0{,}723\bar{\lambda}_d = 0{,}699$

Step 3 As the reduction factor for the buckling of the stiffener is $\chi_d = 0{,}699 < 1$, iterations are required to refine the value of the reduction factor. The iterations are carried out based on the reduced design strength, $\sigma_{com,Ed,i} = \chi_{d,i-1}f_{yb}/\gamma_{M0}$ to obtain new effective widths of the lip and flange in the stiffener and recalculate the critical stress of distortional buckling of the stiffener and thus to obtain new reduction factor. The iteration stops when the reduction factor χ_d converges. The final values obtained after iterations are $b_{e2} = 36{,}1$, $c_{eff} = 19{,}0$ and $\chi_d = 0{,}689$.

- Effective width of the web
  The position of the initial neutral axis (web is assumed as fully effective) with regard to the flange in compression is given by

$h_c = \dfrac{c_p\left(h_p - \frac{c_p}{2}\right) + b_{p2}h_p + \frac{h_p^2}{2} + \frac{c_{eff}^2 \chi_d}{2}}{c_p + b_{p2} + h_p + b_{e1} + (b_{e2} + c_{eff})\chi_d} = 106\,\text{mm}$

The stress ratio thus is:

$\psi = -\dfrac{h_p - h_c}{h_c} = -\dfrac{200 - 106}{106} = -0{,}89$

The corresponding buckling coefficient is calculated by (Eq. (11.13a))

$k_\sigma = 7{,}81 - 6{,}29\psi + 9{,}78\psi^2 = 21{,}2$

The relative slenderness thus is:

$$\bar{\lambda}_{h,red} = \bar{\lambda}_h = \sqrt{\frac{f_{yb}}{\sigma_{cr,h}}} = \sqrt{\frac{390}{1051}} = 0,986$$

where $\sigma_{cr,h} = \dfrac{\pi^2 E}{12(1-\nu^2)} \dfrac{k_\sigma}{(h_p/t)^2} = \dfrac{3,14^2 \times 210000}{12(1-0,3^2)} \dfrac{21,2}{(200/2)^2} = 402\,\text{N/mm}^2$

The width reduction factor thus is (Eq. (11.14a))

$$\rho = \frac{\bar{\lambda}_{h,red} - 0,055(3+\psi)}{\bar{\lambda}_{h,red}^2} = \frac{0,986 - 0,055(3 - 0,89)}{0,986^2} = 0,895$$

The effective width of the zone in compression of the web is:

$$h_{eff} = \rho h_c = 0,895 \times 106 = 94,9\,\text{mm}$$

Part of the effective width near the flange is (see Table 11.1):

$$h_{e1} = 0,4 h_{eff} = 0,4 \times 94,9 = 37,9\,\text{mm}$$

Part of the effective width near the neutral axis is:

$$h_{e2} = 0,6 h_{eff} = 0,6 \times 94,9 = 56,8\,\text{mm}$$

Thus,

$$h_1 = h_{e1} = 37,9\,\text{mm}$$
$$h_2 = (h_p - h_c) + h_{e2} = (200 - 106) + 56,8 = 151\,\text{mm}$$

The effective widths of the web obtained above are based on the position of the initial neutral axis (web is assumed as fully effective). To refine the result iterations are required which is based on the newly obtained effective widths, h_{e1} and h_{e2}, to determine the new position of the neutral axis. The stress ratio, buckling coefficient, relative slenderness, width reduction factor and effective widths of the web thus are re-calculated according to the new position of the neutral axis. Iteration continues until it converges. The final values obtained after iterations are $h_{e1} = 37,9\,\text{mm}$, $h_{e2} = 56,8\,\text{mm}$ and $h_2 = 149\,\text{mm}$ (see Fig. 11.22).

- Effective properties of the section (see Fig. 11.23)
  Effective cross-section area:

$$A_{eff} = t[c_p + b_{p2} + h_2 + h_1 + b_{e1} + (b_{e2} + c_{eff})\chi_d] = 698\,\text{mm}^2$$

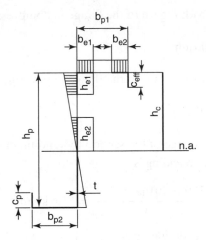

FIGURE 11.22 Symbols used for representing the dimensions of the effective cross-section

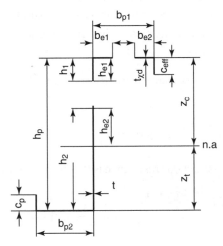

FIGURE 11.23 Symbols used for representing the properties of the effective cross-section

Position of the neutral axis with regard to the flange in compression:

$$z_c = \frac{t\left[c_p\left(h_p - \frac{c_p}{2}\right) + b_{p2}h_p + h_2\left(h_p - \frac{h_2}{2}\right) + \frac{h_1^2}{2} + \frac{c_{eff}^2 \chi_d}{2}\right]}{A_{eff}} = 108\,\text{mm}$$

Position of the neutral axis with regard to the flange in tension:

$$z_t = h_p - z_c = 92\,\text{mm}$$

Second moment of the effective sectional area:

$$I_{eff,y} = \frac{t(h_1^3 + h_2^3 + c_p^3 + \chi_d c_{eff}^3)}{12} + \frac{t^3(b_{p2} + b_{e1} + b_{e2}\chi_d^3)}{12}$$

$$+ c_p t\left(z_t - \frac{c_p}{2}\right)^2 + b_{p2}t z_t^2 + h_2 t\left(z_t - \frac{h_2}{2}\right)^2 + h_1 t\left(z_c - \frac{h_1}{2}\right)^2$$

$$+ b_{e1}t z_c^2 + b_{e2}(\chi_d t)z_c^2 + c_{eff}(\chi_d t)\left(z_c - \frac{c_{eff}}{2}\right)^2$$

$$= 4340000\,\text{mm}^4$$

Effective section modulus with regard to the flange in compression:

$$W_{\text{eff},y,\text{c}} = \frac{I_{\text{eff},y}}{z_{\text{c}}} = 40100\,\text{mm}^3$$

Effective section modulus with regard to the flange in tension:

$$W_{\text{eff},y,\text{t}} = \frac{I_{\text{eff},y}}{z_{\text{t}}} = 47200\,\text{mm}^3$$

The design value of the resistance of the section to bending moment about the y-axis due to local and distortional buckling is

$$M_{\text{c,Rd}} = \frac{f_{\text{yb}} W_{\text{eff},y,\text{c}}}{\gamma_{\text{M0}}} = \frac{390 \times 40100 \times 10^{-6}}{1,0} = 15,6\,\text{kNm}$$

The load table of a section is defined based on the specified type of connection and given span lengths. Assume that the section is used as a single span of 5 m long, both ends are butted to rafter beam using cleats, the section is subjected to gravity loading and is fully restrained laterally.

The design value of the resistance to gravity load on the span due to local and distortional buckling is:

$$P_{\text{c,Rd}} = \frac{8 M_{\text{c,Rd}}}{l} = \frac{8 \times 15,6}{5} = 25,0\,\text{kN}$$

The design value of the plastic shear resistance is (Section 6.2.6 in EN 1993-1-1 (2005)):

$$V_{\text{pl,Rd}} = \frac{A_{\text{v}} \left(f_{\text{yb}}/\sqrt{3} \right)}{\gamma_{\text{M0}}} = \frac{\frac{200 \times 2 \times 390}{\sqrt{3}} \times 10^{-3}}{1,0} = 90,1\,\text{kN}$$

The design value of the shear buckling resistance is (Section 6.1.5 in EN 1993-1-3 (2006)):

$$V_{\text{b,Rd}} = \frac{\frac{h_{\text{p}}}{\sin\phi} t f_{\text{bv}}}{\gamma_{\text{M0}}} = \frac{\frac{200}{\sin 90°} \times 2 \times \frac{0,48 \times 390}{1,49} \times 10^{-3}}{1,0} = 50,3\,\text{kN}$$

in which,

$$f_{\text{bv}} = 0,58 f_{\text{yb}} \qquad \text{if } \bar{\lambda}_{\text{w}} \leq 0,83$$
$$f_{\text{bv}} = 0,48 f_{\text{yb}}/\bar{\lambda}_{\text{w}} \quad \text{if } 0,83 < \bar{\lambda}_{\text{w}}$$

$$\bar{\lambda}_{\text{w}} = 0,346 \frac{h_{\text{p}}}{t} \sqrt{\frac{f_{\text{yb}}}{E}} = 0,346 \times \frac{200}{2} \times \sqrt{\frac{390}{210000}} = 1,49$$

So, the design shear resistance is

$$V_{\text{c,Rd}} = \min(V_{\text{pl,Rd}}, V_{\text{b,Rd}}) = 50,3\,\text{kN}$$

Since the maximum shear force for the simply supported beam occurs at the support, the value of which is only half of the span load, it is obvious that for the present case the design load is controlled by the design value of the resistance due to local and distortional buckling. Therefore the ultimate design load for the section is 25 kN.

Note that, the load factors used for permanent actions and for variable actions are different. The design load for any combinations of dead and imposed loads thus should satisfy

$$\gamma_G P_G + \gamma_Q P_Q \leq P_{c,Rd} = 25\,\text{kN}$$

where $\gamma_G = 1,35$ and $\gamma_Q = 1,5$ are the load factors for dead and imposed loads, P_G and P_Q are the dead and imposed loads, respectively.

The deflection check is usually done by using computer software which is normally provided by the manufacturer. The allowable deflection is determined based on individual cases and is not specified in most design standards.

11.9 CHAPTER CONCLUSIONS

This chapter has described the mechanism of failures of the cold-formed steel sections and the corresponding methods for analyses and principles for design. The focuses have been placed on the analyses of local, distortional and lateral–torsional buckling of the sections for these are the main differences between the cold-formed steel section and the hot rolled steel section. For the cold-formed steel sections the most important properties are the effective section properties. For summarizing the calculation procedure a flowchart of calculation of effective section properties is given in Figs. 11.24 and 11.25.

The design of a cold-formed steel section is much more complicated than that of a hot rolled section. This is partly because the section involves elements which have large width-to-thickness ratios and thus are easier to buckle locally and distortionally, and partly because the section is locally restrained by its supported trapetzoidal sheeting, which complicates the loading system and generates different stresses in restrained and free flanges. Simplified design methods may be used for channel, zed and sigma purlin systems that have no anti-sag bars, no use of sleeves and overlapping between two adjacent sections (see Annex E in EN 1993-1-3 (2006)).

It is well known that for a short beam the design is normally controlled by the bending moment and/or shear force, while for a long beam it is usually controlled by the deflection and lateral–torsional buckling. Thus, purlins are usually designed to be continuous by using sleeves or overlapping in order to satisfy deflection limits and anti-sag bars to prevent twisting during erection and to stabilize the lower flange against wind uplift. For simply supported members, it is the sagging (positive) moment conditions that determine the capacity of the member. For continuous members over one or more internal supports, moments are to be determined elastically. Plastic hinge analysis is not permitted because the slender sections are not able to maintain their full

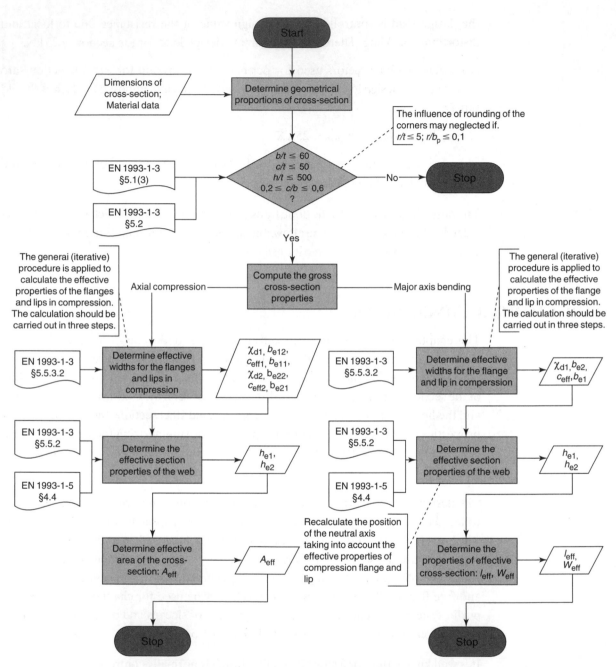

FIGURE 11.24 The flowchart of calculation of effective section properties for cold-formed steel sections under compression or bending (copy from www.access-steel.com by permission)

moment capacity when rotations exceed the point at which the section reaches yield. By utilizing the flexibility of sleeved or overlapping purlins at the supports, some elastic redistribution of moment may be achieved, and hence lead to more efficient design of the member. However, when the purlins are sleeved or overlapped, the design of the

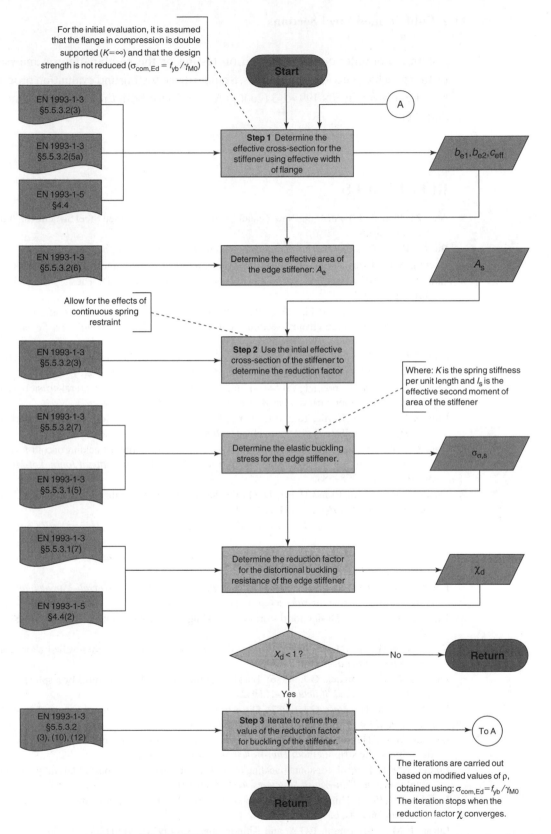

FIGURE 11.25 The flowchart of iterative procedure for calculation of reduction factor due to distortion buckling (copy from www.access-steel.com by permission)

system is normally developed based on testing or the combination of analysis and testing to achieve economic solutions. Standardized testing and evaluation procedures have been given in EN 1993-1-3 (2006) (Annex A) for both the cold-formed members and sheeting.

REFERENCES

AS/NZS 4600 (1996). Australia/New Zealand Standard for Cold-formed Steel Structures. Standards Australia, Sydney.

Bulson, P.S. (1970). *The stability of flat plates*. Chatto & Windus, London.

Cheung, Y.K. (1976). *Finite strip methods in structural analysis*. Pergramon Press, New York.

Chu, X.T. (2004). *Failure Analysis of Cold-formed Steel Sections*. Ph.D. Thesis, Aston University, Birmingham. UK.

Chu, X.T., Kettle, R. and Li, L.Y. (2004a). Lateral–torsion buckling analysis of partial-laterally restrained thin-walled channel-sections beams, *Journal of Constructional Steel Research*, **60(8)**, 1159–1175.

Chu, X.T., Li, L.Y. and Kettle, R. (2004b). The effect of warping stress on the lateral torsion buckling of cold-formed zed-purlins, *Journal of Applied Mechanics*, **71(5)**, 742–744.

Chu, X.T., Rickard, J. and Li, L.Y. (2005a). Influence of lateral restraint on lateral–torsional buckling of cold-formed steel purlins. *Thin-Walled Structures*, **43**, 800–810.

Chu, X.T., Ye, Z.M, Kettle, R. and Li, L.Y. (2005b). Buckling behaviour of cold-formed channel sections under uniformly distributed loads. *Thin-Walled Structures*, **43**, 531–542.

Chu, X.T., Ye, Z.M, Li, L.Y. and Kettle, R. (2006). Local and distortional buckling of cold-formed zed-section beams under uniformly distributed transverse loads, *International Journal of Mechanical Sciences*, **48(8)**, 378–388.

Davies, J.M., Leach, P. and Heinz, D. (1993). Second-order generalized beam theory, *Journal of Constructional Steel Research*, **31**, 221–241.

EN 1993-1-1 (2005). *Eurocode 3: Design of steel structures – Part 1–1: General rules and rules for buildings*. BSI.

EN 1993-1-3 (2006). *Eurocode 3: Design of steel structures – Part 1–3: General – Cold formed thin gauge members and sheeting*. BSI.

EN 1993-1-5 (2006). *Eurocode 3: Design of steel structures – Part 1–5: General – Strength and stability of planar plated structures without transverse loading*. BSI.

Hancock, G.J. (1997). Design for distortional buckling of flexural members, *Thin-Walled Structures*, **27(1)**, 3–12.

Kesti, J. and Davies, J.M. (1999). Local and distortional buckling of thin-walled short columns, *Thin-Walled Structures*, **34**, 115–134.

Lau, S.C.W. and Hancock, G.J. (1986). Buckling of thin flat-walled structures by a spline finite strip method, *Thin-Walled Structures*, **4**, 269–294.

Lau, S.C.W. and Hancock, G.J. (1987). Distortional buckling formulas for channel columns, *Journal of Structural Engineering, ASCE*, **113(5)**, 1063–1078.

Lau, S.C.W. and Hancock, G.J. (1989). Inelastic buckling analyses of beams, columns and plates using the spline finite strip method, *Thin-Walled Structures*, **7**, 213–238.

Li, L.Y. (2004). Lateral–torsional buckling of cold-formed zed-purlins partial-laterally restrained by metal sheeting, *Thin-Walled Structures*, **42(7)**, 995–1011.

Loughlan, J. (1993). Thin-walled cold-formed sections subjected to compressive loading, *Thin-Walled Structures*, **16(1–4)**, 65–109.

Lucas, R.M., Al-Bermani, F.G.A. and Kitipornchai, S. (1997a). Modelling of cold-formed purlin-sheeting systems – Part 1. Full model, *Thin-Walled Structures*, **27(3)**, 223–243.

Lucas, R.M., Al-Bermani, F.G.A. and Kitipornchai, S. (1997b). Modelling of cold-formed purlin-sheeting systems – Part 2. Simplified model, *Thin-Walled Structures*, **27**(**4**), 263–286.

Oden, J.T. (1967). *Mechanics of elastic structures*. McGraw-Hill Book Company, New York.

Rhodes, J. and Lawson, R.M. (1992). *Design of structures using cold-formed steel sections*. SCI Publication, 089. SCI.

Schafer, B.W. (1997). *Cold-formed Steel Behaviour and Design: Analytical and Numerical Modelling of Elements and Members with Longitudinal Stiffeners*. Ph.D. Thesis. Cornell University.

Timoshenko, S.P. and Gere, J.M. (1961). *Theory of elastic stability*. McGraw-Hill Book Company, New York.

Vlasov, V.Z. (1961). *Thin-walled elastic beams*. Israel Program for Scientific Translations, Jerusalem, Israel.

Von Karman, T., Sechler, E.E. and Donnell, L.H. (1932). The strength design of thin plates in compression, *Transactions ASME*, **54**, 53–55.

Vrany, T. (2002). Torsional restraint of cold-formed beams provided by corrugated sheeting for arbitrary input variables. *Proceedings of Eurosteel*. Coimbra, cmm, 2002, 734–742.

Walker, A.C. (1975). *Design and analysis of cold-formed sections*. International Textbook Company Ltd, London.

Winter, G. (1968). Thin walled structures – theoretical solutions and test results, *Preliminary Publications 8th Congress IABSE*, 101–112.

Ye, Z.M., Kettle, R., Li, L.Y. and Schafer, B. (2002). Buckling behaviour of cold-formed zed-purlins partially restrained by steel sheeting, *Thin-Walled Structures*, **40**, 853–864.

Ye, Z.M., Kettle, R. and Li, L.Y. (2004). Analysis of cold-formed zed-purlins partially restrained by steel sheeting, *Computer and Structures*, **82**, 731–739.

Zienkiewicz, O.C. and Taylor, R.L. (2000). *The finite element method*, Butterworth Heinemann, Oxford.

Leg length (mm)	Design strength per unit length ($F_{w,Rd}$) in KN/mm	
	Steel grade S275 $f_u = 430\,MPa$ $\beta_w = 0,85$	Steel grade S355 $f_u = 510\,MPa$ $\beta_w = 0,90$
4	0,654	0,733
5	0,818	0,916
6	0,981	1,099
8	1,308	1,466
10	1,636	1,832
12	1,963	2,199
15	2,453	2,748
18	2,944	3,298
20	3,271	3,664
22	3,598	4,031
25	4,089	4,580

Notes:

(1) $F_{w,Rd} = f_u/(3^{1/2}\,\beta_w\gamma_{M2}) \times 0,7 \times$ (leg length) for equal leg lengths and $\gamma_{M2} = 1,25$.

(2) Steel to Standard 10025-2 with appropriate electrodes (BSEN 499).

Bolt diameter d (mm)	Reduced area of bolt A_s (mm²)	Tension stress using reduced area $F_{t, Rd}$ (kN)	Single shear stress using reduced area $F_{v,Rd}$ (kN)	Single shear stress using gross area $F_{v,Rd}$ (kN)
(M12)	84,3	24,3	16,2	21,7
M16	157	45,2	30,1	38,6
M20	245	70,6	47,0	60,3
(M22)	303	87,3	58,2	73,0
M24	353	101,7	67,8	86,9
(M27)	459	132,2	88,1	109,9
M30	561	161,6	107,7	135,7
(M33)	694	199,9	133,2	164,2
M36	817	235,3	156,9	195,4

Notes:

(1) $F_{v,Rd} = 0,6 \, (A \text{ or } A_s) \, f_{ub}/\gamma_{M2}$

(2) $F_{t,Rd} = 0,9 \, A_s f_{ub}/\gamma_{M2}$

(3) $f_{ub} = 400 \, \text{MPa}; \, \gamma_{M2} = 1,25$

(4) Bolt sizes in brackets are not preferred.

(5) Shear values for other classes of bolt are multiplied by a factor

Class	4.6	4.8	5.6	5.8	6.8	8.8	10.9
Factor	1,0	0,83	1,25	1,04	1,25	2,0	2,08

(6) Tension values for other classes of bolt are multiplied by a factor

Class	4.6	4.8	5.6	5.8	6.8	8.8	10.9
Factor	1,0	1,0	1,25	1,25	1,5	2,0	2,5

Bolt size d (mm)	Reduced area of bolt A_s (mm^2)	Preload force class 8.8 $F_{p,C}$ (kN)	Preload force class 10.9 $F_{p,C}$ (kN)	Slip resistance class 8.8 $F_{s,Rd}$ (kN)	Slip resistance class 10.9 $F_{s,Rd}$ (kN)
(M12)	84,3	47,2	59,1	18,9	23,6
M16	157	87,9	109,9	35,2	44,0
M20	245	137,2	171,5	54,9	68,6
M22	303	169,7	212,1	67,9	84,8
M24	353	197,7	247,1	79,1	98,8
M27	459	257,0	321,3	102,8	128,5
M30	561	314,2	392,7	125,7	157,1
M36	817	457,5	571,9	183,0	228,8

Notes:

(1) Preload $F_{p,C} = 0,7 f_{ub} A_s$ where $f_{ub} = 800$ (Class 8.8) and 1000 (Class 10.9) MPa.

(2) Design slip resistance $F_{s,Rd} = k_s n \mu F_{p,C} / \gamma_{M3}$ where $k_s = 1, n = 1, \mu = 0,5$ and $\gamma_{M3} = 1,25$

(3) Bolt sizes in brackets are not preferred.

Beams, columns, joists and tees

Nominal flange widths (mm)	*Spacings in millimetres*				*Recommended diameter of bolt (mm)*	*Actual b_{min} (mm)*
	S_1	S_2	S_3	S_4		
419–368	140	140	75	290	24	362
330–305	140	120	60	240	24	312
330–305	140	120	60	240	20	300
292–203	140	–	–	–	24	212
190–165	90	–	–	–	24	162
152	90	–	–	–	20	150
146–114	70	–	–	–	20	130
102	54	–	–	–	12	98
89	50	–	–	–	–	–
76	40	–	–	–	–	–
64	34	–	–	–	–	–
51	30	–	–	–	–	–

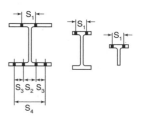

Angles

Nominal leg length (mm)	Spacing of holes						Maximum diameter of bolt		
	S_1 (mm)	S_2 (mm)	S_3 (mm)	S_4 (mm)	S_5 (mm)	S_6 (mm)	S_1 (mm)	S_2 and S_3 (mm)	S_4, S_5 and S_6 (mm)
200	–	75	75	55	55	55	–	30	20
150	–	55	55	–	–	–	–	20	–
125	–	45	50	–	–	–	–	20	–
120	–	45	50	–	–	–	–	20	–
100	55	–	–	–	–	–	24	–	–
90	50	–	–	–	–	–	24	–	–
80	45	–	–	–	–	–	20	–	–
75	45	–	–	–	–	–	20	–	–
70	40	–	–	–	–	–	20	–	–
65	35	–	–	–	–	–	20	–	–
60	35	–	–	–	–	–	16	–	–
50	28	–	–	–	–	–	12	–	–
45	25	–	–	–	–	–	–	–	–
40	23	–	–	–	–	–	–	–	–
30	20	–	–	–	–	–	–	–	–
25	15	–	–	–	–	–	–	–	–

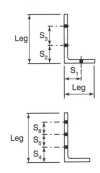

Channels

Nominal flange width (mm)	S_1 (mm)	Recommended diameter of bolt (mm)
102	55	24
89	55	20
76	45	20
64	35	16
51	30	10
38	22	–

TABLE 15 Slenderness correction factor, n, for members with applied loading substantially concentrated within the middle fifth of the unrestrained length.

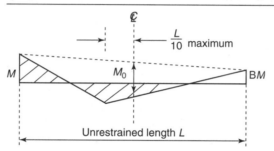

Note 1. All hogging moments are +ve.
Note 2. β is defined in Table 18.
Note 3. M_0 is the mid-length moment on a simply supported span equal to the unrestrained length (see Table 17).

$\gamma = M/M_0$	β positive					0.0	β negative				
	1.0	0.8	0.6	0.4	0.2		−0.2	−0.4	−0.6	−0.8	−1.0
+50.00	1.00	0.96	0.92	0.87	0.82	0.77	0.72	0.67	0.66	0.66	0.65
+10.00	0.99	0.99	0.94	0.90	0.85	0.80	0.75	0.69	0.68	0.68	0.67
+5.00	0.98	0.98	0.97	0.93	0.89	0.84	0.79	0.73	0.71	0.70	0.70
+2.00	0.96	0.95	0.95	0.95	0.94	0.94	0.89	0.84	0.79	0.77	0.76
+1.50	0.95	0.95	0.94	0.94	0.93	0.93	0.92	0.90	0.85	0.80	0.80
+1.00	0.93	0.92	0.92	0.92	0.92	0.91	0.91	0.91	0.91	0.92	0.92
+0.50	0.90	0.90	0.90	0.89	0.89	0.89	0.89	0.89	0.88	0.88	0.88
0.00	0.86	0.86	0.86	0.86	0.86	0.86	0.86	0.86	0.86	0.86	0.86
−0.10	0.85	0.85	0.85	0.85	0.85	0.86	0.86	0.86	0.86	0.86	0.86
−0.20	0.83	0.83	0.83	0.84	0.84	0.85	0.85	0.85	0.86	0.86	0.86
−0.30	0.81	0.82	0.82	0.83	0.83	0.84	0.85	0.85	0.86	0.86	0.87
−0.40	0.79	0.80	0.81	0.81	0.82	0.83	0.84	0.85	0.85	0.86	0.87
−0.50	0.77	0.78	0.79	0.80	0.82	0.83	0.85	0.86	0.86	0.87	0.88
−0.60	0.62	0.66	0.72	0.77	0.80	0.82	0.84	0.85	0.86	0.87	0.88
−0.70	0.56	0.56	0.61	0.67	0.73	0.79	0.83	0.85	0.87	0.88	0.89
−0.80	0.56	0.53	0.54	0.59	0.65	0.71	0.77	0.83	0.89	0.90	0.90
−0.90	0.59	0.57	0.54	0.53	0.57	0.64	0.71	0.77	0.84	0.88	0.91
−1.00	0.62	0.58	0.54	0.52	0.54	0.59	0.66	0.72	0.80	0.85	0.92

(Continued)

TABLE 15 Continued

−1.10	0.66	0.62	0.57	0.54	0.54	0.57	0.63	0.68	0.76	0.83	0.89
−1.20	0.70	0.66	0.60	0.55	0.54	0.55	0.60	0.65	0.73	0.80	0.87
−1.30	0.73	0.69	0.63	0.57	0.55	0.54	0.57	0.61	0.69	0.77	0.83
−1.40	0.74	0.70	0.64	0.58	0.56	0.54	0.55	0.60	0.66	0.74	0.81
−1.50	0.75	0.70	0.64	0.59	0.56	0.54	0.55	0.59	0.65	0.73	0.80
−1.60	0.76	0.72	0.65	0.60	0.57	0.55	0.55	0.58	0.64	0.72	0.80
−1.70	0.77	0.74	0.66	0.61	0.58	0.56	0.55	0.58	0.63	0.70	0.78
−1.80	0.79	0.77	0.68	0.63	0.59	0.56	0.56	0.57	0.62	0.69	0.76
−1.90	0.80	0.79	0.69	0.64	0.60	0.57	0.56	0.57	0.61	0.67	0.75
−2.00	0.81	0.81	0.70	0.65	0.61	0.58	0.56	0.56	0.60	0.66	0.74
−5.00	0.93	0.89	0.83	0.77	0.72	0.67	0.64	0.61	0.60	0.62	0.65
−50.00	0.99	0.95	0.90	0.86	0.79	0.74	0.70	0.67	0.64	0.63	0.65
Infinity	1.00	0.96	0.91	0.86	0.82	0.77	0.72	0.68	0.65	0.65	0.65

Note 4. The values of n in this table apply only to members of UNIFORM section.
Note 5. Values for intermediate values of β and γ may be interpolated.

TABLE 16 Slenderness correction factor, n, for members with applied loading other than as for Table 15.

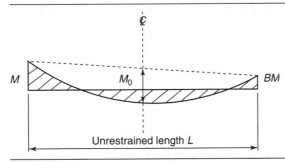

Note 1. All hogging moments are +ve.
Note 2. β is defined in Table 18.
Note 3. M_0 is the mid-length moment on a simply supported span equal to the unrestrained length (see Table 17).

	β positive						β negative				
$\gamma = M/M_0$	1.0	0.8	0.6	0.4	0.2	0.0	−0.2	−0.4	−0.6	−0.8	−1.0
+50.00	1.00	0.96	0.92	0.87	0.83	0.77	0.72	0.67	0.66	0.66	0.65
+10.00	0.99	0.98	0.95	0.91	0.86	0.81	0.76	0.70	0.68	0.68	0.67
+5.00	0.99	0.98	0.97	0.94	0.90	0.85	0.80	0.75	0.71	0.70	0.70
+2.00	0.98	0.98	0.97	0.96	0.94	0.92	0.90	0.86	0.82	0.78	0.76
+1.50	0.97	0.97	0.97	0.96	0.95	0.93	0.92	0.89	0.86	0.83	0.79
+1.00	0.97	0.97	0.97	0.96	0.96	0.95	0.94	0.93	0.93	0.91	0.89
+0.50	0.96	0.96	0.96	0.96	0.96	0.95	0.94	0.94	0.94	0.93	0.92
0.00	0.94	0.94	0.94	0.94	0.94	0.94	0.94	0.94	0.94	0.94	0.94

(Continued)

TABLE 16 Continued

−0.10	0.93	0.93	0.93	0.93	0.94	0.94	0.94	0.94	0.94	0.94	0.94
−0.20	0.92	0.92	0.92	0.92	0.93	0.93	0.93	0.93	0.94	0.94	0.93
−0.30	0.91	0.91	0.92	0.92	0.93	0.93	0.93	0.93	0.94	0.94	0.94
−0.40	0.90	0.90	0.91	0.91	0.92	0.92	0.92	0.92	0.93	0.93	0.93
−0.50	0.89	0.90	0.91	0.91	0.92	0.92	0.92	0.92	0.92	0.92	0.92
−0.60	0.71	0.77	0.84	0.87	0.89	0.91	0.92	0.92	0.92	0.92	0.92
−0.70	0.57	0.64	0.70	0.77	0.82	0.87	0.89	0.91	0.92	0.92	0.91
−0.80	0.47	0.52	0.59	0.67	0.73	0.80	0.86	0.90	0.92	0.92	0.92
−0.90	0.47	0.46	0.50	0.58	0.65	0.73	0.80	0.87	0.90	0.90	0.90
−1.00	0.50	0.48	0.46	0.51	0.58	0.66	0.73	0.81	0.87	0.89	0.89
−1.10	0.54	0.51	0.48	0.49	0.54	0.61	0.69	0.77	0.83	0.87	0.88
−1.20	0.57	0.54	0.50	0.47	0.51	0.56	0.64	0.73	0.80	0.84	0.87
−1.30	0.61	0.56	0.52	0.47	0.49	0.53	0.61	0.70	0.77	0.82	0.86
−1.40	0.64	0.59	0.55	0.49	0.48	0.51	0.58	0.67	0.74	0.79	0.85
−1.50	0.67	0.62	0.57	0.51	0.47	0.49	0.56	0.64	0.71	0.77	0.84
−1.60	0.69	0.64	0.59	0.52	0.48	0.50	0.55	0.63	0.69	0.76	0.83
−1.70	0.71	0.66	0.60	0.54	0.50	0.51	0.55	0.61	0.68	0.74	0.82
−1.80	0.74	0.69	0.62	0.55	0.51	0.51	0.54	0.60	0.66	0.73	0.81
−1.90	0.76	0.71	0.63	0.57	0.53	0.52	0.54	0.58	0.65	0.71	0.80
−2.00	0.78	0.73	0.65	0.58	0.54	0.53	0.53	0.57	0.63	0.70	0.79
−5.00	0.91	0.86	0.80	0.74	0.70	0.65	0.62	0.59	0.58	0.61	0.67
−50.00	0.99	0.95	0.89	0.84	0.79	0.74	0.70	0.66	0.63	0.62	0.65
Infinity	1.00	0.96	0.91	0.86	0.82	0.77	0.72	0.68	0.65	0.65	0.65

Note 4. The values of n in this table apply only to members of UNIFORM section.
Note 5. Values for intermediate values of β and γ may be interpolated.

TABLE 17 Moment diagram between adjacent points of lateral restraint.

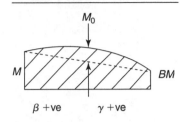

M_0

M BM

β +ve γ +ve

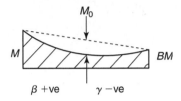

M_0

M BM

β +ve γ −ve

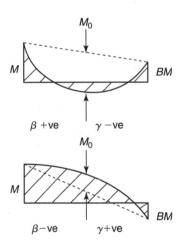

M_0

M BM

β +ve γ −ve

M_0

M BM

β −ve γ +ve

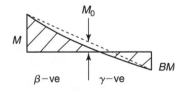

M_0

M BM

β −ve γ −ve

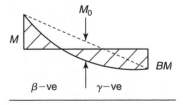

M_0

M BM

β −ve γ −ve

Index